Methods in Molecular Biology

Series Editor
John M. Walker
School of Life and Medical Sciences
University of Hertfordshire
Hatfield, Hertfordshire, AL10 9AB, UK

For further volumes:
http://www.springer.com/series/7651

ADP-ribosylation and NAD+ Utilizing Enzymes

Methods and Protocols

Edited by

Paul Chang

Department of Biology, Koch Institute for Integrative Cancer Research, Massachusetts Institute of Technology, Cambridge, MA, USA; Ribon Therapeutics, Lexington, MA, USA

Humana Press

Editor
Paul Chang
Department of Biology
Koch Institute for Integrative Cancer Research
Massachusetts Institute of Technology
Cambridge, MA, USA

Ribon Therapeutics
Lexington, MA, USA

ISSN 1064-3745 ISSN 1940-6029 (electronic)
Methods in Molecular Biology
ISBN 978-1-4939-8587-6 ISBN 978-1-4939-8588-3 (eBook)
https://doi.org/10.1007/978-1-4939-8588-3

Library of Congress Control Number: 2018949643

Printed on acid-free paper

This Humana Press imprint is published by the registered company Springer Science+Business Media, LLC part of Springer Nature.
The registered company address is: 233 Spring Street, New York, NY 10013, U.S.A.

Preface

This volume is focused on mono-ADP-ribosylation and enzymes that utilize NAD+ including the sirtuins, poly(ADP-ribosyl)transferases (PARPs), and bacterial and eukaryotic ADP-ribosyltransferases. Together these enzymes play important roles in aging, bacterial pathogenicity, and human disease. The last several years have seen tremendous progress in our understanding of mono-ADP-ribosylation and NAD+. This is in large part due to key technological advances and growing interest in the field. Advances in mass spectrometry, chemistry, molecular biology, and high-throughput screening have all contributed to this growth.

These chapters were designed to bring you up to date with the current technologies. They provide detailed descriptions of the key assays and protocols used to study ADP-ribosylation and NAD+ utilizing enzymes. I hope you find these chapters useful for your studies. Thanks to all of the authors and copyeditors for your efforts.

Lexington, MA, USA *Paul Chang*

Contents

Contributors

JEANNETTE ABPLANALP • *Department of Molecular Mechanisms of Disease, University of Zurich, Zurich, Switzerland; Molecular Life Science PhD Program of the Life Science Zurich Graduate School, Zurich, Switzerland*

RACHY ABRAHAM • *Department of Molecular Microbiology and Immunology, Bloomberg School of Public Health, Johns Hopkins University, Baltimore, MD, USA*

SAHIL ADRIOUCH • *Institute of Immunology, University Medical Center Hamburg-Eppendorf, Hamburg, Germany; Faculty of Medicine, University of Rouen, Rouen, France*

IVAN AHEL • *Sir William Dunn School of Pathology, University of Oxford, Oxford, UK*

PETER BANNAS • *Institute of Immunology, University Medical Center Hamburg-Eppendorf, Hamburg, Germany; Department of Radiology, University Medical Center Hamburg-Eppendorf, Hamburg, Germany*

JOSEPH T. BARBIERI • *Microbiology and Immunology, Medical College of Wisconsin, Milwaukee, WI, USA*

FLORIAN J. BOCK • *Cancer Research UK Beatson Institute, Institute of Cancer Sciences, University of Glasgow, Glasgow, UK*

JUAN JOSÉ BONFIGLIO • *Max Planck Institute for Biology of Ageing, Cologne, Germany*

NADY BRAIDY • *Centre for Healthy Brain Ageing, School of Psychiatry, The University of New South Wales, Sydney, NSW, Australia*

MAREIKE BÜTEPAGE • *Institute of Biochemistry and Molecular Biology, Medical School, RWTH Aachen University, Aachen, Germany*

XIAOLU A. CAMBRONNE • *Vollum Institute, Oregon Health and Science University, Portland, OR, USA; Department of Molecular Biosciences, University of Texas at Austin, Austin, TX, USA*

GIULIANA CATARA • *Institute of Protein Biochemistry, National Research Council, Naples, Italy*

PAUL CHANG • *Department of Biology, Koch Institute for Integrative Cancer Research, Massachusetts Institute of Technology, Cambridge, MA, USA; Ribon Therapeutics, Lexington, MA, USA*

CHEN CHEN • *Microbiology and Immunology, Medical College of Wisconsin, Milwaukee, WI, USA*

MICHAEL S. COHEN • *Program in Chemical Biology, Department of Physiology and Pharmacology, Oregon Health and Science University, Portland, OR, USA*

THOMAS COLBY • *Max Planck Institute for Biology of Ageing, Cologne, Germany*

DANIELA CORDA • *Institute of Protein Biochemistry, National Research Council, Naples, Italy*

LAURA ECKEI • *Institute of Biochemistry and Molecular Biology, Medical School, RWTH Aachen University, Aachen, Germany*

DMITRI V. FILIPPOV • *Gorlaeus Laboratories, Leiden Institute of Chemistry, Universiteit Leiden, Leiden, The Netherlands*

Bryan A. Gibson • *Laboratory of Signaling and Gene Regulation, Cecil H. and Ida Green Center for Reproductive Biology Sciences, University of Texas Southwestern Medical Center, Dallas, TX, USA; Division of Basic Research, Department of Obstetrics and Gynecology, University of Texas Southwestern Medical Center, Dallas, TX, USA*

Richard H. Goodman • *Vollum Institute, Oregon Health and Science University, Portland, OR, USA*

Diane E. Griffin • *Department of Molecular Microbiology and Immunology, Bloomberg School of Public Health, Johns Hopkins University, Baltimore, MD, USA*

Giovanna Grimaldi • *Institute of Protein Biochemistry, National Research Council, Naples, Italy*

Giulia Grimaldi • *Department of Nutrition, Institute of Basic Medical Sciences, University of Oslo, Oslo, Norway*

Friedrich Haag • *Institute of Immunology, University medical Center Hamburg-Eppendorf, Hamburg, Germany*

Teemu Haikarainen • *Faculty of Biochemistry and Molecular Medicine, Biocenter Oulu, University of Oulu, Oulu, Finland*

Jun Young Hong • *Department of Chemistry and Chemical Biology, Cornell University, Ithaca, NY, USA*

Ann-Katrin Hopp • *Department of Molecular Mechanisms of Disease, University of Zurich, Zurich, Switzerland; Molecular life Science PhD Program of the Life Science Zurich Graduate School, Zurich, Switzerland*

Michael O. Hottiger • *Department of Molecular Mechanisms of Disease, University of Zurich, Zurich, Switzerland*

Duen-Wei Hsu • *Department of Biochemistry, University of Oxford, Oxford, UK*

Dan Huang • *Laboratory of Signaling and Gene Regulation, Cecil H. and Ida Green Center for Reproductive Biology Sciences, University of Texas Southwestern Medical Center, Dallas, TX, USA; Division of Basic Research, Department of Obstetrics and Gynecology, University of Texas Southwestern Medical Center, Dallas, TX, USA; Department of Cardiovascular Diseases, Clinical Center of Human Gene Research, Union Hospital, Tongji Medical College, Huazhong University of Science and Technology, Wuhan, Hubei Province, P.R. China*

David Hutin • *Department of Pharmacology and Toxicology, University of Toronto, Toronto, ON, Canada*

Myron K. Jacobson • *Department of Pharmaceutical Sciences, University of North Texas System College of Pharmacy, University of North Texas Health Science Center, Fort Worth, TX, USA; Niadyne Pharma, Inc., Fort Worth, TX, USA*

Elaine L. Jacobson • *Niadyne Pharma, Inc., Fort Worth, TX, USA*

Gytis Jankevicius • *Sir William Dunn School of Pathology, University of Oxford, Oxford, UK*

Ilsa T. Kirby • *Program in Chemical Biology, Department of Physiology and Pharmacology, Oregon Health and Science University, Portland, OR, USA*

Hans A. V. Kistemaker • *Gorlaeus Laboratories, Leiden Institute of Chemistry, Universiteit Leiden, Leiden, The Netherlands*

Friedrich Koch-Nolte • *Institute of Immunology, University Medical Center Hamburg-Eppendorf, Hamburg, Germany*

Anna-Lena Kolb • *Department of Biochemistry, University of Oxford, Oxford, UK*

W. Lee Kraus • *Laboratory of Signaling and Gene Regulation, Cecil H. and Ida Green Center for Reproductive Biology Sciences, University of Texas Southwestern Medical*

Center, Dallas, TX, USA; Division of Basic Research, Department of Obstetrics and Gynecology, University of Texas Southwestern Medical Center, Dallas, TX, USA

SARAH KRIEG • *Institute of Biochemistry and Molecular Biology, Medical School, RWTH Aachen University, Aachen, Germany*

NICHOLAS D. LAKIN • *Department of Biochemistry, University of Oxford, Oxford, UK*

LARI LEHTIÖ • *Faculty of Biochemistry and Molecular Medicine, Biocenter Oulu, University of Oulu, Oulu, Finland*

ANTHONY K. L. LEUNG • *Department of Biochemistry and Molecular Biology, Bloomberg School of Public Health, Johns Hopkins University, Baltimore, MD, USA; Department of Oncology, School of Medicine, Johns Hopkins University, Baltimore, MD, USA; Department of Molecular Biology and Genetics, School of Medicine, Johns Hopkins University, Baltimore, MD, USA*

JINYU LI • *College of Chemistry, Fuzhou University, Fuzhou, China*

KEN Y. LIN • *Laboratory of Signaling and Gene Regulation, Cecil H. and Ida Green Center for Reproductive Biology Sciences, University of Texas Southwestern Medical Center, Dallas, TX, USA; Division of Basic Research, Department of Obstetrics and Gynecology, University of Texas Southwestern Medical Center, Dallas, TX, USA*

HENING LIN • *Department of Chemistry and Chemical Biology, Cornell University, Ithaca, NY, USA; Department of Chemistry and Chemical Biology, Howard Hughes Medical Institute, Cornell University, Ithaca, NY, USA*

YUANCHENG LU • *Department of Genetics, Paul F. Glenn Center for the Biology of Aging, Harvard Medical School, Boston, MA, USA*

MARCO LUKOWIAK • *Department of Neurology, University Medical Center Hamburg-Eppendorf, Hamburg, Germany*

BERNHARD LÜSCHER • *Institute of Biochemistry and Molecular Biology, Medical School, RWTH Aachen University, Aachen, Germany*

TIM MAGNUS • *Department of Neurology, University Medical Center Hamburg-Eppendorf, Hamburg, Germany*

MIRKO M. MAKSIMAINEN • *Faculty of Biochemistry and Molecular Medicine, Biocenter Oulu, University of Oulu, Oulu, Finland*

MASATO MASHIMO • *Faculty of Pharmaceutical Sciences, Department of Pharmacology, Doshisha Women's College of Liberal Arts, Kyotanabe, Kyoto, Japan*

IVAN MATIC • *Max Planck Institute for Biology of Ageing, Cologne, Germany*

JASON MATTHEWS • *Department of Pharmacology and Toxicology, University of Toronto, Toronto, ON, Canada; Department of Nutrition, Institute of Basic Medical Sciences, University of Oslo, Oslo, Norway*

ROBERT LYLE MCPHERSON • *Department of Biochemistry and Molecular Biology, Bloomberg School of Public Health, Johns Hopkins University, Baltimore, MD, USA*

STEPHAN MENZEL • *Institute of Immunology, University Medical Center Hamburg-Eppendorf, Hamburg, Germany*

RORY K. MORGAN • *Program in Chemical Biology, Department of Physiology and Pharmacology, Oregon Health and Science University, Portland, OR, USA*

JOEL MOSS • *Pulmonary Branch, National Heart, Lung, and Blood Institute, National Institutes of Health, Bethesda, MD, USA*

SUDARSHAN MURTHY • *Faculty of Biochemistry and Molecular Medicine, Biocenter Oulu, University of Oulu, Oulu, Finland*

SHAO-EN ONG • *Department of Pharmacology, University of Washington, Seattle, WA, USA*

CATHERINE J. PEARS • *Department of Biochemistry, University of Oxford, Oxford, UK*
MELANIJA POSAVEC MARJANOVIĆ • *University of Zagreb, Zagreb, Croatia*
ALINA RAKHIMOVA • *Department of Biochemistry, University of Oxford, Oxford, UK*
BJÖRN RISSIEK • *Department of Neurology, University Medical Center Hamburg-Eppendorf, Hamburg, Germany*
RYAN A. ROGGE • *Laboratory of Signaling and Gene Regulation, Cecil H. and Ida Green Center for Reproductive Biology Sciences, University of Texas Southwestern Medical Center, Dallas, TX, USA; Division of Basic Research, Department of Obstetrics and Gynecology, University of Texas Southwestern Medical Center, Dallas, TX, USA*
GIULIA ROSSETTI • *Computational Biomedicine, Institute for Advanced Simulation IAS-5 and Institute of Neuroscience and Medicine INM-9, Jülich, Germany; Jülich Supercomputing Centre, Jülich, Germany; Department of Oncology, Hematology and Stem Cell Transplantation, Medical School, RWTH Aachen University, Aachen, Germany*
MICHAEL B. SCHULTZ • *Department of Genetics, Paul F. Glenn Center for the Biology of Aging, Harvard Medical School, Boston, MA, USA*
DAVID A. SINCLAIR • *Department of Genetics, Paul F. Glenn Center for the Biology of Aging, Harvard Medical School, Boston, MA, USA; Department of Pharmacology, School of Medical Sciences, The University of New South Wales, Sydney, NSW, Australia*
REBECCA SMITH • *Department of Physiological Chemistry, Biomedical Center Munich, Luwig-Maximilians-Universität München, Planegg-Martinsried, Germany*
EASWARAN SREEKUMAR • *Department of Molecular Microbiology and Immunology, Bloomberg School of Public Health, Johns Hopkins University, Baltimore, MD, USA; Viral Disease Biology Program, Rajiv Gandhi Centre for Biotechnology (RGCB), Thiruvananthapuram, Kerala, India*
LINDA A. STEVENS • *Pulmonary Branch, National Heart, Lung, and Blood Institute, National Institutes of Health, Bethesda, MD, USA*
MELISSA L. STEWART • *Vollum Institute, Oregon Health and Science University, Portland, OR, USA*
GYULA TIMINSZKY • *Department of Physiological Chemistry, Biomedical Center Munich, Luwig-Maximilians-Universität München, Planegg-Martinsried, Germany; Institute of Genetics, Biological Research Centre of the Hungarian Academy of Sciences, Szeged, Hungary*
SEIJI URA • *Department of Biochemistry, University of Oxford, Oxford, UK*
CARMEN VALENTE • *Institute of Protein Biochemistry, National Research Council, Naples, Italy*
PATRICIA VERHEUGD • *Institute of Biochemistry and Molecular Biology, Medical School, RWTH Aachen University, Aachen, Germany*
JIM VOORNEVELD • *Gorlaeus Laboratories, Leiden Institute of Chemistry, Universiteit Leiden, Leiden, The Netherlands*
ANA B. A. WALLIS • *Department of Biochemistry, University of Oxford, Oxford, UK*
ZHIZHI WANG • *Department of Biological Structure, University of Washington, Seattle, WA, USA*
WENQING XU • *Department of Biological Structure, University of Washington, Seattle, WA, USA*
XIAOYU ZHANG • *Department of Chemistry and Chemical Biology, Cornell University, Ithaca, NY, USA*

Part I

Introduction

Chapter 1

Vitamin B3 in Health and Disease: Toward the Second Century of Discovery

Myron K. Jacobson and Elaine L. Jacobson

Abstract

This introductory chapter briefly reviews the history, chemistry, and biochemistry of NAD (the term NAD as it is used here refers to both oxidized and reduced forms of the molecule) consuming ADP-ribose transfer enzymes as components of the involvement of vitamin B3 in health and disease.

Key words Pellagra, Vitamin B3, Nicotinic acid, Nicotinamide, Nicotinamide riboside, ADP-ribose

1 Introduction

The history of vitamin B3 and B3-derived molecules impacting human health and disease dates back a bit more than a century. During the first half century of discovery, parallel research paths leading to landmark discoveries in both biochemistry and public health set the stage for understanding the biology that is the subject of this monograph. In 1906, the laboratory of Arthur Harden described the presence of a heat-stable low-molecular-weight factor that was required for alcoholic fermentation [1]. We now know that NAD was a key component of this factor. In 1915, Joseph Goldberger of the US Public Health Service was charged with determining the cause of the killer disease pellagra that was ravishing the southern regions of the United States. The story of Goldberger's heroic efforts is described in a monograph by Elizabeth W. Etheridge titled "The Butterfly Caste" [2]. The puzzling cause of pellagra was resolved in 1937 when the laboratory of Conrad Elvehjem reported that the simple pyridine derivatives nicotinic acid and nicotinamide could cure the canine equivalent of pellagra [3] and soon thereafter the effective treatment of human pellagra established a landmark in public health in the United States. The two vitamin forms were named niacin and niacinamide, respectively, to avoid public concern about the name nicotinic acid,

Paul Chang (ed.), *ADP-ribosylation and NAD+ Utilizing Enzymes: Methods and Protocols*, Methods in Molecular Biology, vol. 1813, https://doi.org/10.1007/978-1-4939-8588-3_1,

and they were collectively referred to as vitamin B3. In the same time frame as the discovery of B3, research in the laboratories of Otto Warburg and Hans von Euler led to the identification of NAD as the major bioactive form of B3 and its involvement in cellular oxidation/reduction reactions central to energy metabolism [4]. Their work represented a landmark in biochemistry.

The second half century of vitamin B3 research can be considered to have begun with the discovery starting in the 1960s of multiple classes of NAD-consuming enzymes with functions well beyond the simple catabolism of NAD. NAD consumption in the nucleus generating polymers of ADP-ribose was described at nearly the same time by the laboratories of Paul Mandel [5], Osamu Hayaishi [6], and Takashi Sugimura [7]. Concurrently, the Alvin Pappenheimer laboratory discovered that diphtheria toxin inhibits protein synthesis by catalyzing the transfer of a single ADP-ribose moiety of NAD to protein synthesis elongation factor-2 [8]. Subsequently, other bacterial toxins and later endogenous enzymes that catalyze transfer of single ADP-ribose moieties to cellular acceptors were discovered [9]. Still later, two additional classes of NAD-consuming enzymes were discovered: cyclic ADP-ribose synthases in the laboratory of Hon Cheung Lee [10] and NAD-dependent protein deacetylases called sirtuins in the laboratory of Leonard Guarente [11]. The second half century also led to the discovery by the Charles Brenner laboratory of a third form of vitamin B3, nicotinamide riboside [12].

The vitamin B3 metabolome of animal cells that has emerged over the first century of discovery is shown in Fig. 1. It is important to note that the complexity of the metabolome is greater than shown as the figure does not include cellular compartmentalization. NAD occupies a central role in cellular homeostasis as the nexus of energy regeneration, NADP generation to support cellular reductive needs, and multiple classes of ADP-ribose transfer reactions involved in cell signaling mechanisms with fundamental roles in cellular homeostasis that include the maintenance of genomic integrity and cellular adaptation to changes in energy status, with accompanying implications for health and disease.

The reaction catalyzed by all ADP-ribose transfer enzymes, shown in Fig. 2, demonstrates an underlying simplicity in the basic chemistry that nevertheless results in great biochemical diversity due to the wide range of acceptors utilized by the different classes of ADP-ribose transferases. This diversity can be attributed to the versatility of ADP-ribose as a group transfer agent. The ADP-ribose (ADPR) transferome (Fig. 3) shows cellular nucleophiles that have been identified to date that serve as ADP-ribose acceptors. While much focus of current research is on protein acceptors, DNA, RNA, and NAD itself have been identified as acceptors.

The versatility of ADP-ribose as a transfer agent dictates the need for the development and application of new experimental

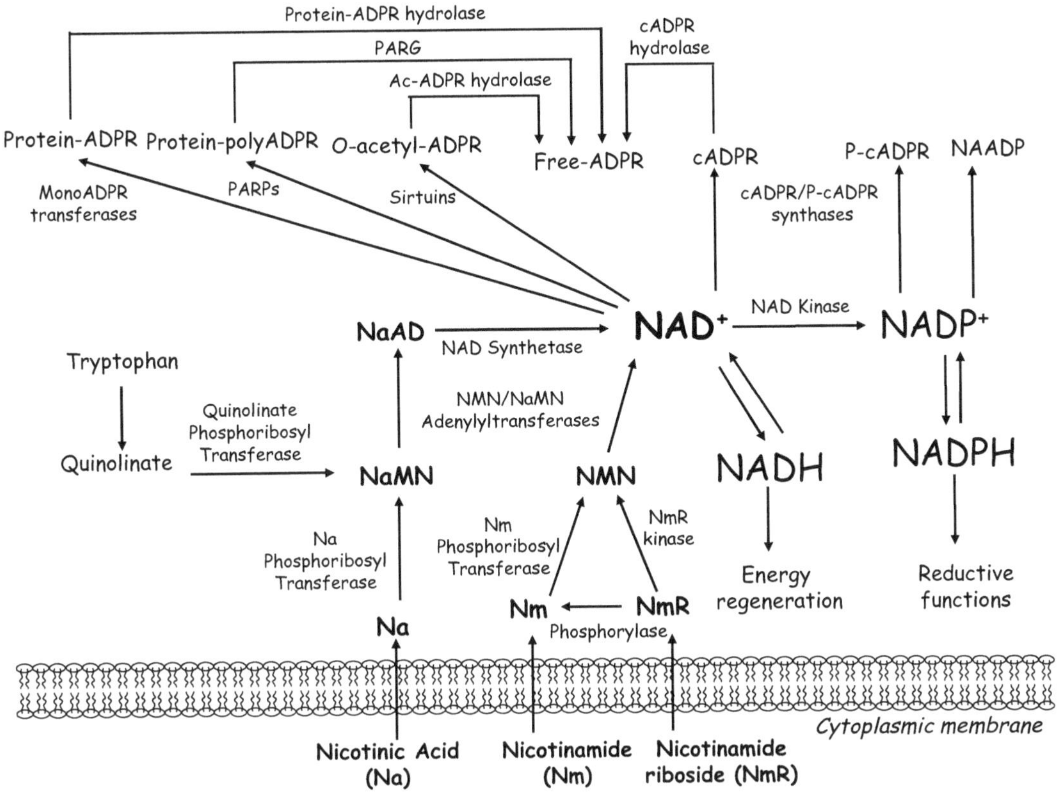

Fig. 1 An overview of the vitamin B3 metabolome in animal cells showing the integration of biosynthetic pathways, classes of NAD-consuming reactions involved in cell signaling, and the involvement in cellular energy regeneration and support of reductive functions

methods to assay NAD-utilizing enzymes, and numerous new methods are described in this monograph. Endogenous monomeric ADP-ribose transfer reactions particularly are incompletely understood, and new methods described here should prove valuable for a better understanding of this aspect of NAD metabolism. The chemical complexity of ADP-ribose and ADP-ribose polymers provides functional roles via non-covalent interactions with other cellular components, and methods for the study of this aspect of ADP-ribose metabolism described here should prove valuable.

The continual consumption of NAD by ADP-ribose transfer reactions dictates that cells continually replenish cellular NAD content, making an understanding of the pathways of NAD biosynthesis and their regulation an important aspect of the role of vitamin B3 in health and disease. Current evidence indicates that the de novo biosynthetic pathway from tryptophan makes a minor contribution to human tissue NAD content [13], making it likely that most tissues derive NAD by salvage pathways. Each vitamin B3 form is converted to NAD utilizing distinct enzymatic steps, which, together with the presence of biosynthetic enzymes in multiple cell

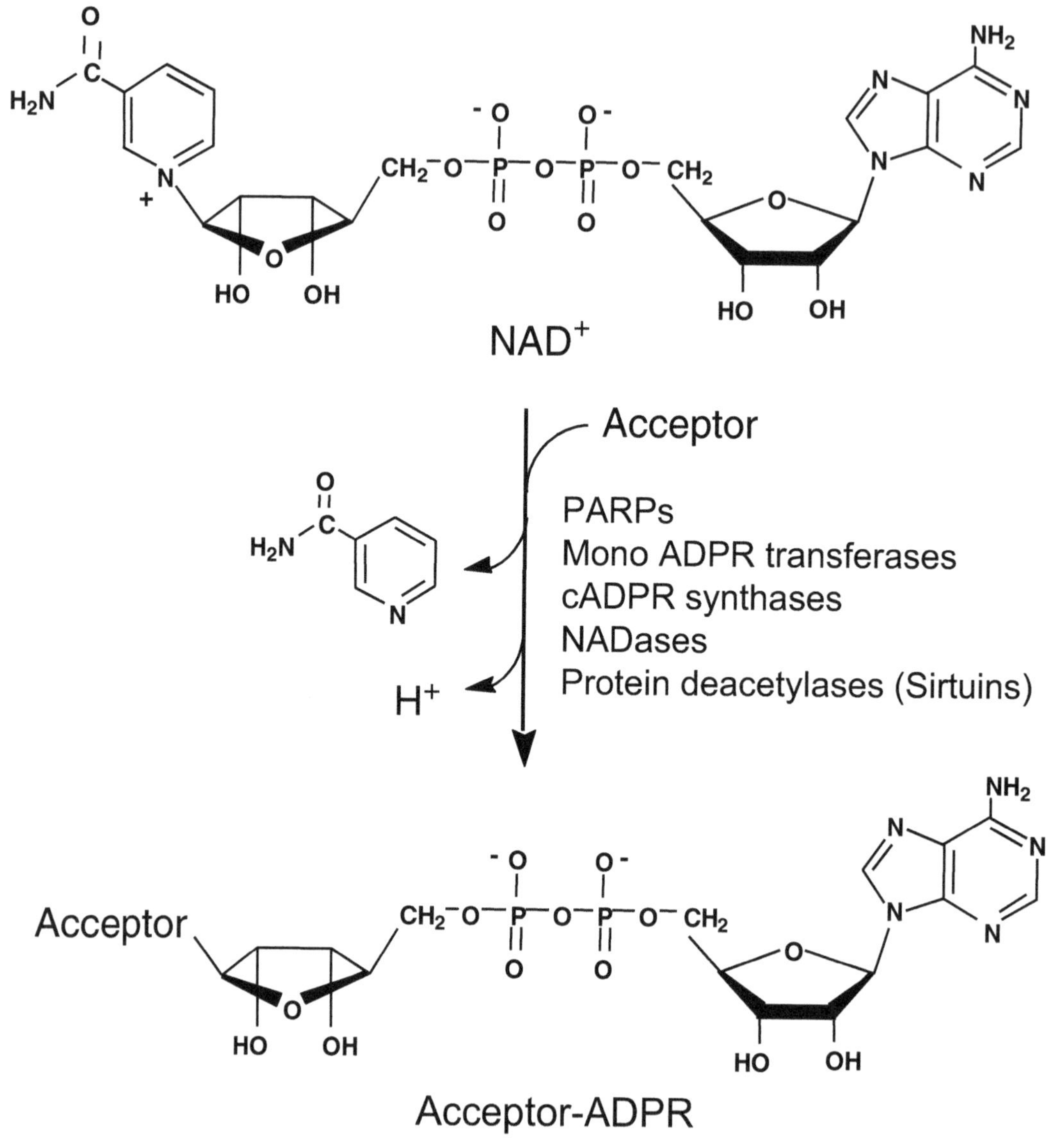

Fig. 2 The basic reaction catalyzed by cellular ADP-ribose transfer enzymes

compartments [14], allows for the potential independent regulation of the three salvage pathways. The regulation of NAD content is of interest as NAD content decline is a feature of many age-related diseases and conditions [15] and NAD replenishment treats and prevents many of these [16]. The ability to determine NAD content and its redox state in different cellular compartments will greatly contribute to our understanding of NAD metabolism, and a method described here should yield valuable information in this regard.

In his excellent review titled "PARPs and ADP-Ribosylation: 50 Years… and Counting" [17], Lee Kraus documents many major

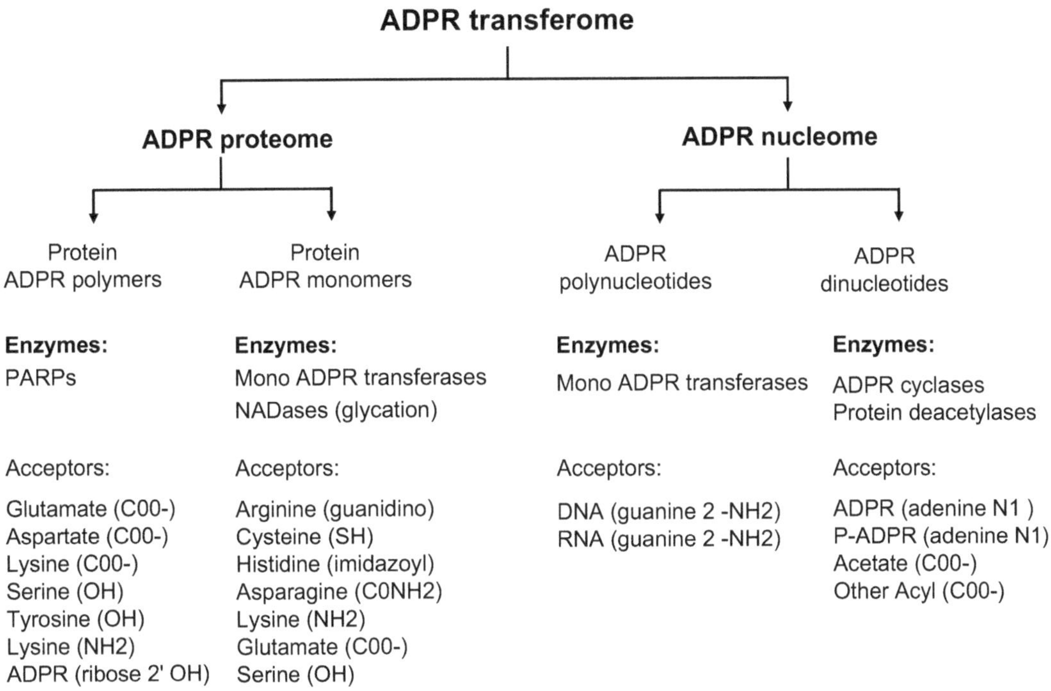

Fig. 3 The ADP-ribose (ADPR) transferome showing the wide range of cellular nucleophiles that have been shown to be modified by NAD-derived ADP-ribose

milestones between the first detection of ADP-ribose polymers in 1964 and the approval of the first PARP inhibitor for cancer therapy in 2014. Most milestones were the result of the development and application of new methods in chemistry, biochemistry, molecular biology, structural biology, genetics, genomics, and proteomics, leading finally to translation to the clinic.

Learning about nature is often limited by our ability to observe. There is much yet to learn as evidenced by the fact that B3 was shown to cure pellagra in the 1930s, yet molecular evidence as to how each of the classical symptoms of the four Ds of pellagra, diarrhea, dermatitis, dementia, and death is prevented by B3 is still incompletely understood. The methods described in this monograph will provide new tools to continue to unravel the remarkable involvement of vitamin B3 in health and disease.

References

1. Harden A, Young WJ (1906) The alcoholic fermentation of yeast-juice. Proc R Soc Lond B Biol Sci 77:405–420
2. Etheridge EW (1972) The butterfly caste: a social history of pellagra in the South. Greenwood Publishing Company, Westport
3. Elvehjem CA et al (1937) Relation of nicotinic acid and nicotinic acid amide to canine black tongue. J Am Chem Soc 59(9):1767–1768
4. Schlenk F (1984) The dawn of nicotinamide coenzyme research. Trends Biochem Sci 9(6):286–288

5. Chambon P, Weill JD, Mandel P (1963) Nicotinamide mononucleotide activation of new DNA-dependent polyadenylic acid synthesizing nuclear enzyme. Biochem Biophys Res Commun 11:39–43
6. Nishizuka Y et al (1967) Studies on the polymer of adenosine diphosphate ribose. I. Enzymic formation from nicotinamide adenine dinuclotide in mammalian nuclei. J Biol Chem 242(13):3164–3171
7. Fujimura S et al (1967) Polymerization of the adenosine 5′-diphosphate-ribose moiety of nicotinamide-adenine dinucleotide by nuclear enzyme. I. Enzymatic reactions. Biochim Biophys Acta 145(2):247–259
8. Goor RS, Pappenheimer AM Jr, Ames E (1967) Studies on the mode of action of diphtheria toxin. V. Inhibition of peptide bond formation by toxin and NAD in cell-free systems and its reversal by nicotinamide. J Exp Med 126(5):923–939
9. Moss J, Stanley SJ (1981) Amino acid-specific ADP-ribosylation. Identification of an arginine-dependent ADP-ribosyltransferase in rat liver. J Biol Chem 256(15):7830–7833
10. Clapper DL et al (1987) Pyridine nucleotide metabolites stimulate calcium release from sea urchin egg microsomes desensitized to inositol trisphosphate. J Biol Chem 262(20):9561–9568
11. Vaziri H et al (2001) hSIR2(SIRT1) functions as an NAD-dependent p53 deacetylase. Cell 107(2):149–159
12. Bieganowski P, Brenner C (2004) Discoveries of nicotinamide riboside as a nutrient and conserved NRK genes establish a Preiss-Handler independent route to NAD+ in fungi and humans. Cell 117(4):495–502
13. Fu CS et al (1989) Biochemical markers for assessment of niacin status in young men: levels of erythrocyte niacin coenzymes and plasma tryptophan. J Nutr 119(12):1949–1955
14. Dolle C et al (2013) NAD biosynthesis in humans—enzymes, metabolites and therapeutic aspects. Curr Top Med Chem 13(23):2907–2917
15. Verdin E (2015) NAD(+) in aging, metabolism, and neurodegeneration. Science 350(6265):1208–1213
16. Canto C, Menzies KJ, Auwerx J (2015) NAD(+) metabolism and the control of energy homeostasis: a balancing act between mitochondria and the nucleus. Cell Metab 22(1):31–53
17. Kraus WL (2015) PARPs and ADP-Ribosylation: 50 years … and counting. Mol Cell 58(6):902–910

Part II

ADP-Ribose Binding Domains as Tools

Chapter 2

Monitoring Poly(ADP-Ribosyl)ation in Response to DNA Damage in Live Cells Using Fluorescently Tagged Macrodomains

Rebecca Smith and Gyula Timinszky

Abstract

Poly(ADP-ribosyl)ation (PARylation) is a dynamic posttranslational modification that is added and removed rapidly at sites of DNA damage. PARylation is important for numerous aspects of DNA repair including chromatin decondensation and protein recruitment. Visualization of PARylation levels after DNA damage induction is generally obtained using traditional immunofluorescent techniques on fixed cells, which results in limited temporal resolution. Here, we describe a microscopy-based method to track ADP-ribosylation at break sites. This method relies on DNA damage induction using a 405 nm FRAP laser on Hoechst-treated cells expressing GFP-tagged PAR-binding proteins, such as macrodomains where the recruitment of the PAR-binder to sites of DNA damage gives an indication of PARylation levels.

Key words DNA damage response, Poly(ADP-ribosyl)ation, PARP1, macrodomain, Live-cell imaging, Microirradiation

1 Introduction

Poly(ADP-ribosyl)ation is one of the earliest detectable posttranslational modifications that occur during the DNA damage response [1]. It is highly dynamic, being generated at break sites by activated PARP1 and rapidly degraded by PARG [1, 2].

The most popular method to visualize cellular PAR levels relies on antibodies that bind PAR, such as the most frequently used 10H mouse monoclonal antibody. Yet, there is ever the desire to observe and study dynamic cellular events as they occur in their native environment: in a live cell. The antibody-based detection of PARylation can also be used to study PAR dynamics in cells in response to genotoxic stimuli. However, the changes in PAR levels are measured in populations of cells fixed at specific time intervals. Although this allows us to measure the amount of PAR over time at the population level, it does not facilitate measurement and visualization of local changes in PAR levels within the same cell over time.

Paul Chang (ed.), *ADP-ribosylation and NAD+ Utilizing Enzymes: Methods and Protocols*, Methods in Molecular Biology, vol. 1813, https://doi.org/10.1007/978-1-4939-8588-3_2,

Time-lapse imaging—taking a series of images over a period of time—can provide the means of capturing PAR dynamics in living cells but requires an easy to detect probe, based on light absorption-emission for light microscopy that either binds PAR or is incorporated into PAR as "PAR itself." The latter approach was successfully used by loading cells with a fluorescent NAD^+ analogue that readily accumulates at the sites of DNA damage induced by laser microirradiation [3, 4]. The use of cell-permeable, fluorescent NAD^+ analogues opens exciting, new avenues to study ADP-ribosylation. However, generating such molecules requires expertise in organic chemistry, which, without the commercialization of such analogues, will limit its use in the wider scientific community.

A good probe should be nontoxic and easy to deliver into cells and should not interfere with the metabolism of its target, in this case PAR. Fluorescent protein-tagged PAR-binding protein domains largely fulfill the above criteria. Being genetically encoded, they do not require any special delivery method other than delivering the plasmid DNA encoding the probe into the cell via conventional transfection. Of the several ADP-ribose-binding protein domains identified to date, such as the macrodomain, the PBZ (PAR-binding zinc finger) domain, the oligonucleotide-/oligosaccharide-binding fold, and the WWE domain [1, 2, 5, 6], macrodomains are the most frequently utilized PAR-binding protein modules to detect or enrich ADP-ribosylated proteins.

In this chapter, we provide a detailed method to examine PAR dynamics upon DNA damage in live cells through the recruitment of fluorescently tagged macrodomains to sites of laser-induced DNA damage. Our goal is to provide the reader with a method that is easy to implement on a range of microscopes.

2 Materials

2.1 Cell Culture Reagents

1. Humidified 37 °C incubator with 5% CO_2.
2. Exponentially growing human U-2 OS (*ATCC HBT-96*) cells.
3. Dulbecco's Modified Eagle Medium (with 4.5 g/L glucose) supplemented with 10% fetal bovine serum, 2 mM glutamine, 100 μg/mL penicillin, and 100 U/mL streptomycin.
4. Phenol-Red-free Leibovitz's L-15 medium supplemented with 10% fetal bovine serum, 2 mM glutamine, 100 μg/mL penicillin, and 100 U/mL streptomycin.
5. Lab-Tek II chambered cover glass.
6. Phosphate-buffered saline (PBS).
7. Trypsin-EDTA.
8. Hemocytometer for cell counting.

2.2 Transfections and Treatments

1. eYFP- human macroH2A1.1 macrodomain (amino acids 162–372) [7], monomeric EGFP-human macroD2 macrodomain (amino acids 7–243) [8], or monomeric EGFP-human PARP14 macrodomain (amino acids 998–1196 of PARP14 as described earlier [9] were amplified and inserted into pEGFP-C1 using *BsrG*I and *Xba*I).
2. Xfect transfection reagent.
3. Hoechst 33342 (10 mg/mL).

2.3 Image Acquisition

1. A Zeiss AxioObserver Z1 microscope equipped with a Yokogawa CXU-X1 confocal scanner unit, an AxioCam HRm CCD camera, and a Zeiss C-Apochromat ×63/1.2 water-immersion objective.
2. RAPP DL-405 nm diode laser coupled through the epifluroscence backboard of the microscope.
3. Compact power and energy meter console (Thorlabs) and standard photodiode power sensor, S120VC (Thor Labs).
4. Calibration slide (optional).

2.4 Image Analysis

1. ImageJ—including "StackReg" plugin.
2. Microsoft Excel.

3 Methods

3.1 Cell Culture

Cells are seeded into an 8-well Lab-Tek chambered cover glass prior to transfection allowing for image acquisition on an inverted confocal microscope. The following experiment has been optimized for U-2 OS cells. Cell numbers and incubation times after transfection may vary depending on cell type used and may require optimization.

Cells are cultured in Dulbecco's Modified Eagle Medium (with 4.5 g/L glucose) supplemented with 10% fetal bovine serum, 2 mM glutamine, 100 μg/mL penicillin, and 100 U/mL streptomycin in a 37 °C humidified incubator with 5% CO_2:

1. Aspirate medium from a flask of cells that are approximately 80% confluent.
2. Add prewarmed PBS to the flask to wash the cells.
3. Aspirate PBS.
4. Add enough prewarmed trypsin-EDTA solution to the flask to cover the cells and leave on the cells for approximately 30 s.
5. Observe cells under a basic inverted light microscope. When cells are rounded, gently tap the flask to detach them from the flask.

6. Resuspend cells in fresh prewarmed complete medium. Add at least double the volume of the trypsin-EDTA solution used to inactivate trypsin.
7. Count a small aliquot of cell suspension and determine the cell density.
8. Dilute cells to 70,000 cells/mL and aliquot 300 μL of cell suspension into each well of an 8-well Lab-Tek II chambered cover glass.
9. Return the Lab-Tek to the incubator, and incubate overnight to allow the cells to attach to the glass.

3.2 Transfection

The transfection reaction detailed below is enough for two wells (plus coverage for pipetting errors) of an 8-well Lab-Tek II chambered cover glass using Xfect transfection reagent. Alternative transfection reagents can be also used after optimization:

1. Aliquot 75 μL of room temperature Xfect reaction buffer into a 1.5 mL tube.
2. Add 1 μg of pmEGFP macrodomain to the buffer and mix gently (*see* **Notes 1** and **2**).
3. Add 0.3 μL of Xfect transfection reagent to the DNA-buffer mixture and vortex for 10 s.
4. Incubate the DNA-buffer-polymer mixture for 15 min at room temperature.
5. Aspirate medium from the wells, and add 140 μL of fresh complete DMEM to each well.
6. Add 30 μL of reaction mixture dropwise to each well.
7. Return the Lab-Tek to the incubator for 4 h.
8. Aspirate medium, and add 300 μL fresh complete DMEM to the cells before returning the Lab-Tek to the incubator.
9. Incubate cells for 24 h.

3.3 Microscopy, DNA Damage Induction, and Image Acquisition

This method describes DNA damage induction and image acquisition to observe fluorescently tagged macrodomain proteins to sites of DNA damage. The following experiment has been optimized using a Zeiss AxioObserver Z1 confocal spinning-disk microscope equipped with an AxioCam HRm CCD camera (Zeiss), Yokogawa CXU-X1 confocal scanner, and a Zeiss C-Apochromat ×63/1.2 water-immersion objective. DNA damage is completed by microirradiating a selected area in the nucleus of Hoechst-presensitized cells with a RAPP DL-405 nm diode laser, which is coupled through the epifluorescence backboard of the microscope. This method can be easily modified for implementation on any FRAP-capable spinning-disk or laser-scanning confocal microscope.

3.3.1 Microscope Setup

1. Turn on the microscope, including the lasers that are required for image acquisition and microirradiation.
2. Turn on the heating system (*see* **Note 3**).
3. Allow lasers to warm for at least 1 h prior to beginning the experiment (*see* **Note 4**).

3.3.2 Calibration of the FRAP Laser

This step will vary depending on the setup of the microscope and the type of FRAP system. Calibration of the FRAP laser will ensure accurate and reproducible DNA damage induction. The following instructions are for calibration of RAPP 405 nm diode laser. Ensure you follow the instruction applicable to your individual FRAP system. For systems that do not require calibration (e.g., point-scanning confocal microscopes), skip ahead to Subheading 3.3.3.

1. Select the ×63 objective, and apply water as the immersion fluid.
2. Place the calibration slide in the slide holder. The coverslip that is the same thickness as the Lab-Tek cover glass should be facing the objective.
3. Focus on the grid or fluorescent marker.
4. Begin imaging with low exposure times (10 ms) with the imaging laser set to the lowest power.
5. On the RAPP software, select the 405 nm laser in the calibration drop-down menu.
6. Click on center spot to ensure the laser is targeted in the middle of the objective.
7. Open the 405 nm shutter, and turn on the laser. You will see a small spot in the center of the region of interest. Adjust the fine focus until this spot is the smallest it can be.
8. Switch the laser off once the spot is in focus.
9. In the calibration menu, select automatic calibration of the 405 nm laser, and allow calibration to continue to completion.
10. Click center spot to once ensure the laser is centered on the objective.
11. Remove the calibration slide and the immersion fluid.
12. Take the power meter, and place the sensor securely over the objective.
13. Switch on the 405 nm laser, and adjust the power percentage until you have determined the appropriate level (*see* **Notes 5** and **6**).
14. Switch the laser off, and close the 405 nm laser shutter.

3.3.3 DNA Damage and Image Acquisition

DNA damage is inflicted on Hoechst-treated cells using irradiation with 405 nm light. The rapid nature of PARyation and PAR signaling means images should be acquired with a minimum of 5 s between each image to ensure good temporal resolution.

1. One hour prior to imaging, replace the medium on the cells with Leibovitz medium containing 0.3 μg/mL Hoechst (*see* **Note 7**).
2. Immediately prior to imaging, remove Hoechst-containing Leibovitz medium, and replace with fresh Leibovitz medium without Hoechst.
3. Add immersion fluid to the ×63 objective, and place the Lab-Tek on the slide holder.
4. Find the focus of cells using bright-field or GFP epifluorescence (*see* **Note 8**).
5. On the Zeiss software, use continuous mode to focus on cells.
6. On the RAPP software, use the line function to draw a line across the nucleus. Adjust the duration time of the line and the percentage of the 405 nm laser (*see* **Note 9**).
7. Copy this line to the other nuclei within the field of view.
8. Upload the run to the RAPP hardware and ensure the shutter is open.
9. On the Zeiss software set the experimental details (*see* **Note 10**). The total duration of the experiment should be set to 10 min with images collected every 5 s (*see* **Note 11**).
10. Begin image acquisition using the 488 nm laser for excitation of the eGFP-tagged macrodomain.
11. Capture at least three images before triggering the microirradiation.
12. When the time series is completed, export data as an OME-TIF or as individual .tif images.

3.4 Image Analysis

The image analysis requires the image processing software ImageJ. Ensure this is installed on your computer before proceeding with the image analysis. Here we describe a simple method for quantifying the recruitment of GFP-tagged macrodomain to the site of DNA damage.

1. Open ImageJ.
2. Open "Set Measurements" found in the Analyze drop-down menu.
3. Select "Mean gray value," and set the number of decimal places to 3.
4. Open ROI manager found in Analyze → Tools → ROI manager.
5. In ImageJ, open the OME-TIF file, or import the individual TIF images using the import image sequence found in File → Import → Image sequence.
6. Select the square tool, and draw a square around an individual nucleus.

7. Duplicate this area using the duplicate stack function (Ctrl + Shift + D)—make sure to click the duplicate stack button.
8. Register the stack by going to Plugins → Registration → StackReg. Select "Rigid Body" in the "Transformation" drop-down menu and select OK (*see* **Note 12**).
9. Go to Image → Lookup Tables → and select HiLo (*see* **Note 13**).
10. Define an area of background using the polygon tool, and add this area to the ROI manager with the "Add" button.
11. Define the nucleus using the polygon tool, and add this area to the ROI manager.
12. Define the area that was irradiated with the 405 nm laser, and add to the ROI manager (*see* **Note 14**).
13. With the image stack still open, select the three areas in the ROI manager, and use "Multi Measure" found under the "More" drop-down menu, and measure the gray values within the areas.
14. Copy the data from the Results pop-up window, and paste into an Excel spreadsheet. The recruitment of the macrodomain to the site of damage can be calculated as Damage fluorescence—Background fluorescence/Nucleus fluorescence—Background fluorescence and normalized to the time point before microirradiation.
15. Complete **steps 6–14** for each nucleus that has been irradiated.
16. Plot the average of the normalized recruitment with error bars showing standard error of the mean (SEM).

4 Notes

1. Fluorescent proteins in nature usually form dimers or tetramers. The formation of dimers between the fluorescent tags might have adverse effects on the behavior of the protein of interest—in this case the PAR-binding probe—and should be avoided. Because of this, we suggest using true monomeric fluorescent proteins or the A206K mutation that makes GFP and its variants monomeric [10–12].
2. The following should be considered when selecting an ADP-ribose-binding probe. What is its substrate specificity? What is its affinity for the substrate? Does it have enzymatic activity toward its substrate? There is an array of ADP-ribose-binding protein domains; however, not all are equally suitable for monitoring PARylation. The WWE domain only binds PAR and not mono-ADP-ribose [13, 14]; however, it binds PAR

very tightly, and in our experience, its overexpression can alter intracellular PAR dynamics. For this reason, it should be used with caution. Some of the macrodomains were reported to bind only mono-ADP-ribosylated proteins [8, 9], yet, some of them have enzymatic activity as well. When it comes to monitoring ADP-ribosylation, we suggest using the macrodomain of the macroH2A1.1 histone variant (Fig. 1a). This macrodomain binds both mono-ADP-ribose and PAR and does not have an extremely tight affinity for ADP-ribose [7, 15]. The recruitment of other mono-ADP-ribose-specific macrodomains (Fig. 1b, c) differs from that of the macroH2A1.1 macrodomain (Fig. 1a). Some of the differences could be related to their binding specificity; however, the experimental evidence is still largely missing.

3. To reduce the amount of stress cells are under while image acquisition occurs, it is important to maintain the temperature to 37 °C. This is normally obtained through an environmental chamber encompassing the microscope. The heating system will also help prevent large drifts in focus that may occur during image acquisition and increase the reliability of results.
4. Lasers for imaging and FRAP should be allowed to warm for at least 1 h prior to the start of the experiment. This will ensure that the laser power has reached a steady state and will not change throughout the course of the experiment. Failure to obtain this steady power state will result in an increase in the variability of data.
5. The energy to induce DNA damage is suggested at around 1 $\mu J/\mu m^2$ where the energy is determined through the power of the laser and the time taken to draw the line, and μm^2 is the area of the line that is drawn. For the setup described, this corresponds to a 15 μm line illuminated for 200 ms with a 405 nm laser set to 35 μW (Fig. 1). The spot size this laser produces is approximately 1 μm. Systems that have a larger spot size will require a combination of higher laser power and longer irradiation times to induce comparable amount of DNA damage.
6. Varying the power of the laser used to FRAP will lead to different amounts of DNA damage induction, which will alter the recruitment of the GFP-tagged macrodomain. To determine the appropriate amount of 405 nm laser power that is required to induce DNA damage, we suggest carrying out a preliminary experiment where the power of the laser and/or the time of irradiation with the 405 nm laser is varied and recruitment of the macrodomain in each setting assessed. An example of such an experiment can be seen in Fig. 1. Altering the laser power as well as the irradiation time changes the peak recruitment of the GFP-tagged macrodomains as well as their release dynamics.

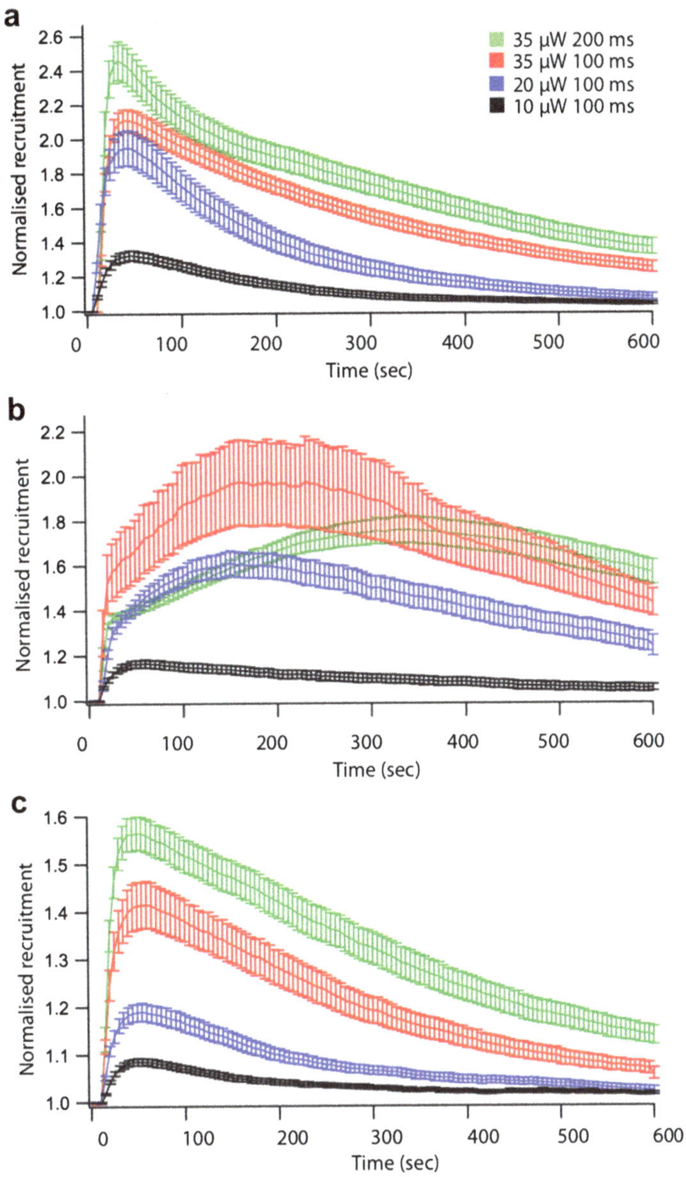

Fig. 1 Varying the amount of 405 nm laser power or illumination alters macrodomain recruitment dynamics. Cells expressing (**a**) the macrodomain of mH2A1.1, (**b**) the macrodomain of macroD2, or (**c**) the macrodomain 2 of PARP14 were irradiated with a 405 nm laser over different laser powers or with different illumination times (green, 35 μW, 200 ms; red, 35 μW, 100 ms; blue, 20 μW, 100 ms; black, 10 μW, 100 ms). The recruitment of the macrodomain at the damage site is normalized to before damage induction. Error bars show SEM

7. We find that concentrations of Hoechst higher than 0.3 μg/mL show cytotoxicity and should be avoided. The concentration could be lowered to 0.1 μg/mL; however, this will also alter the amount of DNA damage induced with the 405 nm

laser irradiation. Cells can also be sensitized through BrdU incorporation. BrdU is typically used at a concentration of 10 μM in the growth medium for 24 h prior to imaging to allow essentially all cells to replicate DNA and thus incorporate BrdU. However, care should be taken when BrdU incorporation is combined with transfection as the transfection might alter cell cycle progression and thus interfere with uniform BrdU labeling of the cell population [16]. While the current method specifically describes how to induce DNA damage in Hoechst sensitized cells with 405 nm irradiation, DNA damage can be induced with a variety of lasers also without the requirement for presensitization [17].

8. Refrain from using UV illumination during the initial focusing step as this may induce DNA damage in the Hoechst-presensitized cells and alter PARylation dynamics/macrodomain recruitment.
9. Make sure that the 405 nm laser is used for "FRAP" and the 488 nm for imaging the GFP-tagged macrodomain.
10. We have described this experiment using GFP-tagged macrodomains. These fusion proteins generally have small molecular weights and therefore express very well. With a transient transfection, there will likely be cells with various amounts of protein expression. We did not detect differences in DNA damage-induced ADP-ribosylation levels when comparing cells that express varying amounts of the GFP-tagged macrodomain of the macroH2A1.1 histone variant (Fig. 2). Nevertheless, we suggest choosing cells with similar expression levels. Importantly, it is necessary to choose cells that have fluorescence levels up to a maximum of 25% of the dynamic range of the camera. When DNA damage is induced, the recruitment of GFP-tagged macrodomains will result in localized areas with high fluorescence. If the original fluorescence is too high, information may be lost when the camera becomes saturated. We find that cells with very high expression levels often have abnormal nuclear morphology and should be avoided.
11. Microscopy is always a compromise between the emitted signal and the adverse effects of illumination, such as cytotoxicity and photobleaching. The cytotoxicity may vary between cell lines. Rounded-up cells, fragmented mitochondria, or the appearance of intracellular vesicles upon imaging could indicate cytotoxicity. The intensity of lasers used should be kept to a minimum level that allows for a good fluorescent signal emitted by the probe (GFP). Furthermore, the intervals the images are taken during the time-lapse should be kept to the minimum that still captures the dynamics of PARylation

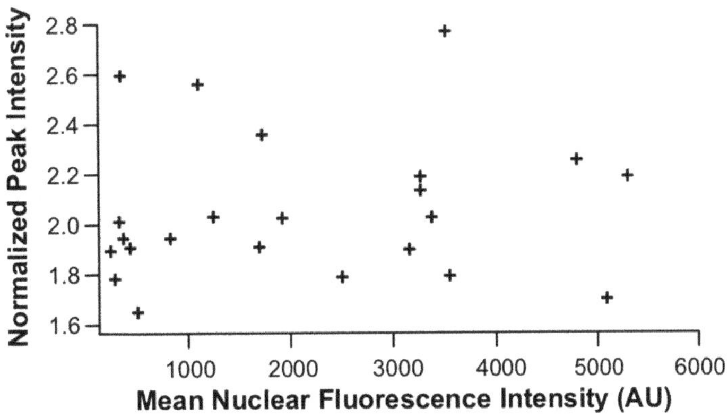

Fig. 2 Overall expression level of macrodomains does not affect peak recruitment levels. Nuclear fluorescence was plotted against the normalized peak recruitment value of the macroH2A1.1 macrodomain

required for the planned experiment. The above considerations are true for photobleaching as well. Enhanced GFP is quite photostable, and we experience negligible photobleaching during our typical experiments. There is a vast array of fluorescent proteins that might be used instead of GFP (for an interactive website visit: http://www.fpvis.org); their photostability but also brightness can differ from that of GFP, and imaging parameters may require optimization.

12. During the 10-min acquisition, there might be rotation/movement of the nuclei. Registration will ensure that the nucleus is in the same position in each frame allowing for accurate measurements to be acquired.
13. This will indicate areas that have high gray values in red and low/no gray values in blue. The registration step will result in areas at the edge of the stack of images that do not have gray values and will appear as blue regions. It is important to avoid these areas when selecting the background region for measurements (Fig. 3)
14. As all nuclei are irradiated with the same sized line, the measured damaged area should also be the same in each nucleus. An alternative strategy to identify the damaged area is through coexpression of a photoactivatable fluorescent protein-tagged chromatin marker such as PATagRFP-H2B with GFP macrodomain. Illumination with 405 nm light in these cells that have been presensitized with Hoechst will induce damage and simultaneously mark the area that has been damaged [18]. The PATagRFP images can then be used for the identification of the irradiated area.

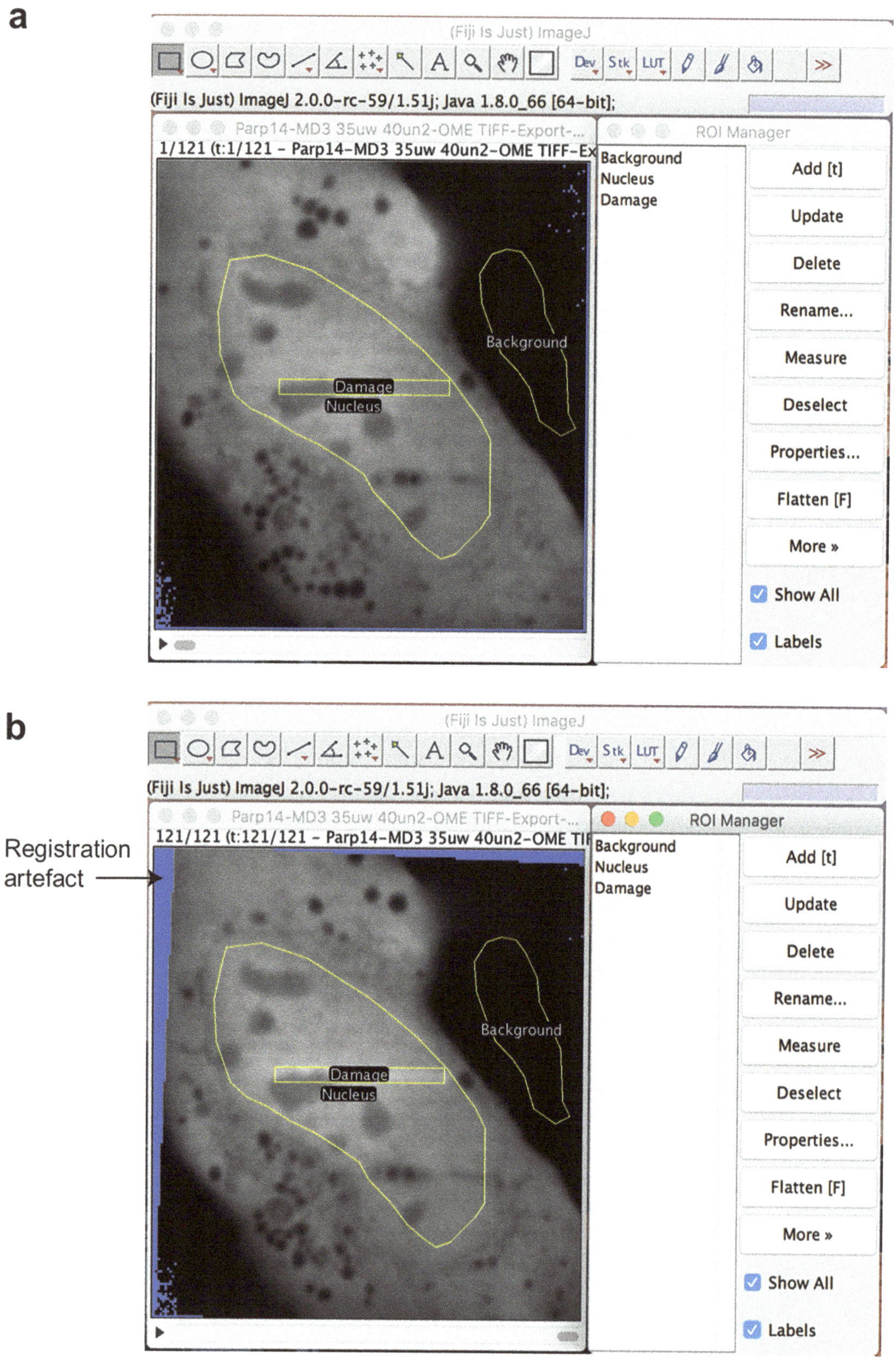

Fig. 3 Registration of image stacks can produce artifacts that should be avoided during analysis. (**a**) Nuclei are registered to time 0 and show no registration artifacts. (**b**) At later time points, registration artifacts (in blue) become apparent and should be avoided when defining areas to be measured. Areas for the background, nucleus, and damage have been defined in both images

Acknowledgments

This work was supported by the Worldwide Cancer Research grant (#14-1315) and the Deutsche Forschungsgemeinschaft grant (TI 817/2-1) to G.T. The authors declare no competing financial interests.

References

1. Barkauskaite E, Jankevicius G, Ladurner AG, Ahel I, Timinszky G (2013) The recognition and removal of cellular poly(ADP-ribose) signals. FEBS J 280(15):3491–3507. https://doi.org/10.1111/febs.12358
2. Barkauskaite E, Jankevicius G, Ahel I (2015) Structures and mechanisms of enzymes employed in the synthesis and degradation of PARP-dependent protein ADP-ribosylation. Mol Cell 58(6):935–946. https://doi.org/10.1016/j.molcel.2015.05.007
3. Buntz A, Wallrodt S, Gwosch E, Schmalz M, Beneke S, Ferrando-May E, Marx A, Zumbusch A (2016) Real-time cellular imaging of protein poly(ADP-ribos)ylation. Angew Chem Int Ed Engl 55(37):11256–11260. https://doi.org/10.1002/anie.201605282
4. Wallrodt S, Buntz A, Wang Y, Zumbusch A, Marx A (2016) Bioorthogonally functionalized NAD(+) analogues for in-cell visualization of poly(ADP-ribose) formation. Angew Chem Int Ed Engl 55(27):7660–7664. https://doi.org/10.1002/anie.201600464
5. Zhang F, Shi J, Bian C, Yu X (2015) Poly(ADP-ribose) mediates the BRCA2-dependent early DNA damage response. Cell Rep 13(4):678–689. https://doi.org/10.1016/j.celrep.2015.09.040
6. Zhang F, Chen Y, Li M, Yu X (2014) The oligonucleotide/oligosaccharide-binding fold motif is a poly(ADP-ribose)-binding domain that mediates DNA damage response. Proc Natl Acad Sci U S A 111(20):7278–7283. https://doi.org/10.1073/pnas.1318367111
7. Timinszky G, Till S, Hassa PO, Hothorn M, Kustatscher G, Nijmeijer B, Colombelli J, Altmeyer M, Stelzer EH, Scheffzek K, Hottiger MO, Ladurner AG (2009) A macrodomain-containing histone rearranges chromatin upon sensing PARP1 activation. Nat Struct Mol Biol 16(9):923–929. https://doi.org/10.1038/nsmb.1664
8. Jankevicius G, Hassler M, Golia B, Rybin V, Zacharias M, Timinszky G, Ladurner AG (2013) A family of macrodomain proteins reverses cellular mono-ADP-ribosylation. Nat Struct Mol Biol 20(4):508–514. https://doi.org/10.1038/nsmb.2523
9. Forst AH, Karlberg T, Herzog N, Thorsell AG, Gross A, Feijs KL, Verheugd P, Kursula P, Nijmeijer B, Kremmer E, Kleine H, Ladurner AG, Schuler H, Luscher B (2013) Recognition of mono-ADP-ribosylated ARTD10 substrates by ARTD8 macrodomains. Structure 21(3):462–475. https://doi.org/10.1016/j.str.2012.12.019
10. Campbell RE, Tour O, Palmer AE, Steinbach PA, Baird GS, Zacharias DA, Tsien RY (2002) A monomeric red fluorescent protein. Proc Natl Acad Sci U S A 99(12):7877–7882. https://doi.org/10.1073/pnas.082243699
11. Zacharias DA, Violin JD, Newton AC, Tsien RY (2002) Partitioning of lipid-modified monomeric GFPs into membrane microdomains of live cells. Science (New York, NY) 296(5569):913–916. https://doi.org/10.1126/science.1068539
12. Snapp EL, Hegde RS, Francolini M, Lombardo F, Colombo S, Pedrazzini E, Borgese N, Lippincott-Schwartz J (2003) Formation of stacked ER cisternae by low affinity protein interactions. J Cell Biol 163(2):257–269. https://doi.org/10.1083/jcb.200306020
13. Wang Z, Michaud GA, Cheng Z, Zhang Y, Hinds TR, Fan E, Cong F, Xu W (2012) Recognition of the iso-ADP-ribose moiety in poly(ADP-ribose) by WWE domains suggests a general mechanism for poly(ADP-ribosyl)ation-dependent ubiquitination. Genes Dev 26(3):235–240. https://doi.org/10.1101/gad.182618.111
14. Zhang Y, Liu S, Mickanin C, Feng Y, Charlat O, Michaud GA, Schirle M, Shi X, Hild M, Bauer A, Myer VE, Finan PM, Porter JA, Huang SM, Cong F (2011) RNF146 is a poly(ADP-ribose)-directed E3 ligase that regulates axin degradation and Wnt signalling. Nat Cell Biol 13(5):623–629. https://doi.org/10.1038/ncb2222
15. Kustatscher G, Hothorn M, Pugieux C, Scheffzek K, Ladurner A (2005) Splicing regu-

lates NAD metabolite binding to histone macroH2A. Nat Struct Mol Biol 12(7):624–625. https://doi.org/10.1038/nsmb956

16. McBroom CA, Sheinin R (1991) Effect of transfection manipulations on mouse cell cycle progression. Biochem Cell Biol 69(9):665–669

17. Kong X, Mohanty SK, Stephens J, Heale JT, Gomez-Godinez V, Shi LZ, Kim JS, Yokomori K, Berns MW (2009) Comparative analysis of different laser systems to study cellular responses to DNA damage in mammalian cells. Nucleic Acids Res 37(9):e68. https://doi.org/10.1093/nar/gkp221

18. Sellou H, Lebeaupin T, Chapuis C, Smith R, Hegele A, Singh HR, Kozlowski M, Bultmann S, Ladurner AG, Timinszky G, Huet S (2016) The poly(ADP-ribose)-dependent chromatin remodeler Alc1 induces local chromatin relaxation upon DNA damage. Mol Biol Cell 27(24):3791–3799. https://doi.org/10.1091/mbc.E16-05-0269

Chapter 3

In Vitro Techniques for ADP-Ribosylated Substrate Identification

Giovanna Grimaldi, Giuliana Catara, Carmen Valente, and Daniela Corda

Abstract

ADP-ribosylation is a post-translational modification of proteins that has required the development of specific technical approaches for the full definition of its physiological roles and regulation. The identification of the enzymes and specific substrates of this reaction is an instrumental step toward these aims. Here we describe a method for the separation of ADP-ribosylated proteins based on the use of the ADP-ribose-binding *macro* domain of the thermophilic protein Af1521, coupled to mass spectrometry analysis for protein identification. This method foresees the coupling of the *macro* domain to resin, an affinity-based pull-down assay, coupled to a specificity step resulting from the clearing of cell lysates with a mutated *macro* domain unable to bind ADP-ribose. By this method both mono- and poly-ADP-ribosylated proteins have been identified.

Key words ADP-ribosylation, *Macro* domain, Af1521 *macro*-domain purification, DMP cross-linker, *Macro*-domain-based pulldown, PARP, ART, ADP-ribosylated substrates, ADP-ribose, PAR

1 Introduction

Post-translational modifications of proteins represent key mechanisms in the regulation of cellular events [1–3]. We report here a method for the identification of the substrates of the ADP-ribosylation reaction, originally developed with the aim of elucidating the protein modification/function relationship [4].

The ADP-ribosylation field has suffered for many years a lack of specific tools to identify the modified proteins, this being particularly the case for the substrates of the mono-ADP-ribosylation (MARylation) enzymes [(ADP-ribosyltransferases cholera toxin-like (ARTCs), ADP-ribosyltransferases diphtheria toxin-like (ARTDs), poly-ADP-ribose-polymerases (PARPs); [5]]. Instead, the poly-ADP-ribosylated (PARylated) substrate identification was initially based on the use of the monoclonal 10H antibody, which binds poly-ADP-ribose (PAR) composed of at least 10 ADP-ribose units [6]. Due to its strong affinity toward PAR, 10H antibody is

Paul Chang (ed.), *ADP-ribosylation and NAD+ Utilizing Enzymes: Methods and Protocols*, Methods in Molecular Biology, vol. 1813, https://doi.org/10.1007/978-1-4939-8588-3_3, © Springer Science+Business Media, LLC, part of Springer Nature 2018

routinely used to enrich PARylated proteins [7–9], thus allowing only the study of the events catalyzed by the poly-ADP-ribose-polymerases of the PARP family. In the case of the MARylated substrates, attempts to develop antibodies specific for the detection of mono-ADP-ribose have been made during the years, but they had a poor application in the field [10–13]. This is due to the fact that these antibodies could detect specific proteins modified in vitro by either cellular enzymes or toxins, while they did not recognize well-known cellular MARylated proteins, possibly due to the fact that the specific epitope included crucial residues in the region of the bound ADP-ribose of discrete substrates [10–13].

Taking advantage of the protein module- the *macro* domain- of the thermophilic protein Af1521 from *Archaeoglobus fulgidus* that recognizes monomeric and/or polymeric forms of ADP-ribose [14], we successfully explored the possibility that this could recognize also the ADP-ribose bound to proteins and, more specifically, the modified proteins isolated from cell lysates [4]. This *macro* domain has been extensively characterized and has been shown to be a highly specific binder of ADP-ribose, with a K_d of 0.13 μM. It binds the distal ribose of the ADP-ribose moiety and thus can recognize both mono- and poly-ADP-ribose [14].

Here we describe the affinity purification method we developed for the study of both the mono- and poly-ADP-ribosylation reactions; this is based on a *macro*-domain-based pull-down assay (to enrich for ADP-ribosylated proteins) and then followed by mass spectrometry (MS) analysis to identify new substrates of this post-translational modification [4].

Following this initial Af1521 *macro*-domain-based method, other ADP-ribose binding modules have been investigated for the identification of ADP-ribosylated proteins. Among these, the catalytically inactive E756D mutant of the poly-ADP-ribose glycohydrolase (PARG-DEAD, a distant member of the *macro*-domain family [15, 16]) was employed in a substrate-trapping strategy [8]. The transfected GFP-tagged PARG-DEAD able to bind ADP-ribosylated proteins (without hydrolyzing PAR) was immunoprecipitated and the substrates identified by liquid chromatography coupled to tandem MS (LC-MS/MS). A comparison of this affinity purification method with the Af1521 *macro* domain and with the 10H antibody affinity-based enrichment indicated a good overlap of substrate recognition with both (96% and 73% overlapping, respectively) [8]. At the same time, these methods showed a degree of specificity toward the substrates identified following genotoxic stress: PARG-DEAD preferentially recovered nuclear proteins; Af1521 *macro* domain under these conditions mainly identified nuclear substrates; 10H antibody preferentially isolated mitochondrial proteins. This may derive from differences in the primary structures of these ADP-ribose-binding modules leading to different affinities for the modified peptides and for the binding to either mono- or poly-ADP-ribosylated substrates. Accordingly, the three *macro* domains (*macro* 1–3) of the mono-ADP-ribosyltransferase PARP14 were shown to specifically interact with MARylated PARP10 and with one of its substrate, the small GTPase Ran (but not with poly-ADP-ribose polymers) [17]. Instead, the *macro* domain of human histone H2A.1, involved in chromatin rearrangement after DNA damage, was shown to pull down PARylated proteins including PARP1, Ku70/80, and DNA-PKc [18]. Finally, the WWE domain from RNF146/Iduna recognizes *iso*-ADP-ribose [19], the minimal structural unit of PAR, and thus can be used to isolate and identify PARylated proteins [20].

The *macro*-domain-based pull-down method detailed below foresees the use of both the wild-type Af1521 *macro* domain and its mutant version (the G42E mutant) unable to bind ADP-ribose and thus used as control of specificity. The *macro*-domain proteins were covalently immobilized onto glutathione Sepharose 4B beads using the dimethyl pimelimidate (DMP) imidoester cross-linker. DMP was selected based on its efficient cross-linking activity under mild conditions of pH and temperature, a useful feature to preserve the native structure of the *macro* protein(s). The use of these *macro*-domain-bound beads allows the increase of the molar excess of the *macro* modules *versus* the total cellular ADP-ribosylated proteins, which improves the recovery in the pull-down assay.

The mutated *macro*-domain-bound beads can be used either to preclear the lysate that in turn is incubated with the wild-type *macro*-domain-bound beads [4] or in parallel [21]. Proteins bound by the G42E *macro* domain have to be considered as non-specific background binders. The modified proteins are then eluted and analyzed by MS.

Of note, the described experimental protocol is performed under denaturing conditions; only those proteins covalently attached to ADP-ribose moieties are captured (thus discriminating between non-covalent and covalent binding to ADP-ribose or PAR).

The detailed procedure to obtain the *macro*-domain protein(s) and the pull-down assay procedure are described in the next paragraph.

2 Materials

It is recommended to use ultrapure chemicals and Milli-Q water for preparing all buffers and solutions.

2.1 Expression and Purification of Af1521 Macro Domain(s)

1. Luria–Bertani liquid medium (LB): 1% (w/v) tryptone, 0.5% (w/v) yeast extract, 1% NaCl (w/v), pH 7.0 (volume, 1.5 L). Add about 200 mL of water to a 2 L graduated glass beaker. Weigh 15 g of tryptone, 15 g of NaCl, and 7.5 g of yeast extract, and transfer to the beaker (*see* **Note 1**). Add water to a

volume of 1 L, and shake until the solutes have dissolved. Adjust the pH to 7.0 with 5 N NaOH (*see* **Note 2**). Make up the volume to 1.5 L with water. Sterilize by autoclaving for 20 min at 15 psi (1.05 kg/cm^2) on liquid cycle.

2. STE buffer: 150 mM NaCl, 20 mM Tris–HCl, 1 mM ethylenediaminetetraacetic acid (EDTA), pH 8.0 (volume, 100 mL). Weigh 0.88 g of NaCl and 0.24 g of Tris–HCl; add about 70 mL of water and 1 μL of 0.1 M EDTA stock solution. Adjust the pH to 8.0, and make up to 100 mL with water. EDTA stock solution (0.1 M) is prepared dissolving 1.86 g of EDTA disodium salt in water; make the volume to 50 mL and store at 4 °C.
3. Phosphate-buffered saline (PBS) (volume, 1 L): 10 mM PO_4^{3-}, 137 mM NaCl, 2.7 mM KCl, pH 7.4. Weigh 1.42 g of Na_2HPO_4, 0.24 g of KH_2PO_4, 8 g of NaCl, and 0.2 g of KCl, and start with 800 mL of distilled water to dissolve all salts. Adjust the pH to 7.4 with HCl. Make up to 1 L with water (*see* **Note 3**).
4. Washing buffer: 1× PBS supplemented before use with 1 mM EDTA and 1 mM dithiothreitol (DTT) (volume, 200 mL). Add 2 μL of 0.1 M EDTA stock solution and 2 μL of 0.1 M DTT stock solution (*see* **Note 4**) to 200 mL of 1× PBS, and finally add protease inhibitors (*see* **Note 5**).

2.2 Cross-Linking of Af1521 Macro Domain(s) with Dimethyl Pimelimidate (DMP)

1. 0.2 M disodium tetraborate solution, pH 8.6 (volume, 50 mL) (*see* **Note 6**). Weigh 3.68 g of disodium tetraborate decahydrate, shake in 30 mL of water until the solute is dissolved, and make up to 50 mL with water. In parallel, weigh 0.62 g of boric acid (H_3BO_3), shake in 30 mL of water until the solute is dissolved, and make up to 50 mL with water. Adjust the pH of the disodium tetraborate solution from ~9.0–9.5 to 8.0 using boric acid solution while stirring.
2. 0.2 M ethanolamine solution, pH 8.3 (volume, 200 mL) (*see* **Note 7**). To obtain 0.2 M ethanolamine, pH 8.3, add 2.55 mL of ethanolamine (98%) to ~150 mL of water. Adjust the pH to 8.3 with NaOH, and bring to 200 mL final volume with water.
3. 20 mM DMP in 0.2 M disodium tetraborate solution. The dimethyl pimelimidate (DMP) is unstable in solution because the imidate moiety is easily hydrolyzed, and thus it is directly added as a powder to the suspension where the reaction will then occur. Here, it is important to verify that the pH of the suspension (after DMP addition) is between 8.0 and 9.0 as the cross-linking efficiency is greatly reduced out of this pH range. Add tenfold molar excess of DMP to the protein(s) when the protein concentration is above 5 mg/mL. Weigh 0.026 g of DMP × 2HCl (with spacer arm length of 9.2 Å) in 5 mL of 0.2 M disodium borate solution, pH 8.6 (*see* **Note 8**).

4. 20 mM reduced glutathione solution (volume, 80 mL). Dissolve 0.5 g of reduced glutathione in 50 mL of 20 mM Tris–HCl, pH 8.0. Adjust the pH to 8.0, and bring to 80 mL final volume (*see* **Note 9**).
5. 0.1 M glycine solution, pH 2.4 (volume, 50 mL). Mix 5.55 g glycine to 30 mL of water. Adjust the pH to 2.4 using 1 M HCl, and bring the volume to 50 mL with water.
6. Glutathione-S-transferase (GST) elution buffer: 100 mM Tris–HCl pH 8.0, 20 mM reduced glutathione, and 5 mM DTT (*see* **Notes 4** and **9**).

2.3 Macro-Domain-Based Pull-Down Assay

1. 1× PBS solution: prepared as described in Subheading 2.1.
2. RIPA lysis buffer (50 mL): 100 mM Tris–HCl, pH 7.5, 150 mM NaCl, 1% (w/v) Igepal, 0.5% (w/v) sodium deoxycholate, 0.1% sodium dodecyl sulfate (SDS), protease inhibitors.

   To obtain 1× RIPA lysis buffer solution, dilute the single components to the above indicated final concentrations starting from the following stock solutions: 1 M Tris–HCl, pH 7.5, 5 M NaCl, 10% (w/v) Igepal, 3% (w/v) sodium deoxycholate, and 10% (w/v) SDS. Add protease inhibitors immediately before use (*see* **Note 10**).
3. 100 mM Tris–HCl solution, pH 7.5 (50 mL): prepare starting from 1 M Tris–HCl, pH 7.5 stock solution.
4. Store all buffers at 4 °C.

3 Methods

Here we describe an affinity purification method that, combined with LC-MS/MS analysis, allows the identification of both the mono- and poly-ADP-ribosylated proteins from cell lysates. The procedure has been set using total cell lysates from HeLa cells, but it can be applied to several cell lines, treated according to the experimental needs.

The following section can be divided into two main parts:

1. Production of the cross-linked GST-Af1521-bound beads [Expression, Purification, and Cross-linking of Af1521 Macro domain(s): Subheadings 3.1 and 3.2].
2. Af1521 *macro*-domain-based pull-down assay from total cell lysates [Subheading 3.3]. A scheme of the procedure is reported in Fig. 1.

We suggest to perform the incubation of total cell lysates in a sequential manner, first with GST-Af1521 G42E-bound beads and then with GST-Af1521 WT-bound beads in order to get a pre-clearing step ("sequential pull-down," see Subheading 3.3.2).

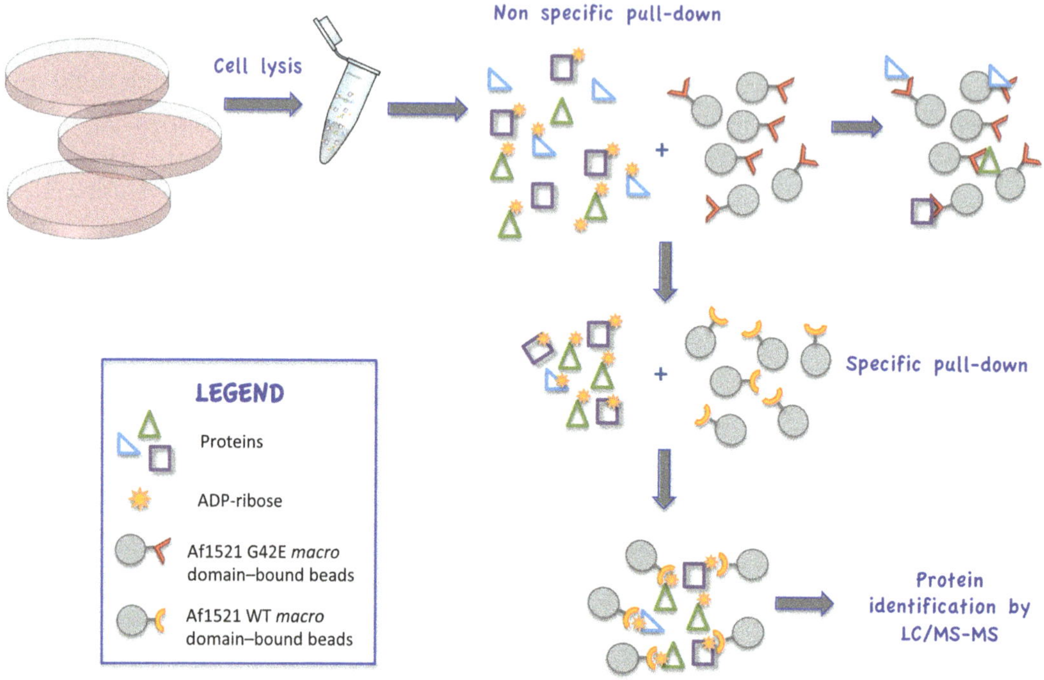

Fig. 1 Schematic representation of the *macro*-domain-based pull-down assay using Af1521 G42E- and Af1521 WT-bound beads according to the "sequential pull-down" reported procedure. After cell lysis, the solubilized proteins are first incubated with the Af1521 G42E-mutated *macro*-domain-bound beads to remove the non-specific binding (Af1521 G42E, non-specific pull-down) and then with the Af1521 WT *macro*-domain-bound beads to recover the ADP-ribosylated proteins (Af1521 WT, specific pull-down). The eluted proteins from the Af1521 WT *macro*-domain-bound beads are then identified by LC-MS/MS

Alternatively, the incubation of cell lysates with the *macro*-domain cross-linked beads (G42E and the WT *macro*-domain proteins) can be performed in parallel, splitting the cell lysates into two aliquots: one incubated with GST-Af1521 G42E-bound beads and the other one with GST-Af1521 WT-bound beads ("parallel pull-down," see Subheading 3.3.3). In this procedure, the eluted proteins from the GST-Af1521 G42E-bound beads have to be considered non-specific (*i.e.*, not ADP-ribose-mediated), and thus they do not represent ADP-ribosylated substrates.

3.1 Expression and Purification of Af1521 Macro Domain(s)

The following procedure is used to purify GST wild-type Af1521 *macro* domain (GST-Af1521 WT) and its G42E mutant (GST-Af1521 G42E), and it leads to a final yield of 2–5 mg of the recombinant proteins from 500 mL of culture:

1. Inoculate *Escherichia coli* BL21-DE3 transformed with pGEX4T1-Af1521 wild-type *macro* domain (or pGEX4T1-Af1521 G42E-mutant *macro* domain) in 50 mL of LB containing 100 μg/mL ampicillin into a glass flask of 250 mL, and incubate overnight at 37 °C under continuous shaking (200 rpm).

2. Dilute the culture 1:10 in 500 mL of the same medium, and monitor the OD_{600} until it reaches 0.6 (usually it takes ~1 h) under continuous shaking (200 rpm).
3. Induce bacteria with the addition of 0.2 mM isopropil-β-D-1-tiogalattopiranoside (IPTG) (*see* **Note 11**) for 3.5 h at 20 °C under continuous shaking (200 rpm).
4. Cool the culture on ice, and centrifuge at 5000 × *g* in a pre-cooled JA14 rotor for 10 min at 4 °C.
5. Discard the supernatant, and resuspend the pellet in 25 mL of cooled STE buffer supplemented, immediately prior to use, with 1 mM DTT, 0.5 mg/mL lysozyme (*see* **Note 12**), and the cocktail of protease inhibitors. Make sure that the lysozyme is completely dissolved in the STE buffer before the addition to the bacterial pellet. Gently resuspend the pellet in this complete STE buffer using 10 mL pre-cooled plastic pipet on ice to avoid protein degradation. Incubate the suspension in a 50 mL conical tube on a rotating wheel for 30 min at 4 °C.
6. Freeze the suspension by immersion in liquid nitrogen, and store at −80 °C overnight or for a few days (*see* **Note 13**).
7. Thaw the suspension by transferring it to a 20 °C water bath; once the suspension is completely thawed, add DTT to a final concentration of 1 mM, 1% (w/v) Triton X-100 (using 10% Triton X-100 stock solution, *see* **Note 14**) and protease inhibitors.
8. Incubate the lysate with gentle agitation at 4 °C for 20 min before sonication on ice 3× 30 s (amplitude 30%, tip with diameter of 6.4 mm) (*see* **Note 15**).
9. Centrifuge the lysate at 20,000 × *g* in a JA20 rotor for 15 min at 4 °C. Save 10 μL of lysate as input (10 μL input/25 mL of total cell lysate) to monitor protein purification.
10. Recover the supernatant, and add it to a 50 mL conical tube where 0.5 mL of bed volume of glutathione Sepharose 4B beads was previously sedimented (*see* **Note 16**). Incubate the suspension on a rotating wheel for 2 h at 4 °C, and then centrifuge at 500 × *g* for 5 min at 4 °C to sediment the beads.
11. Wash the beads 6× with 15 mL washing buffer (centrifuging at 500 × *g* for 5 min at 4 °C). Save 10 μL of bed bead volume to monitor and quantify the yield and quality of purified protein.

Results from a representative purification are shown in Fig. 2.

3.2 Cross-Linking of Af1521 Macro Domain(s) with Dimethyl Pimelimidate (DMP)

1. Wash both the 0.5 mL bed volume of GST-Af1521 WT-bound beads and the 0.5 mL bed volume of GST-Af1521 G42E-bound beads (obtained as described in Subheading 3.1) twice with 15 mL of 0.2 M disodium borate, pH 8.6, and centrifuge at 1000 × *g* for 5 min at room temperature.

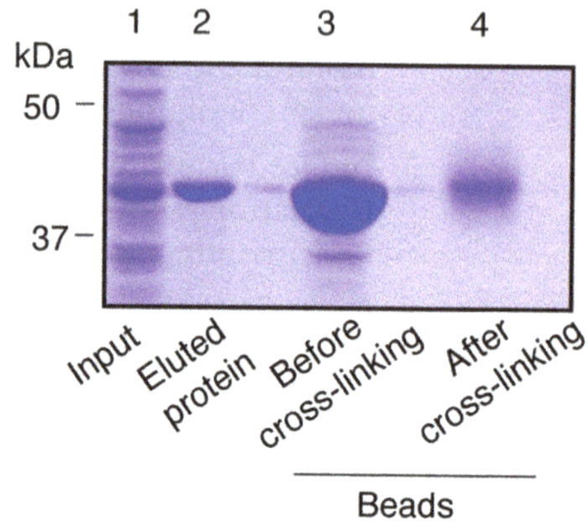

Fig. 2 Samples obtained from a representative purification of recombinant Af1521 *macro*-domain proteins analyzed by SDS-PAGE and stained with Coomassie blue dye. *Lane 1* shows 10 μL of 25 mL of cell lysate from *E. coli* BL21-DE3 transformed with pGEX4T1-Af1521 WT *macro* domain after IPTG induction (Input). *Lane 2* shows 8 μg of WT *macro*-domain protein eluted from glutathione Sepharose beads (eluted protein). *Lanes 3 and 4* show 10 μL of Af1521 WT *macro*-domain-bound beads before (*lane 3*) and after (*lane 4*) cross-linking. Of note, the reduced amount of the eluted WT *macro*-domain protein after cross-linking (*lane 4* versus *lane 3*) correlates with the efficiency of its covalent binding to the beads. Molecular weight standards (kDa) are indicated on the left of the panel

2. Resuspend the bed beads in 1 mL of 0.2 M sodium borate, pH 8.6, and dissolve the DMP cross-linker powder directly into the suspension(s) to a final concentration of 20 mM. Incubate on a rotating wheel for 30 min at room temperature. Save 10 μL of the bed volume of both GST-Af1521 WT- and GST-Af1521 G42E-bound beads to monitor the cross-linking efficiency.
3. Centrifuge at 1000 × *g* for 5 min at room temperature and remove the supernatant.
4. Stop the reaction with the addition of 14 mL of 0.2 M ethanolamine, pH 8.2. Rotate 5 min at room temperature, centrifuge at 1000 × *g* for 5 min and remove the supernatant.
5. Incubate the beads with 10 mL of 0.2 M ethanolamine, pH 8.2, on a rotating wheel for 1 h at room temperature.
6. At the end of this incubation, save 10 μL of the bed volume of both *macro* domains bound to the beads to monitor the cross-linking efficiency.
7. Centrifuge at 1000 × *g* for 5 min at room temperature and remove the supernatant.
8. Wash the beads 3× with 15 mL 1× PBS (*see* **Note 17**), 3× with 15 mL of reduced glutathione solution, once again in 15 mL 1× PBS (centrifuging at 1000 × *g* for 5 min at room temperature),

and finally with 15 mL of 0.1 M glycine, pH 2.4, to remove the proteins that were not covalently bound to the beads.

9. Finally, wash the cross-linked matrices twice with 15 mL PBS centrifuging at 1000 × *g* for 5 min at room temperature.
10. Store at 4 °C the GST-Af1521 WT and GST-Af1521 G42E *macro*-domain cross-linked beads (*see* **Note 18**) in 15 mL of 1× PBS supplemented with 0.02% (w/v) sodium azide (*see* **Note 19**).

The efficiency of the cross-linking can be monitored by analyzing samples removed at each step by SDS-PAGE, followed by Coomassie blue staining.

3.3 Af1521 Macro-Domain-Based Pull-Down Assay

8 mg cell lysate is generally used to identify ADP-ribosylated substrates by LC/MS-MS. The cell lysates can be obtained, according to the experimental procedures of the users, starting from:

1. Untreated and/or stimulated cells.
2. Cells overexpressing the enzyme of interest.
3. Isolated cell fractions (*see* **Note 20**).

3.3.1 Preparation of Cell Lysates

8 mg cell lysate is generally recovered from 80% confluent cells plated in 10–15 culture dishes of 10 cm.

1. Plate and culture adherent cells to approximately 80% confluence. Cells should be in the log phase of growth and healthy (*see* **Note 21**).
2. Remove the culture growth medium and keep dishes on ice for all steps.
3. Carefully and gently wash the cell monolayer with 10 mL/dish of pre-cooled 1× PBS. Remove 1× PBS and repeat the washing procedure 3×.
4. Add 0.5 mL of RIPA lysis buffer (supplemented with protease inhibitors before use, *see* **Note 5**)/dish, and carefully swirl to homogenously distribute the buffer.
5. Scrape cells with a cell scraper or with a silicone spatula, and then transfer the cell lysate into a clear tube.
6. Incubate the cell lysate at 4 °C on a rotating wheel for 20 min.
7. Centrifuge at 15,000 × *g* for 15 min at 4 °C.
8. Recover the supernatant (avoiding the pellet) into a new tube, and evaluate the protein concentration by the bicinchoninic acid method (*see* **Notes 22** and **23**).

3.3.2 Macro-Domain-Based Pull-down ("Sequential Pull-down")

All the incubation steps with GST-Af1521-bound beads have to be performed at 4 °C (*see* **Note 24**):

1. Incubate 8 mg of cell lysate (obtained as described in Subheading 3.3.1) with 100 μL of the cross-linked GST-Af1521

G42E-bound beads (10 μg/μL binding efficiency) for 8 h at 4 °C (*see* **Note 25**), on a rotating wheel.

2. Following this incubation, centrifuge the mixtures at 500 × *g* for 5 min to recover the proteins that did not bind to the GST-Af1521 G42E-bound beads; store the beads at 4 °C in 3 mL of ice-cold 100 mM Tris–HCl, pH 7.5.
3. Incubate the unbound proteins to the GST-Af1521 G42E-bound beads with 100 μL of the cross-linked GST-Af1521 WT-bound beads (10 μg/μL binding efficiency) for a further 8 h at 4 °C (*see* **Note 26**).
4. Following this incubation, centrifuge the samples at 500 × *g* for 5 min and recover the unbound proteins.
5. Wash both GST-Af1521 G42E-bound and GST-Af1521 WT-bound beads eight times with RIPA lysis buffer and further eight times with 100 mM Tris–HCl buffer, pH 7.5, always centrifuging at 500 × *g* for 5 min at 4 °C.
6. At the end of these washing steps, add 100 μL Laemmli sample buffer, boil 5 min at 95 °C, mix to completely recover the proteins bound to the beads, and boil for a further 5 min at 95 °C.
7. Centrifuge at 500 × *g* for 5 min to recover the eluted proteins.
8. Analyze the eluted proteins by SDS/PAGE.
9. Gels can be stained (*e.g.*, GelCode Blue Stain reagent) and further processed for LC-MS/MS identification (*see* **Note 27**).

The most appropriate procedure to process the samples for MS identification should be defined according to the MS strategy adopted for protein identification. A schematic representation of the pull-down procedure is reported in Fig. 1.

Alternatively, for routine experiments, the eluted proteins separated by SDS-PAGE can be transferred onto nitrocellulose for Western blotting analysis using specific antibodies (*see* **Note 28**).

Here, as an example, we report the ADP-ribosylation of endogenous PARP1 and PARP12 under physiological/growth conditions (Fig. 3). To verify the specific ADP-ribose binding of a modified protein, the pull-down experiment was performed in the presence of free ADP-ribose in the cell lysates (at 20 mM final concentration) that competes for the *macro*-domain binding and thus decreases the protein binding to the *macro* domain [4].

3.3.3 Macro-Domain-Based Pull-down ("Parallel Pull-down")

1. Incubate 8 mg of cell lysate (obtained as described in Subheading 3.3.1) with 100 μL of the cross-linked GST-Af1521 G42E-bound beads (10 μg/μL binding efficiency) and further 8 mg of cell lysate with 100 μL of the cross-linked GST-Af1521 WT-bound beads (10 μg/μL binding efficiency) for 8 h at 4 °C (*see* **Notes 25** and **26**), on a rotating wheel.

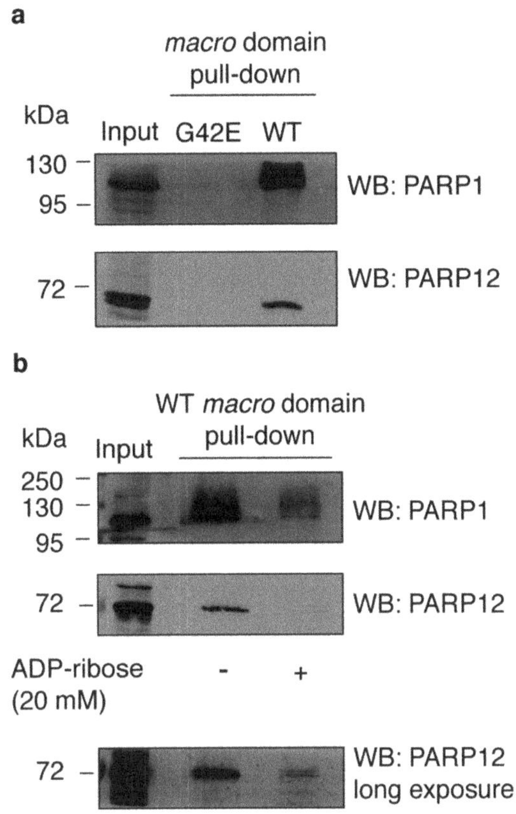

Fig. 3 *Macro* domain-based pull-down assay. Panel (**a**) shows *macro*-domain-based pull-down of untreated HeLa cell lysate (input) first pre-cleared with Af1521 G42E-bound beads and then incubated with Af1521 WT-bound beads. The eluted ADP-ribosylated proteins were separated by SDS-PAGE, transferred onto nitrocellulose, and revealed by Western blotting (WB) with anti-PARP1- and anti-PARP12-specific antibodies (as indicated). *Lane G42E* shows the eluted proteins from Af1521 G42E-bound beads. *Lane WT* shows the eluted proteins after sequential incubation of Af1521 G42E-bound beads followed by Af1521 WT-bound beads. Panel (**b**) shows *macro*-domain-based pull-down of untreated HeLa cell lysate (input) incubated with Af1521 WT-bound beads alone (–) or in the presence of 20 mM free ADP-ribose (+). The eluted ADP-ribosylated proteins were separated by SDS-PAGE, transferred onto nitrocellulose, and revealed by Western blotting (WB) with anti-PARP1 and anti-PARP12 specific antibodies (as indicated in the upper panels). In the presence of 20 mM ADP-ribose, a minimal binding of PARP12 to Af1521 WT-bound beads was detected only after long exposure (1 h, lowest panel). Molecular weight standards (kDa) are indicated on the left of the panels

2. Following this incubation, centrifuge the samples at 500 × *g* for 5 min and recover the unbound proteins.
3. Wash both GST-Af1521 G42E-bound and GST-Af1521 WT-bound beads eight times with RIPA lysis buffer and further eight times with 100 mM Tris–HCl buffer, pH 7.5, always centrifuging at 500 × *g* for 5 min at 4 °C.

4. At the end of these washing steps, add 100 μL Laemmli sample buffer, boil 5 min at 95 °C, mix to completely recover the proteins bound to the beads, and boil for a further 5 min at 95 °C.
5. Centrifuge at 500 × *g* for 5 min to recover the eluted proteins.
6. Analyze the eluted proteins by SDS/PAGE.
7. Gels can be stained (*e.g.*, GelCode Blue Stain reagent) and further processed for LC-MS/MS identification (*see* **Note 27**).

4 Notes

1. The presence of water at the bottom of the glass beaker helps to dissolve the solutes even though the complete solubilization requires at least 15 min at room temperature.
2. First, concentrated NaOH (5 N) can be used to reduce the difference from the starting pH to the required pH. When a pH close to the required one is obtained, it is advised to use a series of NaOH solutions (*e.g.*, 1 and 0.5 N) with lower ionic strengths to avoid a sudden increase in pH above the required pH.
3. It is good practice to prepare a 10× stock solution of PBS. The pH of this stock solution is ~6.8, but when diluted with water to 1× PBS, it should change to 7.4, and the correct pH value should be directly measured using a pH meter and adjusted using HCl or NaOH. The concentrated stock solutions of PBS may precipitate when stored at 4 °C; if this is the case, the PBS should be kept at room temperature before use, until the precipitate completely dissolves.
4. DTT is not stable in solution. Only freshly made DTT solutions should be used. It is good practice to prepare a 1 M DTT stock solution in water dissolving 1.55 g of DTT powder in a final volume of 10 mL. Store at −20 °C in 20 μL aliquots. This stock solution is then diluted 1:10 in water, immediately prior to add it at a 1 mM final concentration in the washing or STE buffer.
5. The simplest way to prepare a protease inhibitor solution is to use protease inhibitor cocktail tablets commercially available and able to inhibit a broad spectrum of serine, cysteine, and metalloproteases as well as calpains. They are formulated to give a ready-to-use protease inhibitor solution upon dissolution in a specified quantity of distilled water. Each tablet is sufficient for a defined volume of solution according to the manufacturer instructions. The tablets are available for the standard volumes of 100, 200, 500, and 1000 mL, and based on this, each tablet can be directly dissolved in 100, 200, 500,

and 1000 mL of buffer solution. It is also possible to prepare a 100× stock solution in water or 100 mM PBS, pH 7.0. This stock solution is stable for 1–2 weeks, stored at 4 °C, or for 12 weeks at −20 °C.

6. It is recommended to use disodium tetraborate decahydrate ($Na_2B_4O_7 \times 10H_2O$) that is completely soluble in water compared to the anhydrous sodium salt.
7. To avoid fast changes in the pH value, use a series of NaOH solutions with lower ionic strengths (*e.g.*, 1 and 0.5 N) to carefully adjust the pH to the required value of 8.3.
8. DMP powder is stored at 4 °C. Equilibrate the DMP vial to room temperature before opening and weighting it, to avoid condensation onto the product. Typically equilibration requires at least 15 min.
9. It is recommended to prepare the reduced glutathione solution immediately before use and carefully adjust the pH to the 8.0 required value.
10. Depending on the experimental aims, the general PARP inhibitor PJ34 is added at a final concentration of 10 μM to avoid lysis-induced PARP activation [22].
11. It is good practice to prepare a 1 M IPTG stock solution in water dissolving 7.15 g of IPTG powder in 20 mL of water and to vortex until it is completely dissolved. This may take some time since IPTG is not very soluble in water. Make the final volume to 30 mL, mix again, aliquot in 1 mL (×30) in autoclaved tubes, and store at −20 °C for several months.
12. The lyophilized powder of lysozyme is stored at −20 °C, and it requires at least 10 min to be fully equilibrated before being weighted at room temperature. The weighted lysozyme is then directly added into the STE buffer.
13. The process of the flash freezing in liquid nitrogen is an optional step although it facilitates the break of the bacterial cells; we suggest including this step in the procedure.
14. It is good practice to prepare 10% (w/v) stock solution of Triton X-100 in water and store it in a 50 mL conical tube at room temperature.
15. We suggest to increase the sonication cycles up to five times if your solution is not completely clear.
16. Of note, the glutathione Sepharose 4B matrix is composed of 45–165 μm diameter wet spherical beads with space arm of 12 atoms. It is stored at 4 °C. Immediately before use, the bottle of glutathione Sepharose 4B beads has to be gently shaken in order to resuspend the matrix (supplied as ~75% slurry in 20% ethanol, as preservative). Use a pipet to take the required bead volume and transfer it to a 10 mL conical tube. Usually, dis-

pense 1.33 mL of the ~75% slurry glutathione Sepharose 4B beads per 1 mL of bed volume required. Sediment the beads by centrifugation at 500 × *g* for 5 min at 4 °C. Carefully discard the supernatant, and wash the beads by the addition of 10 mL cold 1× PBS per mL of bed volume dispensed. Invert 10 times to mix, centrifuge at 500 × *g* for 5 min at 4 °C, and discard the supernatant. Repeat this step three times to completely remove the ethanol and a further three times with the washing buffer to equilibrate the beads. After removing the buffer, resuspend the equilibrated beads in 2 mL of washing buffer, and transfer an equal amount of bed volume in two 50 mL conical tubes. Directly add 25 mL of each bacterial cell lysate.

17. The procedure can be stopped at this step storing the beads overnight at 4 °C.
18. *Macro*-domain-bound beads are stable in 1× PBS [supplemented with 0.02% (w/v) sodium azide] at 4 °C for at least 1 year.
19. It is good practice to prepare 5% (w/v) stock solution of sodium azide in water and store it at room temperature.
20. Cell fractionation can be obtained according to routine, described procedures that will be selected by users according to their experimental aims.
21. If you use suspension culture, grow cells to a density of 1–2 million cells/mL, before pelleting by centrifugation at 300 × *g* for 5 min at room temperature.
22. The use of bicinchoninic acid method is preferred since SDS in the lysis buffer can interfere with Bradford assay.
23. When using HeLa cells, 2 × 10 cm tissue culture dishes can be lysed in 0.8 mL RIPA lysis buffer to recover a lysate of ~2 μg/μL protein concentration.
24. It should be noted that the Af1521 protein bears a hydrolase activity, as indicated in in vitro experiments performed at 30 ° C, and it removes ADP-ribose groups from modified acidic residues (glutamate/aspartate) [23, 24]; however, at low temperatures (4 ° C), its hydrolase activity is negligible, whereas the ADP-ribose binding capacity is maintained [21, 25].
25. It is important to equilibrate the cross-linked GST-Af1521 G42E-bound beads with RIPA lysis buffer before incubation with cell lysates. Wash the bead bed volume by the addition of 1 mL of pre-cooled RIPA lysis buffer. Invert 10 times to mix, centrifuge at 500 × *g* for 5 min at 4 °C, and discard the supernatant. Repeat this step three times.
26. The cross-linked GST-Af1521 WT-bound beads were previously equilibrated as above.

27. GelCode Blue Stain reagent does not require an organic solvent for destaining (differently from other staining methods); this makes this reagent highly compatible with MS analysis and N-terminal sequence analysis.
28. For routine experiments, from 0.8 to 1.5 mg of cell lysates can be used (although the appropriate amount of cell lysates can be determined by users according to the experimental conditions and experimental aims). In this case, 20 μL of the cross-linked *macro*-domain-bound beads can be used. The samples are processed as described in the Subheadings 3.3.2 and 3.3.3 (except for eight washes with RIPA lysis buffer and eight washes with 100 mM Tris–HCl buffer, pH 7.5, that can be both reduced to three) and the eluted proteins separated by SDS-PAGE, transferred onto nitrocellulose for Western blotting analysis with specific antibodies.

Acknowledgments

This study was supported by the Italian Association for Cancer Research (AIRC, Milan, Italy, IG10341 and 14675 to D.C.), TRansforming IDEas in Oncological research award (TRIDEO, AIRC-Fondazione Cariplo, Milan, Italy, IG17524 to C.V.), the PNR-CNR Aging Program, the Flag project Nanomax, and the POR project OcKey. G.G. received a fellowship from the Italian Foundation for Cancer Research (FIRC, Milan, Italy).

References

1. Jensen ON (2006) Interpreting the protein language using proteomics. Nat Rev Mol Cell Biol 7(6):391–403
2. Corda D, Di Girolamo M (2003) Functional aspects of protein mono-ADP-ribosylation. EMBO J 22(9):1953–1958
3. Hottiger MO (2015) Nuclear ADP-ribosylation and its role in chromatin plasticity, cell differentiation, and epigenetics. Annu Rev Biochem 84:227–263
4. Dani N et al (2009) Combining affinity purification by ADP-ribose-binding macro domains with mass spectrometry to define the mammalian ADP-ribosyl proteome. Proc Natl Acad Sci U S A 106(11):4243–4248
5. Hottiger MO, Hassa PO, Luscher B, Schuler H, Koch-Nolte F (2010) Toward a unified nomenclature for mammalian ADP-ribosyltransferases. Trends Biochem Sci 35(4):208–219
6. Kawamitsu H et al (1984) Monoclonal antibodies to poly(adenosine diphosphate ribose) recognize different structures. Biochemistry 23(16):3771–3777
7. Gagne JP et al (2008) Proteome-wide identification of poly(ADP-ribose) binding proteins and poly(ADP-ribose)-associated protein complexes. Nucleic Acids Res 36(22):6959–6976
8. Gagne JP et al (2012) Quantitative proteomics profiling of the poly(ADP-ribose)-related response to genotoxic stress. Nucleic Acids Res 40(16):7788–7805
9. Isabelle M, Gagne JP, Gallouzi IE, Poirier GG (2012) Quantitative proteomics and dynamic imaging reveal that G3BP-mediated stress granule assembly is poly(ADP-ribose)-dependent following exposure to MNNG-induced DNA alkylation. J Cell Sci 125(Pt 19):4555–4566
10. Eide B, Gierschik P, Spiegel A (1986) Immunochemical detection of guanine nucleotide binding proteins mono-ADP-ribosylated by bacterial toxins. Biochemistry 25(21): 6711–6715

11. Meyer T, Hilz H (1986) Production of anti-(ADP-ribose) antibodies with the aid of a dinucleotide-pyrophosphatase-resistant hapten and their application for the detection of mono(ADP-ribosyl)ated polypeptides. Eur J Biochem 155(1):157–165—
12. Osago H, Terashima M, Hara N, Yamada K, Tsuchiya M (2008) A new detection method for arginine-specific ADP-ribosylation of protein—a combinational use of anti-ADP-ribosylarginine antibody and ADP-ribosylarginine hydrolase. J Biochem Biophys Methods 70(6):1014–1019
13. Schwab CJ, Colville MJ, Fullerton AT, McMahon KK (2000) Evidence of endogenous mono-ADP-ribosylation of cardiac proteins via anti-ADP-ribosylarginine immunoreactivity. Proc Soc Exp Biol Med 223(4):389–396
14. Karras GI et al (2005) The macro domain is an ADP-ribose binding module. EMBO J 24(11):1911–1920
15. Slade D et al (2011) The structure and catalytic mechanism of a poly(ADP-ribose) glycohydrolase. Nature 477(7366):616–620
16. Wang Z, Gagne JP, Poirier GG, Xu W (2014) Crystallographic and biochemical analysis of the mouse poly(ADP-ribose) glycohydrolase. PLoS One 9(1):e86010
17. Forst AH et al (2013) Recognition of mono-ADP-ribosylated ARTD10 substrates by ARTD8 macrodomains. Structure 21(3):462–475
18. Timinszky G et al (2009) A macrodomain-containing histone rearranges chromatin upon sensing PARP1 activation. Nat Struct Mol Biol 16(9):923–929
19. Wang Z et al (2012) Recognition of the iso-ADP-ribose moiety in poly(ADP-ribose) by WWE domains suggests a general mechanism for poly(ADP-ribosyl)ation-dependent ubiquitination. Genes Dev 26(3):235–240
20. Bartolomei G, Leutert M, Manzo M, Baubec T, Hottiger MO (2016) Analysis of chromatin ADP-ribosylation at the genome-wide level and at specific loci by ADPr-ChAP. Mol Cell 61(3):474–485
21. Jungmichel S et al (2013) Proteome-wide identification of poly(ADP-ribosyl)ation targets in different genotoxic stress responses. Mol Cell 52(2):272–285
22. Bilan V et al (2017) New quantitative mass spectrometry approaches reveal different ADP-ribosylation phases dependent on the levels of oxidative stress. Mol Cell Proteomics 16(5):949–958
23. Jankevicius G et al (2013) A family of macrodomain proteins reverses cellular mono-ADP-ribosylation. Nat Struct Mol Biol 20(4):508–514
24. Rosenthal F et al (2013) Macrodomain-containing proteins are new mono-ADP-ribosylhydrolases. Nat Struct Mol Biol 20(4):502–507
25. Larsen SC et al (2017) Proteome-wide identification of in vivo ADP-ribose acceptor sites by liquid chromatography-tandem mass spectrometry. Methods Mol Biol 1608:149–162

Chapter 4

Assessment of Intracellular Auto-Modification Levels of ARTD10 Using Mono-ADP-Ribose-Specific Macrodomains 2 and 3 of Murine Artd8

Mareike Bütepage, Sarah Krieg, Laura Eckei, Jinyu Li, Giulia Rossetti, Patricia Verheugd, and Bernhard Lüscher

Abstract

Mono-ADP-ribosylation is a posttranslational modification, which is catalyzed in cells by certain members of the ADP-ribosyltransferase diphtheria toxin-like family (ARTD) of ADP-ribosyltransferases (aka PARP enzymes). It involves the transfer of a single residue of ADP-ribose (ADPr) from the cofactor NAD^+ onto substrate proteins. Although 12 of the 17 members of the ARTD family have been defined as mono-ARTDs in in vitro assays, relatively little is known about their exact cellular functions. A major challenge is the detection of mono-ADP-ribosylated (MARylated) proteins in cells as no antibodies are available that detect exclusively MARylated proteins. As an alternative to classical antibodies, the MAR-specific binding domains macro2 and macro3 of Artd8 can be utilized alone or in combination, to demonstrate intracellular auto-modification levels of ARTD10 in cells in both co-immunoprecipitation and co-localization experiments. Here we demonstrate that different macrodomain constructs of human ARTD8 and murine Artd8, alone or in combination, exert differences with regard to their interaction with ARTD10 in cells. Precisely, while the macrodomains of murine Artd8 interacted with ARTD10 in cells in a MARylation-dependent manner, the macrodomains of human ARTD8 interacted with ARTD10 independent of its catalytic activity. Moreover, we show that a combination of macro2 and macro3 of murine Artd8 was recruited more efficiently to ARTD10 during co-localization experiments compared to the single domains. Therefore, murine Artd8 macrodomain constructs can serve as a tool to evaluate intracellular ARTD10 auto-modification levels using the described methods, while the human ARTD8 macrodomains are less suited because of ADPr-independent binding to ARTD10. Protocols for co-immunoprecipitation and co-localization experiments are described in detail.

Key words ARTD8/PARP14, ARTD10/PARP10, Co-immunoprecipitation, Immunofluorescence, Macrodomain, Modeling, Mono-ADP-ribosylation, Structure prediction

---

Patricia Verheugd and Bernhard Lüscher contributed equally to this work

Paul Chang (ed.), *ADP-ribosylation and NAD+ Utilizing Enzymes: Methods and Protocols*, Methods in Molecular Biology, vol. 1813, https://doi.org/10.1007/978-1-4939-8588-3_4, © Springer Science+Business Media, LLC, part of Springer Nature 2018

1 Introduction

ADP-ribosylation is a posttranslational modification (PTM) of proteins, which is catalyzed in cells by enzymes of the ADP-ribosyltransferase diphtheria toxin-like (ARTD) family of ADP-ribosyltransferases (aka PARP enzymes). Seventeen ARTDs exist, and the active family members transfer ADP-ribose (ADPr) from their cofactor β-NAD^+ onto substrate proteins with release of nicotinamide. They can be subclassified into enzymes that transfer a single ADPr residue onto their substrates (mono-ADP-ribosylation or MARylation) and those that are able to form ADPr polymers on their substrates (poly-ADP-ribosylation or PARylation). One member of the ARTD family, ARTD13, is catalytically inactive. PARylation is a well-studied PTM, playing a role in DNA damage repair, chromatin and transcriptional regulation, and signaling [1]. However, the majority of ARTD enzymes are restricted to MARylation. Subclassification of ARTDs into those enzymes that MARylate and those that PARylate has been based on three characteristic amino acids in the catalytic center of the enzymes (H-Y-E in poly-ARTDs and H-Y-E variants in mono-ARTDs) [2]. However, recent findings indicate that this does not provide a reliable classification. In vitro biochemical assays, in which radioactively labeled β-NAD^+ is added to an ARTD enzyme, resulting in auto-modification or in labeling of a substrate and subsequent resolution of the proteins by SDS-PAGE and autoradiography, together with the analysis of MAR and PAR chains upon release from substrates and their visualization on polyacrylamide gels to determine chain length, allows to classify ARTDs more reliably. Mono-ARTDs typically show one distinct labeled protein on SDS-PAGE and release of MAR, while poly-ARTDs resolve as a smear of labeled bands and release PAR chains. ARTD3 and ARTD4 contain the H-Y-E motif and were therefore proposed to have PARylation activity but have been classified as mono-ARTDs based on in vitro auto-modification assays [3–5]. Moreover, ARTD9 possesses a Q-Y-T motif and does not auto-modify and therefore was considered inactive. However, in the presence of a cofactor, ARTD9 is catalytically active [6]. Thus, the most recent census indicates that ARTD3, 4, 7–12 and 14–17 are mono-ARTDs [2, 3, 7].

ARTD10, the founding member of the mono-ARTD subgroup, has been identified as an interaction partner of the oncoprotein MYC in 2005 [8] and was described as a MARylating enzyme in 2008 [7]. Several kinases were identified as potential substrates of ARTD10, and MARylation of GSK3β by ARTD10 negatively regulates its kinase activity [9]. ARTD10 furthermore controls the NF-κB signaling pathway dependent on its catalytic activity and on its ability to bind to K63-linked poly-ubiquitin.

NEMO has been described as a relevant substrate, and its MARylation interferes with its poly-ubiquitination, an essential step for signal propagation and NF-κB activation [10]. ARTD10 expression is induced by interferon stimulation, and it can act as a repressor of viral replication and cellular translation [11, 12], pointing to a potential role of ARTD10 as a modulator of immune responses. In addition, ARTD10-dependent MARylation also controls cell proliferation although relevant substrates have not been identified so far [7, 13]. Taken together, MARylation by ARTD10 participates in the control of a variety of cellular processes, and it is therefore of interest to define how ARTD10's catalytic activity is regulated in cells [14]. However, demonstrating catalytic activity of mono-ARTDs in cells in general and defining MARylation of individual substrates and the functional consequences more specifically has been challenging. While progress has been made with regard to mass spectrometry-based methods for the detection of ADP-ribosylated proteins, still no antibody exists that specifically recognizes MARylated proteins, in contrast to PARylation, which can easily be detected using antibodies that recognize PAR chains of varying length.

Macrodomains are protein domains with a conserved globular fold of ~150 amino acids and typically with affinity for free ADPr, protein-linked ADPr, and/or other ADPr-derivatives, such as *O*-acetyl-ADPr [15]. While several macrodomains act as binding modules for PAR, in addition to a variety of other protein domains recognizing PAR chains [16], only macro2 and macro3 of Artd8 have so far been described as domains binding exclusively MARylated proteins [15–17] (Fig. 1). We have made use of these two binding domains recognizing MARylation to demonstrate that MARylation/auto-modification of ARTD10 occurs in cells [11, 17]. More precisely, it was shown that all three macrodomains of human ARTD8, hereafter referred to as ARTD8macro1, ARTD8macro2, or ARTD8macro3, bind ADPr via a conserved ADPr-binding pocket. However, ARTD8macro1 binds free ADPr with lower affinity than ARTD8macro2 or ARTD8macro3. Consistent with these findings, only ARTD8macro2, ARTD8macro3, and a construct comprising all three macrodomains (ARTD8macro1-3), but not ARTD8macro1 alone, are able to interact with MARylated ARTD10 in in vitro pulldown experiments [17]. MARylation-dependent interaction between macro2, macro3, or macro1-3 and ARTD10 in vitro has also been shown for the macrodomains of murine Artd8. The domains of murine origin have been used to demonstrate intracellular interaction of the macrodomains with MARylated ARTD10 and endogenous RAN [17].

Here we present detailed protocols for assessing auto-modification levels of ARTD10 in cells using co-immunoprecipitation as well as co-localization experiments employing Artd8macro2 and

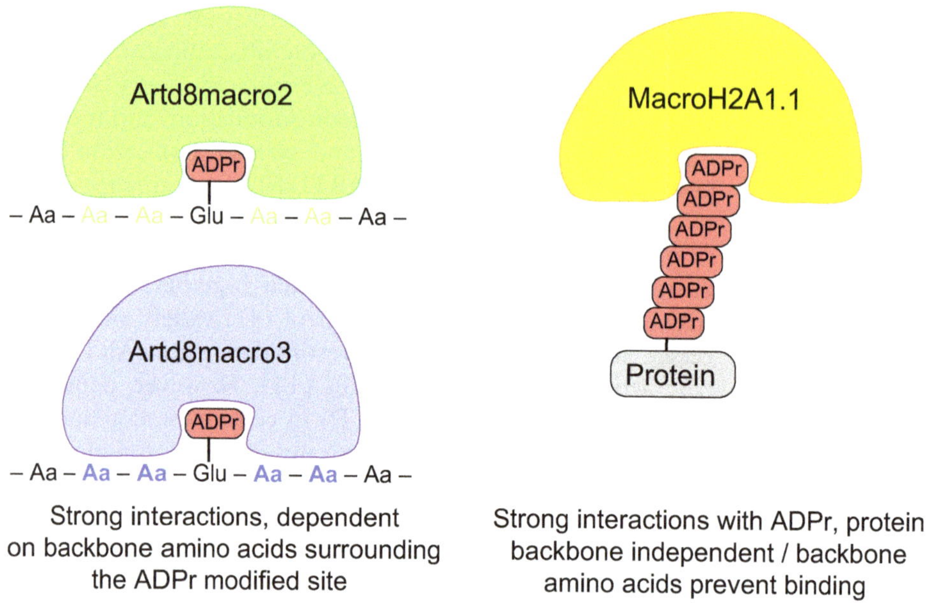

Fig. 1 Model for the interaction of macrodomains with either mono- or poly-ADP-ribose. The macrodomains of murine Artd8, Artd8macro2, and Artd8macro3 have somewhat different ARTD10-binding activities, leading to the suggestion that they interact with MARylation sites that possess distinct amino acid sequences surrounding the modification site. In contrast to MacroH2A1.1, Artd8macro2 and Artd8macro3 do not bind to the terminal ADPr of PAR chains. This indicates that they require additional determinants, i.e., the amino acid backbone of the substrate, for efficient binding. These additional determinants are not accessible in PARylated substrates

Artd8macro3. With the appropriate modifications, these protocols can also be used to address MARylation of other proteins. Briefly, for co-immunoprecipitation experiments, expression plasmids encoding N-terminally EGFP-tagged ARTD8/Artd8 macrodomains and HA-tagged ARTD10 are co-transfected in HEK293 cells (*see* Subheadings 2.1.1 and 3.1.1). Whole-cell lysates are prepared and macrodomains are pulled down using anti-EGFP antibodies coupled to Protein G Sepharose beads (*see* Subheadings 2.1.2 and 3.1.2). Coprecipitated proteins, such as MARylated ARTD10, are detected by standard Western blot analysis using specific antibodies against the protein of interest (*see* Subheadings 2.1.3 and 3.1.3), whereupon the amount of detected protein should correlate with its intracellular MARylation levels (Fig. 2). Suitable controls should always be included into the analysis, such as the use of ADPr-binding-deficient macrodomain mutants, to exclude the possibility of ADPr-independent interactions.

For co-localization studies, we routinely use HeLa Flp-In T-REx cell lines stably expressing ARTD10 or ARTD10-G888W, a catalytically inactive mutant, upon addition of doxycycline [13, 17]. When overexpressed, ARTD10 accumulates in highly dynamic cytoplasmic dot-like structures [18]. EGFP-tagged Artd8macro2

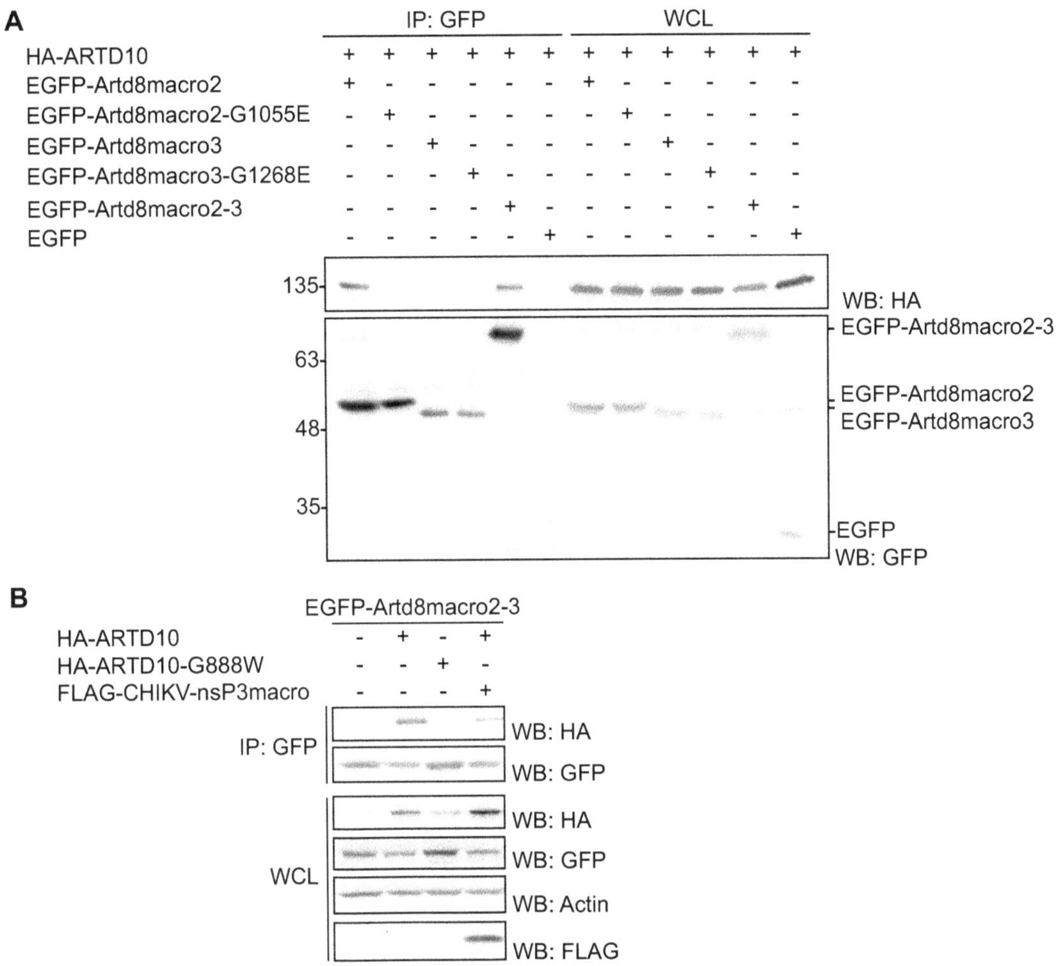

Fig. 2 Macrodomains of murine Artd8 interact with ARTD10 dependent on MARylation. (**a**) HEK293 cells were co-transfected with plasmids encoding HA-ARTD10 together with EGFP-tagged Artd8macro2 (WT/G1055E mutant), Artd8macro3 (WT/G1268E mutant) or Artd8macro2-3. EGFP-containing protein complexes were immunoprecipitated (IP) from whole-cell lysates (WCL). ARTD10 and EGFP fusion proteins were analyzed by Western blotting. Artd8macro2 and macro3 interact with ARTD10 in a MARylation-dependent manner. (**b**) HA-ARTD10-WT, together with or without FLAG-CHIKV-nsP3macro, or HA-ARTD10-G888W was transiently expressed in HEK293 cells stably expressing EGFP-Artd8macro2-3. EGFP-containing protein complexes were immunoprecipitated (IP) from whole-cell lysates (WCL). Coprecipitated HA-ARTD10 and EGFP-levels were analyzed by Western blotting. Coprecipitation of ARTD10 with Artd8macro2-3 can be partially reversed by co-expression of the CHIKV-nsP3macro MAR-hydrolase

and Artd8macro3 are recruited into these dots depending on catalytic activity of ARTD10 and ADPr-binding ability of the macrodomains. Thus, the extent to which the macrodomains are recruited into these dots can be assessed by confocal (immuno) fluorescence microscopy. Cells are seeded onto coverslips and transfected with expression plasmids for EGFP-tagged ARTD8/Artd8 macrodomains (*see* Subheadings 2.2.1 and 3.2.1).

Twenty-four h after transfection, ARTD10 expression is induced for additional 24 h. Cells are fixed (*see* Subheadings 2.2.2 and 3.2.2) and stained using ARTD10-specific antibodies (*see* Subheadings 2.2.3 and 3.2.3). Localization of ARTD10 and the macrodomains is analyzed by confocal (immuno)fluorescence microscopy (*see* Subheadings 2.2.4 and 3.2.4) (Fig. 3).

Using the described experimental setups, we typically make the following observations (described in detail in Subheadings 3.1.4 and 3.2.5): In both co-immunoprecipitation and in co-localization experiments, the macrodomains 2 and 3 of murine Artd8 interact with ARTD10 in a strictly ADPr-dependent fashion [17]. Mutations in the ADPr-binding pocket of the macrodomains, which interfere with ADPr binding, prevent interaction with ARTD10 [17]. Intracellular reversal of ARTD10 auto-modification, e.g., by co-expression of the Chikungunya virus (CHIKV) nsP3 macrodomain, results in reduced interaction of the Artd8 macrodomains with ARTD10 [11]. Moreover, we observe that a combination of macro2 and macro3 of murine Artd8 (Artd8macro2-3) is recruited more efficiently to ARTD10 during co-localization experiments compared to the single domains.

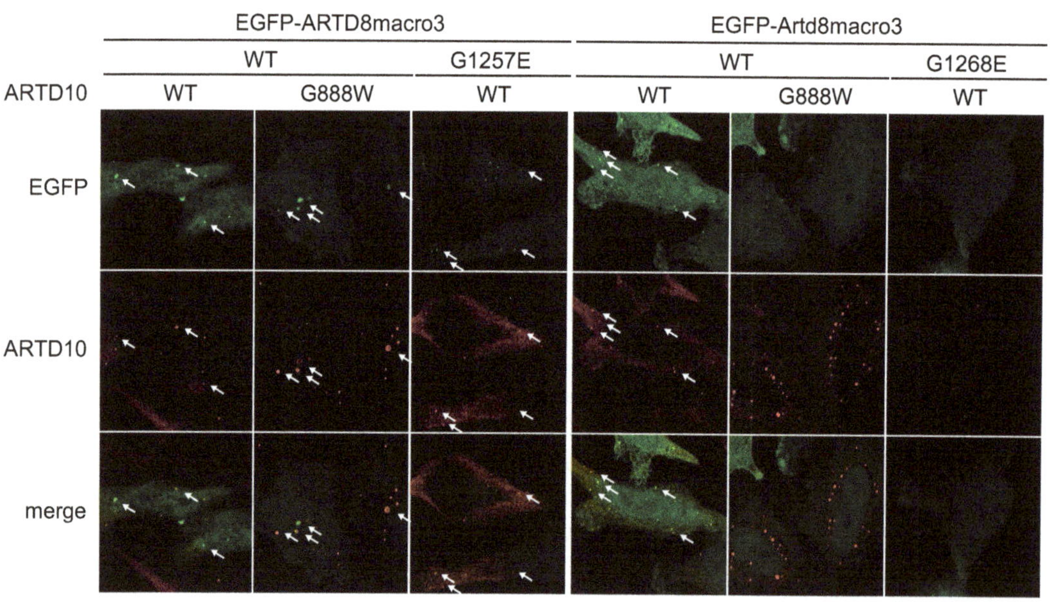

Fig. 3 Macrodomains of human ARTD8 interact with ARTD10 independent of MARylation activity. HeLa Flp-In T-REx cells expressing either ARTD10-WT or ARTD10-G888W upon doxycycline addition were transfected with constructs encoding EGFP-tagged ARTD8macro3, EGFP-tagged Artd8macro3, or the respective ADPr-binding-deficient mutants (ARTD8macro3-G1257E, Artd8macro3-G1268E). 24 h after transfection, ARTD10 expression was induced by addition of 500 ng/mL doxycycline for additional 24 h. Cells were fixed and stained using an ARTD10-specific primary antibody and an Alexa Fluor 555-coupled secondary antibody. EGFP and Alexa Fluor 555 (immuno)stainings were analyzed by confocal fluorescence microscopy. While Artd8macro3 co-localizes with ARTD10 dots (see arrows) in a MARylation-dependent manner, human ARTD8macro3 co-localizes with ARTD10 independent of its catalytic activity. Picture sizes: 50 × 50 μm

In contrast to the macrodomains 2 and 3 of murine Artd8, the macrodomains of human ARTD8 interact with ARTD10 independent of its catalytic activity in co-localization and co-immunoprecipitation experiments. We inferred from in silico structural comparisons that macro3 of human and murine origin exerts sequence variations in two surface-exposed loops in the vicinity of the ADPr-binding pocket, resulting in differences in the electrostatic surface potential of the two domains and in increased flexibility of both loops in human ARTD8 macro3 compared to its murine counterpart (Fig. 4). Keeping in mind that these MAR-specific macrodomains most probably recognize the substrate backbone in addition to the ADPr modification (Fig. 1), these sequence variations in both domains might account for differences in substrate recognition and therefore in interaction with ARTD10.

Altogether this illustrates that the macrodomains 2 and 3 of murine Artd8 can be used to estimate intracellular auto-modification levels of ARTD10 compared to a defined control, either in co-immunoprecipitation experiments, where the amount of coprecipitated ARTD10 correlates with its MARylation status, or in co-localization studies, where the extent of co-localization in ARTD10 dots as well as its background staining can reflect the degree of ARTD10 auto-modification. The described methods can help in the future to define factors (e.g., other proteins, signaling pathways, PTMs or inhibitors) that regulate intracellular ARTD10 auto-modification activity or reverse its auto-modification. Moreover, EGFP-Artd8macro2-3 can be adapted to demonstrate MARylation of other proteins, as shown for NEMO and RAN, by combining EGFP-Artd8macro2-3 with substrate-specific antibodies in, for example, co-immunoprecipitation, co-localization, and proximity ligation experiments.

2 Materials

Reagents and equipment listed below are routinely used in our laboratory. Materials and reagents from other providers should be equally suitable.

2.1 Co-immunoprecipitation of Mono-ADP-Ribosylated ARTD10 with Artd8 Macrodomains

2.1.1 Cell Seeding and Transfection

1. HEK293 cells (ATCC CRL-1573).
2. HEPES buffer: 10 mM HEPES pH 7.3, 142 mM NaCl, 6.7 mM KCl, autoclaved.
3. HBS buffer: 21 mM HEPES pH 6.95, 137 mM NaCl, 5 mM KCl, 0.7 mM Na_2HPO_4, sterile filtered (*see* **Note 1**).
4. 2.5 M $CaCl_2$, sterile filtered.
5. Plasmids:

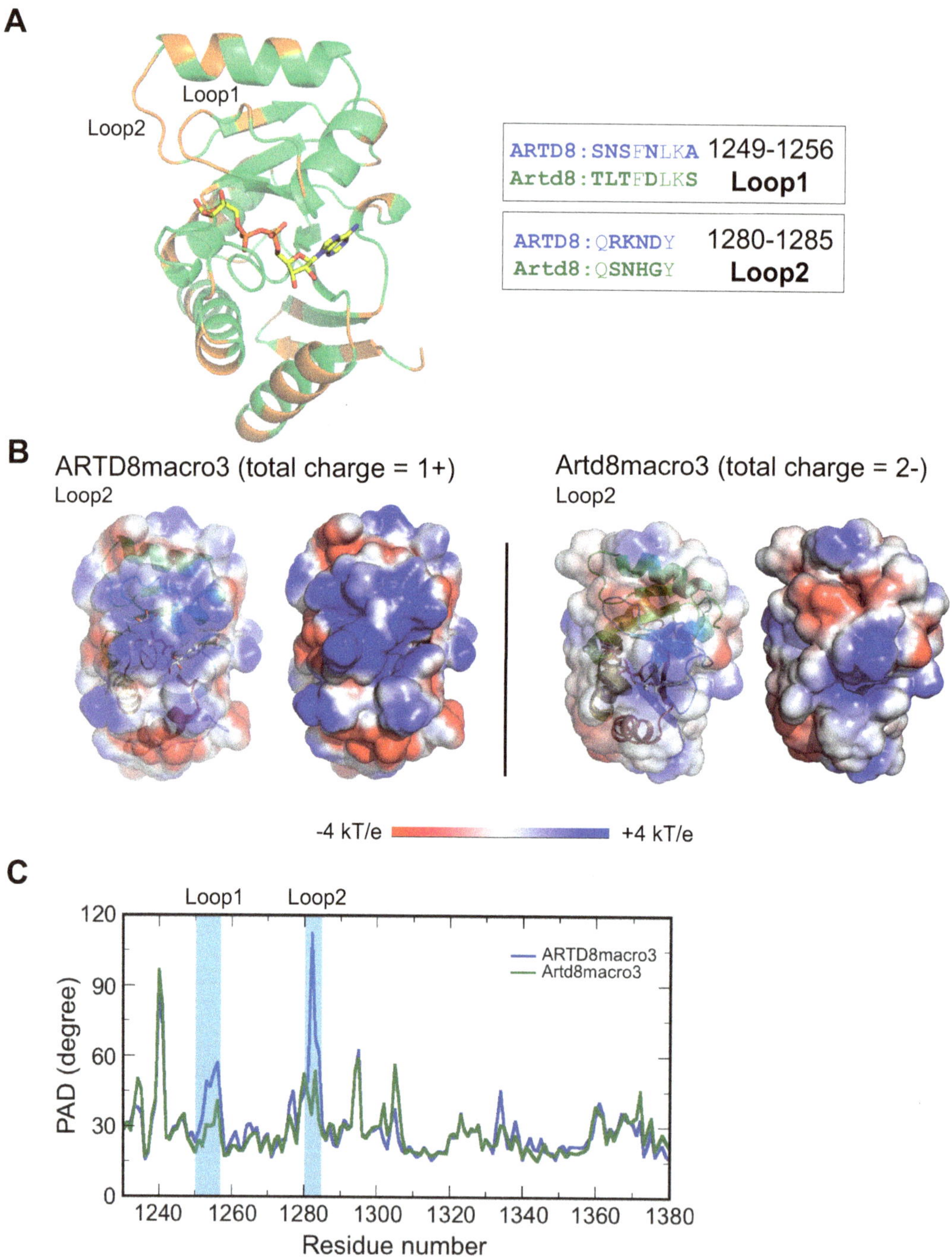

Fig. 4 Structural differences between murine and human ARTD8macro3. (**a**) Sequence variations in murine Artd8macro3 (colored in orange) projected onto the structure of human ARTD8macro3 (colored in green; PDB ID: 4ABK [17]). (**b**) Electrostatic potential (±4 kT/e) rendered on the surfaces of proteins from the most representative structures in molecular dynamics simulations. (**c**) PAD analysis [19] to characterize the local flexibility of each residue across molecular dynamics simulations. The index goes from 0 to 180°: The higher the index is, the higher is the local flexibility of the residue

Eukaryotic expression plasmids of N-terminally HA-tagged ARTD10: pEVRF0-HA-ARTD10 [7], pEVRF0-HA-ARTD10-G888W [7].

6. Eukaryotic expression plasmids of N-terminally EGFP-tagged macrodomains of human/murine ARTD8/Artd8 (based on pEGFP-C1): GW-pEGFP-Artd8macro2 [17], GW-pEGFP-Artd8macro2-G1055E [17], GW-pEGFP-Artd8macro3 [17], GW-pEGFP-Artd8macro3-G1268E, GW-pEGFP-Artd8macro2-3.

   Eukaryotic expression plasmid of N-terminally FLAG-tagged macrodomain of CHIKV-nsP3: GW-pcDNA3-FLAG-CHIKV-nsP3macro [11].

2.1.2 Preparation of Cell Lysates and Co-immunopre-cipitation

1. PBS: 140 mM NaCl, 2.6 mM KCl, 2 mM Na_2HPO_4, and 1.45 mM KH_2PO_4.
2. Lysis buffer: 50 mM Tris pH 7.5, 150 mM NaCl, 1 mM EDTA, 10% (v/v) glycerol, 1 mM DTT, 1% (v/v) NP-40, 1× protease inhibitor cocktail (P8340, Sigma), and 100 μM sodium vanadate.
3. Protein G Sepharose 4 FastFlow.
4. Anti-GFP antibody (9F9, Rockland).
5. Four times sample buffer (SB): 320 mM Tris–HCl pH 6.8, 40% (v/v) glycerol, 20% (w/v) SDS, 0.5% (w/v) Bromophenol blue, and 200 mM β-mercaptoethanol.

2.1.3 SDS-PAGE and Western Blot Analysis

1. Lower Tris buffer (separating gel): 1.5 M Tris–HCl pH 8.8, 0.4% SDS.
2. Upper Tris buffer (stacking gel): 0.5 M Tris–HCl pH 6.8, 0.4% SDS.
3. 20% (w/v) sodium dodecyl sulfate (SDS).
4. 20% (w/v) ammonium persulfate (APS).
5. *N*,*N*,*N*′,*N*′-tetraethylmethylenediamine (TEMED).
6. 30% acrylamide/bis-acrylamide 4K Mix 37.5:1.
7. Running buffer (Laemmli): 25 mM Tris, 250 mM glycine, 0.1% (w/v) SDS.
8. Isopropanol.
9. Mini-PROTEAN 3 gel electrophoresis system: casting stand, casting frames, glass plates, and combs (Bio-Rad).
10. Protein molecular weight standards.
11. Semidry transfer buffer: 25 mM Tris, 192 mM glycine, and 20% (v/v) methanol.
12. Blotting paper.
13. Nitrocellulose membrane.

14. Semidry transfer blotter.
15. PBS-T: 0.05% (v/v) Tween-20 in PBS.
16. Ponceau S solution: 0.05% (w/v) Ponceau S, 1% acetic acid, in H_2O.
17. Blocking solution: 5% nonfat dry milk powder (w/v) in PBS-T.
18. Antibodies diluted as indicated in PBS-T: anti-GFP (9F9, Rockland, 1:5000), anti-HA (3F10, Roche, 50–200 ng/mL) or anti-HA (16B12, Covance, 1:1000) (*see* **Note 2**), anti-FLAG (M2, Sigma-Aldrich, 1:5000), anti-actin (C4, MP Biomedicals, 1:500) antibody, anti-mouse IgG-HRP conjugate, and anti-rat IgG-HRP conjugate.
19. ECL Western blotting substrate.
20. Luminescence imaging device.

2.2 Co-localization of ARTD10 with Artd8 Macrodomains

2.2.1 Cell Seeding and Transfection

1. Sterile glass coverslips ⌀18 mm.
2. Cell lines: HeLa Flp-In T-REx ARTD10 and HeLa Flp-In T-REx ARTD10 G888W [13].
3. 12-well plates.
4. Transfection reagents: HBS buffer and 2.5 M $CaCl_2$ (*see* Subheading 2.1.1).
5. Plasmids:

   Eukaryotic expression plasmids of N-terminally EGFP-tagged macrodomains of human/murine ARTD8/Artd8 (based on pEGFP-C1): GW-pEGFP-ARTD8macro3, GW-pEGFP-ARTD8macro3-G1257E, GW-pEGFP-Artd8macro3 [17], GW-pEGFP-Artd8macro3-G1268E, GW-pEGFP-Artd8macro1-3 [17], and GW-pEGFP-Artd8macro2-3.

   Eukaryotic expression plasmids of N-terminally EGFP-tagged macrodomains of murine Artd8 (based on pcDNA5/FRT/TO): pcDNA5/FRT/TO-EGFP-Artd8macro2-3, pcDNA5/FRT/TO-EGFP-Artd8macro2-3-G1055E/G1268E.

   Eukaryotic expression plasmid of N-terminally FLAG-tagged macrodomain of CHIKV-nsP3: GW-pcDNA3-FLAG-CHIKV-nsP3macro [11].
6. HEPES buffer (*see* Subheading 2.1.1).
7. Doxycycline, 1 mg/mL in H_2O.

2.2.2 Fixation

1. PBS (*see* Subheading 2.1.2).
2. Formaldehyde solution: 3.7% (w/v) paraformaldehyde in PBS, heat to 60 °C while stirring until the paraformaldehyde has dissolved completely. After the solution has cooled down, aliquot and store it at −20 °C (*see* **Note 3**).

2.2.3 Permeabilization, Blocking, and Immunostaining

1. Permeabilization solution: 0.1% Triton-X-100 (v/v) in PBS.
2. Blocking solution: 1% (w/v) bovine serum albumin (BSA), 0.1% (v/v) Triton-X-100 in PBS.
3. Antibody dilution buffer: 0.2% BSA in PBS.
4. Anti-ARTD10 antibody (5H11) [17], final dilution 1:50 in antibody dilution buffer.
5. Goat anti-rat IgG (H+L)-Alexa Fluor 555 conjugate: final dilution 1:1000 in antibody dilution buffer (2 μg/mL).
6. Hoechst 33258, final concentration of 1 μg/mL in H_2O.
7. Mowiol 4-88: add 2.4 g Mowiol to 6 mL glycerol and stir to mix. Add 6 mL H_2O and stir for 2 h at room temperature. Mix with 12 mL 0.2 M Tris pH 8.5 and incubate for 10 min at 50–60 °C to dissolve. Centrifuge for 15 min at 5000 × *g*. Store in aliquots at −20 °C. Warm to 37 °C before use.
8. Microscope glass slides.

2.2.4 Acquiring Images

1. LSM710 confocal microscope (Zeiss) equipped with a Plan-Apochromat 63×/1.40 Oil DIC M27 objective, an AxioCam (Zeiss), a UV 405 nm diode, an argon laser, a DPSS 561 nm laser, and beam splitter filters.
2. Immersion oil.

3 Methods

3.1 Co-immunoprecipitation of Mono-ADP-Ribosylated ARTD10 with Artd8 Macrodomains

3.1.1 Cell Seeding and Transfection

First, ARTD10 (WT and optionally the G888W mutant) has to be co-expressed together with the macrodomains (WT and mutated) to allow the proteins to interact in cells. Therefore, HEK293 cells are seeded and transfected the next day with plasmids encoding for the respective proteins. The use of macrodomains fused to an epitope tag such as EGFP allows for the capturing of macrodomain-containing complexes in later steps of the protocol.

1. Seed 1.5×10^6 HEK293 cells per 10 cm dish. One 10 cm dish will be used per immunoprecipitation (IP). Allow the cells to adhere to the plate overnight.
2. The next day, co-transfect cells with plasmids encoding the desired EGFP-tagged macrodomains together with HA-tagged ARTD10-WT or ARTD10-G888W (*see* **Note 4**).
3. For transfection of one 10 cm cell culture dish, mix 20 μg of the desired plasmid DNA (*see* **Note 5**) in a reaction tube, and dilute in 950 μL HBS buffer; add 50 μL of 2.5 M $CaCl_2$ dropwise and vortex briefly. Incubate for 15 min at room temperature. Add transfection mixture dropwise to the cells. Gently swirl the plates and incubate at 37 °C for 4–6 h.

4. Wash the cells with pre-warmed HEPES buffer to remove the precipitate (*see* **Note 6**). Incubate the cells for 48 h in a cell culture incubator.

3.1.2 Preparation of Cell Lysates and Co-immunoprecipitation

To demonstrate interaction between the macrodomains and ARTD10 (and additional MARylation substrates), macrodomain-containing complexes need to be immunoprecipitated from lysates of the cells transfected as in Subheading 3.1.1. Addition of EGFP-specific antibodies to the lysates results in the formation of macrodomain-containing immune complexes, which are captured by co-incubation with antibody-binding Protein G Sepharose beads. To remove nonspecific interactions, the beads are washed with lysis buffer. IP samples as well as fractions of the input lysates are prepared for the subsequent SDS-PAGE.

The following steps should be carried out at 4 °C/on ice. Precool all the buffers and centrifuges.

1. Forty-eight h after transfection, wash the cells gently with PBS. Aspirate PBS completely. Add 500 μL of lysis buffer to each 10 cm dish.
2. Scrape cells off the plate with a cell scraper, and transfer lysate to a 1.5 mL reaction tube. Incubate for 30 min on ice.
3. Centrifuge lysates for 20 min at 4 °C at 16,000 × *g* to remove cell debris. Transfer supernatant to a new reaction tube.
4. Mix 25 μL of lysate with 8.3 μL 4× SB, heat for 5 min at 95 °C, and store as input sample (5% input) at −20 °C until gel loading (*see* **Note 7**).
5. Equilibrate Protein G Sepharose beads. Per IP, use 25 μL of 50% bead slurry. Pipette the needed amount of beads for all IPs into one tube, and fill it up with lysis buffer. Mix by inverting several times.
6. Centrifuge for 2 min at 500 × *g* to pellet the beads. Aspirate the supernatant and repeat the washing step.
7. Split the beads to several tubes according to the number of samples. Centrifuge again and aspirate residual buffer.
8. Pipette the remaining cell lysate onto the beads.
9. Add 500 ng of anti-GFP antibody to every tube. Incubate for 1 h at 4 °C under constant rotation.
10. Pellet the beads by centrifugation at 500 × *g* for 2 min.
11. Aspirate supernatant (*see* **Note 8**), and perform three wash steps with 500 μL lysis buffer each (*see* **Note 9**) and centrifugation for 2 min at 500 × *g*.
12. Aspirate supernatant from the beads completely, and resuspend in 30 μL of 1× SB. Heat the samples for 5 min at 95 °C, and spin down before gel loading (*see* **Note 10**).

3.1.3 SDS-PAGE and Semidry Western Blot

The presence of ARTD10 (and other MARylation substrates) in the macrodomain-containing immune complexes is detected by standard Western blots analysis. Therefore, SDS-PAGE and subsequent Western blotting are performed followed by immune detection of the desired antigens in IP and input samples.

1. Cast a 10% SDS acrylamide separating gel: Mix 2.8 mL Lower Tris, 5.6 mL H_2O and 4.15 mL acrylamide/bis-acrylamide. Add 65 μL 20% APS and 12.5 μL TEMED, mix, and immediately pour into the gel plate assembly to ¾ of the volume. Cover the surface with isopropanol, and wait for 15–20 min until the gel has polymerized (*see* **Note 11**).
2. After polymerization of the separating gel, remove the isopropanol carefully and cast the stacking gel on top of the separating gel: Mix 2.5 mL H_2O, 1 mL Upper Tris, and 0.5 mL acrylamide/bis-acrylamide. Add 20 μL 20% APS and 8 μL TEMED, mix well, and pour the mixture on top of the separating gel. Insert the comb, and wait for 15–20 min until the gel has completely polymerized.
3. Remove the comb and residual gel pieces, assemble the electrophoresis chamber, and fill it up with running buffer. Rinse the slots of the gel with running buffer using a syringe.
4. Load a size standard protein ladder and total amounts of IP and input samples.
5. Run the gel at 150–200 V until the loading dye has reached the bottom of the gel.
6. Equilibrate six blotting papers per gel and the blotting membrane in transfer buffer.
7. Disassemble the acrylamide gels and equilibrate them in transfer buffer.
8. Assemble the Western blot in the following order: Cathode (−)—3× blotting paper—gel—membrane—3× blotting paper—anode (+).
9. Remove air bubbles by flattening the blot sandwich.
10. Run the transfer at 1.5–2 mA/cm^2 gel for 1 h 15 min.
11. Disassemble the blot layers and incubate the membrane in Ponceau S solution for a few seconds under agitation to check for equal blotting. Cut the stained membrane horizontally between 75 and 100 kDa size standard bands. Wash the membrane pieces in PBS-T to reverse the Ponceau staining.
12. Incubate the membrane pieces in blocking solution for 30 min at room temperature under constant agitation.
13. Wash the membrane pieces three times for 5 min with PBS-T.

14. Incubate the membrane piece with proteins >75 kDa with the anti-HA antibody and the membrane piece with proteins <75 kDa with the anti-GFP antibody overnight at 4 °C under constant agitation (*see* **Note 12**).
15. Wash the membrane pieces three times for 10 min with PBS-T.
16. Incubate the membrane pieces with the respective secondary antibodies for 1 h at room temperature under constant agitation.
17. Wash the membrane pieces three times for 10 min with PBS-T.
18. Develop the membrane pieces using ECL solutions.
19. Optional: Detect actin levels in the input fractions on the same membrane piece (<75 kDa) repeating the **steps 14–18** using the respective primary and secondary antibodies as an additional input control.

3.1.4 Data Interpretation

EGFP-tagged Artd8macro2 and Artd8macro3 coprecipitate ARTD10 when co-expressed in HEK293 cells but not catalytically inactive ARTD10-G888W (Fig. 2a) [17]. Mutation of a conserved glycine residue in the ADPr-binding pocket of these macrodomains (Artd8macro2-G1055E and Artd8macro3-G1268E) interferes with ADPr binding. EGFP-tagged Artd8macro2 and three mutants do no longer interact with ARTD10 in co-immunoprecipitation experiments (Fig. 2a) [17]. Furthermore, reversibility of ARTD10 auto-modification in cells resulting from co-expression of hydrolases such as the CHIKV-nsP3 macrodomain can also be demonstrated using the described method (Fig. 2b) [11], indicating that the amount of ARTD10 pulled down with the macrodomains correlates with its intracellular MARylation status. It should however be noted that Artd8macro2 and Artd8macro3 alone exert somewhat distinct binding toward ARTD10 (Fig. 2a) [17]. We have proposed that these macrodomains require an interaction with the substrate backbone in addition to the substrate-bound ADPr. This would explain not only the fact that they apparently do not interact with (the terminal ADPr residue of) PAR but also the differential binding of macro2 and macro3 toward one specific MARylated protein such as ARTD10 and possibly reflects that these macrodomains interact with MARylated proteins in a site-specific manner (Fig. 1) [17].

3.2 Co-localization of ARTD10 with Artd8 Macrodomains

3.2.1 Cell Seeding and Transfection

First, ARTD10 (WT/G888W mutant) has to be co-expressed in cells together with the macrodomains (WT/mutants) to allow the proteins to interact in cells. Because the samples need to be stained for immunofluorescence microscopy in later steps of the protocol, cells are grown directly on glass coverslips. HeLa cells, which stably and inducibly express ARTD10 (WT/G888W mutant), are transfected with plasmids encoding for EGFP-tagged macrodomains, which allows for the direct detection of macrodomains via EGFP fluorescence. Twenty-four h after transfection, ARTD10 expression is induced resulting in co-expression of both proteins.

1. Seed 5 × 10^4 HeLa Flp-In T-REx ARTD10 cells per well of a 12-well plate in 1 mL cell culture medium on sterile glass coverslips (*see* **Note 13**). Allow the cells to adhere to the coverslip overnight.
2. Transfect the cells with a plasmid allowing for expression of the desired EGFP-tagged macrodomain using calcium phosphate precipitation. Mix the needed amounts of plasmid DNA (2 μg DNA in total per well; *see* **Note 14**), and dilute it in 95 μL HBS buffer per well. Add 5 μL 2.5 M $CaCl_2$ and vortex briefly. Incubate the transfection mix for 15 min at room temperature. Add the mixture dropwise to the cells. Swirl the plate gently and incubate overnight in the cell culture incubator.
3. The next day, wash cells with pre-warmed HEPES buffer to remove the precipitate. Add 1 mL fresh medium to each well and 500 ng/mL doxycycline to the desired wells to induce ARTD10 expression. Incubate the cells in the cell culture incubator for 24 h.

3.2.2 Fixation

Cells are fixed to preserve cellular morphology, the subcellular architecture, and antigens for antibody binding by chemical crosslinking with formaldehyde.

The following steps should be carried out at room temperature unless otherwise noted:

1. Wash the cells twice with PBS.
2. Remove PBS and add 500 μL formaldehyde solution per well. Incubate for 20 min.
3. Remove formaldehyde solution and wash cells extensively with PBS (*see* **Note 15**).

3.2.3 Permeabilization, Blocking, and Immunostaining

To make the cells accessible for antibodies, they are permeabilized by incubation with detergent. Samples are blocked to minimize unspecific binding of the antibodies. Next, samples are stained with ARTD10-specific primary antibodies and Alexa Fluor 555-coupled secondary antibodies before mounting the coverslips on microscope glass slides.

1. Wash the cells for 5 min with 500 μL permeabilization solution.
2. Incubate the cells for 30 min in blocking solution.
3. Aspirate blocking solution and incubate coverslips with primary antibody for 1 h at 37 °C (*see* **Note 16**).

4. Wash three times with PBS.
5. From now on, avoid exposing the samples, antibodies, and Hoechst stain to direct light to prevent photo-bleaching.
6. Incubate coverslips with secondary antibody for 1 h.
7. Wash three times with PBS and once with H_2O (*see* **Note 17**).
8. Incubate coverslips for 5 min with Hoechst 33258 to stain nuclei (optional).
9. Wash coverslips three times with H_2O.
10. Mount coverslips using 10 μL Mowiol on microscope glass slides (*see* **Note 18**). Allow the mounting medium to solidify for several hours. Store the samples at 4 °C protected from light until image acquirement.

3.2.4 Acquiring Images

EGFP and Alexa Fluor 555 signals are acquired from a single focal plane using confocal laser-scanning microscopy. The microscope settings for image acquirement will depend on the type of confocal microscope and software used as well as on the quality of the samples.

1. Spot a drop of immersion oil on the objective before placing the glass slide on top.
2. The settings for imaging with the Zeiss LSM 710 confocal microscope and the Zen2012 software, as applied in Figs. 3 and 5, are as follows: Plan-Apochromat 63×/1.40 Oil DIC M27 objective, picture size 1024 × 1024 px, pixel dwell time 6.30 μs, 1.5-fold digital zoom (corresponding picture size of 89.9 × 89.9 μm).
3. Optional Excitation of nuclear Hoechst staining is performed using the 405 nm laser line (UV 405 nm diode). Fluorescence is detected using a beam splitter filter (MBS 405) with the detection range set from 415 to 487 nm.
4. EGFP fluorescence is excited using the 488 nm laser line (argon, 25 mW, 2% output), and detected using a beam splitter filter (MBS 488/561/633) with the detection range set from 493 to 552 nm. Alexa Fluor 555 fluorescence is excited using the 561 nm laser line (DPSS, 20 mW, 2% output), and detected using a beam splitter filter (MBS 488/561/633) with the detection range set from 562 to 715 nm (*see* **Note 19**).
5. For co-localization experiments, each individual channel needs to be adjusted to the same optical slice thickness. The pinhole therefore should be set to 1 AU for one channel and adjusted to the same corresponding optical slice thickness in the other channels (*see* **Note 20**).
6. Settings should be kept equal for all images belonging to a single experiment.

3.2.5 Data Interpretation

EGFP-tagged Artd8macro2 and Artd8macro3 are recruited into ARTD10 dots in HeLa Flp-In T-REx cells that express ARTD10, but not into ARTD10-G888W dots upon induction in HeLa Flp-In T-REx cells (Fig. 3) [17]. Of note is that ARTD10-G888W forms dots at least as efficient as ARTD10, indicating that this behavior is independent of MARylation. ADPr-binding-deficient EGFP-tagged Artd8macro2-G1055E [17] or EGFP-tagged Artd8macro3-G1268E (Fig. 3) no longer co-localize with ARTD10 dots. In contrast to murine Artd8macro3, human ARTD8macro3 co-localizes with both ARTD10 and ARTD10-G888W dots (Fig. 3). To exclude the possibility that the macrodomain recognizes a MARylated site on ARTD10 that does not originate from ARTD10 auto-modification but from MARylation by another mono-ARTD enzyme in cells, we created an ARTD8macro3 mutant (G1257E; corresponding to Artd8macro2-G1055E), which is supposed to interfere with ADPr binding. However, this mutant displays the same behavior in cells as the wild-type macrodomain (Fig. 3). Thus, although interaction of human ARTD8 macrodomains with ARTD10 in vitro is MARylation/auto-modification dependent, this is not simply transferrable to the *in-cell* situation. This might be explained by sequence variations in two surface-exposed loops, loop 1 and loop 2, between macro3 of human ARTD8 and murine Artd8, which show an overall sequence identity of 70.4%. The two loops are in the vicinity of where the protein backbone of the MARylated substrate is postulated to interact with macro3 (Fig. 4a). These sequence variations in addition to others, which are exposed on the macrodomains, result in marked differences in their electrostatic surface potential (Fig. 4b) and in increased flexibility of both loops in ARTD8macro3 compared to Artd8macro3 as inferred from molecular dynamics simulations and PAD (protein angular dispersion) analysis (Fig. 4c) [19]. Thus, these two loops might be important for substrate recognition, in addition to ADPr binding, and could therefore determine binding site specificity. With regard to its interaction with MARylated ARTD10, the altered surface potential in ARTD8macro3 might prevent binding of the MARylated site on ARTD10 that is recognized by its murine counterpart and favor a MARylation-independent interaction at a different site, a suggestion that is consistent with the ADPr-independent interaction of human ARTD8macro3 with ARTD10. Furthermore, differential modification sites produced by ARTD10 during in vitro ADP-ribosylation assays and in cells could account for the differences seen in in vitro and in *in-cell* experiments. We suppose that the role of ARTD8 and ARTD10, in addition to other mono-ARTDs, in innate immunity has dictated a different evolutionary development of the macrodomains. While maintaining the binding to ADPr, the substrate specificity may be distinct. Thus, although our findings are rather black and white regarding

ARTD10 binding, it is well possible that the human ARTD8macro2 and 3 domains will recognize distinct MARylated substrates that have not been identified yet. We therefore conclude that the macrodomains of human ARTD8 are not reliable as a tool for assessing the intracellular auto-modification status of ARTD10. For this purpose, only the macrodomains 2 and 3 of murine Artd8 should be used.

We have observed previously that the interaction between ARTD10 and a construct comprising all three macrodomains of murine Artd8, macro1-3, seems to be stronger than the interaction with the single macrodomains [17]. In co-localization experiments, we observed that while Artd8macro3 co-localizes with ARTD10 dots with an additional substantial background distribution throughout the rest of the cell (Figs. 3 and 5a), the combined macrodomain construct is recruited more efficiently into ARTD10 dots with a reduction in cytoplasmic background staining (Fig. 5a) [17]. This effect is even more pronounced when, instead of a macro1-3 construct, a macro2-3 construct is co-expressed together with ARTD10, resulting in a strong formation of dots by EGFP-Artd8macro2-3 that all co-localize with ARTD10 dots and almost no residual EGFP background staining, with all these constructs being expressed equally (not shown). The strong co-localization of EGFP-Artd8macro2-3 with ARTD10 dots might be explained by an increased avidity that results from the combination of two macrodomains in one construct and/or with macro2 and macro3 potentially recognizing different auto-modification sites on ARTD10, resulting in cooperative binding and thus highly efficient co-localization with ARTD10 dots. The recruitment of EGFP-Artd8macro2-3 to ARTD10 dots as well as its distribution throughout the rest of the cell (cytoplasmic background) seems to be directly correlated to the ability of the macrodomains to interact with ARTD10-linked mono-ADPr in cells. This is concluded from different control experiments. EGFP-Artd8macro2-3 is evenly distributed throughout the cell when co-expressed with ARTD10-G888W, compatible with its defect to auto-modify and hence a lack of EGFP-Artd8macro2-3 binding sites (Fig. 5). Moreover, the cellular distribution of an ADPr-binding-deficient mutant,

Fig. 5 (continued) with constructs encoding EGFP-Artd8macro2-3, EGFP-Artd8macro2-3-G1055E/G1268E (Artd8macro2-3-GE/GE) or FLAG-CHIKV-nsP3macro. 24 h after transfection, ARTD10 expression was induced by addition of 500 ng/mL doxycycline for additional 24 h. Cells were fixed and stained using an ARTD10-specific primary antibody and an Alexa Fluor 555-coupled secondary antibody. EGFP and Alexa Fluor 555 (immuno)stainings were analyzed by confocal fluorescence microscopy. The strong recruitment of EGFP-Artd8macro2-3 into ARTD10-WT dots and its reduced cytoplasmic expression induced by co-expression of ARTD10-WT are reversed by co-expression of the CHIKV-nsP3 macrodomain hydrolase or introduction of mutations in the ADPr-binding cleft

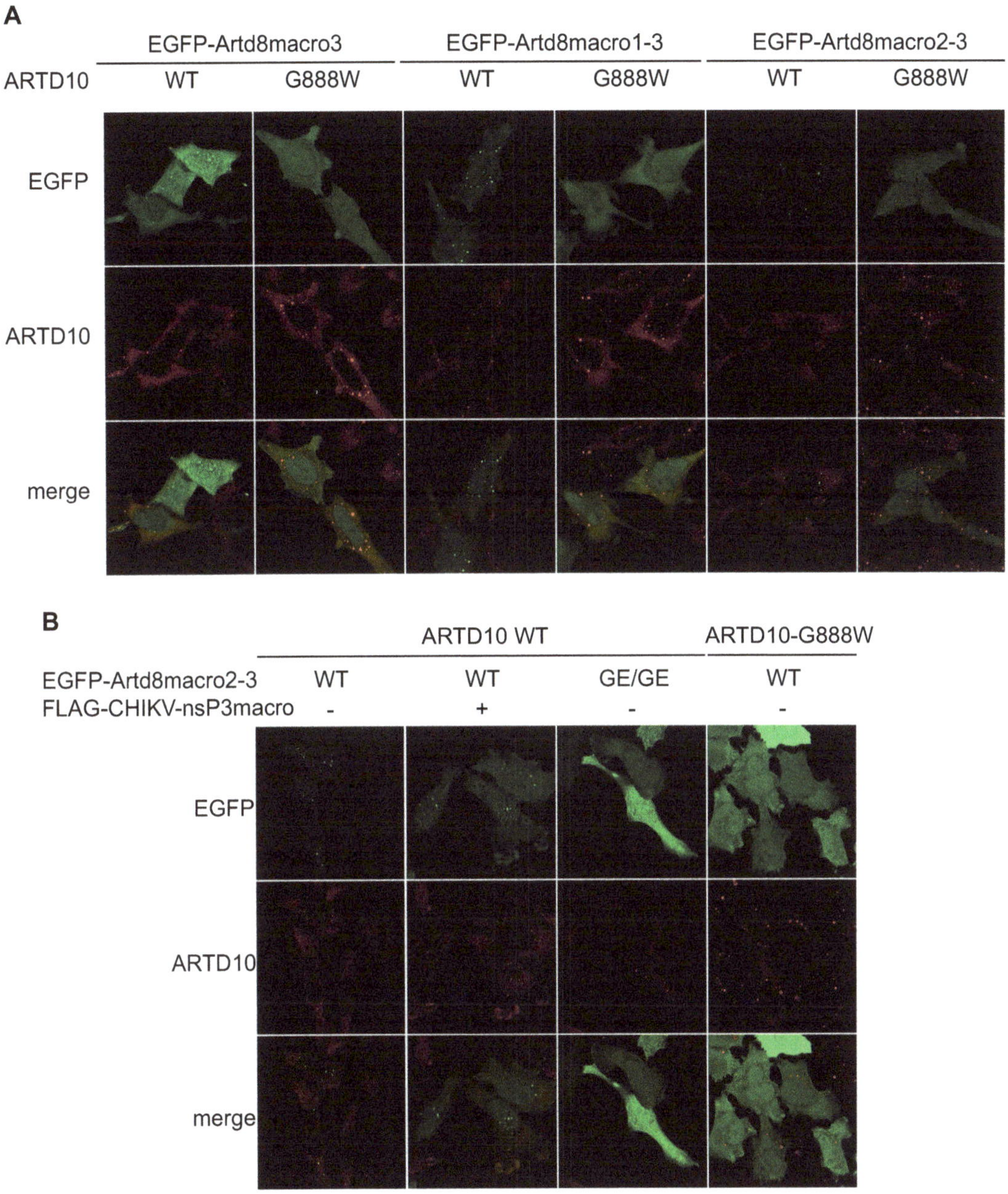

Fig. 5 A combined macro2 and macro3 construct of murine Artd8 is recruited more efficiently to ARTD10 dots than the single macro3 construct. (**a**) HeLa Flp-In T-REx cells expressing either ARTD10-WT or ARTD10-G888W upon doxycycline addition were transfected with constructs encoding EGFP-Artd8macro3, EGFP-Artd8macro1-3, or EGFP-Artd8macro2-3. 24 h after transfection, ARTD10 expression was induced by addition of 500 ng/mL doxycycline for additional 24 h. Cells were fixed and stained using an ARTD10-specific primary antibody and an Alexa Fluor 555-coupled secondary antibody. EGFP and Alexa Fluor 555 (immuno)stainings were analyzed by confocal fluorescence microscopy. Upon co-expression with ARTD10, Artd8macro3 alone is recruited into ARTD10 dots with residual background staining, while the combined EGFP-fusion constructs of Artd8macro1-3 or macro2-3 are recruited more efficiently into ARTD10 dots with a reduction in the residual cytoplasmic EGFP signal. (**b**) HeLa Flp-In T-REx cells stably expressing either ARTD10-WT or ARTD10-G888W were transfected

EGFP-Artd8macro2-3-G1055E/G1268E, is not affected by ARTD10 induction (Fig. 5b). Also, the staining of ARTD10 dots by EGFP-Artd8macro2-3 is reduced in parallel to an increase in broad, unfocused distribution of the fusion protein when the CHIKV-nsP3 macrodomain, which exerts robust MAR hydrolase activity toward ARTD10, is co-expressed (Fig. 5b) [11]. Together, these findings provide evidence that the relative co-localization of ARTD10 and EGFP-Artd8macro2-3 is a direct readout of ARTD10 MARylation.

4 Notes

1. The pH of the buffer should be carefully and exactly adjusted to 6.95 to ensure optimal transfection results.
2. Alternatively, use an antibody directed against ARTD10.
3. Frozen aliquots of formaldehyde solution should be warmed to 37 °C before use until any precipitate has dissolved completely. Formaldehyde is toxic. Solutions should be made and handled in a fume hood.
4. Optional: To assess reversal of ARTD10 auto-modification by MAR hydrolases, such as CHIKV-nsP3macro, co-transfect a hydrolase-encoding plasmid at this stage, e.g., GW-pcDNA3-FLAG-CHIKV-nsP3macro (*see* Subheading 2.1.1). In case it is desired to demonstrate MARylation of other proteins (that are not sufficiently expressed at the endogenous level for easy detection), plasmids encoding these proteins can be co-transfected during this step accordingly.
5. The amounts of each plasmid used for co-transfection should be tested individually and adjusted to reach adequate expression levels of each protein. Fill up with empty vector to 20 μg plasmid DNA per 10 cm dish in total.
6. The cells can be incubated with HEPES buffer for up to 5 min at 37 °C. Control removal of precipitate with a light microscope.
7. Optional (*see* **Note 4**): In case FLAG-CHIKV-nsP3macro is co-transfected, FLAG-CHIKV-nsP3macro expression levels should be detected in the input fractions. FLAG-CHIKV-nsP3macro is best resolved on a 15% SDS acrylamide gel, while ARTD10 and the Artd8 macrodomains can be resolved on the same 10% SDS acrylamide gel. Make sure to take off enough of input lysate to load two gels or use gradient gels.
8. The IP supernatant can be kept and stored at −20 °C for further analyses.

9. Increasing the salt concentration of the buffer during the washing step can help to prevent nonspecific interactions.
10. Samples can be stored at −20 °C but should be heated again shortly at 95 °C before gel loading.
11. Optional (*see* **Notes 4** and 7): Cast an additional 15% SDS acrylamide gel to resolve FLAG-CHIKV-nsP3macro. Mix 2.8 mL Lower Tris, 3.5 mL H_2O and 6.25 mL acrylamide/bis-acrylamide. Add 65 μL 20% APS and 12.5 μL TEMED to start polymerization.
12. Optional (*see* **Notes 4**, 7 and **11**): Incubate the additional Western blot for detection of FLAG-CHIKV-nsP3macro with an anti-FLAG antibody (*see* Subheading 2.1.3).
13. Non-sterile coverslips can be sterilized directly in the wells by washing them for a few minutes with 70% ethanol followed by washing with PBS twice.
14. We usually transfect 0.5 μg of macrodomain plasmid filled up with 1.5 μg of empty vector. In case intracellular removal of auto-modification of ARTD10 by hydrolases such as the CHIKV-nsP3 macrodomain should be investigated, co-transfect a hydrolase-encoding plasmid together with the Artd8 macrodomain construct, and fill up with less empty vector accordingly. Plasmid amounts should in general be carefully titrated and adjusted to reach adequate expression levels of the proteins.
15. Fixed cells can be stored in PBS at 4 °C for a few days until staining.
16. Spot 50 μL of antibody solution per sample on Parafilm, and transfer the coverslips cells up on the antibody solution. Incubation should be performed in a humid chamber to prevent evaporation of the antibody solution.
17. Washing with H_2O is important to avoid the formation of salt crystals during the following steps in the protocol.
18. Mowiol can be pre-warmed to 37 °C. Cut the tip before pipetting and avoid air bubble formation.
19. Optimize image brightness and signal to noise ratio by adjusting the master gain and the laser power if necessary.
20. For each individual channel, the setting of 1 airy unit (1 AU) corresponds to the optimal pinhole diameter that ensures confocal acquisition (the majority of out-of-focus light is filtered out, and the best signal-to-noise ratio is reached). However, setting every channel to a pinhole size of 1 AU can result in a varying optical slice thickness. For co-localization experiments, make sure to adjust the pinhole size so that each channel has the same optical slice thickness (e.g., 1 μm).

Acknowledgments

We thank G. Müller-Newen and the core facility "Immunohistochemistry and Confocal Microscopy Facility" of the Medical School of the RWTH Aachen University for support. Our work was supported by grants from the German Science Foundation (DFG LU 466/16-1), the IZKF Aachen (O2-1-2014) of the Medical School of the RWTH Aachen University, the Start-Up program of the Excellence Initiative of the RWTH Aachen University (StUpPD_119_13), and by the START program of the Faculty of Medicine, RWTH Aachen University (117/15 and 121/17).

References

1. Hottiger MO (2015) Nuclear ADP-ribosylation and its role in chromatin plasticity, cell differentiation, and epigenetics. Annu Rev Biochem 84:227–263. https://doi.org/10.1146/annurev-biochem-060614-034506
2. Hottiger MO, Hassa PO, Luscher B, Schuler H, Koch-Nolte F (2010) Toward a unified nomenclature for mammalian ADP-ribosyltransferases. Trends Biochem Sci 35(4):208–219. https://doi.org/10.1016/j.tibs.2009.12.003
3. Vyas S, Matic I, Uchima L, Rood J, Zaja R, Hay RT, Ahel I, Chang P (2014) Family-wide analysis of poly(ADP-ribose) polymerase activity. Nat Commun 5:4426. https://doi.org/10.1038/ncomms5426
4. Loseva O, Jemth AS, Bryant HE, Schuler H, Lehtio L, Karlberg T, Helleday T (2010) PARP-3 is a mono-ADP-ribosylase that activates PARP-1 in the absence of DNA. J Biol Chem 285(11):8054–8060. https://doi.org/10.1074/jbc.M109.077834
5. Kickhoefer VA, Siva AC, Kedersha NL, Inman EM, Ruland C, Streuli M, Rome LH (1999) The 193-kD vault protein, VPARP, is a novel poly(ADP-ribose) polymerase. J Cell Biol 146(5):917–928
6. Yang CS, Jividen K, Spencer A, Dworak N, Ni L, Oostdyk LT, Chatterjee M, Kusmider B, Reon B, Parlak M, Gorbunova V, Abbas T, Jeffery E, Sherman NE, Paschal BM (2017) Ubiquitin modification by the E3 Ligase/ADP-ribosyltransferase Dtx3L/Parp9. Mol Cell 66(4):503–516.e505. https://doi.org/10.1016/j.molcel.2017.04.028
7. Kleine H, Poreba E, Lesniewicz K, Hassa PO, Hottiger MO, Litchfield DW, Shilton BH, Luscher B (2008) Substrate-assisted catalysis by PARP10 limits its activity to mono-ADP-ribosylation. Mol Cell 32(1):57–69. https://doi.org/10.1016/j.molcel.2008.08.009
8. Yu M, Schreek S, Cerni C, Schamberger C, Lesniewicz K, Poreba E, Vervoorts J, Walsemann G, Grotzinger J, Kremmer E, Mehraein Y, Mertsching J, Kraft R, Austen M, Luscher-Firzlaff J, Luscher B (2005) PARP-10, a novel Myc-interacting protein with poly(ADP-ribose) polymerase activity, inhibits transformation. Oncogene 24(12):1982–1993. https://doi.org/10.1038/sj.onc.1208410
9. Feijs KL, Kleine H, Braczynski A, Forst AH, Herzog N, Verheugd P, Linzen U, Kremmer E, Luscher B (2013) ARTD10 substrate identification on protein microarrays: regulation of GSK3beta by mono-ADP-ribosylation. Cell Commun Signal 11(1):5. https://doi.org/10.1186/1478-811x-11-5
10. Verheugd P, Forst AH, Milke L, Herzog N, Feijs KL, Kremmer E, Kleine H, Luscher B (2013) Regulation of NF-kappaB signalling by the mono-ADP-ribosyltransferase ARTD10. Nat Commun 4:1683. https://doi.org/10.1038/ncomms2672
11. Eckei L, Krieg S, Butepage M, Lehmann A, Gross A, Lippok B, Grimm AR, Kummerer BM, Rossetti G, Luscher B, Verheugd P (2017) The conserved macrodomains of the non-structural proteins of Chikungunya virus and other pathogenic positive strand RNA viruses function as mono-ADP-ribosylhydrolases. Sci Rep 7:41746. https://doi.org/10.1038/srep41746
12. Atasheva S, Frolova EI, Frolov I (2014) Interferon-stimulated poly(ADP-ribose) polymerases are potent inhibitors of cellular transla-

tion and virus replication. J Virol 88(4):2116–2130. https://doi.org/10.1128/JVI.03443-13
13. Herzog N, Hartkamp JD, Verheugd P, Treude F, Forst AH, Feijs KL, Lippok BE, Kremmer E, Kleine H, Luscher B (2013) Caspase-dependent cleavage of the mono-ADP-ribosyltransferase ARTD10 interferes with its pro-apoptotic function. FEBS J 280(5):1330–1343. https://doi.org/10.1111/febs.12124
14. Kaufmann M, Feijs KL, Luscher B (2015) Function and regulation of the mono-ADP-ribosyltransferase ARTD10. Curr Top Microbiol Immunol 384:167–188. https://doi.org/10.1007/82_2014_379
15. Feijs KL, Forst AH, Verheugd P, Luscher B (2013) Macrodomain-containing proteins: regulating new intracellular functions of mono(ADP-ribosyl)ation. Nat Rev Mol Cell Biol 14(7):443–451. https://doi.org/10.1038/nrm3601
16. Verheugd P, Butepage M, Eckei L, Luscher B (2016) Players in ADP-ribosylation: readers and erasers. Curr Protein Pept Sci 17(7):654–667
17. Forst AH, Karlberg T, Herzog N, Thorsell AG, Gross A, Feijs KL, Verheugd P, Kursula P, Nijmeijer B, Kremmer E, Kleine H, Ladurner AG, Schuler H, Luscher B (2013) Recognition of mono-ADP-ribosylated ARTD10 substrates by ARTD8 macrodomains. Structure 21(3):462–475. https://doi.org/10.1016/j.str.2012.12.019
18. Kleine H, Herrmann A, Lamark T, Forst AH, Verheugd P, Luscher-Firzlaff J, Lippok B, Feijs KL, Herzog N, Kremmer E, Johansen T, Muller-Newen G, Luscher B (2012) Dynamic subcellular localization of the mono-ADP-ribosyltransferase ARTD10 and interaction with the ubiquitin receptor p62. Cell Commun Signal 10(1):28. https://doi.org/10.1186/1478-811x-10-28
19. Caliandro R, Rossetti G, Carloni P (2012) Local fluctuations and conformational transitions in proteins. J Chem Theory Comput 8(11):4775–4785. https://doi.org/10.1021/ct300610y

Chapter 5

Biochemical and Biophysical Assays of PAR-WWE Domain Interactions and Production of *iso*-ADPr for PAR-Binding Analysis

Zhizhi Wang and Wenqing Xu

Abstract

The poly(ADP-ribose) polymerase (PARP) family of proteins utilize NAD^+ as the substrate to modify protein acceptors with either mono(ADP-ribose) (MAR) or poly(ADP-ribose) (PAR). MAR and PAR have been shown to regulate distinct cellular processes. *Iso*-ADP-ribose (*iso*-ADPr) is the smallest internal PAR structural unit containing the characteristic ribose-ribose glycosidic bond formed during poly(ADP-ribosyl)ation. The WWE domain of RNF146 specifically recognizes the *iso*-ADPr moiety in PAR but does not interact with MAR. This provides a way to distinguish PAR from MAR modification and to isolate PARylated proteins. *Iso*-ADPr can be used to detect the PAR-specific binding properties of interested proteins. Here we describe the detailed method to generate and purify *iso*-ADPr and its use in PAR-binding analysis through isothermal titration calorimetry (ITC) analysis.

Key words PARP, PAR, PARylation, ADP-ribose, MAR, *iso*-ADPr, WWE domain

1 Introduction

Poly(ADP-ribosyl)ation (PARylation) is a key protein posttranslational modification catalyzed by PARP1–5 [1–4] and is involved in many cellular processes, including DNA damage repair, gene transcription, protein degradation, cell proliferation, and apoptosis [5–9]. In contrast, mono-ADP-ribosylation (MARylation) is catalyzed by distinct enzymes and has distinct roles in many biological processes including signal transduction and intracellular trafficking [10–13]. MARylation is also an approach used by various fungal and bacterial pathogens to take control of the host cell [13–15].

Iso-ADPr is the internal structural unit of PAR containing the characteristic ribose-ribose glycosidic bond. The RNF146 WWE domain and *iso*-ADPr was first shown forming a tight complex by our lab [16]. Because *iso*-ADPr is formed during PARylation, it is a unique tool to distinguish PAR-binding proteins from MAR-binding ones [16–19]. Purified *iso*-ADPr has also been used for

Paul Chang (ed.), *ADP-ribosylation and NAD+ Utilizing Enzymes: Methods and Protocols*, Methods in Molecular Biology, vol. 1813, https://doi.org/10.1007/978-1-4939-8588-3_5,

structural studies [16, 20, 21]. Importantly, we found that the RNF146 WWE domain binds to *iso*-ADPr with high affinity but has no detectable binding with MAR [16]. Therefore, this WWE domain has also been used as a tool to isolate PARylated proteins [22]. In this chapter, we describe the detailed method of semi-biosynthesis and purification of *iso*-ADPr and provide an example for analyzing protein-PAR interaction using *iso*-ADPr.

2 Materials

Prepare all solutions using Milli-Q water with a resistivity of 18.2 MΩ cm (at 25 °C) and reagents at analytical grade. Solutions should be filtered using a 0.22 μm pore size filter before use and stored under ambient conditions.

2.1 PARylation Reaction

1. Stock solutions: 1 M $MgCl_2$; 1 M Tris-HCl pH 8.0; 5 M NaCl; 1 M DTT.
2. Dissolve buffer: 50 mM Tris-HCl pH 8.0, 50 mM NaCl, 5 mM DTT.
3. β-nicotinamide adenine dinucleotide (β-NAD), 45 mM in dissolve buffer.
4. Histone Type IIA (from calf thymus, Sigma), 6 mg/mL in dissolve buffer.
5. Histone H1, 20 mg/mL (Millipore).
6. Activator DNA, 5'-GGAATTCC-3′, 10 mg/mL in ddH_2O.
7. Human PARP1 HSA enzyme, 10,000 units/mL (Trevigen).

2.2 Purification of PAR Polymer

1. 100% w/v trichloroacetic acid (TCA) in ddH_2O.
2. 100% EtOH; hydrochloric acid (HCl), 12.1 N; chloroform; 100% isopropanol.
3. Proteinase K.
4. 10% w/v sodium dodecyl sulfate (SDS) in ddH_2O.
5. Stock solutions: 5 M NaOH; 1 M NaAc/HAc buffer, pH 5.0.
6. The phenol-chloroform-isoamyl alcohol mixture, 25:24:1 v/v (phenol saturated with Tris-HCl pH 8.0).
7. 70% v/v EtOH in ddH_2O.

2.3 PAR Polymer Digestion by Phosphodiesterase

Venom phosphodiesterase I (PDE),100 units/mL in 50 mM Tris-HCl pH 8.0, 50 mM NaCl.

2.4 Iso-ADPr Purification

1. Stock solution: 1 M Tris-HCl pH 7.5; 1 M NH_4HCO_3.
2. Buffer A, 5 mM Tris-HCl pH 7.5; Buffer B, 5 mM Tris-HCl pH 7.5, 1 M NaCl; gel filtration buffer, 50 mM NH_4HCO_3.

3. 1 mL HiTrap Q HP column (GE Healthcare); Superdex 75 10/300 GL column (GE Healthcare); P-1 pump (GE Healthcare); AKTA FPLC (GE Healthcare); SpeedVac concentrator.

2.5 RNF146 WWE Domain Purification

The cDNA encoding human RNF146 WWE domain (residues 100–183, NGEYAWYYEGRNGWWQYDERTSRELEDAFSKGKKN TEMLIAGFLYVADLENMVQYRRNEHGRR RKIKRDIIDIPKKGVAGLRLD) is cloned into pGEX-4T1 with an N-terminal GST tag and a TEV cleavage site (ENLYFQG) in between.

2.6 Isothermal Titration Calorimetry (ITC)

1. Stock solutions: 1 M HEPES pH 7.5; 1 M TCEP.
2. MicroCal VP-ITC (Malvern).
3. ITC buffer: 20 mM HEPES pH 7.5, 150 mM NaCl, and 1 mM TCEP.

3 Methods

3.1 PARylation

The in vitro PARylation reaction is set up using PARP1 as enzyme. Prepare 15 mL PARylation reaction in a 50 mL Falcon centrifuge tube.

1. Add components in the following order (final concentration indicated):

   20 mM $MgCl_2$, 50 mM Tris-HCl pH8.0, 50 mM NaCl, 10 mM DTT, 1.5 mM β-NAD, 0.1 mg/mL histone H1, 0.1 mg/mL histone Type IIA, 66.7 μg/mL activator DNA, 300 units/mL PARP1 (*see* **Note 1**).
2. Mix reaction well by inverting the centrifuge tube.
3. Incubate reaction at room temperature for 2 h on a laboratory rotisserie (*see* **Note 2**).

3.2 PAR Polymer Purification

The reaction is stopped by ice-cold TCA. The white pellet is washed with ice-cold ethanol and further dissolved in buffer containing protease K and SDS. PAR polymer is detached from protein using NaOH and purified by phenol/chloroform extraction and isopropanol precipitation.

3.2.1 TCA Precipitation of PARylation Reaction

1. Prechill PARylation reaction tube on ice after incubation.
2. Prechill 100% w/v TCA on ice.
3. Prechill 100% EtOH on ice.
4. Add 5 mL 100% w/v ice-cold TCA to the 15 mL reaction solution.
5. Mix reaction well by inverting the centrifuge tube. Incubate on ice for 1 h.

6. Spin at 4 °C for 15 min/3000 *g*. Carefully decant the supernatant after centrifugation.
7. Wash the white pellets by 12 mL ice-cold 100% EtOH.
8. Mix reaction well by inverting the centrifuge tube.
9. Spin at 4 °C for 15 min/3000 *g*. Carefully decant the supernatant.
10. Air-dry the white pellets in the fume hood.

3.2.2 Separation of PAR Polymer from Protein

1. Add 10 mL buffer containing 50 mM Tris-HCl pH 8.0, 50 mM NaCl, 0.2 mg/mL proteinase K, and 0.1% w/v SDS to the white pellets. Incubate at 37 °C overnight.
2. Add 500 μL of 5 M NaOH to the 10 mL overnight solution. Mix thoroughly and incubate at 37 °C for 3 h.
3. Add 200 μL of HCl (12.1 N) to the solution. Mix well (*see* **Note 3**).
4. Add equal volume (10.7 mL) of phenol-chloroform-isoamyl alcohol mixture (25:24:1 v/v) to the solution. Mix thoroughly at room temperature (*see* **Note 4**).
5. Spin at 4 °C for 10 min/3000 *g*. Carefully pipette out the upper phase (aqueous) to a new 50 mL Falcon centrifuge tube.
6. Wash the aqueous phase solution with equal volume (10.7 mL) of chloroform. Mix thoroughly at room temperature (*see* **Note 4**).
7. Spin at 4 °C for 10 min/3000 *g*. Carefully pipette out the upper phase (aqueous phase, 10 mL) to a new 50 mL Falcon centrifuge tube.

3.2.3 Isopropanol Precipitation of PAR Polymer

1. Prechill 100% isopropanol on ice.
2. Prechill 70% v/v EtOH on ice.
3. Add 4.3 mL of 1 M NaAc/HAc pH 5.0 to the 10 mL aqueous phase solution to final concentration of 0.3 M NaAc/HAc, pH ~5.2.
4. Add 0.7 volume (10 mL) of ice-cold 100% isopropanol. Mix thoroughly and incubate at −20 °C for 2 h to completely precipitate PAR polymer (*see* **Note 5**).
5. Aliquot the isopropanol precipitation solutions to 1.5 mL Eppendorf centrifuge tubes.
6. Spin at 4 °C for 30 min/20,000 *g*. Carefully decant the supernatant (*see* **Note 6**).
7. Wash the pellets with 100 μL ice-cold 70% v/v EtOH each Eppendorf centrifuge tube.
8. Spin at 4 °C for 15 min/20,000 *g*. Carefully remove the supernatant by pipette.

9. Air-dry the pellets in the fume hood.
10. Dissolve all the pellets in 1 mL ddH_2O (*see* **Note 7**).

3.3 PAR Digestion by PDE

The purified PAR polymer is hydrolyzed into *iso*-ADPr by PDE.

1. Prepare 1.6 mL PDE digestion of PAR reaction in a 1.7 mL Eppendorf centrifuge tube.
2. Add components in the following order (final concentration indicated): 50 mM Tris-HCl pH 8.0, 50 mM NaCl, 1 mL PAR in ddH_2O from previous step, 50 units of PDE, and 15 mM $MgCl_2$ (*see* **Note 8**). Mix well and incubate at room temperature overnight on a laboratory rotisserie.

3.4 Iso-ADPr Purification

iso-ADPr is further purified by anion-exchange chromatography and gel filtration.

3.4.1 Anion-Exchange Chromatography for Iso-ADPr

1. Spin the PDE digestion reaction at 4 °C for 15 min/20,000 *g*. Carefully pipette out the supernatant (1.6 mL) to a new 50 mL Falcon centrifuge tube.
2. Dilute the reaction solution by Buffer A to final volume 50 mL.
3. Load the diluted sample to 1 mL HiTrap Q HP column by P1 pump.
4. Run anion-exchange chromatography at 1 mL/min flow rate with a liner gradient 0–25% Buffer B in 25 min. Collect the fractions for the main UV peak.
5. Aliquot *iso*-ADPr after anion-exchange chromatography to new Eppendorf centrifuge tubes (500 μL each).
6. Concentrate to ~500 μL total by SpeedVac concentrator (*see* **Note 9**).

3.4.2 Gel Filtration for Iso-ADPr

1. Run gel filtration for *iso*-ADPr against the gel filtration buffer on Superdex 75 10/300 GL column at flow rate of 0.6 mL/min. Collect the fractions (500 μL each) of the main UV peak (*see* **Note 10**).
2. Dry the fractions with *iso*-ADPr in 50 mM NH_4HCO_3 by SpeedVac as much as possible (*see* **Note 11**).
3. Redissolve *iso*-ADPr with 100 μL ddH_2O. Measure concentration; freeze and store at −80 °C (*see* **Note 12**).

3.5 RNF146 WWE Domain Purification

The GST-RNF146(WWE) fusion protein is overexpressed in *E. coli* BL21 (DE3) cells and grown in LB media. Bacteria cell pellets are lysed by sonication. GST fusion proteins are eluted from Glutathione Sepharose 4B beads. GST tag is removed by TEV at 4 °C overnight. Then the RNF146(WWE) protein is further purified by a cation exchange column and finally purified by a Superdex 200 column on FPLC. The peak fractions are pooled and concentrated to ~10 mg/mL in a buffer containing 20 mM Tris-HCl pH 7.5, 100 mM NaCl, and 2 mM DTT.

3.6 ITC

ITC is performed to analyze the interaction and binding stoichiometry between *iso*-ADPr and WWE domain proteins. Since the *iso*-ADPr moiety is formed during PARylation, a specific interaction between the target proteins with *iso*-ADPr indicates its interaction with PAR polymers.

1. Exchange buffer for purified WWE domain protein to ITC buffer by a Superdex 75 10/300 GL column to a final concentration of ~20 μM.
2. Dilute *iso*-ADPr by ITC buffer to ~500 μM.
3. Run ITC experiment using a MicroCal VP-ITC at 16 °C. A typical titration consisted of injecting 30–40 of 5 μL aliquots of the ligand (*iso*-ADPr) into the WWE domain protein sample (1.4218 mL sample well) at time intervals of 5 min, to ensure that the titration peak reached the baseline.
4. Analyze the ITC data using the software Origin 7.0 provided by the manufacturer. Fit data using a one-site model. As an example, the ITC result of the RNF146(WWE) and *iso*-ADPr is shown in Fig. 1.
5. ITC analysis using ADPr (MAR) should be performed as a control for PAR binding.

4 Notes

1. Add the right amount of ddH_2O first and then the other components in the order listed earlier. Always keep PAPR1 enzyme at −20 °C, and add it to the reaction last.
2. White precipitate appears slowly during the incubation of the PARylation reaction.
3. The final pH after adding HCl is around 8.5.
4. The phenol-chloroform-isoamyl alcohol mixture and chloroform should be added in the fume hood.
5. In order to cool down the isopropanol precipitation solution faster, 15 min of −80 °C incubation is recommended prior to −20 °C incubation.
6. There are white round pellets at the bottom of the Eppendorf centrifuge tubes.
7. The concentration of PAR polymer can be measured by NanoDrop and calculated based on its extinction coefficient of 13,500 cm^{-1} M^{-1} at 258 nm [23]. The UV peak should be

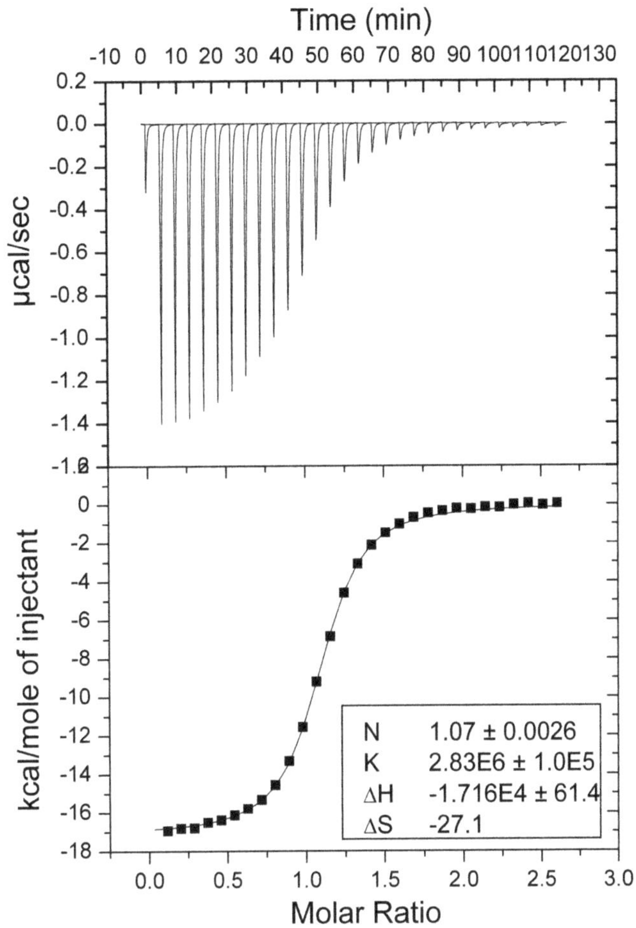

Fig. 1 ITC analysis of the interaction between RNF146(WWE) and iso-ADPr. ITC data showed the 1:1 interaction between the human RNF146 WWE domain and iso-ADPr with a binding affinity (Kd) of 353 nM

at ~260 nm. UV absorbance ratios of A_{260nm}/A_{280nm} and A_{260nm}/A_{230nm} both should be over 4.

8. Add the right amount of ddH_2O first and then the other components in the order listed earlier. Add $MgCl_2$ to the reaction last, or it will precipitate with PAR polymer.
9. The concentration of *iso*-ADPr can be measured by NanoDrop and calculated based on its extinction coefficient of 13,500 cm^{-1} M^{-1}. The UV peak should be at ~260 nm. UV absorbance ratios of A_{260nm}/A_{280nm} and A_{260nm}/A_{230nm} both should be close to 6 and 5, respectively.
10. Discard the late fractions that overlap with the salt peak.
11. NH_4HCO_3 is volatile.

12. The concentration of *iso*-ADPr can be measured by NanoDrop and calculated based on its extinction coefficient of 13,500 cm^{-1} M^{-1}. The UV peak should be at ~259 nm. UV absorbance ratios of A_{260nm}/A_{280nm} and A_{260nm}/A_{230nm} both should be over 6 and 5, respectively. The identity and purity can be further confirmed by LC-MS. The molar yield from β-NAD to *iso*-ADPr is about 20%.

Acknowledgment

This work was supported by NIH grant R01 GM099766 to W.X.

References

1. Barkauskaite E, Jankevicius G, Ahel I (2015) Structures and mechanisms of enzymes employed in the synthesis and degradation of PARP-dependent protein ADP-ribosylation. Mol Cell 58(6):935–946. https://doi.org/10.1016/j.molcel.2015.05.007
2. Hottiger MO, Hassa PO, Luscher B, Schuler H, Koch-Nolte F (2010) Toward a unified nomenclature for mammalian ADP-ribosyltransferases. Trends Biochem Sci 35(4):208–219. https://doi.org/10.1016/j.tibs.2009.12.003
3. Daniels CM, Ong SE, Leung AK (2015) The promise of proteomics for the study of ADP-ribosylation. Mol Cell 58(6):911–924. https://doi.org/10.1016/j.molcel.2015.06.012
4. Pascal JM, Ellenberger T (2015) The rise and fall of poly(ADP-ribose): an enzymatic perspective. DNA Repair 32:10–16. https://doi.org/10.1016/j.dnarep.2015.04.008
5. Bock FJ, Todorova TT, Chang P (2015) RNA regulation by poly(ADP-ribose) polymerases. Mol Cell 58(6):959–969. https://doi.org/10.1016/j.molcel.2015.01.037
6. Hottiger MO (2015) Nuclear ADP-ribosylation and its role in chromatin plasticity, cell differentiation, and epigenetics. Annu Rev Biochem 84:227–263. https://doi.org/10.1146/annurev-biochem-060614-034506
7. Gupte R, Liu Z, Kraus WL (2017) PARPs and ADP-ribosylation: recent advances linking molecular functions to biological outcomes. Genes Dev 31(2):101–126. https://doi.org/10.1101/gad.291518.116
8. Gibson BA, Kraus WL (2012) New insights into the molecular and cellular functions of poly(ADP-ribose) and PARPs. Nat Rev Mol Cell Biol 13(7):411–424. https://doi.org/10.1038/nrm3376
9. Hsiao SJ, Smith S (2008) Tankyrase function at telomeres, spindle poles, and beyond. Biochimie 90(1):83–92
10. Houtkooper RH, Pirinen E, Auwerx J (2012) Sirtuins as regulators of metabolism and healthspan. Nat Rev Mol Cell Biol 13(4):225–238. https://doi.org/10.1038/nrm3293
11. Fabrizio G, Scarpa ES, Di Girolamo M (2015) State of the art of protein mono-ADP-ribosylation: biological role and therapeutic potential. Front Biosci 20:405–430
12. Butepage M, Eckei L, Verheugd P, Luscher B (2015) Intracellular mono-ADP-ribosylation in signaling and disease. Cell 4(4):569–595. https://doi.org/10.3390/cells4040569
13. Aravind L, Zhang D, de Souza RF, Anand S, Iyer LM (2015) The natural history of ADP-ribosyltransferases and the ADP-ribosylation system. Curr Top Microbiol Immunol 384:3–32. https://doi.org/10.1007/82_2014_414
14. Grimaldi G, Corda D, Catara G (2015) From toxins to mammalian enzymes: the diversity of mono-ADP-ribosylation. Front Biosci 20:389–404
15. Deng Q, Barbieri JT (2008) Molecular mechanisms of the cytotoxicity of ADP-ribosylating toxins. Annu Rev Microbiol 62:271–288. https://doi.org/10.1146/annurev.micro.62.081307.162848
16. Wang Z, Michaud GA, Cheng Z, Zhang Y, Hinds TR, Fan E, Cong F, Xu W (2012) Recognition of

the iso-ADP-ribose moiety in poly(ADP-ribose) by WWE domains suggests a general mechanism for poly(ADP-ribosyl)ation-dependent ubiquitination. Genes Dev 26(3):235–240. https://doi.org/10.1101/gad.182618.111

17. Li M, Lu LY, Yang CY, Wang S, Yu X (2013) The FHA and BRCT domains recognize ADP-ribosylation during DNA damage response. Genes Dev 27(16):1752–1768. https://doi.org/10.1101/gad.226357.113
18. Zhang F, Chen Y, Li M, Yu X (2014) The oligonucleotide/oligosaccharide-binding fold motif is a poly(ADP-ribose)-binding domain that mediates DNA damage response. Proc Natl Acad Sci U S A 111(20):7278–7283. https://doi.org/10.1073/pnas.1318367111
19. Kim IK, Stegeman RA, Brosey CA, Ellenberger T (2015) A quantitative assay reveals ligand specificity of the DNA scaffold repair protein XRCC1 and efficient disassembly of complexes of XRCC1 and the poly(ADP-ribose) polymerase 1 by poly(ADP-ribose) glycohydrolase. J Biol Chem 290(6):3775–3783. https://doi.org/10.1074/jbc.M114.624718
20. Wang Z, Gagne JP, Poirier GG, Xu W (2014) Crystallographic and biochemical analysis of the mouse poly(ADP-ribose) glycohydrolase. PLoS One 9(1):e86010. https://doi.org/10.1371/journal.pone.0086010
21. DaRosa PA, Wang Z, Jiang X, Pruneda JN, Cong F, Klevit RE, Xu W (2015) Allosteric activation of the RNF146 ubiquitin ligase by a poly(ADP-ribosyl)ation signal. Nature 517(7533):223–226. https://doi.org/10.1038/nature13826
22. Bartolomei G, Leutert M, Manzo M, Baubec T, Hottiger MO (2016) Analysis of chromatin ADP-ribosylation at the genome-wide level and at specific loci by ADPr-ChAP. Mol Cell 61(3):474–485. https://doi.org/10.1016/j.molcel.2015.12.025
23. Shah GM, Poirier D, Duchaine C, Brochu G, Desnoyers S, Lagueux J, Verreault A, Hoflack JC, Kirkland JB, Poirier GG (1995) Methods for biochemical study of poly(ADP-ribose) metabolism in vitro and in vivo. Anal Biochem 227(1):1–13. https://doi.org/10.1006/abio.1995.1245

Part III

Enzymatic Assays of NAD+ Utilizing Enzymes

Chapter 6

Assays for NAD⁺-Dependent Reactions and NAD⁺ Metabolites

Michael B. Schultz, Yuancheng Lu, Nady Braidy, and David A. Sinclair

Abstract

Nicotinamide adenine dinucleotide (NAD⁺) is an essential redox cofactor and signaling molecule that controls the activity of enzymes involved in metabolism, DNA repair, and cellular survival, such as the PARPs, CD38, and the sirtuins. Here, we describe three methods for measuring the activity of these enzymes: the etheno-NAD⁺ assay measures NAD⁺ hydrolase activity using an NAD⁺ analog to produce a fluorescent product that is measured in real time; the PNC1 assay converts a native product of NAD⁺ hydrolysis, nicotinamide, into a quantitative fluorescent readout; and liquid chromatography tandem mass spectrometry (LC-MS/MS) is used to characterize the entire NAD⁺ metabolome in a sample. These methods will enable new insights into the roles that NAD⁺ and the enzymes that utilize it play in health and disease.

Key words Aging, NAD⁺ nicotinamide, ADP-ribose, PARP, ARTD, CD38, BST1, Sirtuin, Metabolomics, Mass spectrometry, Metabolism, Epigenetics, HDAC

1 Introduction

The oxidized form of nicotinamide adenine dinucleotide (NAD⁺) is an essential coenzyme for redox reactions and co-substrate for multiple families of proteins that control how a cell responds to its environment. Processes regulated by NAD⁺ include energy metabolism, inflammation, DNA repair, gene expression, and aging. Classically, NAD⁺ acts as a redox intermediate, shuttling the energy generated from the breakdown of glucose (during glycolysis and the TCA cycle) to complex I of the electron transport chain. More recently, NAD⁺ has also come to be appreciated as a signaling molecule. Three major classes of enzymes utilize NAD⁺ for non-redox reactions: PARPs/ARTDs, CD38/BST1, and sirtuins. All three hydrolyze NAD⁺ to produce nicotinamide and some form of ADP-ribose. In this chapter, we will focus on the latter two families (although all assays are amenable to any NAD⁺ hydrolase) in describing three assays to measure the hydrolysis of NAD⁺.

Paul Chang (ed.), *ADP-ribosylation and NAD+ Utilizing Enzymes: Methods and Protocols*, Methods in Molecular Biology, vol. 1813, https://doi.org/10.1007/978-1-4939-8588-3_6, © Springer Science+Business Media, LLC, part of Springer Nature 2018

CD38 and BST1 are membrane proteins that hydrolyze NAD^+ to produce (cyclic-)ADP-ribose. CD38 is a transcriptional target of NF-κB that is upregulated in response to inflammation [1]. It also regulates insulin sensitivity via consumption of NAD^+ [2]. BST1, a homolog of CD38, regulates stem cell proliferation in response to energy availability via cADPR signaling [3] and has been associated with neurological disorders [4].

Sirtuins are NAD^+-dependent lysine de-acylases and ADP-ribosyltransferases. Mammals have seven sirtuins (SIRT1–7), each one with a different spectrum of enzymatic reactions, intracellular localizations, and protein substrates (reviewed in [5]). Notable targets of sirtuin deacylation include histones; transcription factors such as p53 [6], PGC-1α [7], and NF-κB [8]; enzymes such as glutamate dehydrogenase [9] and LKB1 [10]; and signal transduction effectors such as Notch [11] and IRS1 [12]. Sirtuins respond to the availability of NAD^+, which increases during calorie restriction or exercise, thereby enhancing DNA repair, increasing mitochondrial function, and reducing inflammation, among other beneficial effects.

All three families of NAD^+ hydrolases, including PARPs and ARTDs, have two important similarities. First, they consume NAD^+, which needs to be resynthesized via biosynthetic pathways starting with NAD^+ precursors such as nicotinamide riboside or tryptophan. Secondly, the dissociation constants of many of these enzymes for NAD^+ are relatively high (reviewed in [13]). Taken together, these two properties mean that these enzymes are often in competition for the same pool of NAD^+ and that increased activity of one can suppress the activity of others [14, 15].

Three assays that measure the hydrolysis of NAD^+ are described in this chapter. The etheno-NAD^+ assay uses a modified NAD^+ molecule which, when hydrolyzed, produces a fluorescent etheno-ADP-ribose product. By tracking the increase in fluorescence in real-time, the NAD^+ hydrolase activity of a sample can be inferred [16]. This assay is amenable for use with recombinant enzymes and cell or tissue samples.

The PNC1 assay measures the other portion of the hydrolyzed NAD^+ molecule: nicotinamide. In this assay (Fig. 1), an enzyme such as SIRT1 generates nicotinamide from NAD^+ in **step 1**. An enzyme from budding yeast, yPnc1, is then utilized to convert the nicotinamide into nicotinic acid and NH_4^+ (free ammonia) in **step 2**. The ammonia is then reacted with ortho-phthalaldehyde (OPT) and dithiothreitol (DTT) to produce the fluorescent 1-alkylthio-substituted isoindoles [17] in **step 3**. Therefore, the fluorescence detected is proportional to the amount of nicotinamide produced during the NAD^+ hydrolysis reaction.

The final assay described in this chapter, liquid chromatography tandem mass spectrometry (LC-MS/MS), can be used for the quantification of the entire NAD metabolome. This technique has

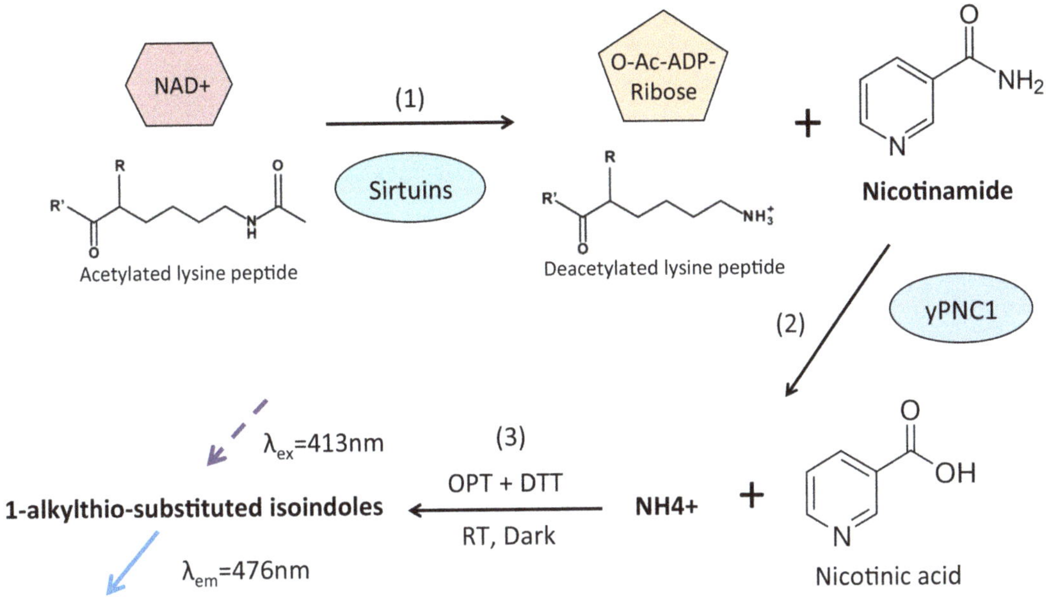

Fig. 1 Overview of the PNC1 assay, with a sirtuin as the NAD^+ hydrolase

become a standard in the metabolomics field and can detect NAD^+ and related metabolites over a wide dynamic range down to low picomolar levels. It displays superior specificity over reverse phase high-performance liquid chromatography (HPLC). Two separation methods are employed to characterize the NAD^+ metabolome: alkaline separation for all metabolites containing a ribose sugar and acid separation for metabolites without a sugar moiety.

It is anticipated that the methods discussed herein will allow for standardization of assays related to NAD^+ across different laboratories and to overcome challenges associated with translation of preclinical studies toward clinical practice.

2 Materials

2.1 Etheno-NAD+ Assay

1. Assay buffer: 250 mM sucrose and 40 mM Tris–HCl, pH 7.4, filtered.
2. Nicotinamide 1,N^6-ethenoadenine dinucleotide (Sigma-Aldrich, N2630), dissolved in assay buffer at 60 mM. Store at −20 °C in small aliquots, to avoid repeated freeze-thaw cycles.
3. NADase enzyme inhibitor:
   - For example, apigenin (Sigma-Aldrich, N3145) for inhibition of CD38. Dissolve in DMSO for a stock concentration of 6 mM.
   - 3-Aminobenzamide (Sigma-Aldrich, A0788) or PJ-34 (Abcam, ab120981) are popular PARP inhibitors.

4. 96-well black half-area flat bottom polystyrene plates (Corning, 3694).
5. Plate reader capable of reading fluorescence (Ex/Em = 310/410).
6. If using recombinant enzymes in vitro:
   - Recombinant enzyme, such as CD38 (R&D Systems, 2404-AC-010) or BST1 (R&D Systems, 4736-AC-010).
7. If using with cell or tissue lysate:
   - NETN lysis buffer (100 mM NaCl, 20 mM Tris–HCl pH 8.0, 0.5 mM EDTA, 0.5% (v/v). NP-40) with protease inhibitors (Sigma-Aldrich, 11873580001).
   - BCA protein quantification assay (Thermo Fisher, 23227).

2.2 PNC1 Assay

1. Purified recombinant SIRT1 (or another nicotinamide-producing enzyme) and yPnc1 enzymes produced by previously described protocols [18].
2. Sirtuins require acetylated peptide substrates with acetylated lysine near the middle of the sequence (e.g., Ac-TARK(ac) STG-NH2) (*see* **Note 4**-for the convenience of the experimenter, notes are grouped by assay are therefore not ordered by when they appear).
3. Reaction buffer: PBS (pH 7.4) supplemented with 1 mM DTT (Thermo Fisher, R0861). Tris buffer (pH 8.0) can be used if PBS is incompatible with the enzyme.
4. OPT Developer Reagent: Dissolve OPT (Sigma-Aldrich, P0657) in pure ethanol (33.3 mM) and dissolve DTT in PBS (14.28 mM). Mix three volumes of ethanol and seven volumes of PBS to make a 30% ethanol/70% PBS (pH 7.4) solution supplemented with 10 mM OPT and 10 mM DTT. Protect from light and store at −20 °C until use.
5. NAD^+ (Sigma-Aldrich, N7004) aliquots prepared at 100 mM in water and stored at −20 °C.
6. Nicotinamide (Sigma-Aldrich, 72,340) stock solution prepared at 100 mM in water, aliquoted and stored at −20 °C.
7. 96-well black polystyrene plates (Corning, 3915) suitable for fluorometry.
8. Plate reader capable of reading fluorescence (Ex/Em = 413/476).
9. 37 °C incubator.
10. Orbital shaker.

2.3 LM/MS/MS Metabolomics

1. Prepare all solutions and buffers using Milli-Q water.
2. Methanol: HPLC-grade 100% methanol (Sigma-Aldrich).
3. Standard curve of NAD^+ and its metabolites:

- Obtain NAD^+, NMN, NAMN, ADP, ATP, NADH, NAAD, $NADP^+$, NADPH, NA, NAM, 1-methyl NAM, and NR (Sigma-Aldrich).
- Prepare a 50 mM solution for each metabolite in water (*see* **Note 10**).
- Mix all stock solutions and add water to make a solution containing 200 μM of each metabolite.
- Perform serial dilutions in water to make standard solutions at the following concentrations: 0, 0.1, 0.2, 0.6, 2, 6, 20, 60, and 200 μM.

4. Fleischmann's active dry yeast extract (Icon Isotopes, Summit, New Jersey) contains ^{13}C isotopes of all metabolites with a sugar moiety.
5. Isotopically enriched internal standards:
   - $^{2}H_4$-NAM, $^{13}C_5$-adenosine, adenosine-3′,5′-cyclic-$^{13}C_5$-monophosphate, and adenosine 5'-$^{13}C_5$-monophosphate (Toronto Research Chemicals).
   - $^{15}N_5$-ADP and adenosine $^{13}C_{10}{}^{15}N_5$ 5′-triphosphate (Sigma-Aldrich).
6. Speed vacuum.
7. Alkaline separation chromatography reagents:
   - Solvent A1: 7.5 mM ammonium acetate with 0.05% (v/v) ammonium hydroxide.
   - Solvent B1: 0.05% (v/v) ammonium hydroxide in acetonitrile.
   - Porous graphitic carbon reversed phase material, Hypercarb LC-MS/MS column (1 mm × 100 mm) (Thermo Fisher) (*see* **Note 11**).
8. Acid separation chromatography reagents (*see* **Note 12**):
   - Solvent A2: 10 mM ammonium acetate with 0.1% formic acid.
   - Solvent B2: 0.1% formic acid in acetonitrile.
   - Hypercarb LC-MS/MS column (2.1 mm × 100 mm) (Thermo Fischer).
9. LC-MS/MS instrument: Thermo Scientific UPLC-MSD assembly consisting of an Accela AS injector, Accela UPLC pump, and a TSQ Vantage bench-top mass spectrometer (Thermo Scientific) equipped with a heated electrospray probe (HESI).
10. Pierce™ Coomassie (Bradford) Protein Assay Kit (Thermo Fischer, 23200) or the BCA protein quantification assay (Thermo Fisher, 23227).

3 Methods

3.1 Etheno-NAD$^+$ Assay

3.1.1 Reagent Preparation

1. Plan the experimental layout. See Fig. 2 for an example layout to generate a dose-response curve with an inhibitor. Include an etheno-NAD$^+$ only control. If using an enzyme inhibitor, also include an etheno-NAD$^+$ with inhibitor (no enzyme) control, and a no-inhibitor control.
2. If using recombinant enzymes in vitro:
   - Prepare recombinant enzyme on ice. Dissolve it in assay buffer to 3× the desired concentration for the final reaction. For example, CD38 should be prepared at a concentration of 0.3 ng/μL, for a final reaction concentration of 0.1 ng/μL (*see* **Note 1** for other enzymes).
   - If using cell or tissue lysates:
   - Collect and homogenize the samples in NETN lysis buffer (or an alternative buffer that will lyse the cells but preserve enzyme activity). The specific method for collection/homogenization will depend on the type of sample.
   - Spin the samples in a refrigerated centrifuge at maximum speed for 10 min to remove insoluble material.
   - Quantify the protein concentration in the supernatants using the BCA protein assay. Normalize the concentration across your samples by adding lysis buffer.
3. Dilute the etheno-NAD$^+$ 100-fold in cold reaction buffer to a concentration of 600 μM. (The final reaction concentration will be 200 μM.) Each single reaction needs 10 μL of diluted etheno-NAD$^+$.
4. If testing an NADase inhibitor, prepare serial dilutions. Dilutions should be prepared in assay buffer to 3× the desired concentration for the final reaction. For example, if the highest concentration of apigenin to be tested is 100 μM, it should be prepared at a concentration of 300 μM (*see* **Note 2** if using DMSO).
5. Allow all reagents to briefly warm up to room temperature.

3.1.2 Perform the Assay

1. Add 10 μL recombinant enzyme/sample lysate to the bottom of each well of the 96-well plate to be used. For the etheno-NAD$^+$ only control and the etheno-NAD$^+$ + inhibitor control wells, instead add 10 μL reaction buffer.
2. Add 10 μL of inhibitor solutions to their respective wells. Add 10 μL of reaction buffer to wells without inhibitor. Wait 10 min to allow the inhibitors to interact with the recombinant enzyme/sample lysate.

	0% activity		Highest[inhibitor]--->Lowest [Inhibitor](μM)						100% activity			
	1	2	3	4	5	6	7	8	9	10	11	12
A	EtNAD only*	EtNAD+Inb**	100	10	1	0.5	0.1	0.01	0***			
B	EtNAD only	EtNAD+Inb	100	10	1	0.5	0.1	0.01	0			
C	EtNAD only	EtNAD+Inb	100	10	1	0.5	0.1	0.01	0			
D	EtNAD only	EtNAD+Inb	100	10	1	0.5	0.1	0.01	0			
E	EtNAD only	EtNAD+Inb	100	10	1	0.5	0.1	0.01	0			
F												
G												
H												

*Etheno-NAD only control (10 μL etheno-NAD + 20 μL reaction buffer)
**Etheno NAD + inhibitor (at maximum concentration) control (10 μL etheno-NAD + 10 μL inhibitor + 10 μL reaction buffer)
***No inhibitor control (10 μL etheno-NAD + 10 μL enzyme/sample + 10 μL reaction buffer)

Fig. 2 Example experimental layout to generate a dose-response curve for an NADase inhibitor (Inb)

3. As quickly as possible, add 10 μL of etheno-NAD+ to the bottom of each well (preferably with a multichannel pipette). Pipette up and down once when adding to ensure mixing.
4. Immediately place the 96-well plate into the plate reader. Read fluorescence (Ex/Em = 310/410) frequently, at least once per minute. Continue the reading for up to 1 h.

3.1.3 Analysis

1. To determine the NADase activity, calculate the slope of the curve for each well (AFU/min). Include only data points that are representative of the initial reaction rate, when the curve is linear (*see* **Note 3**).
2. To remove background, subtract the average slope from the etheno-NAD+ only control from all wells. If necessary, subtract the average slope of the etheno-NAD+ inhibitor control from all wells with inhibitor present (scaled the inhibitor concentration).
3. Use GraphPad Prism (recommended), Microsoft Excel, or similar software to plot the relative NADase activity. If performing a dose response with an inhibitor, GraphPad Prism can be used to calculate the inhibitor's IC_{50}.

3.2 PNC1 Assay

3.2.1 Setup and Reaction

1. Thaw the nicotinamide stock solution on ice and dilute to make 0, 5, 10, 20, 30, 40, and 50 μM standards. Pipette 1 μL of each standard into its respective well of a 96-well plate. (The final nicotinamide concentrations are 100-fold diluted.)
2. Apply 100 μL of the reaction buffer with 1 μg yPNC1 to each standard well and mix by pipetting.

3. Prepare a master-mix on ice. The volume should be enough for assaying two times the number of samples in triplicates, as the extra set will be used to account for background fluorescence. Each single reaction needs the following: acetylated peptide substrate (for sirtuin reactions) (10–30 μM), yPnc1 enzyme (1–2 μg), NAD-consuming enzyme (1–2 μg for SIRT1) (*see* **Note 5**) and reaction buffer (100 μL minus the volume of other components). Mix by gentle vortexing.
4. Divide the master-mix into two tubes. Add NAD^+ (75–200 μM) to one tube, and add an equal volume of distilled water to another tube (*see* **Note 6**). Mix again by gentle vortexing.
5. Add 100 μL of the master-mix to all wells. (All samples should be performed in triplicate.)
6. Test compounds (*see* **Note 7**), such as enzyme activators or inhibitors, can be added into both with and without NAD^+ wells at this stage. Add equal volume of solvent (e.g., DMSO) into control wells (*see* **Note 8**). Mix by pipetting.
7. Gently tap the plate to mix the solutions, and cover with a plate sealer. Incubate at 37 °C on a plate shaker with gentle agitation for 1 h (*see* **Note 9**).

3.2.2 Development

1. Warm OPT Developer Reagent at 42 °C for 15 min (avoid light). Vortex if precipitation of DTT is observed.
2. After the incubation period, quickly add 100 μL of OPT Developer Reagent into each well under low light with multichannel pipette. Gently tap to mix the solutions, cover with a plate sealer and wrap with aluminum foil. Incubate at room temperature for 1 h on a plate shaker with gentle agitation.
3. Read the fluorescence on a plate reader with filters set to excitation ~420 (±10) nm and emission ~460 (±20) nm. Both 0.1 and 1 s read times work well.

3.2.3 Analysis

1. Use the nicotinamide standards to plot a standard curve of fluorescence intensity versus nicotinamide concentration.
2. Calculate the net fluorescence for each reaction condition by subtracting the mean fluorescence of the background control reactions (no NAD^+) from the experimental reaction (with NAD^+), $F_{\text{corrected}} = F_{+\text{NAD}} - \text{F}_{-\text{NADcontrol(mean value)}}$.
3. Net fluorescence can be converted into amounts of nicotinamide using the linear equation generated from the standard curve. More nicotinamide produced during the reaction represents higher enzyme activity.

3.3 NAD Metabolomics

3.3.1 Extraction from Tissue

1. Collect tissue samples, rinse in ice-cold PBS (*see* **Note 13**), and remove excess PBS using a Kim wipe. Snap-freeze tissue samples in liquid nitrogen and store at −80 °C until extraction.
2. Weigh tissue samples when commencing metabolite extraction.
3. Add 500 μL of cold methanol per 10 mg of tissue to a 10 mL Falcon tube on ice.
4. Transfer tissue sample into the tube and keep on ice. Homogenize the frozen tissue sample in methanol using a Polytron homogenizer or TissueLyser (Qiagen) (*see* **Note 14**).
5. Centrifuge the homogenate at 15,000 rpm at 4 °C for 10 min.
6. Carefully remove the supernatant and transfer to a new sample tube.

3.3.2 Extraction from Cultured Cells

1. Culture cells in correctly supplemented media (*see* **Note 15**).
2. Aspirate culture media and wash once with ice-cold PBS.
3. Add 500 μL of prechilled methanol per well for a 6-well plate. (Scale based on surface area for larger plates.)
4. Harvest the cells using a cell scraper and transfer the cell lysate into a 1.5 mL Eppendorf tube on ice (*see* **Note 13**).
5. Homogenize the cells using a syringe with a 23–26 gauge needle.
6. Centrifuge the homogenate at 15,000 rpm at 4 °C for 10 min.
7. Carefully remove the supernatant and transfer to a new sample tube.

3.3.3 Sample Setup

1. Make a 1:40 dilution of Fleischmann's extract in 10 mM ammonium acetate.
2. Mix each point on the standard curve and all experimental samples 1:1 with diluted Fleischmann's extract.
3. Add 2 μL of internal standards to each point on the standard curve and all experimental samples.
4. Dry samples and standards using a speed vacuum.
5. Reconstitute samples in 200 μL of 100 mM ammonium acetate.
6. Transfer reconstituted samples and standards into 200 μL LC/MS-MS glass vials and capped before analysis.

3.3.4 Liquid Chromatography

1. Alkaline separation for detection of eight metabolites (NMN, NAMN, ADP, ATP, NAD^+, NADH, NAAD, NADP+) (*see* **Note 16**):
   - Column: 1 mm × 100 mm Hypercarb.
   - Flow rate: 0.08 mL/min.

- Column temperature: 60 °C.
- Equilibrate the column for 20 min with solvent A1 before the first injection.
- After sample injection, apply solvent B1 on a gradient.

2. Acid separation for detection of three metabolites (NA, NAM, and NR) (*see* **Note 16**):
   - Column: 2.1 mm × 100 mm Hypercarb.
   - Flow rate: 0.2 mL/min.
   - Column temperature: 60 °C.
   - Equilibrate the column with solvent A2 before the first injection.
   - After sample injection, apply solvent B2 on a gradient.

3.3.5 Analysis

1. To confirm the identity of the NAD⁺ metabolite peaks, the results from Q1 ion scanning can be compared to MS/MS spectra from online databases Metlin (http://metlin.scripps.edu/) and Massbank (http://www.massbank.jp/?lang=en). The transitions are shown in Table 1. Peak areas can be integrated using Thermo Scientific Xcalibur™ software (version 2.2, 2011).
2. To quantify metabolites in pmol units, compute the ratio of the integrated peak area for each isotopically labeled metabolite to the peak area of its unlabeled version, for all samples in the standard curve. Interpolate the concentration of each isotopically labeled metabolite (*see* **Note 17**). By then computing the ratios of the peak areas of the unlabeled vs. labeled metabolites in the experimental samples, pmol concentrations of unlabeled metabolites can be inferred.
3. Normalize the metabolite concentrations to the amount of protein present in the tissue or cell culture sample using the Bradford or BCA protein assays or to the number of cells.

4 Notes

Etheno-NAD⁺ Assay

1. The NADase activity of various enzymes can differ drastically. For example, CD38 has >100 fold the activity of BST1. The enzyme concentration and the length of observation time for the reaction may need to be adjusted depending on this activity.
2. If your inhibitor is first dissolved in DMSO, keep the DMSO concentration the same in all samples. For example, if apigenin is first dissolved at 6 mM in 100% DMSO and then diluted in

Table 1

Transitions for NAD+ metabolites using LC-MS/MS

Analyte	Transitions (*m/z*)
NAM	123
$^{2}H_{4}$-NAM	127
Adenosine (ADO)	266
$^{13}C_{5}$-ADO	271
NA	124
NMN	335
NAD	664
NaAD	665
NADH	666
NaMN	336
NADP	744
ADP	426
$^{15}N_{5}$-ADP	431
NADPH	746
ATP	506
$^{13}C_{10}{}^{15}N_{5}$-ATP	521

reaction buffer to 300 μM at 5% DMSO, all serial dilutions should be made in reaction buffer with 5% DMSO. Reaction buffer with 5% DMSO would also be added to the controls without inhibitor.

3. Many plate readers can calculate the slope of the curve for each well. This is recommended, as exporting the raw data and calculating the slope elsewhere can be unwieldy. If this must be done, the use of Pivot tables in Excel or data analysis software such as MATLAB or R is recommended.

PNC1 Assay

4. The peptide length is usually 5–15 amino acids. To reduce background fluorescence, shorter peptides or peptides with fewer aromatic groups are preferred. Also, peptides with hydrophobic groups adjacent to the acetyl–lysine are recommended for STAC-mediated SIRT1 activation [19].
5. For enzymes that have lower activity than SIRT1, the protein amount added should be increased (e.g., SIRT6).

6. NAD^+ could exhibit fluorescence at high concentrations (>200 μM). If high NAD^+ concentrations are required, the background reaction formulation can be altered. Instead of not adding NAD^+, background reactions can include NAD^+ but exclude the enzyme or use the corresponding non-acetylated peptide.
7. Test compounds should not alter yPnc1 activity or diminish the fluorescence signal. This can be discerned by incubating with a nicotinamide standard curve in this assay.
8. DMSO may inhibit the enzyme activity and lower the signal. Therefore, adding equal amount of DMSO in the same concentration to control wells is important. Typically <4% DMSO in the final reaction is desirable.
9. Longer incubation times may be needed if: (1) the enzyme has lower activity; (2) the activity on the assayed substrate is weak; or (3) the nicotinamide standard curve doses are very high.

NAD^+ Metabolomics

10. β-Nicotinamide adenine dinucleotide hydrate (Sigma-Aldrich, N1511) should be used to make NAD^+ standard solution. It may be difficult to dissolve NAD^+ powder in water, and it may be vortexed rigorously until dissolved. Aliquots of NAD^+ and its metabolites should be stored in −80 °C freezer. Avoid multiple freeze/thaw cycles with stock solutions. Frozen stock solutions of NAD^+ metabolites are stable at −80 °C for at least 1 year.
11. Column selection is critical when running LC-MS/MS methodology. The Hypercarb column is suited for measuring NAD^+ and its metabolites in terms of accuracy, reliability, and reproducibility [20].
12. Formic acid is a very strong and corrosive. It should be handled with care. It is recommended that the solution is prepared in the fume hood and using personnel protective equipment.
13. Tissue/cellular collection should be done as rapidly as possible to prevent degradation of the NAD^+ metabolites.
14. Use a homogenizer in the fume hood and wear personal protective equipment, including lab coat and safety goggles.
15. Grow cells in 6-well plates or larger for quantification of NAD^+ metabolites at detectable amounts.
16. The carbohydrate group is absent in NAM, NA, and NR and cannot be labeled by a heavy-labeled glucose. Therefore, NAM, NA, and NR are best quantified using an acidic separation with NAM and NA and $^2H_4$-NAM as internal standard. This method allows the quantification of eight metabolites

using alkaline separation and three metabolites in an acidic separation.

17. Samples are mixed with Fleischmann's extract and internal standards to allow for quantification of each metabolite in pmol units. For instance, if a particular metabolite concentration was between the 20 and 60 μM standards (10 μM and 30 μM after the 1:1 dilution) in the Fleischmann's sample, then in a 2.5 μL injection volume there would be 100 pmol of that metabolite (40 is between 2 and 6; 40 * 2.5 = 100).

Acknowledgments

N.B. is the recipient of the Australian Research Council Discovery Early Career Research Award at the University of New South Wales, Sydney, Australia.

References

1. Kang B-NN, Tirumurugaan KG, Deshpande DA, Amrani Y, Panettieri RA, Walseth TF, Kannan MS (2006) Transcriptional regulation of CD38 expression by tumor necrosis factor-alpha in human airway smooth muscle cells: role of NF-kappaB and sensitivity to glucocorticoids. FASEB J 20:1000–1002
2. Escande C, Nin V, Price NL, Capellini V, Gomes AP, Barbosa MT, O'Neil L, White TA, Sinclair DA, Chini EN (2013) Flavonoid apigenin is an inhibitor of the NAD+ ase CD38: implications for cellular NAD+ metabolism, protein acetylation, and treatment of metabolic syndrome. Diabetes 62:1084–1093
3. Yilmaz ÖHH, Katajisto P, Lamming DW, Gültekin Y, Bauer-Rowe KE, Sengupta S, Birsoy K, Dursun A, Yilmaz VO, Selig M, Nielsen GP, Mino-Kenudson M, Zukerberg LR, Bhan AK, Deshpande V, Sabatini DM (2012) mTORC1 in the Paneth cell niche couples intestinal stem-cell function to calorie intake. Nature 486:490–495
4. Lopatina O, Yoshihara T, Nishimura T, Zhong J, Akther S, Fakhrul AA, Liang M, Higashida C, Sumi K, Furuhara K, Inahata Y, Huang J-JJ, Koizumi K, Yokoyama S, Tsuji T, Petugina Y, Sumarokov A, Salmina AB, Hashida K, Kitao Y, Hori O, Asano M, Kitamura Y, Kozaka T, Shiba K, Zhong F, Xie M-JJ, Sato M, Ishihara K, Higashida H (2014) Anxiety- and depression-like behavior in mice lacking the CD157/BST1 gene, a risk factor for Parkinson's disease. Front Behav Neurosci 8:133
5. Bonkowski MS, Sinclair DA (2016) Slowing ageing by design: the rise of NAD(+) and sirtuin-activating compounds. Nat Rev Mol Cell Biol 17:679–690
6. Vaziri H, Dessain SK, Ng Eaton E, Imai SI, Frye RA, Pandita TK, Guarente L, Weinberg RA (2001) hSIR2(SIRT1) functions as an NAD-dependent p53 deacetylase. Cell 107:149–159
7. Rodgers JT, Lerin C, Haas W, Gygi SP, Spiegelman BM, Puigserver P (2005) Nutrient control of glucose homeostasis through a complex of PGC-1alpha and SIRT1. Nature 434:113–118
8. Yeung F, Hoberg JE, Ramsey CS, Keller MD, Jones DR, Frye RA, Mayo MW (2004) Modulation of NF-kappaB-dependent transcription and cell survival by the SIRT1 deacetylase. EMBO J 23:2369–2380
9. Lombard DB, Alt FW, Cheng H-LL, Bunkenborg J, Streeper RS, Mostoslavsky R, Kim J, Yancopoulos G, Valenzuela D, Murphy A, Yang Y, Chen Y, Hirschey MD, Bronson RT, Haigis M, Guarente LP, Farese RV, Weissman S, Verdin E, Schwer B (2007) Mammalian Sir2 homolog SIRT3 regulates global mitochondrial lysine acetylation. Mol Cell Biol 27:8807–8814
10. Price NL, Gomes AP, Ling AJ, Duarte FV, Martin-Montalvo A, North BJ, Agarwal B, Ye L, Ramadori G, Teodoro JS, Hubbard BP, Varela AT, Davis JG, Varamini B, Hafner A, Moaddel R, Rolo AP, Coppari R, Palmeira CM, de Cabo R, Baur JA, Sinclair DA (2012) SIRT1 is required for AMPK activation and the

beneficial effects of resveratrol on mitochondrial function. Cell Metab 15:675–690

11. Guarani V, Deflorian G, Franco CA, Krüger M, Phng L-KK, Bentley K, Toussaint L, Dequiedt F, Mostoslavsky R, Schmidt MHH, Zimmermann B, Brandes RP, Mione M, Westphal CH, Braun T, Zeiher AM, Gerhardt H, Dimmeler S, Potente M (2011) Acetylation-dependent regulation of endothelial Notch signalling by the SIRT1 deacetylase. Nature 473:234–238
12. Zhang J (2007) The direct involvement of SirT1 in insulin-induced insulin receptor substrate-2 tyrosine phosphorylation. J Biol Chem 282:34356–34364
13. Cantó C, Menzies KJ, Auwerx J (2015) NAD(+) metabolism and the control of energy homeostasis: a balancing act between mitochondria and the nucleus. Cell Metab 22:31–53
14. Bai P, Cantó C, Oudart H, Brunyánszki A, Cen Y, Thomas C, Yamamoto H, Huber A, Kiss B, Houtkooper RH, Schoonjans K, Schreiber V, Sauve AA, Menissier-de Murcia J, Auwerx J (2011) PARP-1 inhibition increases mitochondrial metabolism through SIRT1 activation. Cell Metab 13:461–468
15. Braidy N, Guillemin GJ, Mansour H, Chan-Ling T, Poljak A, Grant R (2011) Age related changes in NAD+ metabolism oxidative stress and Sirt1 activity in wistar rats. PLoS One 6:e19194
16. Muller HM, Muller CD, Schuber F (1983) NAD+ glycohydrolase, an ecto-enzyme of calf spleen cells. Biochem J 212:459–464
17. Sugawara K, Oyama F (1981) Fluorogenic reaction and specific microdetermination of ammonia. J Biochem 89:771–774
18. Hubbard BP, Sinclair DA (2013) Measurement of sirtuin enzyme activity using a substrate-agnostic fluorometric nicotinamide assay. Methods Mol Biol 1077:167–177
19. Hubbard BP, Gomes AP, Dai H, Li J, Case AW, Considine T, Riera TV, Lee JE, E SY, Lamming DW, Pentelute BL, Schuman ER, Stevens LA, Ling AJ, Armour SM, Michan S, Zhao H, Jiang Y, Sweitzer SM, Blum CA, Disch JS, Ng PY, Howitz KT, Rolo AP, Hamuro Y, Moss J, Perni RB, Ellis JL, Vlasuk GP, Sinclair DA (2013) Evidence for a common mechanism of SIRT1 regulation by allosteric activators. Science 339:1216–1219
20. Trammell SA, Brenner C (2013) Targeted, LCMS-based metabolomics for quantitative measurement of NAD(+) metabolites. Comput Struct Biotechnol J 4:e201301012

Chapter 7

Generating Protein-Linked and Protein-Free Mono-, Oligo-, and Poly(ADP-Ribose) In Vitro

Ken Y. Lin, Dan Huang, and W. Lee Kraus

Abstract

ADP-ribosylation is a covalent posttranslational modification of proteins that is catalyzed by various types of ADP-ribosyltransferase (ART) enzymes, including members of the poly(ADP-ribose) polymerase (PARP) family. ADP-ribose (ADPR) modifications can occur as mono(ADP-ribosyl)ation, oligo(ADP-ribosyl)ation, or poly(ADP-ribosyl)ation, depending on the particular ART enzyme catalyzing the reaction, as well as the specific reaction conditions. Understanding the biology of ADP-ribosylation requires facile and robust means of generating and detecting the modification in all of its forms. Here we describe how to generate protein-linked mono(ADP-ribose), oligo(ADP-ribose), and poly(ADP-ribose) (MAR, OAR, and PAR, respectively) in vitro as an automodification of PARPs 1 or 3. First, epitope-tagged PARP-1 (a PARP polyenzyme) and PARP-3 (a PARP monoenzyme) are expressed individually in insect cells using baculovirus expression vectors, and purified using immunoaffinity chromatography. Second, the purified recombinant PARPs are incubated individually in the presence of different concentrations of NAD^+ (as a donor of ADPR groups) and sheared DNA (to activate their catalytic activities) resulting in various forms of auto-ADP-ribosylation. Third, the products are confirmed using ADPR detection reagents that can distinguish among MAR, OAR, and PAR. Finally, if desired, the OAR and PAR can be deproteinized. The protein-linked and free MAR, OAR, and PAR generated in these reactions can be used as standards, substrates, or binding partners in a variety of ADPR-related assays.

Key words ADP-ribose (ADPR), ADPR binding domain (ARBD), ADP-ribosylation, ADP-ribosyltransferase (ART), Automodification, Mono(ADP-ribosyl)ation (MARylation), Nicotinamide adenosine dinucleotide (NAD^+), Oligo(ADP-ribosyl)ation (OARylation), Poly(ADP-ribose) polymerase (PARP), Poly(ADP-ribosyl)ation (PARylation), Posttranslational modification (PTM)

1 Introduction

ADP-ribosylation is a posttranslational modification of proteins that results from the covalent addition of ADP-ribose (ADPR) moieties derived from donor beta-nicotinamide adenine dinucleotide (βNAD^+) on glutamate, aspartate, serine, lysine, arginine, and

Ken Y. Lin and Dan Huang contributed equally to this work

Paul Chang (ed.), *ADP-ribosylation and NAD+ Utilizing Enzymes: Methods and Protocols*, Methods in Molecular Biology, vol. 1813, https://doi.org/10.1007/978-1-4939-8588-3_7,

possibly other amino acid residues [1–4]. The modification is catalyzed by a diverse family of ADP-ribosyltransferase (ART) enzymes, including (1) bacterial toxins (e.g., cholera and diphtheria toxins) [5, 6], (2) ecto-ADP-ribosyltransferases (ectoARTs) [7], (3) members of the sirtuin family [8–10], and (4) members of the poly(ADP-ribose) polymerase (PARP) family [2, 11]. ARTs generally catalyze mono(ADP-ribosyl)ation (MARylation) of their substrate proteins, although some PARPs, such as PARP-1, catalyze oligo- and poly(ADP-ribosyl)ation (OARylation and PARylation, respectively) [2, 5, 7, 12, 13] (Fig. 1). Many ARTs are capable of modifying themselves (i.e., automodification). ADP-ribosylation of proteins has emerged as a dynamic regulator of a wide variety of cellular and molecular processes, including transcription, RNA processing, DNA damage repair, unfolded protein responses, and cellular signaling [2, 15–17].

In addition to ARTs, which add ("write") ADPR, cells express glycohydrolases, which remove ("erase") ADPR, and proteins containing ADPR binding domains (ARBDs), which bind ("read") mono-, oligo-, and/or poly(ADP-ribose) (MAR, OAR, and PAR, respectively) [2, 12, 15, 18–20]. Interestingly, nature has devised protein modules that specifically recognize and bind to various forms of ADP-ribose [18, 20]. The latter (i.e., ARBDs "readers"), which are found in a variety of proteins with diverse functions, including a number of PARP family members [2, 18, 20], include macrodomains and WWE domains [18, 20]. Macrodomains bind MAR, as well as the terminal ADPR moieties in OAR and PAR, allowing them to bind to all three forms of ADPR (i.e., MAR, OAR, and PAR) [21–23]. On the other hand, WWE domains bind the iso-ADPR linkages joining ADPR monomers, limiting their binding to OAR and PAR [24–26]. ARBDs can be used as research tools for the recognition of various forms of ADPR by fusing them to other protein domains or modules [23, 27–32], including the Fc region of rabbit immunoglobulin to generate antibody-like ADPR binding proteins [29, 30].

The field is faced with a number of fundamental questions about ADP-ribosylation. For example, (1) which proteins are substrates (i.e., targets) for ADP-ribosylation?, (2) which specific amino acid residues within those substrate proteins are modified?, (3) what is the nature of the ADPR modification (mono-, oligo-, or poly)?, and (4) how does ADP-ribosylation modulate the activity or function of the substrate proteins? Answering these questions requires a facile and robust means of generating and detecting ADPR in all of its forms. Herein, we describe methods for (1) the expression and purification of recombinant PARP-1 and PARP-3; (2) using these proteins to generate protein-linked MAR, OAR,

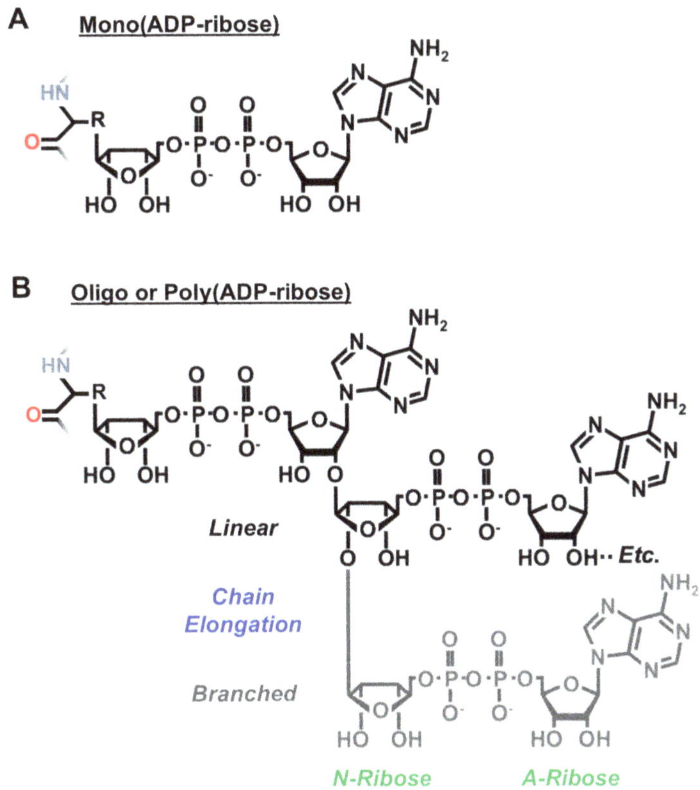

Fig. 1 Chemical structures of protein-linked ADP-ribose. PARP family members and other ADP-ribosyltransferases (ARTs) catalyze the covalent attachment of ADP-ribose units derived from donor NAD^+ molecules on substrate proteins. Common residues of attachment include lysine, glutamate, and aspartate. (**a**) Chemical structure of protein-linked mono(ADP-ribose) (MAR). (**b**) Chemical structure of protein-linked oligo- or poly(ADP-ribose) (OAR and PAR, respectively). Chain elongation can proceed in a linear or branched manner through different ribose-ribose linkages, as shown. For linear elongation, the adenine-proximal ribose (A-ribose) from a terminal ADP-ribose unit is joined through an $\alpha(1 \rightarrow 2)$ *O*-glycosidic bond to the nicotinamide-proximal ribose (N-ribose; relative to the nicotinamide moiety in NAD^+) of the next ADP-ribose unit added to the growing chain. For branched elongation, the glycosidic bond occurs between the N-ribose of a terminal ADP-ribose unit and the N-ribose of the next ADP-ribose unit added to the growing chain. PAR polymers can grow to over 200 units in size (indicated by "etc."), with branching reactions occurring approximately once every 20–50 linear elongation additions [14]

and PAR in vitro as an automodification of PARPs 1 or 3; and (3) deproteinizing the automodified PARP-1 to generate free OAR or PAR (Fig. 2). The protein-linked and free ADPR forms can be used as standards in a variety of ADPR-related assays.

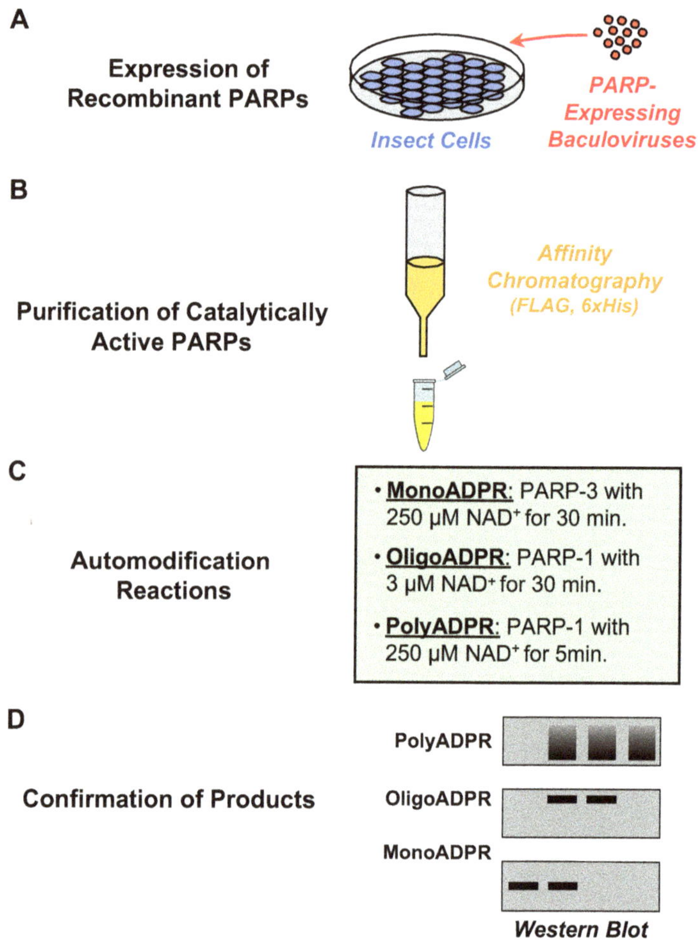

Fig. 2 An approach for generating various forms of protein-linked ADP-ribose. Flow diagram of the steps required to generate and test various forms of protein-linked ADP-ribose using biochemical approaches. (**a**) Expression of recombinant PARPs in insect cells using a baculovirus-based expression system. (**b**) Purification of catalytically active PARPs from insect cell lysates using affinity chromatography. (**c**) Automodification of purified PARPs using controlled reaction conditions. (**d**) Confirmation of the ADPR products by Western blotting using ADPR detection reagents

2 Materials

All reagents should be the highest quality molecular biology and biochemistry grade.

2.1 Expression of Recombinant PARP-1 and PARP-3 Proteins

1. Sf9 insect cells.
2. Appropriate serum-free or serum-containing Sf9 cell growth medium (e.g., Sf-900™ II SFM medium).

3. Recombinant baculoviruses for expressing FLAG epitope-tagged wild-type PARP-1 or PARP-3.
4. Cell culture spinner flasks.
5. Low RPM magnetic cell culture stir plate.
6. 27 °C incubator.
7. Phosphate-buffered saline (PBS), ice cold.

2.2 Purification of Recombinant PARP-1 and PARP-3 Proteins

1. >18 MΩ double distilled H_2O (ddH_2O).
2. Protease inhibitor mix (e.g., Roche cOmplete™ EDTA-free tablets provided in glass vials, which can be dissolved to the appropriate concentration) (*see* **Note 1**).
3. FLAG lysis buffer: 20 mM HEPES (pH 7.9), 0.5 M NaCl, 4 mM $MgCl_2$, 0.4 mM EDTA, 20% glycerol, 250 mM nicotinamide, 2 mM β-mercaptoethanol, and 2× protease inhibitor mix (*see* **Notes 2** and **3**).
4. Dounce homogenizer with a type B (tight) pestle.
5. High-speed centrifuge with an SS-34 rotor (Sorvall) or a comparable centrifuge/rotor pair.
6. FLAG dilution buffer: 20 mM HEPES (pH 7.9), 10% glycerol, 0.02% NP-40 (*see* **Note 3**).
7. Probe-style sonicator.
8. Anti-FLAG M2 agarose resin, 50% slurry.
9. FLAG wash buffer #1: 20 mM HEPES (pH 7.9), 150 mM NaCl, 2 mM $MgCl_2$, 0.2 mM EDTA, 15% glycerol, 0.01% NP-40, 100 mM nicotinamide, 0.2 mM β-mercaptoethanol, 1 mM PMSF, 1 μM aprotinin, 100 μM leupeptin (*see* **Notes 2** and **3**).
10. FLAG wash buffer #2/high salt: 20 mM HEPES (pH 7.9), 1 M NaCl, 2 mM $MgCl_2$, 0.2 mM EDTA, 15% glycerol, 0.01% NP-40, 100 mM nicotinamide, 0.2 mM β-mercaptoethanol, 1 mM PMSF, 1 μM aprotinin, 100 μM leupeptin (*see* **Notes 2** and **3**).
11. FLAG wash buffer #3: 20 mM HEPES (pH 7.9), 150 mM NaCl, 2 mM $MgCl_2$, 0.2 mM EDTA, 15% glycerol, 0.01% NP-40, 0.2 mM β-mercaptoethanol, 1 mM PMSF, 1 μM aprotinin, 100 μM leupeptin (*see* **Notes 2** and **3**).
12. FLAG peptide (Sigma), 5 mg/mL dissolved in Tris buffered saline [TBS: 50 mM Tris–HCl (pH 7.4), 150 mM NaCl] (*see* **Notes 2** and **3**).
13. FLAG elution buffer: Use FLAG wash buffer #3 containing 0.2 mg/mL FLAG peptide (*see* **Notes 2** and **3**).
14. Bradford protein assay solution.

2.3 In Vitro Auto(ADP-Ribosyl)ation Reactions with PARP-1 and PARP-3

1. Purified recombinant full-length FLAG-tagged PARP-1 and PARP-3 (from Subheadings 3.1 and 3.2 below).
2. Sonicated salmon sperm DNA.
3. Bovine serum albumin (BSA).
4. Automodification Buffer: 20 mM HEPES (pH 8.0), 5 mM $MgCl_2$, 5 mM $CaCl_2$, 0.01% NP-40, 25 mM KCl, 1 mM DTT, 0.1 mg/mL sonicated salmon sperm DNA, 0.1 mg/mL BSA.
5. NAD^+, 10 mM, dissolved in 10 mM Tris–HCl (pH 7.9) (*see* **Note 4**).

2.4 Confirmation of Protein-Linked ADPR Products by Western Blotting

1. 10% PAGE-SDS gel and an appropriate electrophoresis apparatus.
2. 4× SDS-PAGE loading buffer: 200 mM Tris–HCl (pH 6.8), 10% SDS, 40% glycerol, 0.04% bromophenol blue, and 400 mM DTT (*see* **Note 5**).
3. Nylon-backed nitrocellulose membrane.
4. Western transfer apparatus.
5. Anti-mono(ADP-ribose) binding reagent (EMD Millipore: MABE1076, RRID: AB_2665469).
6. Anti-poly(ADP-ribose) binding reagent (EMD Millipore: MABE1031; RRID: AB_2665467).
7. Anti-pan(ADP-ribose) binding reagent (EMD Millipore: MABE1016, RRID: AB_2665466).

2.5 Generation and Analysis of Protein-Free OAR and PAR

1. Trichloroacetic acid, 100% (w/v) (*see* **Note 3**).
2. Trichloroacetic acid, 20% (w/v) (*see* **Note 3**).
3. Ethanol, 100%.
4. TE: 10 mM Tris–HCl (pH 8.0), 1 mM EDTA.
5. DNase I, 10 units/μL.
6. SDS, 10% (w/v).
7. Proteinase K, 20 mg/mL.
8. 1 M KOH/50 mM EDTA solution.
9. Hydrochloric acid (HCl).
10. Phenol/chloroform/isoamyl alcohol (25:24:1, v/v).
11. Chloroform.
12. Glycogen, 10 mg/mL dissolved in nuclease-free water.
13. 3 M sodium acetate (pH 5.2).
14. 70% ethanol.
15. Urea-PAGE loading solution: 6 M urea, 25 mM NaCl, 4 mM EDTA, 0.02% (w/v) xylene cyanol, and 0.02% (w/v) bromophenol blue.

16. 5× TBE electrophoresis buffer: 0.45 M Tris base (pH 7.6), 0.45 M boric acid, and 10 mM EDTA.
17. Ammonium persulfate (APS), 10% (w/v) prepared in water (*see* **Note 6**).
18. *N,N,N′,N′*- tetramethylethylenediamine (TEMED).
19. Polyacrylamide gel, 20% (w/v) with a 19:1 (w/v) acrylamide/bisacrylamide ratio: 100 mL of 30% 19:1 (w/v) acrylamide/bisacrylamide, 30 mL 5× TBE electrophoresis buffer, and 10 mL ddH_2O. Add 0.75 mL of 10% APS and 6 μL of TEMED immediately before use to initiate polymerization.
20. Silver stain kit.

3 Methods

3.1 Expression of PARP-1 and PARP-3 in Insect Cells

Insect cells are useful eukaryotic cells for the high-level expression of recombinant proteins, which in many cases retain their natural posttranslational modification state. The protocol below describes how to generate high-titer baculoviruses, which can be used for expressing recombinant full-length FLAG-tagged PARP-1 and PARP-3 in insect cells. The expressed proteins can then be purified as described in Subheading 3.2 below.

1. Grow Sf9 cells using your preferred growth conditions (*see* **Note 7**).
2. Amplify PARP-expressing recombinant baculoviruses at a low multiplicity of infection (MOI) to generate more viruses by infecting Sf9 cells (*see* **Note 7**). Use multiple rounds of low MOI amplifications as needed to generate 75 to 100 mL of high-titer virus. Collect and store the baculovirus stocks at 4 °C protected from light.
3. In parallel to the baculovirus amplification, expand uninfected Sf9 cells in suspension culture so that 250–500 mL of cells in logarithmic growth are available for a large-scale infection (*see* **Note 8**). The cells should be grown at 27 °C in a spinner flask placed on a low RPM magnetic cell culture stir plate.
4. Infect the suspension cultures of Sf9 cells in logarithmic growth from **step 3** with 50–100 mL of freshly amplified high-titer recombinant baculovirus expressing FLAG-tagged PARP-1 or PARP-3 from **step 2**. You will need one 250–500 mL suspension culture for each protein preparation.
5. 48 h after infection, collect the Sf9 cells by centrifuging the culture at 1000 × *g* for 10 min in a large centrifuge bottle (*see* **Note 9**).
6. Gently resuspend the Sf9 cell pellet in 40 mL of ice-cold PBS and transfer the suspension to a 50 mL conical tube.

7. Collect the cells again by centrifuging at 1000 × *g* for 10 min in the 50 mL conical tube. Remove the PBS.
8. Flash-freeze the cell pellets in liquid N_2 and store them at −80 °C.

3.2 Purification of PARP-1 and PARP-3 from Insect Cells

Baculovirus vectors allow for the high-level expression of recombinant affinity-tagged proteins in insect cells. The protocol below describes how to purify recombinant full-length FLAG-tagged PARP-1 and PARP-3 expressed in insect cells, which can be used in the biochemical assays described below, as well as a host of other assays not described here.

1. Thaw PARP-1 or PARP-3-containing Sf9 cell pellets on wet ice.
2. Resuspend the cells in 7 mL of FLAG lysis buffer per 100 mL of initial insect cell culture volume.
3. Lyse cells by dounce homogenization ten times on ice with a tight pestle. Transfer to a 50 mL screw-capped centrifuge tube.
4. Incubate on ice with intermittent gentle mixing to allow salt extraction to occur.
5. Clarify the lysate by centrifugation at 27,000 × *g* at 4 °C for 20 min (*see* **Note 10**). Collect the supernatant and mix with an equal volume of ice-cold FLAG dilution buffer (*see* **Note 11**).
6. Sonicate the lysate briefly using a probe sonicator (15 s, 65% amplitude), while keeping the sample on ice.
7. Clarify the lysate again as in **step 5**.
8. While the sample is being centrifuged, wash and pre-equilibrate anti-FLAG M2 agarose in FLAG wash buffer #1.
9. Mix the clarified lysate with equilibrated anti-FLAG M2 agarose resin (200 μL of a 50% slurry per 250 mL of initial insect cell culture volume) in a screw-capped conical tube.
10. Incubate the lysate with the anti-FLAG M2 agarose resin for 3 h at 4 °C with gentle continuous mixing (e.g., using a nutator).
11. Centrifuge the mixture in the conical tube at 1000 × *g* for 10 min at 4 °C in a benchtop centrifuge.
12. Wash the resin by gently resuspending it in 40 mL of FLAG wash buffer #1, incubating it for 5 min at 4 °C with gentle mixing and collecting it by centrifugation in the conical tube at 1000 × *g* for 10 min at 4 °C in a benchtop centrifuge.
13. Repeat **step 12**.

14. Wash the resin with 25 mL of FLAG wash buffer #2 as described in **step 12**.
15. Repeat **step 14**.
16. Wash the resin with 25 mL of FLAG wash buffer #3 as described in **step 12**.
17. Repeat **step 16**, for a total of six washes with three different buffers (*see* **Note 12**).
18. Elute the FLAG-tagged PARP proteins from the anti-FLAG M2 agarose resin using 200 μL of FLAG elution buffer (*see* **Note 13**). Add the elution buffer to the washed resin and incubate on ice with gentle mixing for 10 min. Collect the resin by centrifugation as described above and carefully remove the supernatant to a new tube on ice, making sure not to bring any resin along with the eluate. Repeat at least three times (*see* **Note 14**).
19. Quantify the eluted protein by using a Bradford protein assay or by comparison to increasing amounts of a known protein (e.g., BSA) run side by side with the elutions on a Coomassie-stained SDS-PAGE gel (*see* **Notes 15** and **16**) (Fig. 3).
20. Aliquot the purified protein, flash-freeze in liquid N_2, and store at −80 °C.

3.3 In Vitro Auto(ADP-Ribosyl)ation Reactions with PARP-1 and PARP-3 to Generate Protein-Linked Mono-, Oligo-, and Poly(ADP-Ribose)

Protein-linked mono-, oligo-, and poly(ADP-ribose) can be generated by taking advantage of the fact that PARP-1 and PARP-3 undergo auto(ADP-ribosylation) when activated by free DNA ends—a process called automodification. At high NAD^+ concentrations, PARP-1 produces poly(ADP-ribose) in its automodification reaction, whereas at lower, more limiting NAD^+ concentrations, PARP-1 produces oligo(ADP-ribose) in its automodification reaction. PARP-3 is a mono(ADP-ribosyl)transferase that produces mono(ADP-ribose) in its automodification reaction. In this section, we describe how to generate protein-linked mono-, oligo-, and poly(ADP-ribose) using purified recombinant PARP-1 and PARP-3 using automodification in vitro.

1. Incubate 500 ng of purified recombinant PARP-1 or PARP-3 at 25 °C in a 100 μL reaction volume in automodification buffer under the following conditions (*see* **Notes 17** and **18**):
   (a) PARP-1 with 250 μM NAD^+ for 5 min to generate poly(ADP-ribose)
   (b) PARP-1 with 3 μM NAD^+ for 30 min to generate oligo(ADP-ribose)
   (c) PARP-3 with 250 μM NAD^+ for 30 min to generate mono(ADP-ribose)

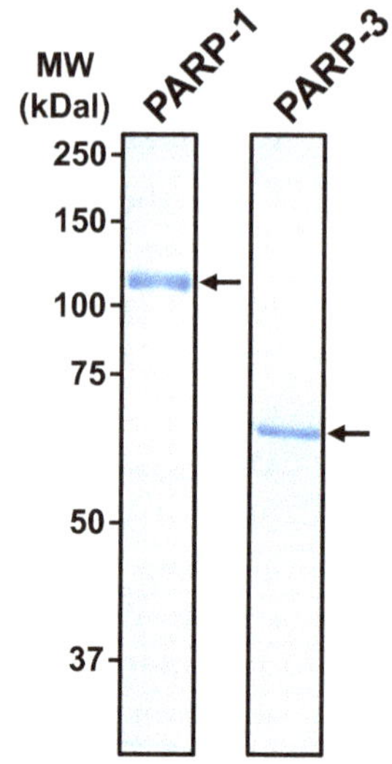

Fig. 3 SDS-PAGE analysis of purified recombinant PARP-1 and PARP-3. Human PARP-1 and PARP-3 were expressed in insect cells using a baculovirus-based expression system and purified from cell lysates using affinity chromatography. Aliquots of the purified proteins were run on a 10% acrylamide-SDS gel, which was stained using Coomassie Brilliant Blue. Molecular weight markers (in kilo-Daltons, kDal) are shown for reference. Black arrows indicate the locations of the purified proteins

2. Stop the reactions by the addition of 4× SDS-PAGE loading buffer, followed by heating at 100 °C for 5 min (*see* **Note 19**).
3. Assess the progress of the autoPARylation reaction with PARP-1 by running 25% of the reaction on a Coomassie-stained SDS-PAGE gel (*see* **Notes 20** and **21**) (Fig. 4).
4. Analyze the protein-linked MAR, OAR, and PAR products by running 5% of the reaction products on Western or dot blots using antibody-like ADP-ribose binding reagents (*see* **Note 22**) (Fig. 5).

3.4 Generation of Protein-Free Oligo(ADP-Ribose) and Poly(ADP-Ribose)

For some applications, it may be useful to separate the OAR or PAR from PARP-1 to use in a protein-free state for some assays [note that MAR (i.e., ADPR) can be purchased in a protein-free state]. This section describes a method for deproteinizing and purifying protein-free OAR or PAR.

1. Perform automodification reactions using purified PARP-1 protein as described in Subheading 3.3, **step 1**. To facilitate visualization of OAR during gel analysis, the reaction can be supplemented with 60 nM ^{32}P-NAD^{+} (*see* **Note 23**).
2. Stop the reaction by precipitation with ice-cold trichloroacetic acid (20%, v/v added from a 100% stock solution) on ice for 15 min.

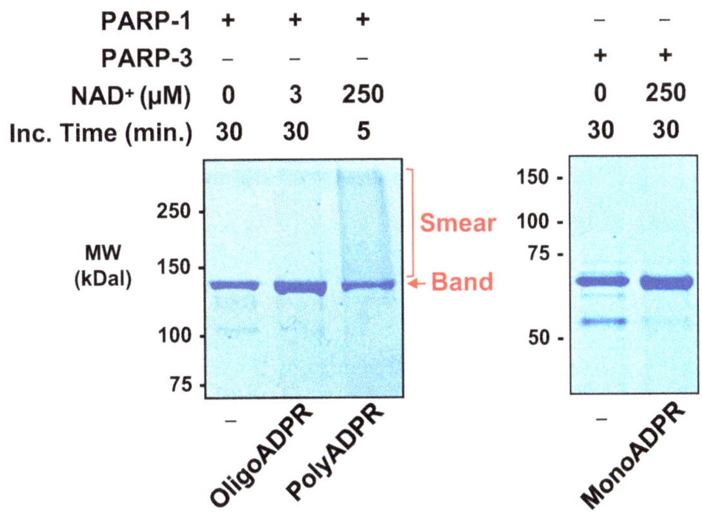

Fig. 4 SDS-PAGE analysis of purified recombinant PARP-1 and PARP-3 subjected to in vitro automodification. Purified recombinant human PARP-1 or PARP-3 (each at 500 ng/μL) was incubated with NAD^+ (concentrations indicated) and sheared salmon sperm DNA (0.1 mg/mL) for the times indicated to promote automodification. Aliquots of the reactions were run on a 10% acrylamide-SDS gel, which was stained using Coomassie Brilliant Blue. Molecular weight markers (in kiloDaltons, kDal) are shown for reference. Incubation of PARP-1 with 250 μM NAD^+ for 5 min produces poly(ADP-ribose), whereas incubation of PARP-1 with 3 μM NAD^+ for 30 min produces oligo(ADP-ribose). Incubation of PARP-3 with 250 μM NAD^+ for 30 min produces mono(ADP-ribose). Poly(ADP-ribosyl)ated PARP-1 exhibits a slower, heterogeneous migration (indicated as smearing, with a subsequent reduction of the primary band), which is not observed with oligo(ADP-ribosyl)ated PARP-1

3. Collect the precipitates by centrifugation at 15,000 × *g* for 15 min at 4 °C in a microcentrifuge.
4. Wash the pellets twice with ice-cold 20% trichloroacetic acid, followed by three times with 100% ethanol. After each wash, centrifuge briefly to make sure that the pellet does not become dislodged.
5. Remove any residual ethanol by air drying, and then dissolve the pellets in 100 μL of TE.
6. Remove the DNA by adding 1 μL of DNase I (10 units/μL), and incubate for 15 min at 37 °C.
7. Bring the reaction to 0.1% SDS from a 10% stock (*see* **Note 24**).
8. Add proteinase K to 0.2 mg/mL from a 20 mg/mL stock, and incubate at 50 °C for 2–3 h or at 37 °C overnight.
9. Add an equal volume of 1 M KOH/50 mM EDTA to detach the polymers from the digested proteins and incubate for 2 h at 60 °C.

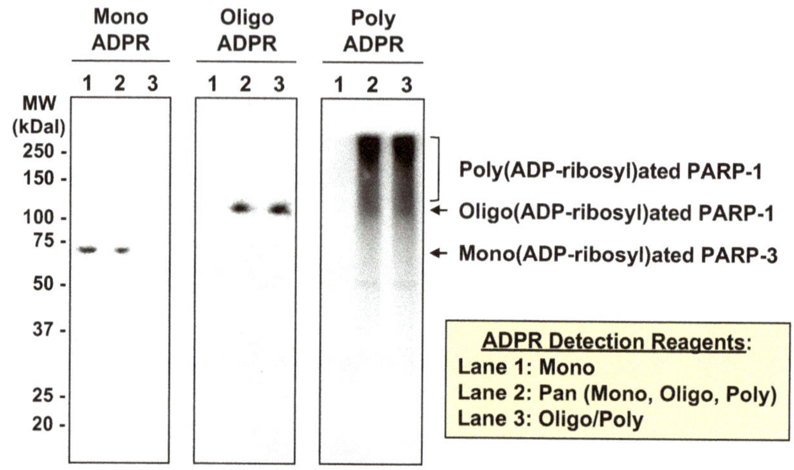

Fig. 5 Western blot analysis of MARylated PARP-3, OARylated PARP-1, and PARylated PARP-1 using antibody-like ADPR detection reagents. Purified recombinant human PARP-1 or PARP-3 was incubated with NAD^+ and sheared salmon sperm DNA under the conditions specified in the legend to Fig. 4. Aliquots of the reactions were run on a 10% acrylamide-SDS gel and subjected to Western blotting using the following antibody-like ADPR detection reagents: (1) anti-mono(ADP-ribose) binding reagent (EMD Millipore: MABE1076), (2) anti-pan(ADP-ribose) binding reagent (EMD Millipore: MABE1016), and (3) anti-poly(ADP-ribose) binding reagent (EMD Millipore: MABE1031). Molecular weight markers (in kiloDaltons, kDal) are shown for reference. The loading in each lane were standardized for the amount of terminal ADP-ribose units using anti-pan(ADP-ribose) binding reagent (EMD Millipore: MABE1016)

10. Following the alkaline treatment, adjust the pH to 8.0 using HCl.
11. Add an equal volume of phenol/chloroform/isoamyl alcohol and vortex for 30 s.
12. Centrifuge the mixture at 15,000 × *g* for 5 min at room temperature in a microcentrifuge to separate the aqueous and organic phases.
13. Transfer the upper aqueous phase to a new microcentrifuge tube.
14. Add an equal volume of chloroform and vortex for 30 s.
15. Repeat **steps 12** and **13**.
16. Estimate the volume of the aqueous phase.
17. Add 1 μL of 10 mg/mL glycogen, 1/10 volume of 3 M sodium acetate (pH 5.2), and 2.5 volumes of ice-cold ethanol. Mix the solution well.
18. Incubate at −20 °C for 1 h (to overnight).
19. Centrifuge at 15,000 × *g* for 30 min at 4 °C in a microcentrifuge.
20. Remove and discard the supernatant without disturbing the pellet.

21. Carefully wash the pellet with 1 mL of 70% ethanol and re-centrifuge at maximum speed for 5 min at room temperature in a microcentrifuge.
22. Repeat **steps 20** and **21**.
23. Air-dry the pellets and dissolve them in small volume of TE. Store the free polymers at −20 °C in aliquots until needed. The protein-free products can be analyzed by dot blotting or urea-PAGE (*see* **Notes 25** and **26**).
24. For dot blotting, combine an aliquot of free OAR or PAR with urea-PAGE loading solution, spot on a nylon-backed nitrocellulose membrane, and blot using ADPR detection reagents (Fig. 6).
25. For PAGE analysis, prepare a 20 cm × 20 cm × 0.15 cm 20% polyacrylamide gel with a 19:1 (w/v) acrylamide/bisacrylamide ratio using the recipe described in Subheading 2.5, **step 19**. Pre-run the gel for 1 h in TBE buffer at 400 V. Replace the TBE buffer and run the free oligomers and polymers at a constant voltage of 400 V. Stain the gel using silver stain or, if the product was ^{32}P-labeled (see **step 1** above), fix and dry the

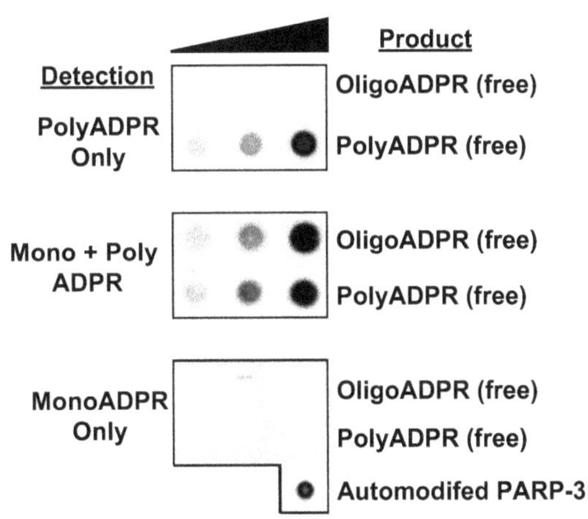

Fig. 6 Dot blot analysis of protein-free OAR and PAR. Purified recombinant human PARP-1 was incubated with NAD$^+$ and sheared salmon sperm DNA under the conditions specified in the legend to Fig. 4 to generate OAR and PAR. The OAR and PAR products were deproteinized, precipitated, pelleted, dissolved, spotted on a nitrocellulose membrane in increasing amounts, and subjected to dot blotting using the following: (1) anti-PAR monoclonal antibody 10H ("PolyADPR Only"), (2) anti-poly(ADP-ribose) binding reagent (EMD Millipore: MABE1031) ("Mono + Poly"), or (3) anti-mono(ADP-ribose) binding reagent (EMD Millipore: MABE1076) ("MonoADPR Only"). For the latter, automodified PARP-3 was spotted as a positive control. The loading in each dot was standardized for the amount of terminal ADP-ribose units using anti-pan(ADP-ribose) binding reagent (EMD Millipore: MABE1016)

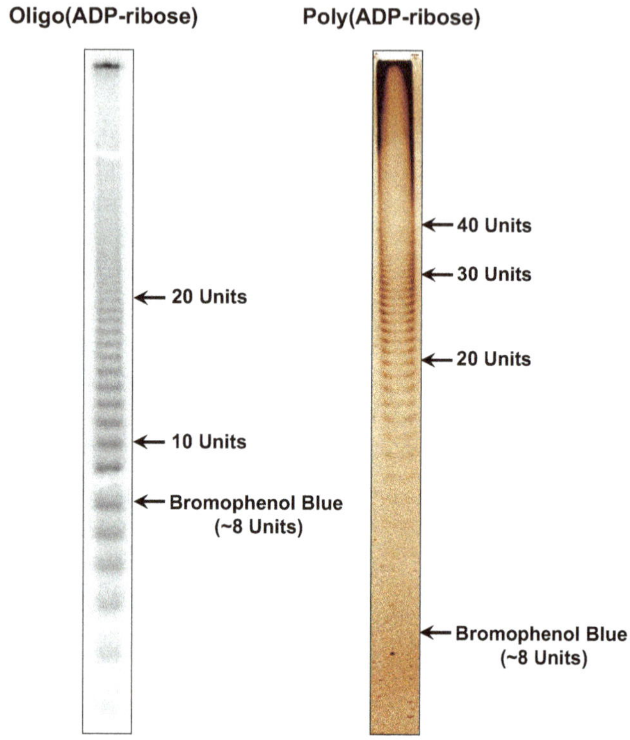

Fig. 7 PAGE analysis of protein-free PAR. Purified recombinant human PARP-1 was incubated with NAD^+ and sheared salmon sperm DNA under the conditions specified in the legend to Fig. 4 to generate oligo(ADP-ribose) (*left*) and poly(ADP-ribose) (*right*). To facilitate visualization of OAR during gel analysis, the PARP-1 automodification reaction was supplemented with 60 nM ^{32}P-NAD^+. The OAR and PAR products were deproteinized and subjected to PAGE analysis with subsequent phosphorimage analysis (for OAR) or silver staining (for PAR). The migration of bromophenol blue (BPB), which runs at the location of chains of ~8 ADPR units [33, 34], was used as a marker

gel, then visualize by autoradiography or by phosphorimage analysis (Fig. 7).

3.5 Using Protein-Linked and Protein-Free Mono-, Oligo-, and Poly(ADP-Ribose)

Protein-linked and protein-free MAR, OAR, and PAR can be used in a variety of assays as standards, substrates, or binding/interaction partners. The reaction conditions described herein generate OAR and PAR with average lengths of ~10 ADPR units and ~30 ADPR units, respectively (*see* **Notes 27** and **28**) (Fig. 7).

4 Notes

1. Make fresh for each use.
2. While this buffer can be made with the base components in advance and stored, the reducing agent (β-mercaptoethanol)

and protease inhibitors (PMSF, aprotinin, leupeptin, or protease inhibitor mix) should be added immediately before use.

3. This buffer/solution should be ice cold for use.
4. NAD^+ is labile. Distribute in small, single-use aliquots after dissolving. Store at −80 °C.
5. DTT is labile. Make the 4× SDS-PAGE loading buffer without DTT, and add the DTT immediately before use from a fresh stock solution. β-mercaptoethanol (0.7 M or 5% of the final sample) can be added immediately before use as an alternative to DTT as the reducing agent.
6. APS is labile. The 10% stock solution should not be stored at 4 °C for more than 2 weeks.
7. Please refer to other sources for detailed descriptions of Sf9 cell culture, as well as baculovirus generation and use.
8. Healthy Sf9 cells in suspension cultures at densities between 5×10^5 and 2×10^6 cells/mL should be in logarithmic growth phase.
9. Expression of the recombinant PARP proteins is usually detectable after 15 h of infection and increases until ~48 h post-infection. We recommend determining the time of maximal protein expression in a time course experiment.
10. 27,000 × *g* can be achieved at 15,000 RPM using an SS-34 rotor in a Sorvall centrifuge.
11. This will bring the NaCl concentration to 250 mM, which is compatible with the FLAG affinity purification.
12. The washes with wash buffer #1, a low salt buffer containing 150 mM NaCl, remove protein contaminants. The wash with 1 M NaCl (FLAG wash buffer #2) removes DNA and RNA contamination from PARP-1 and PARP-3, which can be significant for these nucleic acid-binding proteins. Nicotinamide is included in wash buffers #1 and #2 to inhibit automodification of PARP-1 and PARP-3. Nicotinamide is then omitted in the wash buffer #3.
13. Elute with the same volume of wash buffer as the packed bed volume of beads. For example, if you used 400 μL of a 50% slurry, this will contain a packed bed volume of 200 μL of beads.
14. Multiple elutions should be performed to monitor for efficiency of elution. The peak of elution will usually occur in the first two elutions, but significant useable protein may be eluted out to elutions 4 or 5, depending on the expression and purification conditions.

15. Running the purified proteins (~300 to 500 ng) on a Coomassie-stained SDS-PAGE gel will allow an assessment of the purity and quality of the purified proteins (Fig. 3).
16. The expected yield is ~0.4–0.5 mg/mL in elutions 1 and 2, and ~0.1–0.3 mg/mL in elutions 3 and 4, for a total combined yield of ~250 μg (i.e., 800 μL of a ~0.3 mg/mL solution if all elutions were combined).
17. Sonicated salmon sperm DNA is added because PARP-1 and PARP-3 are DNA-dependent PARPs whose catalytic functions are activated by free DNA ends. The addition of sheared DNA may not be necessary for other PARP family members.
18. We do not routinely quantify the catalytic activity of our PARP-1 and PARP-3 preparations. The conditions described here for automodification have worked well across many different preparations of PARP-1 and PARP-3. If the expected automodification results are not achieved with a particular preparation of PARP-1 or PARP-3, try varying the amount of NAD^+ in the reactions or the length of the incubation.
19. If you plan to generate free OAR or PAR, skip this step and proceed directly to Subheading 3.4 (although, if this is your first time, you should confirm the reaction products as described in Subheading 3.3 before proceeding).
20. This step is optional but recommended for the first time through the protocol.
21. As the reaction progresses, the PARP-1 protein band will smear into a slower-migrating automodified product, an outcome not observed with mono- or oligo(ADP-ribosylated) PARP-1 or mono(ADP-ribosylated) PARP-3 (Fig. 3).
22. The reaction conditions described here generate different levels of protein-linked ADP-ribose units. To achieve equal signals in Western or dot blots, you can standardize each reaction for the amount of terminal ADP-ribose units using anti-pan(ADP-ribose) binding reagent (EMD Millipore: MABE1016) (Fig. 5).
23. Using ^{32}P-NAD^+ in the automodification reaction will generate a radioactively labeled product, which should be handled using standard radioactive material use precautions.
24. SDS helps the proteinase K digest proteins more efficiently.
25. Protein-free OAR can also be generated by controlled digestion of PAR with subsequent HPLC purification to isolate specific species of defined length.
26. Additional information about working with free polymers can be found in refs [34, 35].
27. Free monoADPR can be purchased from Sigma (A0752). Producing free monoADPR by enzymatic means is inefficient.

Free ADPR can be analyzed by thin layer chromatography but is not suitable for analysis by SDS-PAGE or dot blotting.

28. Blotting with the ADPR detection reagents and the 10H monoclonal antibody can be used to confirm the length of the OAR and PAR chains generated under the reaction conditions described herein (Figs. 5 and 6). The PAR chains generated under the reaction conditions described herein have an average length of ~25 to 35 units (Fig. 7).

Acknowledgments

The PARP-related research in the Kraus lab is supported by grants from the National Institutes of Health, NIDDK (DK069710), the Cancer Prevention and Research Institute of Texas (CPRIT) (RP160319), and the Cecil H. and Ida Green Center for Reproductive Biology Sciences Endowments.

References

1. Bonfiglio JJ, Fontana P, Zhang Q et al (2017) Serine ADP-ribosylation depends on HPF1. Mol Cell 65:932–940 e936
2. Gibson BA, Kraus WL (2012) New insights into the molecular and cellular functions of poly(ADP-ribose) and PARPs. Nat Rev Mol Cell Biol 13:411–424
3. Laing S, Unger M, Koch-Nolte F et al (2011) ADP-ribosylation of arginine. Amino Acids 41:257–269
4. Schreiber V, Dantzer F, Ame JC et al (2006) Poly(ADP-ribose): novel functions for an old molecule. Nat Rev Mol Cell Biol 7:517–528
5. Deng Q, Barbieri JT (2008) Molecular mechanisms of the cytotoxicity of ADP-ribosylating toxins. Annu Rev Microbiol 62:271–288
6. Simon NC, Aktories K, Barbieri JT (2014) Novel bacterial ADP-ribosylating toxins: structure and function. Nat Rev Microbiol 12:599–611
7. Glowacki G, Braren R, Firner K et al (2002) The family of toxin-related ecto-ADP-ribosyltransferases in humans and the mouse. Protein Sci 11:1657–1670
8. Hawse WF, Wolberger C (2009) Structure-based mechanism of ADP-ribosylation by sirtuins. J Biol Chem 284:33654–33661
9. Rack JG, Morra R, Barkauskaite E et al (2015) Identification of a class of protein ADP-Ribosylating sirtuins in microbial pathogens. Mol Cell 59:309–320
10. Van Meter M, Mao Z, Gorbunova V et al (2011) Repairing split ends: SIRT6, mono-ADP ribosylation and DNA repair. Aging (Albany NY) 3:829–835
11. Ame JC, Spenlehauer C, De Murcia G (2004) The PARP superfamily. BioEssays 26:882–893
12. Hottiger MO (2016) SnapShot: ADP-ribosylation signaling. Mol Cell 62:472
13. Vyas S, Matic I, Uchima L et al (2014) Family-wide analysis of poly(ADP-ribose) polymerase activity. Nat Commun 5:4426
14. Kiehlbauch CC, Aboul-Ela N, Jacobson EL et al (1993) High resolution fractionation and characterization of ADP-ribose polymers. Anal Biochem 208:26–34
15. Gupte R, Liu Z, Kraus WL (2017) PARPs and ADP-ribosylation: recent advances linking molecular functions to biological outcomes. Genes Dev 31:101–126
16. Luo X, Kraus WL (2012) On PAR with PARP: cellular stress signaling through poly(ADP-ribose) and PARP-1. Genes Dev 26:417–432
17. Ryu KW, Kim DS, Kraus WL (2015) New facets in the regulation of gene expression by ADP-ribosylation and poly(ADP-ribose) polymerases. Chem Rev 115:2453–2481
18. Barkauskaite E, Jankevicius G, Ladurner AG et al (2013) The recognition and removal of cellular poly(ADP-ribose) signals. FEBS J 280:3491–3507

19. Fontana P, Bonfiglio JJ, Palazzo L et al (2017) Serine ADP-ribosylation reversal by the hydrolase ARH3. elife 6
20. Teloni F, Altmeyer M (2016) Readers of poly(ADP-ribose): designed to be fit for purpose. Nucleic Acids Res 44:993–1006
21. Karras GI, Kustatscher G, Buhecha HR et al (2005) The macro domain is an ADP-ribose binding module. EMBO J 24:1911–1920
22. Kustatscher G, Hothorn M, Pugieux C et al (2005) Splicing regulates NAD metabolite binding to histone macroH2A. Nat Struct Mol Biol 12:624–625
23. Timinszky G, Till S, Hassa PO et al (2009) A macrodomain-containing histone rearranges chromatin upon sensing PARP1 activation. Nat Struct Mol Biol 16:923–929
24. Kang HC, Lee YI, Shin JH et al (2011) Iduna is a poly(ADP-ribose) (PAR)-dependent E3 ubiquitin ligase that regulates DNA damage. Proc Natl Acad Sci U S A 108:14103–14108
25. Wang Z, Michaud GA, Cheng Z et al (2012) Recognition of the iso-ADP-ribose moiety in poly(ADP-ribose) by WWE domains suggests a general mechanism for poly(ADP-ribosyl) ation-dependent ubiquitination. Genes Dev 26:235–240
26. Zhang Y, Liu S, Mickanin C et al (2011) RNF146 is a poly(ADP-ribose)-directed E3 ligase that regulates axin degradation and Wnt signalling. Nat Cell Biol 13:623–629
27. Aguilera Gomez A, Van Oorschot MM, Veenendaal T et al (2016) In vivo vizualisation of mono-ADP-ribosylation by dPARP16 upon amino-acid starvation. elife 5:e21475
28. Bartolomei G, Leutert M, Manzo M et al (2016) Analysis of chromatin ADP-ribosylation at the genome-wide level and at specific loci by ADPr-ChAP. Mol Cell 61:474–485
29. Gibson BA, Zhang Y, Jiang H et al (2016) Chemical genetic discovery of PARP targets reveals a role for PARP-1 in transcription elongation. Science 353:45–50
30. Luo X, Ryu KW, Kim DS et al (2017) PARP-1 controls the adipogenic transcriptional program by PARylating C/EBPβ and modulating its transcriptional activity. Mol Cell 65:260–271
31. Martello R, Leutert M, Jungmichel S et al (2016) Proteome-wide identification of the endogenous ADP-ribosylome of mammalian cells and tissue. Nat Commun 7:12917
32. Murawska M, Hassler M, Renkawitz-Pohl R et al (2011) Stress-induced PARP activation mediates recruitment of Drosophila Mi-2 to promote heat shock gene expression. PLoS Genet 7:e1002206
33. Griesenbeck J, Oei SL, Mayer-Kuckuk P et al (1997) Protein-protein interaction of the human poly(ADP-ribosyl)transferase depends on the functional state of the enzyme. Biochemistry 36:7297–7304
34. Haince JF, Poirier GG, Kirkland JB (2004) Nonisotopic methods for determination of poly(ADP-ribose) levels and detection of poly(ADP-ribose) polymerase. Curr Protoc Cell Biol Chapter 18:Unit 18.17
35. Malanga M, Bachmann S, Panzeter PL et al (1995) Poly(ADP-ribose) quantification at the femtomole level in mammalian cells. Anal Biochem 228:245–251

Chapter 8

Methods to Study TCDD-Inducible Poly-ADP-Ribose Polymerase (TIPARP) Mono-ADP-Ribosyltransferase Activity

David Hutin, Giulia Grimaldi, and Jason Matthews

Abstract

TCDD-inducible poly-ADP-ribose polymerase (TIPARP; also known as PARP7 and ARTD14) is a mono-ADP-ribosyltransferase that has emerged as an important regulator of innate immunity, stem cell pluripotency, and transcription factor regulation. Characterizing TIPARP's catalytic activity and identifying its target proteins are critical to understanding its cellular function. Here we describe methods that we use to characterize TIPARP catalytic activity and its mono-ADP-ribosylation of its target proteins.

Key words Immunoprecipitation, TCDD-inducible poly-ADP-ribose polymerase (TIPARP), Protein purification, Biotinylated-NAD^+, ^{32}P-NAD^+

1 Introduction

ADP-ribosylation is a posttranslation modification that plays an important role in numerous cellular responses, including DNA repair, oxidative stress, and immune responses, but also gene transcription, protein degradation, and cellular metabolism [1, 2]. ADP-ribosylation is catalyzed by members of the poly-ADP-ribose polymerase PARP family, also known as ADP-ribosyltransferase diphtheria toxin-like (ARTD) family. PARP family members transfer ADP-ribose from nicotinamide adenine dinucleotide (NAD^+) to their target proteins and in the process release nicotinamide (NAM). The majority of the 17 PARPs in humans transfer one molecule of ADP-ribose (mono-ADP-ribosylation; MARylation) rather than several ADP-ribose moieties, (poly-ADP-ribosylation; PARylation) onto themselves or onto their target proteins [3–5]. Although most of the current understanding of ADP-ribosylation has come from studying PARylation and more specifically PARP1 (ARTD1) activity and function, the recent discovery that the majority of PARP family members exhibit

Paul Chang (ed.), *ADP-ribosylation and NAD+ Utilizing Enzymes: Methods and Protocols*, Methods in Molecular Biology, vol. 1813, https://doi.org/10.1007/978-1-4939-8588-3_8,

mono-ADP-ribosyltransferase activity has led to intense research interest in characterizing the enzymatic activities, target proteins, and cellular functions of MARylating members of the PARP family [6]. The report that macrodomain-containing proteins, including MacroD1, MacroD2, and C6orf130, recognize and hydrolyze mono-ADP-ribose from modified proteins [7, 8] revealed that MARylation is a dynamic modification.

TCDD-inducible poly-ADP-ribose polymerase (TIPARP; also known as PARP7 and ARTD14) is a mono-ADP-ribosyltransferase [9]. TIPARP was first identified as a target gene of the aryl hydrocarbon receptor (AHR) [10] and is most evolutionarily conserved with PARP12 (ARTD12) and PARP13 (ARTD13) [4, 9, 11]. AHR is a ligand-activated transcription factor that is activated by numerous environmental pollutants, dietary ligands, and metabolic breakdown products. TIPARP functions as part of a negative feedback loop, by repressing AHR function through mono-ADP-ribosylation, which is reversed by the ADP-ribosylase, MacroD1 [12]. *Tiparp*$^{-/-}$ mice exhibit an increased sensitivity to AHR-ligand-induced toxicities, supporting the role of TIPARP as a negative regulator of AHR function [12]. TIPARP also plays a role in viral replication and innate immunity and has been reported to influence the pluripotency of embryonic stem cells [13, 14] Moreover, TIPARP is regulated by other transcription factors and signaling pathways, including androgen receptor [15], hypoxia factor 1 α [16], and platelet-derived growth factor (PDGF) [17], and by interferons [18], suggesting that TIPARP has vast cellular roles. Characterizing TIPARP's catalytic activity and identifying the proteins it modifies are critical to understanding its cellular function.

In this chapter, we describe methods that we use to characterize TIPARP catalytic function and its MARylation of AHR. These approaches rely on the expression and purification of TIPARP and AHR from bacteria, and TIPARP activity is assayed in the presence of ^{32}P-NAD^{+} or biotinylated-NAD^{+}. The addition of labeled ADP-ribose onto TIPARP or its target proteins is visualized after SDS-PAGE and/or immunoblotting. MARylation results in a very slight upward shift in the migration of modified proteins after SDS-PAGE, with the proteins running very similar to their predicted molecular weights. However, these techniques are done in vitro and do not address whether TIPARP can modify itself or target proteins, such as AHR, in cells and under more physiological conditions. To determine this, we use a combination of immunoprecipitation of TIPARP or a target protein followed by detection of ADP-ribose using an anti-PAN-ADPr binding reagent. By immunoprecipitating the proteins of interest, we increase the specificity and interpretation of the data, since the anti-PAN-ADPr recognizes both mono-ADP-ribosylated and poly-ADP-ribosylated peptide sequences.

2 Materials

Prepare all solutions using ultrapure water with the exception of the running and western blot transfer buffers that are prepared using pure water. Prepare and store all solutions at room temperature unless indicated otherwise.

2.1 Purification of GST-Tagged Proteins

1. One Shot BL21 DE3 star chemically competent *E. coli*.
2. Plasmid DNA is purified from DH5α bacteria culture by plasmid prep kit.
3. pGEX-4T1-TIPARP (human TIPARP full-length sequence). pGEX-4T1-$AHR_{430–848}$ (human AHR partial sequence encoding amino acids 430–848) [12]. The pGEX-4T1-$AHR_{430–848}$ also contains a 6× histidine tag at the C-terminal end, which was found to improve stability.
4. SOC Medium.
5. LB-ampicillin culture medium: To a 1 L Erlenmeyer flask, add 10 g tryptone, 5 g yeast extract, 10 g NaCl, 1 mL NaOH. Autoclave for 45 min at 121 °C. After autoclaving, freshly add 100 μg/mL ampicillin before inoculation with bacteria culture.
6. LB plate with ampicillin: To a 1 L Erlenmeyer flask, add 10 g tryptone, 5 g yeast extract, 10 g NaCl, 1 mL NaOH, 15 g bactoagar. Autoclave 45 min at 121 °C. After autoclaving place the flask in a water bath at 60 °C, add 100 μg/mL ampicillin when the medium is at 60 °C, and pour into petri dishes. Let them solidify at room temperature. Store at 4 °C and protect from light.
7. 1 M isopropyl β-D-1-thiogalactopyranoside (IPTG). Prepared as 1 mL aliquots and stored at −20 °C.
8. Innova® 40 incubator (New Brunswick).
9. Amerex Gyromax 737R (Amerex Instruments).
10. Disposable 25 mL elution column with stopcock (Bio-Rad).
11. Branson Ultrasonics Sonifier™ S-450 Digital Ultrasonic Cell Disruptor/Homogenizer (Branson Ultrasonics).
12. Glutathione (GSH) Sepharose High Performance (GE Healthcare Life Sciences): Glutathione Sepharose is washed with lysis buffer three times and resuspend to make a 50% slurry.
13. Lysis buffer: 20 mM Tris–HCl pH 8.0, 500 mM NaCl, 1 mM EDTA, 1% Triton X-100, 10% glycerol, 2 mM DTT, 1× protease inhibitor cocktail (PIC), 200 μg/mL lysozyme, (Sigma-Aldrich), 400U DNase I (Roche). DTT, PIC, and DNase I should be freshly added before use. Store buffers at 4 °C.

14. Wash buffer: 20 mM Tris–HCl pH 8.0, 500 mM NaCl, 1 mM EDTA, 1% Triton X-100, 10% glycerol, 2 mM DTT. Store at 4 °C.
15. High-salt wash buffer: 20 mM Tris–HCl pH 8.0, 1 M NaCl, 1 mM EDTA, 1% Triton X-100, 10% glycerol, 2 mM DTT. Store at 4 °C.
16. Elution buffer: 100 mM Tris–HCl pH 8.0, 150 mM NaCl, 0.2% Tween 20, 10% glycerol, 40 mM GSH. Prepare and store at 4 °C without GSH. Freshly add GSH to elution buffer on the day of the experiment. Be sure to adjust the pH to 8.0 after the addition of GSH.
17. Amicon Ultra centrifugal filter units MWCO 50 kDa (Merck).
18. Bicinchoninic acid (BCA) assay (Pierce).
19. SimplyBlue Safe Stain (Thermo Scientific).

2.2 32P-NAD Assay

1. 20× PARP buffer: 1 M Tris–HCl pH 8.0, 4 mM DTT, 80 mM $MgCl_2$. Store at −20 °C.
2. 500 μM β-NAD^+ (Sigma-Aldrich). Store at −80 °C.
3. GST-TIPARP: purified with GST protein purification protocol.
4. Target purified protein for hetero-ribosylation testing. For example, GST-$AHR_{430–848}$ (amino acids 430–848), purified using the GST protein purification protocol, purchased commercially or by other established protein purification protocols [12].
5. ^{32}P-NAD^+ (Perkin-Elmer).
6. 5× SDS sample buffer: 250 mM Tris–HCl, pH 6.8, 10% SDS, 50% (v/v) glycerol, 0.1% (w/v) bromophenol blue. Add DTT to a final concentration of 100 mM before use.
7. GelCode Blue Stain Reagent (Pierce).

2.3 Biotinylated-NAD+ Assay

1. 20× PARP buffer 1 M Tris-Base, pH 8.0, 4 mM DTT, 80 mM $MgCl_2$. Store at −20 °C.
2. Biotinylated-NAD^+ (500 μM; Trevigen).
3. GST-TIPARP: purified with GST protein purification protocol.
4. Target purified protein for hetero-ribosylation testing. For example, GST-AHR (amino acids 430–848) purified using the GST protein purification protocol, purchased commercially or by other established protein purification protocols [12].
5. 5× SDS sample buffer: 250 mM Tris–HCl, pH 6.8, 10% SDS, 50% (v/v) glycerol, 0.1% (w/v) bromophenol blue. Add DTT to a final concentration of 100 mM.
6. Horseradish peroxidase streptavidin (Maravai LifeSciences).

2.4 35S-Methionine Protein Shift Assay

1. TNT Coupled Reticulocyte Lysate System (Promega).
2. ^{35}S-methionine (Perkin-Elmer).
3. Activated DNA (Sigma-Aldrich).
4. β-NAD^{+} (Sigma-Aldrich).
5. 5× SDS sample buffer: 250 mM Tris–HCl, pH 6.8, 10% SDS, 50% (v/v) glycerol, 0.1% (w/v) bromophenol blue. Freshly add DTT to a final concentration of 100 mM.
6. KODAK™ ENLIGHTNING™ Rapid Autoradiography Enhancer (Perkin-Elmer).
7. Model 583 gel dryer (Bio-Rad).

2.5 Anti-PAN-ADPr Detection of ADP-Ribosylated Proteins in Cells

2.5.1 Transfection

1. Plasmid DNA is purified from DH5α bacteria culture by plasmid isolation kit.
2. Transfection reagent: Lipofectamine 2000 Transfection Reagent (Thermo Scientific) Opti-MEM™ Reduced Serum Media ((Thermo Scientific).
3. COS-1 cells (adherent fibroblast from *Cercopithecus aethiops* kidney).
4. Culture media: Dulbecco's Modified Eagle's Medium 4.5 g/L glucose with 10% fetal bovine serum, 1% penicillin/streptomycin.

2.5.2 Immunoprecipitation

1. Lysis and wash buffer: 500 mM NaCl, 1% NP40, 20 mM HEPES pH 7.4.
2. Protease inhibitor cocktail (Roche).
3. Diagenode Bioruptor Plus (Diagenode).
4. Antibodies: rabbit polyclonal anti-AHR (H-211; Santa Cruz), mouse monoclonal anti-GFP (3E6; Life Technologies).
5. Protein A beads or protein G dynabeads (Life Technologies): Protein A is used for polyclonal antibodies raised in rabbits, and protein G is used with monoclonal antibodies raised in mice.
6. 5× SDS sample buffer: 250 mM Tris–HCl, pH 6.8, 10% SDS, 50% (v/v) glycerol, 0.1% (w/v) bromophenol blue. Freshly add DTT to a final concentration of 100 mM.

2.5.3 SDS-Polyacrylamide Gel

1. Running buffer: 25 mM Tris base, 192 mM glycine, 0.1% SDS, and water up to 1 L.
2. Mini-PROTEAN Tetra cell system and Mini-Protean TGX Precast 10% and 4–15% acrylamide 10-well gels (Bio-Rad).
3. 5× SDS sample buffer: 250 mM Tris–HCl, pH 6.8, 10% SDS, 50% (v/v) glycerol, 0.1% (w/v) bromophenol blue. Add fresh DTT for a final concentration of 100 mM.

2.5.4 Immunoblotting

1. Polyvinylidene difluoride (PVDF) membrane.
2. Western blot transfer buffer: 25 mM Tris base, 192 mM glycine, 20% methanol.
3. Antibodies: mouse monoclonal anti-GFP (JL8; Clontech), rabbit polyclonal anti-AHR (H-211), anti-PAN ADP-ribose reagent (EMD Millipore).
4. 10× Tris-buffered saline (TBS): 1.5 M NaCl, 0.1 M Tris–HCl, pH 7.4.
5. TBS containing 0.1% Tween 20 (TBST).
6. Blocking buffer: 5% skim milk powder in TBST.
7. Mini-PROTEAN Tetra Blotting Module.

3 Methods

3.1 ADP-Ribosylation Assay Using GST-TIPARP and ^{32}P-NAD

The addition of labeled ADP from the catalysis of ^{32}P-NAD or biotinylated-NAD by TIPARP is an effective means to determine its in vitro catalytic activity and to identify its protein targets. However, a source of TIPARP protein, such as bacterial overexpressed, is required. We have found that overexpression of TIPARP in bacteria as a GST fusion rather than a 6× histidine fusion protein improved its solubility. Overexpressing GST-TIPARP at 16 °C and ensuring that the purification is done at 4 °C and keeping the purified protein on ice improves its stability and functionality. Unlike PARylation in which hundreds of ADP-ribose molecules can be added to the enzyme causing a large upward shift in its mobility in SDS-PAGE (PARP1), large upward shifts in enzyme migration are not observed by MARylating PARP family members. Therefore, the presence of a signal at the appropriate molecular weight after incubation of purified GST-TIPARP, target protein (AHR) in the presence of ^{32}P-NAD, or biotinylated-NAD indicates active enzyme or modified target protein (Fig. 1). Truncated and mutant variants of TIPARP can also be tested in this in vitro ADP-ribosylation assay.

3.1.1 Bacterial Culture and Protein Purification

1. Transform BL21 DE3 star competent cells with plasmid: Mix 10 ng of plasmid (pGEX-4T1-TIPARP) or (pGEX-4T1-$AHR_{430-848}$) with 50 μL of BL21 DE3 star bacteria (*see* **Note 1**), and incubate on ice for 30 min. Incubate the mixture for 30 s at 42 °C and place it back on ice for 1 min. Add 500 μL of SOC medium and incubate for 1 h at 37 °C at 250 rpm (*see* **Note 2**). Plate 100 μL of the transformed culture on LB plates containing ampicillin. Incubate at 37 °C overnight.
2. Pick a colony from the plate and grow it in 3 mL of LB-amp for 8 h at 37 °C and 250 rpm.

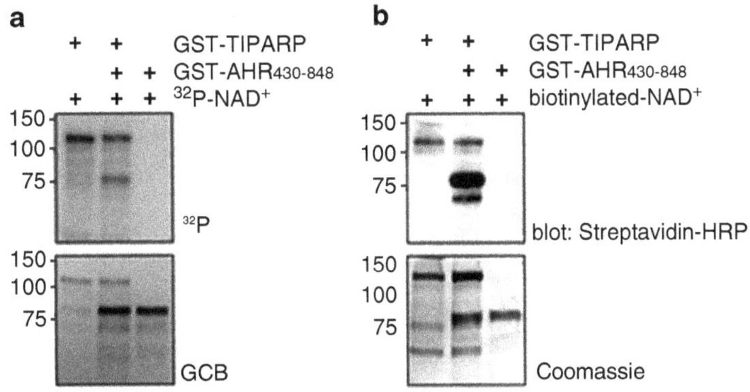

Fig. 1 Characterization of TIPARP mono-ADP-ribosyltransferase activity using ^{32}P-NAD$^+$ and biotinylated-NAD$^+$. (**a**) GST-tagged TIPARP protein (500 ng) was incubated with 2 μg of GST-AHR$_{430–848}$ and 2 μCi ^{32}P-NAD$^+$ for 20 min at room temperature. Mono-ADP-ribosylation of proteins was detected by autoradiography after SDS-PAGE and transfer to PVDF membrane. GelCode Blue (GCB) staining of the PVDF membrane prior to autoradiography was used to visualize protein loaded. (**b**) GST-tagged TIPARP protein (1 μg) was incubated with 1 μg of GST-AHR$_{430–848}$ and 25 μM biotinylated-NAD$^+$ for 30 min at room temperature. Fifteen percent of the reaction was separated by SDS-PAGE and transferred to a PVDF membrane. The membranes were blocked in 3% BSA and incubated with streptavidin-HRP before developing with ECL. The remaining 85% of the reaction was separated by SDS-PAGE and stained with Coomassie Blue to visualize the proteins. The data indicate that TIPARP mono-ADP-ribosylates itself and AHR

3. Transfer 500 μL of the previous culture to 50 mL LB-amp media in a sterile 250 mL Erlenmeyer flask, and incubate overnight at 37 °C and 250 rpm.
4. The following morning, add the appropriate amount of the overnight culture to each of six 2-L flasks containing 500 mL of LB-amp such that the optical density at 600 nm (OD_{600}) is 0.1. Incubate the flasks at 37 °C with 250 rpm. When the optical density reaches 0.4–0.6, place the flasks on ice in the cold room for at least 2 h. At the end of the day, add IPTG (final 0.5 mM) and incubate at 16 °C and 170 rpm overnight (*see* **Note 3**).
5. All of the following steps must be performed in the cold room to avoid the degradation of TIPARP.
6. Transfer the cultures into six centrifuges bottles, centrifuge at 5000 × *g* for 15 min at 4 °C.
7. During centrifugation step prepare 100 mL of lysis buffer.
8. After the centrifugation, discard the supernatant and save the pellet. To each pellet add 10 mL of lysis buffer and pipet to resuspend the pellet. Transfer 30 mL of lysate to each of two ultracentrifuges tubes. Incubate the tubes for 20 min on ice.

9. After incubation, sonicate the bacteria on ice with 40% of amplitude for 30 s followed by 30 s of cooling to avoid overheating of the protein. Repeat the sonication two more times. The following sonication conditions are based on a Branson Ultrasonics Sonifier™ S-450 Digital Ultrasonic Cell Disruptor/Homogenizer.
10. Seal the tubes tightly with parafilm, and rotate them slowly at 4 °C for 20 min.
11. Prepare GSH beads: 800 μL of 50% slurry is prepared for each ultracentrifuge tube.
12. Adjust the volumes of the tubes using lysis buffer to balance and ultracentrifuge for 20 min at 4 °C and 20,000 *g*.
13. Transfer each supernatant to a 50 mL conical tube. Add the beads to each tube, and incubate for 1 h by rotating slowly on wheel at 4 °C.
14. Centrifuge the tube at 1000 *g* for 5 min at 4 °C.
15. Aspirate the supernatant and remove as much as possible without disturbing the beads. Wash the beads in each tube in the following order:

    Wash 1: 50 mL of wash buffer and incubate for 5 min with rotation.

    Wash 2: 50 mL of was buffer and incubate for 5 min with rotation.

    Wash 3: 50 mL of high-salt lysis buffer and incubate for 5 min with rotation.

    Wash 4: 50 mL of wash buffer and incubate for 5 min with rotation.
16. After the last centrifugation, remove the supernatant leaving 10 mL of lysis buffer.
17. Resuspend the beads and transfer the 10 mL of bead slurry to a disposable 25 mL elution column with stopcock. Allow the flow through to drop into the discard tray.
18. Once all lysis buffer has passed through, add 5 mL of elution and incubate for 20 min in the cold room.
19. Place the elution column over a 50 mL conical tube, remove the plug, collect the eluate, and place on ice.
20. Transfer the eluate to a concentration column Amicon Ultra centrifugal filter units MWCO 50 kDa. Centrifuge at 1000 *g* at 4 °C for 15 min or until a volume of 300–400 μL is reached (*see* **Note 4**).
21. Add 5 mL of fresh elution buffer (without glutathione) to the concentration column and centrifuge as in **step 20**.
22. Mix the concentrated protein, transfer it from the concentration column to a fresh 1.5 mL tube, and place on ice.

23. Determine the protein concentration using a bicinchoninic acid (BCA) assay according to the manufacturer's instructions.
24. To analyze the concentration and purity load 10 µL of eluate and 0.5, 1, 1.5, and 2 µg of BSA. After the electrophoresis, stain the gel with SimplyBlue Safe Stain or Coomassie Blue.

3.1.2 ^{32}P-NAD Assay

The use of ^{32}P-NAD$^+$ must be done following radioisotope safety protocols in a safe location and with adequate personal protective equipment.

1. Set up the assay reaction as follows: 0.5 µg GST-TIPARP protein, 2 µg GST-AHR$_{430–848}$ protein, 1.5 µL 20× PARP buffer, 2 µCi ^{32}P-NAD$^+$, and water to a final volume of 30 µL.
2. After the addition of ^{32}P-NAD$^+$, mix the reaction thoroughly by pipetting. Allow the reaction to proceed at room temperature for 20 min.
3. Stop the reaction by adding 10 µL 5× SDS sample buffer and heating to 95 °C for 5 min.
4. Let the reaction mix cool down and collect the precipitate by centrifugation.
5. Load 15 µL of the reaction mix onto an SDS acrylamide gel, and run at 100 V for 90 min or until the dye front reaches the bottom of the gel.
6. Activate a PVDF membrane by soaking in an appropriate volume of 100% methanol for 30 s. Place membrane in cold transfer buffer until needed.
7. Transfer protein to the PVDF membrane at 100 V for 60 min at room temperature with an ice pack and on a stir plate with stirring.
8. After transfer, quickly wash the membrane with water for 1 min. Continue to hydrate the membrane by swirling briefly in water. Repeat this with additional 3–4 changes of water; hydration of membrane is crucial for the following staining.
9. Stain membrane with GelCode Blue (GCB) Stain Reagent for 20 min at room temperature.
10. Destain membrane with a solution of 50% methanol and 1% acetic acid, replacing the solution several times until protein bands become clear.
11. Let the membrane dry in a fume hood.
12. Expose membrane to film and develop. It is recommended to do several exposures (30 min, 1 h, 2 h). Older batches of ^{32}P-NAD$^+$ may be used, but longer exposure times may be required (overnight). Note that the half-life of ^{32}P is 14 days.

3.1.3 Biotin-NAD$^+$ ADP-Ribosylation Assay

1. Set up the assay reaction as follows: 1 μg GST-TIPARP protein, 1 μg GST-$AHR_{430–848}$ protein, 1.5 μL 20× PARP buffer, 1.5 μL biotinylated-NAD$^+$ of 500 μM, and water to a final volume of 30 μL.
2. Incubate the reaction at room temperature for 30 min. Stop the reaction by adding 10 μL 5× SDS sample buffer and heating to 95 °C for 5 min.
3. Load 5 μL of the reaction mix onto an SDS acrylamide gel and 25 mL of the sample on a duplicate gel that will be stained with Coomassie Blue after SDS-PAGE. Run gels at 100 V for 90 min or until the dye front reaches the bottom of the gel.
4. For the gel loaded with 5 μL of the reaction, transfer the proteins by western blotting onto a PVDF membrane as described above. For the gel loaded with 25 μL of the reaction stain with Coomassie Blue for 30 min followed by destaining with repeated changes of the destain solution (50% methanol and 10% acetic acid).
5. Block the membrane for 30 min in 3% bovine serum albumin (BSA) in TBST.
6. Incubate the membrane for 30 min with a 1:5000 dilution of horseradish peroxidase streptavidin in TBST.
7. After thoroughly washing the membrane with TBST, expose it to an enhanced chemiluminescence (ECL) substrate and detect the horseradish peroxidase enzyme activity either by exposing it to film or with an appropriate imaging system.

3.2 ^{35}S Methionine Protein Shift Assay

The ^{32}P-NAD$^+$ or biotin-NAD$^+$ ADP-ribosylation assays described above require purified protein, which can be challenging given the relative insolubility of TIPARP. An alternative approach is to use ^{35}S methionine in vitro translation from a plasmid backbone as a source of TIPARP. The ^{35}S methionine-labeled protein can be incubated with and without NAD$^+$. The incorporation of ADP-ribose causes a small shift or upward smear to indicate MARylation, which differs from large upward shift in mobility observed with PARylation (Fig. 2). A signal will be observed under all conditions, but a small shift in mobility will be observed in the presence of NAD$^+$. The added advantage of this approach for TIPARP is that no protein tag, such as GST, is required. The use of ^{35}S methionine must be done following radioisotope safety protocols in a safe location and with adequate personal protective equipment.

1. Prepare in vitro transcribed and translated full-length TIPARP, TIPARP H532A catalytic mutant protein from pcDNA3.1 or full-length PARP1 from pCMV-AC-PARP1 [9] using the TNT Coupled Reticulocyte Lysate System, and ^{35}S-methionine according to the manufacturer's instructions. We use 1 μg of DNA template.

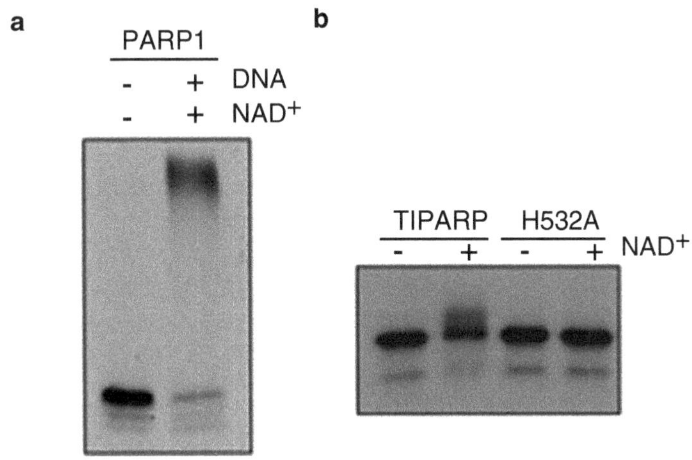

Fig. 2 Protein mobility shift assay using in vitro translated protein in the presence and absence of NAD+. Comparisons of modified and unmodified ^{35}S-labeled PARP1 and TIPARP. In vitro translated (**a**) PARP1 was incubated with activated DNA in the presence or absence of NAD+. (**b**) TIPARP or TIPARP catalytic point mutant (H532A) were incubated with or without 500 μM NAD+. Proteins were resolved SDS-PAGE and visualized by autoradiography. The arrow denotes shift in molecular weight of PARP1 due to the addition of poly-ADP-ribose. Note the slight shift in the mobility of TIPARP indicative of mono-ADP-ribosylation. No shift in mobility was observed for the TIPARP H532A catalytic mutant

2. Set up the ADP-ribosylation assay reaction as follows: 5% or 2.5 μL of ^{35}S-labeled TIPARP or PARP1 protein, 1.5 μL 20× PARP buffer, presence or absence of 1.5 μL NAD+ 500 μM. For PARP1, add 1 μg of activated DNA. Adjust the total volume to 30 μL and incubate for 30 min at 30 °C.
3. Reactions are stopped by the addition of 10 μL 1× SDS sample buffer and heating to 95 °C for 5 min.
4. Load 15 μL of the reaction mix onto 10% SDS-PAGE gel, and run at 100 V for 90 min or until the dye front reaches the bottom of the gel.
5. Incubate gel in KODAK™ ENLIGHTNING™ Rapid Autoradiography Enhancer for 30 min with gentle rocking.
6. Then incubate the gel in gel drying solution (10% (v/v) ethanol and 5% acetic acid) for 20 min with gentle rocking.
7. Dry gel at 80 °C for 2 h using a Model 583 gel dryer.
8. Visualize proteins by autoradiography. Exposure times will vary depending age of the ^{35}S. We start with 2–24 h.

3.3 Analysis of ADP-Ribosylation with Anti-PAN-ADPr Reagent

Characterizing TIPARP activity and identifying its cellular targets is critical to understanding its function. The use of anti-PAN-ADPr reagent, which is a macrodomain-rabbit IgG fusion protein that recognizes mono- and poly-ADP-ribose, represents an important tool for identifying ADP-ribosylation in intact cells or cell extracts. Since we have been unable to identify a suitable antibody to detect TIPARP, GFP-TIPARP is transfected into mammalian cells, such has COS1 followed by immunoprecipitation of cell extracts using anti-GFP antibody or antibody raised against protein modified by TIPARP (Fig. 3). This in cell ADP-ribosylation assays using the anti-PAN-ADPr reagent has a few advantages over in vitro ADP-ribosylation assay including: (1) TIPARP is overexpressed in mammalian cells rather than bacteria, (2) no additional NAD^+ above the cellular concentrations is added to the assay, and (3) the assay also takes into account subcellular localization differences between TIPARP and other proteins, which might influence the ability of TIPARP to modify those proteins in cells compared with in vitro ADP-ribosylation assays.

3.3.1 Transfection

1. Plate COS-1 cells (*see* **Note 5**) in 2 mL of Dulbecco's Modified Eagle's Medium (DMEM) high glucose per well of a 6-well plate at a seeding density of 1×10^5 cells/mL. Incubate cells for 24 h at 37 °C.

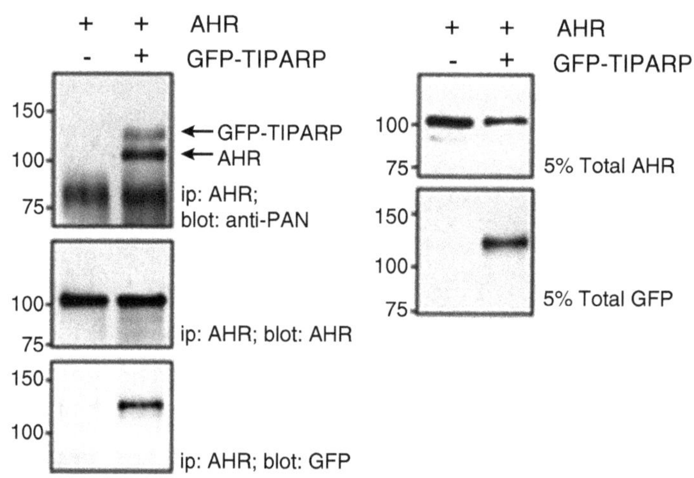

Fig. 3 TIPARP mono-ADP-ribosylates AHR using anti-PAN-ADPr reagent. COS-1 cells were transfected with full-length human AHR in the presence or absence of GFP-TIPARP. Cell extracts were prepared and immunoprecipitated with an anti-AHR antibody (H-211; Santa Cruz). Immunoprecipitated proteins were resolved by SDS-PAGE and electrophoretically transferred to PVDF membrane. After blocking, the membrane was probed with anti-PAN-ADPr reagent for 1 h at room temperature, washed, and probed with an anti-rabbit linked HRP antibody before developing with ECL. The membrane was stripped using Restore™ PLUS Western Blot Stripping Buffer (Thermo Scientific) and probed with the indicated antibodies. The 5% total input samples were also resolved by SDS-PAGE and probed with the indicated antibodies

2. On the following day, use the ration of 2 μL of Lipofectamine 2000 per 1 μg of DNA when preparing the transfection reactions. Prepare one microcentrifuge tube with appropriate amount of Lipofectamine 2000 in 50 μL Opti-MEM and another microcentrifuge tube with the desired amount DNA in 50 μL Opti-MEM. Mix the tube together and add dropwise onto cultured cells. Transfect 2 μg of GFP-TIPARP in the presence or absence of 1 μg of target protein (*see* **Note 6**).
3. Incubate cells for 24 h at 37 °C.

3.3.2 Immunoprecipitation

1. Remove DMEM medium from the wells and wash the cells with 2 mL of cold (4 °C) PBS.
2. Resuspend cells in 250 μL of lysis buffer and incubate on dry ice for 5 min.
3. Thaw the lysate at room temperature and sonicate (Diagenode Bioruptor Plus) using the following conditions: “low power” 4 cycles 15 s on, 30 s off.
4. Centrifuge the lysate for 10 min at 10,000 *g* at 4 °C, and separate the supernatant into two microfuge tubes, one with 50 μL (will be used as input) and one with 200 μL (will be used for immunoprecipitation). Keep the input on ice until Subheading 3.3.3, **step 1**. Proceed with the immunoprecipitation with the remaining lysate.
5. During the centrifugation step, prepare the beads: Aliquot 100 μL of the bead slurry to a 1.5 mL microcentrifuge tube. Centrifuge for few seconds to concentrate the beads at the bottom of the tube, and place the tube on a Dynabeads Magnetic Particle Concentrator. When the beads are concentrated on the side of the tube, aspirate the supernatant and wash the beads with 1 mL of lysis buffer. Repeat three times. After the last wash, resuspend the beads in 100 μL of lysis buffer.
6. Add 2 μg of antibody and 20 μL of beads in 200 μL of lysate. Incubate for 2 h at 4 °C with gentle agitation.
7. Pulse centrifuge the tube for a few seconds to concentrate the beads at the bottom of the tube. Place the tube on a Dynabeads Magnetic Particle Concentrator, and let the beads concentrate on the side of the tube. Aspirate the supernatant. Wash the beads with 1 mL of lysis buffer rotating for 5 min at 4 °C. Repeat the wash three times.
8. During the wash step, prepare the elution buffer.
9. After the last wash, remove the supernatant and add 45 μL of 1× SDS sample buffer. Vortex to mix the beads and 1× SDS sample buffer. Incubate for 5 min at 95 °C. Place the tube on a Dynabeads Magnetic Particle Concentrator and transfer the supernatant to a new microcentrifuge tube.

3.3.3 SDS-Polyacrylamide Gel Electrophoresis

1. Take 10 μL from the input (Subheading 3.3.2, **step 4**), add 2.5 μL 5× SDS sample buffer, and incubate 5 min at 95 °C
2. Unpack a precast 10% or 4–15% TGX minigel and wash the wells with water. Prepare a tank for electrophoresis with the gel and fill it with 1× running buffer. Load 20 μL of the immunoprecipitated sample (Subheading 3.3.2 **step 9**) and 10 μL of the input sample (Subheading 3.3.3 **step 1**). Run the gel at 100 V for 90 min for optimal separation of AHR (96 kDa) from GFP-TIPARP (predicted 106 kDa but runs approximately 130 kDa) (*see* **Note 7**).

3.3.4 Immunoblot

1. When the electrophoresis is complete, uncast the gel and incubate in cold transfer buffer for 5 min. Cut the PVDF membrane approximately the same size as the gel. Activate PVDF membrane by immersing it in 100% methanol for 30 s.
2. Incubate the membrane in transfer buffer for 5 min with agitation.
3. Make a blotting sandwich with one layer of foam, one layer of filter paper, the gel and the membrane, one layer of filer paper, and one layer of foam. Place it in a tank with an ice pack and fill with transfer buffer. Transfer for 100 V for 60 min with constant stirring.
4. After the transfer incubate the membrane overnight in blocking buffer at 4 °C.
5. The next day, pour off blocking buffer and incubate the membrane for 1 h in a dilution of 1:2000 of anti-PAN ADPr reagent in TBST at room temperature.
6. Wash the membrane three times for 10 min in TBST.
7. Incubate the membrane for 1 h in a 1:4000 dilution of anti-rabbit.
8. Wash the membrane three times for 10 min in TBST.
9. Air-dry the membrane, incubate enhanced chemiluminescence (ECL) substrate according to the manufacturer's instructions, and incubate for 5 min. Detect the horseradish peroxidase enzyme activity either by exposing it to film or with an appropriate imaging system.
10. The membrane can be stripped and blocked again before incubation with another antibody.

4 Notes

1. Although we have not done an exhaustive comparison among different bafcterial strains for protein overexpression, BL21 DE3 star cells give slight improved yields of soluble GST-TIPARP compared with BL21 DE3 and BL21 cells. However,

the overall yields of soluble GST-TIPARP are relatively low (approximately 0.20–0.4 mg/L) but sufficient to characterize its enzymatic function. Attempts to purify 6× histidine-tagged TIPARP using Ni-NTA Agarose were unsuccessful because of protein insolubility.

2. The incubator shaking settings depending on the unit. We use a New Brunswick Innova® 40 incubator for routine bacterial culturing for 1–3 L of culture and for incubation during plasmid transformation.
3. Because of the relatively low yield of soluble GST-TIPARP protein, we generally express the protein in a total 3 L of LB medium separated into 6 × 2 liter Erlenmeyer flasks. Six 2 L flasks fit well in an Innova® 40 incubator. If a larger expression of GST-TIPARP and more flasks or large flask are needed requiring a different incubator, then shaking speed of the incubator should be adjusted accordingly. From a starting culture of OD_{600} of 0.1 at a speed setting of 250 rpm, it takes about 2–3 h to reach an OD_{600} between 0.4 and 0.6. For the overnight incubation at 16 °C at 170 rpm, we use a Amerex Gyromax 737R.
4. Since GST-TIPARP has predicted molecular weight of 103 kDa, we concentrate the protein using an Amicon Ultra centrifugal filter unit with a MWCO 50 kDa. We have also purified various deletion mutants of TIPARP and other smaller GST fusion proteins. During purification of TIPARP deletion mutants, we choose the MWCO of the Amicon Ultra centrifugal filter units according to our needs.
5. For the immunoprecipitation assays, we transfect GFP-TIPARP [9] because this allows us to easily visualize transfection efficiency using a fluorescent microscope. We have also found that we can achieve higher levels of transfection with GFP-tagged protein compared with an equal transfection of FLAG or 3× FLAG-TIPARP. It is not clear if the large GFP tag stabilizes TIPARP protein, but in our hands GFP-TIPARP expresses at higher levels than untagged or FLAG-tagged TIPARP.
6. For the transient transfection assay, other easily transfected cell lines, such as HEK293 or cell line of interest could certainly be used. We routinely use COS-1 cells because of their high transfection efficiency and ability to express high levels of transfected TIPARP.
7. TIPARP has a predicted molecular weight of 76 kDa. However, we have noted that in vitro translated, GFP- and GST-tagged version of TIPARP (human and mouse) migrate approximately 20–25 kDa higher than their predicted molecular weights. This shift in migration is lost with the deletion of the first 1–199 amino acids of human TIPARP [9], suggesting that the relatively uncharacterized N-terminus of TIPARP affects its migration in SDS-PAGE.

References

1. Welsby I, Hutin D, Leo O (2012) Complex roles of members of the ADP-ribosyl transferase super family in immune defences: looking beyond PARP1. Biochem Pharmacol 84(1):11–20. https://doi.org/10.1016/j.bcp.2012.02.016
2. Feijs KL, Verheugd P, Luscher B (2013) Expanding functions of intracellular resident mono-ADP-ribosylation in cell physiology. FEBS J 280(15):3519–3529. https://doi.org/10.1111/febs.12315
3. Vyas S, Matic I, Uchima L, Rood J, Zaja R, Hay RT, Ahel I, Chang P (2014) Family-wide analysis of poly(ADP-ribose) polymerase activity. Nat Commun 5:4426. https://doi.org/10.1038/ncomms5426
4. Hottiger MO, Hassa PO, Luscher B, Schuler H, Koch-Nolte F (2010) Toward a unified nomenclature for mammalian ADP-ribosyltransferases. Trends Biochem Sci 35(4):208–219. https://doi.org/10.1016/j.tibs.2009.12.003
5. Krishnakumar R, Kraus WL (2010) The PARP side of the nucleus: molecular actions, physiological outcomes, and clinical targets. Mol Cell 39(1):8–24. https://doi.org/10.1016/j.molcel.2010.06.017
6. Vyas S, Chang P (2014) New PARP targets for cancer therapy. Nat Rev Cancer 14(7):502–509. https://doi.org/10.1038/nrc3748
7. Rosenthal F, Feijs KL, Frugier E, Bonalli M, Forst AH, Imhof R, Winkler HC, Fischer D, Caflisch A, Hassa PO, Luscher B, Hottiger MO (2013) Macrodomain-containing proteins are new mono-ADP-ribosylhydrolases. Nat Struct Mol Biol 20(4):502–507. https://doi.org/10.1038/nsmb.2521
8. Feijs KL, Kleine H, Braczynski A, Forst AH, Herzog N, Verheugd P, Linzen U, Kremmer E, Luscher B (2013) ARTD10 substrate identification on protein microarrays: regulation of GSK3beta by mono-ADP-ribosylation. Cell Commun Signal 11(1):5. https://doi.org/10.1186/1478-811X-11-5
9. MacPherson L, Tamblyn L, Rajendra S, Bralha F, McPherson JP, Matthews J (2013) 2,3,7,8-Tetrachlorodibenzo-p-dioxin poly(ADP-ribose) polymerase (TiPARP, ARTD14) is a mono-ADP-ribosyltransferase and repressor of aryl hydrocarbon receptor transactivation. Nucleic Acids Res 41(3):1604–1621. https://doi.org/10.1093/nar/gks1337
10. Ma Q, Baldwin KT, Renzelli AJ, McDaniel A, Dong L (2001) TCDD-inducible poly(ADP-ribose) polymerase: a novel response to 2,3,7,8-tetrachlorodibenzo-p-dioxin. Biochem Biophys Res Commun 289(2):499–506
11. Vyas S, Chesarone-Cataldo M, Todorova T, Huang YH, Chang P (2013) A systematic analysis of the PARP protein family identifies new functions critical for cell physiology. Nat Commun 4:2240. https://doi.org/10.1038/ncomms3240
12. Ahmed S, Bott D, Gomez A, Tamblyn L, Rasheed A, MacPherson L, Sugamori KS, Cho T, Yang Y, Grant DM, Cummins CL, Matthews J (2015) Loss of the mono-ADP-ribosyltransferase, TIPARP, increases sensitivity to dioxin-induced steatohepatitis and lethality. J Biol Chem 290(27):16824–16840. https://doi.org/10.1074/jbc.M115.660100
13. Roper SJ, Chrysanthou S, Senner CE, Sienerth A, Gnan S, Murray A, Masutani M, Latos P, Hemberger M (2014) ADP-ribosyltransferases Parp1 and Parp7 safeguard pluripotency of ES cells. Nucleic Acids Res 42(14):8914–8927. https://doi.org/10.1093/nar/gku591
14. Yamada T, Horimoto H, Kameyama T, Hayakawa S, Yamato H, Dazai M, Takada A, Kida H, Bott D, Zhou AC, Hutin D, Watts TH, Asaka M, Matthews J, Takaoka A (2016) Constitutive aryl hydrocarbon receptor signaling constrains type I interferon-mediated antiviral innate defense. Nat Immunol 17(6):687–694. https://doi.org/10.1038/ni.3422
15. Bolton EC, So AY, Chaivorapol C, Haqq CM, Li H, Yamamoto KR (2007) Cell- and gene-specific regulation of primary target genes by the androgen receptor. Genes Dev 21(16):2005–2017
16. Dave KA, Whelan F, Bindloss C, Furness SG, Chapman-Smith A, Whitelaw ML, Gorman JJ (2009) Sulfonation and phosphorylation of regions of the dioxin receptor susceptible to methionine modifications. Mol Cell Proteomics 8(4):706–719. https://doi.org/10.1074/mcp.M800459-MCP200
17. Chen WV, Delrow J, Corrin PD, Frazier JP, Soriano P (2004) Identification and validation of PDGF transcriptional targets by microarray-coupled gene-trap mutagenesis. Nat Genet 36(3):304–312
18. Atasheva S, Akhrymuk M, Frolova EI, Frolov I (2012) New PARP gene with an anti-alphavirus function. J Virol 86(15):8147–8160. https://doi.org/10.1128/JVI.00733-12

Chapter 9

Dictyostelium as a Model to Assess Site-Specific ADP-Ribosylation Events

Anna-Lena Kolb, Duen-Wei Hsu, Ana B. A. Wallis, Seiji Ura, Alina Rakhimova, Catherine J. Pears, and Nicholas D. Lakin

Abstract

The amoeba *Dictyostelium discoideum* is a single-cell organism that can undergo a simple developmental program, making it an excellent model to study the molecular mechanisms of cell motility, signal transduction, and cell-type differentiation. A variety of human genes that are absent or show limited conservation in other invertebrate models have been identified in this organism. This includes ADP-ribosyltransferases, also known as poly-ADP-ribose polymerases (PARPs), a family of proteins that catalyze the addition of single or poly-ADP-ribose moieties onto target proteins. The genetic tractability of *Dictyostelium* and its relatively simple genome structure makes it possible to disrupt PARP gene combinations, in addition to specific ADP-ribosylation sites at endogenous loci. Together, this makes *Dictyostelium* an attractive model to assess how ADP-ribosylation regulates a variety of cellular processes including DNA repair, transcription, and cell-type specification. Here we describe a range of techniques to study ADP-ribosylation in *Dictyostelium*, including analysis of ADP-ribosylation events in vitro and in vivo, in addition to approaches to assess the functional roles of this modification in vivo.

Key words ADP-ribosylation, PARP, ADP-ribosyltransferase, *Dictyostelium*, DNA repair

1 Introduction

Studying a variety of biological phenomena in relatively simple genetic model organisms has been instrumental in providing novel concepts and insights into how these processes work in humans. However, while ADP-ribosyltransferases (ARTs) that catalyze mono- and poly-ADP-ribosylation are present in a variety of eukaryotic model organisms, in many instances the complement of these enzymes is not as extensive as in humans [1, 2]. This makes it difficult to use these powerful experimental systems to unravel the complexities of how a variety of ARTs converge to regulate different biological processes. For example, DNA damage-responsive ARTs are absent in yeast, the most commonly used genetic model to study DNA repair, prohibiting the use of this experimental

Paul Chang (ed.), *ADP-ribosylation and NAD+ Utilizing Enzymes: Methods and Protocols*, Methods in Molecular Biology, vol. 1813, https://doi.org/10.1007/978-1-4939-8588-3_9,

system to assess how ADP-ribosylation regulates DNA repair and other processes.

Of the 17 human ARTs identified, PARP1, PARP2, PARP5a (Tankyrase 1), and PARP5b (Tankyrase 2) catalyze poly-ADP-ribose chains. All other ARTs catalyze mono-ADP-ribosylation, with the exception of PARP9 and PARP13, which are inactive [3]. Recently, we and others found that vertebrate DNA repair factors lost in certain invertebrate models are unusually conserved in the genetically tractable amoeba *Dictyostelium*, including ARTs that catalyze mono-ADP-ribosylation [4–12]. Similar to humans, two ARTs are required for tolerance of cells to DNA single-strand breaks (SSBs), while a third that catalyzes mono-ADP-ribosylation (Adprt1a) is required to promote repair of DNA double-strand breaks (DSBs) by NHEJ [10–13]. The mechanistic basis of how ARTs regulate the repair process is similarly conserved, with a variety of ADP-ribose interaction domains being required to promote the assembly of repair factors at DNA lesions [12, 14, 15]. This striking conservation of ART function in the *Dictyostelium* DNA damage response suggests that this organism will be a useful model to study how ADP-ribosylation regulates other cellular processes.

Dictyostelium is a soil-dwelling single-celled haploid eukaryote that feeds on bacteria. However, if the food source becomes depleted, cells aggregate to undergo a series of carefully programmed differentiation steps that generate a multicellular fruiting body of spores supported by a stalk of vacuolated cells [16]. This simple developmental program has resulted in *Dictyostelium* being a much-used model to study cell fate and differentiation, chemotaxis, cell motility, and signal transduction. This has led to the development of a number of techniques that makes this organism a powerful system to study processes regulated by ADP-ribosylation. For example, large numbers of cells can be grown easily in the laboratory, providing significant amounts of material for cell biology, biochemical, or proteomics analysis. The fully sequenced and annotated genome and associated resources (dictybase.org), in addition to the haploid nature of the *Dictyostelium* genome, make gene disruption and replacement strategies by targeted homologous recombination relatively straightforward. Additionally, libraries of *Dictyostelium* mutants can be generated easily to employ in genetic or chemical screens [17–21]. The relatively simple genome organization of *Dictyostelium* also makes it possible to genetically alter certain genes that are challenging to manipulate in vertebrates. For example, the high copy number arrays of histone genes in vertebrates make it difficult to disrupt specific sites in these genes to assess how histone ADP-ribosylation events regulate a variety of cellular processes in vivo. In contrast, *Dictyostelium* contain single-copy histone genes [22, 23], making this approach tractable [13].

Here we describe techniques to study ADP-ribosylation in *Dictyostelium*, including the analysis of ADP-ribosylation events in vitro and in vivo, in addition to gene disruption strategies to assess the functional roles of ARTs and ADP-ribosylation in vivo. Given the expertise of our laboratory, we concentrate on the use of these assays to assess the role of ARTs and histone ADP-ribosylation in DNA repair. However, given *Dictyostelium* is an established model to study a variety of pathways including gene expression, cell-type specification, and cell motility, these techniques can equally be applied to study how ADP-ribosylation regulates processes other than genome stability.

2 Materials

All solutions are made in double-distilled water (or otherwise purified water).

2.1 Buffers, Solutions, and General Materials

1. 10× Phosphate-*buffered* saline (PBS): 1.37 M NaCl, 27 mM kCl, 100 mM Na_2HPO_4, 18 mM KH_2PO_4, (pH 7.4).
2. 1× Phosphate-*buffered* saline with Tween20 (PBST): Dilute 10× PBS tenfold in water; add 0.1% Tween® 20 (pH 7.4).
3. 10× Tris-buffered saline (TBS): 24.8 mM Tris–HCl, 1.37 M NaCl, (pH 7.4).
4. 1× Tris-buffered saline with Tween20 (TBST): Dilute 10× TBS tenfold in water; add 0.1% Tween® 20 (pH 7.4).
5. Blasticidin S: 10 mg/mL stock, made up in water. Stored at −20 °C. Final concentration for selection is 10 μg/mL.
6. H50 buffer: 50 mM kCl, 20 mM Hepes, 10 mM NaCl, 5 mM $NaHCO_3$, 1 mM $NaH_2PO_4 \cdot H_2O$, 1 mM $MgSO_4.7H_2O$ (pH 7.0). Filter sterilize. For long-term storage, keep at −20 °C.
7. HL5: 5 g/L proteose peptone, 5 g/L thiotone E peptone, 10 g/L glucose, 5 g/L yeast extract, 0.35 g/L $Na_2HPO_4 \cdot 7H_2O$, 0.35 g/L KH_2PO_4, 0.05 g/L dihydrostreptomycin sulfate (pH 6.5).
8. KK2: 19 mM KH_2PO_4, 3.6 mM K_2HPO_4.
9. Mounting media containing 4′,6-diamidino-2-phenylindole (DAPI).
10. NEB-IF: 10 mM PIPES (pH 6.8), 300 mM sucrose, 3 mM $MgCl_2$, 20 mM NaCl, 0.5% Triton, use while cold, store at 4 °C, and use within 24 h.
11. NEB-HAE: 50 mM Tris pH 8.0, 10 mM NaCl, 3 mM $MgCl_2$, 3 mM $CaCl_2$, 0.5 M sorbitol, 0.6% Triton X 100, complete protease inhibitor cocktail (EDTA-free, Roche, two mini tablets per 10 mL), phosphatase inhibitor cocktail 2 and 3 (Sigma,100 μL from each cocktail for 10 mL of NLB), 10 μM benzamide, and 200 μM DEA.

12. Nuclear lysis buffer (NLB): 50 mM Hepes, 150 mM NaCl, 1 mM EDTA, phosphatase inhibitor cocktail 2 and 3 (Sigma, 100 μL from each cocktail for 10 mL of NLB), complete protease inhibitor tablets (EDTA-free, Roche, 2 mini tablets per 10 mL), 10 μM benzamide, 200 μM DEA.
13. Nuclear lysis buffer* (NLB*): NLB supplemented with 10 μM benzamide and 200 μM DEA.
14. SDS loading buffer: 25 mM Tris (pH 6.8), 10% glycerol, 2% SDS, 0.1% bromophenol blue, 100 mM dithiothreitol. Aliquot and store at −20 °C.
15. SM agar: 1% peptone, 56 mM glucose, 0.1% yeast extract, 16 mM KH_2PO_4, 5.5 mM $K_2HPO_4{\cdot}3H_2O$, 4 mM $MgSO_4$, 1.7% agar (autoclaved). Pour into Petri dishes to agar depth of about 1.5 cm. Keep the Petri dish lid off the plate while the agar sets to avoid the condensation.
16. BugBuster protein extraction reagent (EMD Millipore): recommended for gentle extraction of proteins from bacteria to avoid contamination with chromosomal DNA.
17. 1× ADP-ribosylation buffer: 50 mM Tris–HCl pH 8, 75 μM NAD^+, 2 mM $MgCl_2$.

2.2 Antibodies and ADP-Binding Reagents

1. Antibody solution for IF: Serum from the animal in which the secondary antibody was raised is diluted to 10% in 1× TBS. Prior to preparing the antibody solution, centrifuge the serum at full speed for 3 min to remove debris which appears as a pellet and surface scum. The antibody solution should be made with the serum appearing between the two debris fractions.
2. Primary antibody for Western blotting: Diluted either in 1× PBST or in 5% milk in 1× TBST for Western blotting. We have successfully used the following antibodies: α-γH2AX (Abcam, ab11174), immunofluorescence and Western blot (1:1000); α-poly(ADP-ribose) (PAR; polyclonal, Trevigen, 4336-BPC-100), immunofluorescence (1:300), Western blot (1:1000); α-H3 (Abcam, ab12079–100), Western blot (1:2000); α-Actin (sc-1615), Western blot (1:1000); HRP-conjugated streptavidin (Sigma, RPN1231), Western blot (1:4000); α-His (Sigma, **SAB1305538**), Western blot (1:2000).
3. ADP-binding reagents: Anti-ADP-ribose binding reagents are useful for the detection of ADP-ribosylated proteins on membranes or on fixed cells, similar to antibody-based Western blotting and immunofluorescence. Poly-ADP-ribose binding reagent (Millipore; MABE1031) and pan-ADP-ribose binding reagent (Millipore; MABE1016) are used 1:1000 diluted in 5% milk in 1× TBST.
4. Secondary antibodies and other detection agents for Western blotting: Antibodies conjugated to a fluorescent tag (e.g.,

TRITC or FITC) for immunofluorescence (diluted 1:80 in antibody solution) or to horseradish peroxidase for Western blotting (diluted 1:5000 in 1× PBST or in 5% milk in 1× TBST). HRP-conjugated streptavidin (Sigma; diluted 1:4000 in 1× PBST).

2.3 Cells and Cell Culture

1. *Dictyostelium discoideum* Ax2 cells (available from the *Dictyostelium* stock center via dictyBase.org) are grown either in shaking suspension or adhered to the surface of a plastic dish in HL5 medium at 22 °C. Genetically modified strains are grown in presence of blasticidin (5–10 μg/mL).
2. *Klebsiella aerogenes* (*Ka*) suspension: Grow lawn of *Ka* on SM agar plate. Scrape some bacteria using 200μl pipette tip into 1 mL of KK2, and vortex to remove clumps. Bacteria available from *Dictyostelium* stock center (dictyBase.org).

2.4 Plasmids

1. pLPBLP or equivalent plasmid containing blasticidin S resistance cassette: The pLPBLP plasmid can be obtained from the stock center accessed via dictyBase.org. This plasmid can be used to generate *Dictyostelium* disruption strains by insertion of DNA sequences of the gene of interest flanking the blasticidin S resistance cassette.
2. pJET 1.2: (CloneJET PCR Cloning Kit, Thermo Scientific) is used for subcloning.
3. pDXA-3C: Obtained from the stock center accessed via dictyBase.org. This plasmid is used as extrachromosomal expression vector under control of the actin 15 promoter, carrying a C-terminal c-Myc-tag and a neomycin resistance cassette [24, 25].
4. pREP: A plasmid for co-transformation with pDXA-3C expression vector. The plasmid can be obtained from the stock center accessed via dictyBase.org.

2.5 Equipment

1. Biorad 0.1 cm gene pulser cuvette (product code: 165-2089).
2. Biorad Gene Pulser Xcell electroporator.
3. Vacuum-Gel dryer.
4. Parafilm.
5. Phosphorimager.
6. Microscope: 1×71 Olympus microscope at ×100 magnification with an Olympus lens and immersion oil (Lenzol): C10600-10B-H Hamamatsu Photonics camera with HCImage Acquisition (Hamamatsu Photonics) image software or equivalent.
7. PCR machine.
8. Western blot developer machine.
9. Western blotting: Biorad Mini-PROTEAN® 3 Cell system or equivalent.
10. X-ray films.

3 Methods

3.1 ADP-Ribosylation Assays In Vitro

Analysis of ADP-ribosylation in vitro is able to define the type of modification catalyzed by a given ART (e.g., poly- or mono-ADP-ribosylation), in addition to identifying potential sites on substrates modified by these enzymes [26]. Findings from these approaches can then guide the analysis of ADP-ribosylation events in vivo. While these approaches have been developed during the study of human ARTs, they are transferable to the analysis of ARTs from other organisms, including *Dictyostelium* [13]. Given ARTs ADP-ribosylate themselves, the type of modification catalyzed can be identified by assessing the auto-modification status of ARTs following incubation with NAD+, in addition to sheared DNA that is a requirement for activation of DNA damage-responsive ARTs [3]. To identify potential substrates, a similar workflow is employed, with the exception that reactions additionally include the putative substrate being analyzed.

3.1.1 ART Auto-ADP-Ribosylation Assays

These assays involve incubating a purified recombinant ART in a reaction buffer containing NAD^+ and sheared DNA fragments containing DNA breaks that activate the activity of the enzyme. Inclusion of biotin-conjugated or radioactive NAD^+ allows detection of the auto-modified ART following SDS-PAGE and Western blotting or autoradiography, respectively.

Detection of ART Auto-ribosylation Using Streptavidin Conjugated NAD^+

1. Recombinant His-tagged ART is expressed and purified from bacteria using standard protocols. In order to minimize activation of ART by bacterial DNA released by lysis procedures such as sonication, we recommend a gentler lysis procedure using BugBuster.
2. Set up a 20 μL reaction on ice in 1× ADP-ribosylation buffer including 25 μM biotinylated NAD^+ either with or without 5 μg/mL of sheared salmon sperm DNA to activate the enzymatic activity of the ART. Add a final concentration of 1 μM recombinant ART to start the reaction (*see* **Note 1**).
3. Incubate the mixture at room temperature for 30 min.
4. Quench the reaction by adding 20 μL 2× SDS loading buffer, and boil the mixture for 5 min (*see* **Note 2**).
5. Resolve the samples by SDS-PAGE using gels with the appropriate concentration of acrylamide depending on the molecular weight of the ART. For ARTs between 80 and 100 kDa, an 8% gel is used. Transfer the proteins from the gel onto a PVDF membrane using standard Western blotting procedures.
6. Block the membrane within 1× PBST containing 1% BSA for 2 h at room temperature.

7. Wash the membrane three times for 10 min with 1× PBST.
8. Incubate the membrane with a 1:4000 dilution of HRP-conjugated streptavidin in 1× PBST.
9. Wash the membrane three times for 10 min with 1× PBST, and detect ADP-ribosylated proteins by ECL.

Detection of ART Auto-ribosylation Using Radioactive NAD^+

This protocol is essentially the same as described in Subheading 3.1.1 with the exception that streptavidin-conjugated NAD^+ is substituted for radioactive NAD^+ and detection of ADP-ribosylated proteins is detected by autoradiography. All stated concentrations are used as final concentration in the reaction.

1. Set up a 20 μL reaction on ice in 1× ADP-ribosylation buffer including 100 nM ^{32}P-NAD^+ either with or without 5 μg/mL of sheared salmon sperm DNA to activate the enzymatic activity of the ART. Add 1 μM of recombinant ART to the mix to start the reaction.
2. Incubate the mixture at room temperature for 30 min.
3. Quench the reaction by adding an equal volume of 2× SDS loading buffer, and incubate at 70 °C for 10 min. This condition of protein denaturation was used to minimize the risk of spilling the radioactive probe due to overheating. Alternatively, insert a very small hole in the lid of the Eppendorf tube containing the radioactive sample using a needle prior to adding 2× SDS loading buffer, and boil it for 5 min with a lead pig above the tube to shield the radioactivity.
4. Centrifuge the sample, and subject the proteins to SDS-PAGE using gels with the appropriate concentration of acrylamide depending on the molecular weight of the ART (*see* Subheading 3.1.1).
5. Place the gel on Whatman paper, and cover with cling film, smoothing out as many wrinkles as possible. Vacuum dry the gel for 1 h at 80 °C, or longer if required, and detect the auto-ribosylated proteins by autoradiography (*see* **Note 3**).

3.1.2 ART-Mediated ADP-Ribosylation of Substrates In Vitro

Investigation of ADP-ribosylation of substrates is performed and processed similar to detecting auto-ribosylation described above, with the exception that the substrate being tested is added to the reaction mixture prepared in **step 1** of the protocols described in Subheading 3.1.1. Inclusion of biotin-conjugated or radioactive NAD^+ allows detection of the modified substrate following SDS-PAGE and Western blotting or autoradiography, respectively. The amount of substrate included in the reactions can vary and needs to be determined experimentally, although we have found that 100 ng of protein is usually sufficient. The resolution of SDS-PAGE needs to be adjusted by altering the concentration of acrylamide in gels depending on the molecular weight of the substrate. The use of recombinant proteins mutated at potential ADP-ribosylation sites can be compared with wild-type proteins in these assays to assess whether a specific site is ADP-ribosylated by a given ART.

3.1.3 Differentiate Between Mono- and Poly-ADP-Ribosylation In Vitro

While poly-ADP-ribosylation was the original modification identified to be catalyzed by ARTs, more recently it has become apparent that the majority of these enzymes catalyze mono-ADP-ribosylation [3, 27]. The ability of a given ART to catalyze mono- or poly-ADP-ribosylation, either *in cis* or *in trans*, can be defined by assessing whether the modification is removed by enzymes that degrade poly-ADP-ribose chains (PARG) [28, 29] or mono-ADP-ribose moieties (MacroD1) [30–32]. An alternative technique is to assess whether the modified substrate is recognized by ADP-ribose interaction domains that specifically recognize poly-ADP-ribose chains or mono- and poly-(ADP-ribose) structures (Pan-ADP-ribose).

Differential Degradation of ADP-Ribose Modifications Using MacroD1 and PARG

This technique is very similar to those described in Subheading 3.1.1, with the exception that the ADP-ribosylation is terminated prior to addition of enzymes that remove ADP-ribose modifications. Enzymes that remove either mono-ADP-ribose (MacroD1) or poly-ADP-ribose (PARG) are then added to the reaction prior to resolution and detection of proteins by SDS-PAGE and Western blotting. The removal of the modification by either MacroD1 or PARG indicates whether the protein is MARylated or PARylated, respectively.

1. Set up a 20 μL reaction on ice in 1× ADP-ribosylation buffer including 25 μM biotinylated NAD^+ either with or without 5 μg/mL of sheared salmon sperm DNA to activate the enzymatic activity of the ART. Add the substrate of interest (if being analyzed) followed by recombinant ART (1 μM) to start the reaction.
2. Incubate the mixture at room temperature for 30 min.
3. Terminate the ADP-ribosylation reaction by adding 10 mM of the ART inhibitor 3-aminobenzamide.
4. Add 0.4 μM MacroD1 protein or 9 nM PARG enzyme to separate reactions, and incubate at 37 °C for 30 min.
5. Quench the reaction by adding an equal volume of 2× SDS loading buffer, and boil the mixture for 5 min.
6. Assess the ADP-ribosylation status by Western blot as described in Subheading 3.1.1.

Differential Detection of Modifications Using ADP-Ribose Recognition Domains

In this assay, ADP-ribosylation assays are performed as in Subheading 3.1.1, with the exception that only unlabeled NAD^+ is included in reactions. Following SDS-PAGE, ADP-ribosylated proteins are detected by employing reagents that bind either poly-ADP-ribose or pan-ADP-ribose (i.e., both mono- and poly-(ADP-ribose) structures). Recognition of the substrate by both pan- and poly-ADP-ribose binding reagents indicates poly-ADP-ribosylation, while recognition by exclusively pan-ADP-ribose binding reagent indicates mono-ADP-ribosylation.

1. Set up a 20 μL reaction on ice in 1× ADP-ribosylation buffer including an additional 25 μM NAD^+ either with or without 5 μg/mL of sheared salmon sperm DNA to activate the enzymatic activity of ART. Add the substrate (if being analyzed) followed by 1 μM recombinant ART to start the reaction.
2. Incubate the mixture at room temperature for 30 min.
3. Quench the reaction by adding an equal volume of 2× SDS loading buffer, and boil the mixture for 5 min.
4. Separate the sample using SDS-PAGE. The percentage of acrylamide in the gel depends on the molecular weight of the substrate and/or ART being analyzed (*see* Subheading 3.1.1).
5. Block the membrane with 5% milk in 1× TBST for 1 h at room temperature.
6. Incubate the membrane with poly-ADP-ribose binding reagent (Millipore; MABE1031), which recognizes poly-ADP-ribose chains, or pan-ADP-ribose binding reagent (Millipore; MABE1016), which binds to mono- and poly-ADP-ribose chains, in 5% milk in 1× TBST for 1 h at room temperature (*see* **Note 4**).
7. Wash the membrane three times for 10 min with 1× TBST.
8. Incubate the membrane with anti-rabbit secondary antibody in 5% milk in 1× TBST for 1 h at room temperature.
9. Wash the membrane three times for 10 min each with 1× TBST, and detect ADP-ribosylated protein by ECL.

3.2 ADP-Ribosylation Detection In Vivo

While the analysis of ADP-ribosylation events in vitro can inform on potential targets for a given ART, in addition to the sites on the protein modified, it is important to assess whether these events similarly occur in vivo. Reagents that recognize proteins only when modified with poly-ADP-ribose chains or mono-ADP-ribose moieties allow the detection of ADP-ribosylation events in vivo either directly by immunofluorescence microscopy or by Western blotting following fractionation of cellular compartments or purification of proteins. Given our interest in DNA damage-responsive ARTs, we describe procedures to assess ADP-ribosylation events associated with DNA damage using microscopy to assess nuclear ADP-ribosylation or biochemical fraction of chromatin and histones from cells. However, these techniques are easily transferable to other cellular fractionation and protein purification procedures, broadening the use of these approaches to other biological contexts.

3.2.1 Immunofluorescence

Immunofluorescence is a commonly used technique to assess the cellular localization of proteins. It is a particularly powerful technique to assess the accumulation of DNA repair proteins into punctate nuclear foci following DNA damage, an observation that reflects the accumulation of DNA repair factors at DNA damage sites. This technique works particularly well when employing reagents that recognize posttranslational modifications, including reagents that recognize ADP-ribosylated proteins (Fig. 1). Cells are left untreated or exposed to the appropriate DNA-damaging agent. Prior to fixing cells, a mild detergent extraction is employed to reduce the background signal, revealing extraction-resistant signals that represent modifications at sites of DNA damage. ADP-ribosylation is subsequently detected by immunofluorescence using either ADP-ribosylation specific antibodies or ADP-ribose interaction domains that specifically recognize poly-ADP-ribose chains or mono- and poly-(ADP-ribose) structures (Pan-ADP-ribose).

Preparation of Coverslips

1. Dilute exponentially growing *Dictyostelium* cells to 1×10^6 cells/mL in HL5, and add 1 mL of cells to each well of a 24-well plate, with each well containing a clean, uncoated coverslip. Allow cells to settle on coverslips for 1 h, followed by washing the coverslips gently with 1 mL HL5 to remove nonadherent cells.
2. Incubate the cells in the 24-well plate with a DNA-damaging agent. The optimum concentration and incubation time to induce ADP-ribosylation differs between DNA-damaging agents and needs to be experimentally determined. A guide to DNA-damaging agents employed in *Dictyostelium* is provided in Table 1.
3. Place the coverslip cell side up on Parafilm, and incubate it for 5 min with NEB-IF (do not let the coverslip dry out at any stage, and add and remove liquid with care). For a 10 mm diameter coverslip, 40–60 μL liquid is sufficient (*see* **Note 5**).
4. Transfer the coverslip into a fresh well of a 24-well plate containing 1 mL of 1× TBS.
5. For the fixation of cells, remove the 1× TBS and add 1 mL 70% ethanol to the 24-well plate, and incubate it for 5 min, followed by rinsing with 100% methanol.
6. Gently rinse the coverslip three times with 1 mL 1× TBS in the 24-well plate. At this stage, coverslips can be stored for up to 4 days at 4 °C in 1 mL of 1× TBS and processed later.

Immuno-detection

1. Transfer coverslips cell side up to fresh Parafilm, and incubate with blocking solution for 1 h at room temperature.
2. Gently replace the blocking solution with antibody solution containing the appropriate ADP-ribose detection reagent. Incubate the coverslip for 1 h at room temperature.

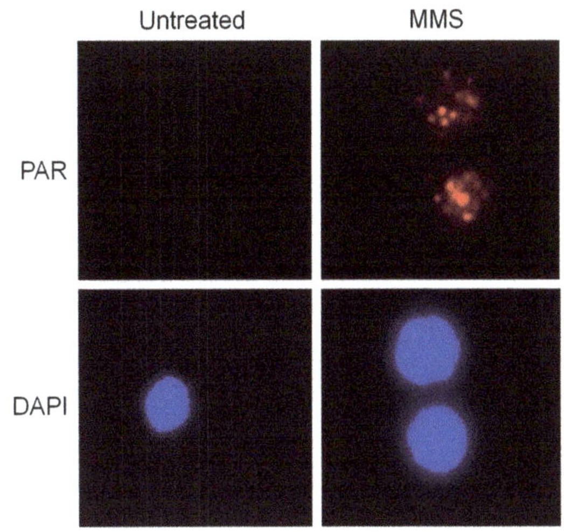

Fig. 1 Visualization of nuclear ADP-ribosylation by immunofluorescence microscopy. Ax2 cells were left untreated or exposed to 1 mM MMS and processed for immunofluorescence microscopy as described in protocol 3.2.1. ADP-ribosylation was detected using an antibody that recognizes proteins that are modified with poly-ADP-ribose chains

3. Transfer coverslips to a fresh 24-well plate, and gently rinse three times in 1 mL of 1× TBS.
4. Transfer coverslips, cell side up, to fresh Parafilm, and incubate with antibody solution containing secondary antibody for 1 h at room temperature in the dark.
5. Repeat washing as in **step 3**.
6. Transfer the coverslips, cell side down, onto a glass slide containing a small volume of mounting media with DAPI, and seal the edges of the coverslip with clear nail polish. Allow nail polish to dry for around 30 min before microscopy analysis. Coverslips can be stored at 4 °C in the dark, but the signal will fade with time.
7. Approximately 200 cells are counted per condition in three independent experiments. Cells that show staining over the untreated control are considered as "positive" and may be categorized based on the level of staining. The percentage of "positive" nuclei is determined as proportion of the total number of counted nuclei (visualized by DAPI).

3.2.2 Chromatin Extraction

The ADP-ribosylation of chromatin-associated proteins can also be monitored by Western blot. In this protocol, cells are left untreated or exposed to an appropriate DNA-damaging agent. A relatively straightforward cellular fractionation procedure is employed to isolate cytosol, nucleosol, and chromatin fractions from cells

Table 1
Agents used to induce different types of DNA damage in Dictyostelium

Agent	Predominant type of DNA damage	Response	Reference
H_2O_2[a]	Single-strand breaks	Sensitivity Gene induction ADP-ribosylation foci formation	[11, 12, 45]
MMS	Alkylated bases leading to single-strand breaks via base excision repair[a]	Gene induction Sensitivity ADP-ribosylation foci formation	[11, 12, 46, 47]
Ionizing radiation	Single- and double-strand breaks	Sensitivity	[48, 49]
Bleomycin	Double-strand breaks	Gene induction Sensitivity H2AX phosphorylation Cell cycle arrest	[6, 7, 48, 50, 51]
Phleomycin	Double-strand breaks	Sensitivity H2AX phosphorylation ADP-ribosylation foci formation	[12, 52]
Cisplatin	Interstrand cross-links[a]	Sensitivity Gene induction Resistance ADP-ribosylation	[8, 15, 19, 20]
UV radiation	Thymidine dimers[a]	Survival Gene induction Cell cycle arrest Spore germination	[45, 46, 49, 53, 54]

This is far from an exhaustive list
[a]High doses of these agents will also lead to generation of double strand

(Fig. 2). Analysis of chromatin fractions by SDS-PAGE and Western blotting using reagents that detect ADP-ribosylation is able to detect ADP-ribosylated proteins in chromatin.

The following protocol is modified from [12]. It is important to note that poly-ADP-ribosylation is a highly dynamic posttranslational modification that can be removed by PARG. Additionally, DNA fragmentation during the extraction procedure can result in activation of DNA damage-responsive ARTs. Therefore, an important addition to these procedures is the inclusion of ART and PARG inhibitors in extraction buffers to prevent ectopic activation of ARTs and removal of PAR chains during the fractionation procedure.

1. Collect 1.5×10^7 exponentially growing cells, spin them down at $1500 \times g$ for 3 min, and resuspend in HL5 at 5×10^6 cells/mL.
2. Treat cells with a DNA-damaging agent while gently agitating. The optimum concentration and incubation time differs between DNA-damaging agents and needs to be experimentally determined (*see* Table 1).
3. Following the treatment, collect the cells by spinning at $1500 \times g$ for 3 min, and wash them with 1 mL of ice-cold KK2.
4. Gently resuspend pellet in ice-cold NLB* supplemented with 0.1% Triton X-100 to a final density of 5×10^6 cells/mL, and incubate for 15 min at 4 °C.
5. Centrifuge at $14{,}000 \times g$ for 3 min at 4 °C to gain a pellet, P1, and supernatant fraction, S1. S1 can be discarded or retained for subsequent analysis.
6. Repeat **steps 4** and **5** with pellet P1.
7. Gently resuspend pellet P1 in the same volume of NLB* as used in **step 4** supplemented with 200 μg/mL RNaseA, and incubate while rotating at room temperature for 30 min.
8. Centrifuge at $14{,}000 \times g$ for 3 min at 4 °C to gain a pellet, P2, and supernatant fraction, S2. Pellet P2 is the detergent-insoluble fraction and includes chromatin and its associated proteins (*see* **Note 6**).

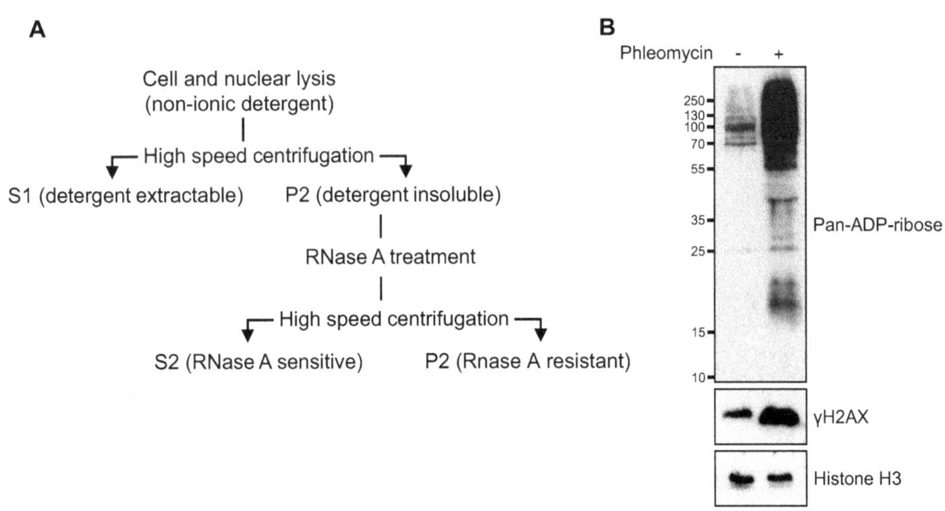

Fig. 2 Detection of ADP-ribosylated proteins in chromatin fractions prepared from cells. Schematic of the chromatin fractionation protocol is presented in (**a**). Fraction P2 which is both detergent and RNaseA resistant, containing chromatin and its associated proteins. ADP-ribosylated proteins can subsequently be detected by SDS-PAGE and Western blot analysis using reagents that detect mono- and poly-ADP-ribosylated targets (**b**). This analysis can also be performed using reagents that detect other types of ADP-ribosylation events

9. Resuspend P2 in 1× SDS loading buffer and analyze fractions by Western blot. When 1.5×10^7 cells were used, P2 can be resuspended in 80 μL of SDS loading buffer, and typically 10 μL were used for Western blot analysis. However, the volume of SDS loading buffer can vary according to the protein of interest and the ease of detection by Western blot. A control Western blot is required to confirm equal loading of chromatin, which can be done using an antibody against a histone protein (e.g., histone H3) or actin. Instead of primary antibodies, α-poly-ADP-ribose binding reagent or α-pan-ADP-ribose binding reagent can be used to detect either mono- and poly- or only mono-ADP-ribosylation. Preparation and processing is similar to a conventional Western blotting when using ADP-ribose binding reagents and is described in Subsection 3.1.3.

3.2.3 Histone Acid Extraction

Histones are an important component of chromatin modified by ARTs [33]. While ADP-ribosylated histones will co-fractionate with chromatin using the techniques described in Subheading 3.2.2, it is also possible to specifically analyze the ADP-ribosylation status of these proteins using an optimized acid-extraction protocol developed to enrich *Dictyostelium* histones from vegetative cells [13, 22]. Essentially, following exposure of cells to DNA damage, chromatin is isolated from cells using a protocol similar to that described in Subheading 3.2.2. However, these samples are subsequently subjected to an acid-extraction step that extracts histone proteins from the chromatin, resulting in a histone-enriched sample. Following resolution of histone extracts by SDS-PAGE, ADP-ribosylation events can be analyzed by Western blot (Fig. 3).

1. Collect 2.1×10^8 exponentially growing cells and centrifuge them at 1500 × *g* for 3 min. Resuspend cells in HL5 at a concentration of 1×10^8–5×10^7 cells/mL.
2. Treat cells with the DNA-damaging agent while gently agitating.
3. Following the treatment, collect the cells by spinning at 1500 × *g* for 3 min. Resuspend the cell in 2 mL ice-cold KK2 and repeat the centrifugation step.
4. Gently resuspend the pellet in NEB-HAE, containing ART and PARG inhibitors, to a final density of 1×10^8 cells/mL. Incubate with rotation for 15 min at 4 °C to promote hypotonic swelling and cell lysis by mechanical shearing.
5. Centrifuge at 2300 × *g* for 5 min to give rise to a pellet containing the nuclei, P1, and a supernatant fraction, S1. S1 can be discarded or retained for subsequent analysis.
6. Gently resuspend P1 in the same volume of NEB-HAE containing 4 M urea and 2% β-mercaptoethanol as used in **step 4,** and incubate while rotating at 4 °C for 15 min.

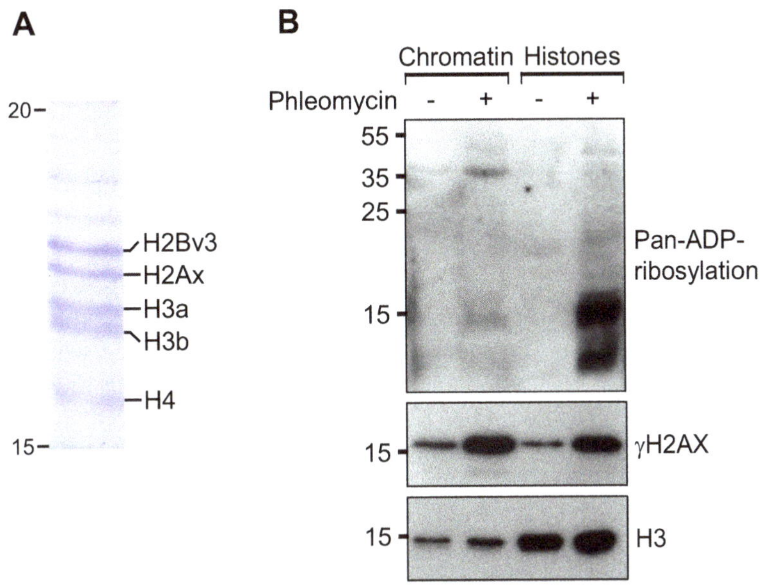

Fig. 3 Detection of ADP-ribosylated proteins in histone-enriched fractions prepared from cells. (**a**) Fractionation of histones from *Dictyostelium* cells using an optimized acid-extraction protocol [13, 22]. Samples were subjected to SDS-PAGE and analyzed by staining with Coomassie Brilliant Blue. (**b**) *Dictyostelium* cells were left untreated or exposed to 300 μg/mL phleomycin for 1 h. Chromatin or histone extracts were prepared as described in Subheadings 3.1.3 and 3.1.3.1, respectively. Samples were subjected to SDS-PAGE and Western blotting using the indicated antibodies

7. Repeat **step 5**.
8. Gently resuspend P1 in 0.4 M HCl at a density of 5×10^8 cell/mL, and incubate on a rotating wheel overnight at 4 °C.
9. Harvest acid-extracted histones by centrifugation at 16,000 × *g* for 15 min at 4 °C which gives rise to a supernatant, S2, containing the histones and a pellet, P2, containing cell debris.
10. Precipitate the histones by adding 6.5 volumes of acetone to S2, and invert the tube several times. Incubate the sample at −20 °C for 2 h.
11. Centrifuge at 16,000 × *g* for 15 min at 4 °C to give rise to a pellet, P3, containing precipitated histones and a supernatant S3.
12. Carefully remove the supernatant and wash P3 twice with ice-cold acetone without disturbing the pellet, and centrifuge as described in **step 11**. The acetone is used to remove the acid from the solution without dissolving the histone pellet.
13. Carefully remove the supernatant and air dry the pellet in the fume hood for 20 min.
14. Resuspend the pellet in 8 M urea supplemented with 5% β-mercaptoethanol.

15. Samples can be diluted with SDS loading buffer and the PARylation status of acid-extracted histones analyzed by Western blotting using ADP-ribose binding reagents as described in Subheading 3.1.3.

3.3 Functional Analysis of ADP-Ribosylation Events

The ability to perform gene disruption and replacement strategies using targeted HR in *Dictyostelium* make this a powerful system to assess the functional consequences of disrupting specific ADP-ribosylation events. Generating gene knockouts is relatively straightforward in *Dictyostelium*, and we have made strains disrupted in specific ARTs to define which enzymes are required to mediate ADP-ribosylation events in response to genotoxic stress [11, 12, 15]. Methods to generate *Dictyostelium* gene disruption strains have been described in detail elsewhere [34, 35]. Therefore, in this section we focus on techniques to define ADP-ribosylation sites in vivo, including knocking-in mutations at ADP-ribosylation sites using gene replacement technologies.

3.3.1 Expression and Purification of Recombinant Proteins

Putative ADP-ribosylation sites on target proteins can be identified in vitro (Subheading 3.1.2), and recent advances in mass spectrometry have begun to define the ADP-ribosylome [36–43]. In both instances, it is important to verify ADP-ribosylation sites by mutating the specific amino acids modified on the protein to confirm this abrogates ADP-ribosylation in vivo. One way this can be achieved is by expressing recombinant proteins mutated a putative ADP-ribosylation sites and comparing their modification status relative to wild-type controls. The inclusion of an epitope tag onto the recombinant proteins can facilitate this analysis either by inducing a shift in the molecular weight of the protein to distinguish it from its endogenous counterpart during Western blot analysis or to allow immunopurification of the recombinant proteins (e.g., *see* [13]). Given inclusion of an epitope tag may interfere with the modification status of a protein, depending on its location, it is recommended that this analysis is performed with both C-terminal and N-terminal tagged proteins. Performing this analysis is a relatively straightforward way to confirm ADP-ribosylation sites prior to committing to knocking in mutations at the relevant endogenous loci.

This methodology first requires cloning of the gene of interest, either wild type or mutated at putative ADP-ribosylation sites, into the pDXA-3C expression vector downstream of a constitutive actin15 promoter [24, 25]. The vector also contains the *Dictyostelium* Ddp2 origin of replication (ORI), conferring the *cis* functions for extrachromosomal replication. Expression of pDXA-3C vector requires the co-transfection with the helper plasmid pREP conferring the Ddp2 ORI *trans* replication function [24, 25]. pDXA-3C also contains a neomycin resistance gene. Therefore, following transfection of cells with pDXA-3C and pREP, incubation in G418 will select for cells that retain the vectors and thus express the gene of interest. A comparison of the ADP-ribosylation status of the recombinant wild-type and mutant proteins can be monitored by employing methods described in Subheadings 3.2.3 and 3.2.2.

1. The pDXA-3C vector contains an ATG start codon upstream of the multiple cloning site. The cloning strategy should be designed to insert the gene of interest and associated epitope tag in frame with this codon. Amplify the entire gene of interest by PCR from AX2 genomic DNA. To facilitate cloning into the pDXA-3C expression vector, the forward primer contains an appropriate restriction enzyme, as, for example, KpnI, followed by a Myc-tag on its 5'end that is in the same reading frame as the gen of interest. The reverse primer should contain another appropriate restriction enzyme site to allow insertion of the gene into pDXA3C. The inclusion of the Myc-tag ensures detecting via Western blotting using a Myc-antibody.
2. Subclone the amplified DNA fragment first into pJET 1.2, and confirm insertion of the PCR product by restriction enzyme analysis and accuracy of the PCR amplification by DNA sequencing using the appropriate primer (pJET cloning kit contains primers flanking the insertion site that can be used for sequencing).
3. Introduce mutations in the potential ADP-ribosylation sites using the QuickChange site-directed mutagenesis kit with primers containing the desired mutation in the center, each complementary to opposite strands of the vector. Confirm the introduction of the relevant mutation by DNA sequencing.
4. Excise the gene of interest containing the mutation with appropriate restriction enzymes from pJET 1.2, and insert it into the extrachromosomal vector pDXA-3C in frame with the ATG start codon present in the vector. This should also be performed with the control wild-type gene.
5. Transfect AX2 cells by electroporation with 2–4 μg of the extrachromosomal vector along with 1 μg of the helper plasmid pREP in H50 buffer using standard procedures [35]. As a control, transfect AX2 cells with pDXA-3 containing the wild-type gene of interest. This is necessary to compare the ADP-ribosylation status of the mutated and non-mutated protein. Transfer the transfected cells in a 10 cm dish, and add 10 mL of HL5. The pDXA3C vector contains a G418-resistance cassette to allow gene expression and selection of transfected cells. The next day add 10 μg/mL G418 to select for cells that have taken up the vector.
6. Once cells become confluent, approximately 5–7 days after the transfection, transfer them into flasks and passage them for another 2–4 days in media containing 10 μg/mL G418 to maintain the selection process.

7. Check the expression level of the Myc-tagged protein by Western blotting. Whole-cell extracts are prepared by pelleting 5×10^6 cells at 1500 × *g* for 3 min. Wash cells by resuspending in 1 mL KK2 before pelleting again at 1500 × *g* for 3 min. Discard the supernatant and resuspend the cell pellet in 30 μL 2× SDS loading buffer. Boil the sample for 5 min and resolve 5 μL of the sample by SDS-PAGE to assess recombinant protein expression levels by Western blotting with a Myc-antibody.
8. Investigate the ADP-ribosylation status of the mutated and non-mutated proteins by performing histone acid extraction (Subheadings 3.2.3) or chromatin preparation (Subheading 3.2.2) and analysis by Western blotting (Subheading 3.1.3).

3.3.2 Mutation of ADP-Ribosylation Sites at Endogenous Loci

Using gene replacement strategies to knock-in mutations at ADP-ribosylation sites in the endogenous loci makes it possible to generate ADP-ribosylation null versions of a given protein, enabling the phenotypic consequences of this mutation to be assessed. The ability to perform gene disruption and replacement strategies in *Dictyostelium* makes this an attractive system to perform these experiments. This is exemplified by the study of histones. The high copy number arrays of histone genes in vertebrates make it difficult to genetically manipulate these loci. In contrast, *Dictyostelium* contains single copies of most histone genes, making their genetic manipulation relatively straightforward [13, 22, 23].

The techniques below describe procedures to generate strains disrupted in site-specific mutations at ADP-ribosylation sites (e.g., *see* Fig. 4). While we focus on the analysis of histone genes, this approach can be employed for other substrates modified by ARTs. Essentially, this methodology involves generating a gene replacement cassette that contains the gene of interest, but with point mutations at putative ADP-ribosylation sites. The gene also contains genomic DNA sequences that flank the gene to allow site-specific recombination at the desired *locus*, thus replacing the wild-type gene with the mutated version. A blasticidin resistance cassette is also engineered into the gene replacement cassette to allow for selection of recombinant strains. Following transfection and selection for recombinant strains, the appropriate gene disruption in selected clones is confirmed by PCR. Having isolated strains with mutations at specific sites within a gene, the impact of ADP-ribosylation of the protein can be assessed using methodologies described in Subheading 3.2 and the impact of a variety of cellular phenotypes assessed.

1. To generate a knock-in mutation, use PCR to amplify the entire coding sequence of the gene of interest in addition to 1 kb of flanking genomic DNA 5′ and 3′ to the gene.

2. Subclone the entire amplified fragment into pJET 1.2 and confirm the integrity of the PCR by DNA sequencing.
3. Excise the selection marker (BSR cassette) from the vector pLPBLP [44] using SmaI, and insert it into a region of the amplified fragment 3′ to the gene of interest using an appropriate blunt end restriction enzyme site. It is important to insert the selection marker into an inert region of the genome and not disrupt any promoter region or gene which might be downstream of the gene being studied. Therefore, insert the BSR cassette 20–200 bp after the stop codon. This also guarantees that 0.7–1 kb of the 3'region of the amplified fragment is still downstream of the selection marker to facilitate gene replacement of the endogenous gene by targeted homologous recombination.
4. Note: If the only restriction enzyme sites available in the 3'region of the gene are facilitating DNA overhangs, mediate DNA end blunting using standard protocols (using Klenow fragment or Mung Bean Nuclease). If there is no restriction enzyme site available, introduce a blunt end restriction enzyme site using the QuikChange method.
5. Introduce the desired mutations using the QuikChange method. If possible, also introduce a novel restriction enzyme site without changing the encoded amino acid sequence to facilitate screening of mutants.
6. Verify the introduced mutation(s) by DNA sequencing.
7. Excise the targeting construct containing the mutated gene, BSR cassette, and flanking genomic DNA sequences from the vector using appropriate restriction enzymes. Purify the DNA fragment using phenol-chloroform extraction.

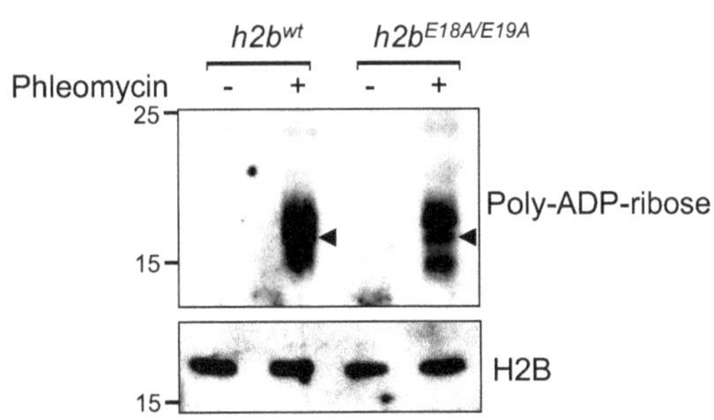

Fig. 4 Analysis of ADP-ribosylation of histones in a strain that has been mutated at E18 and E19 at the endogenous *h2b* gene. *Dictyostelium* strains with either wild-type *h2b* ($h2b^{wt}$) or *h2b* mutated at E18 and E19 ($h2b^{E18AE19A}$) were left untreated or exposed to 300 μg/mL phleomycin for 60 min. Histone-enriched acid extracts were prepared as described in Subheading 3.1.3.1 and Western blotting performed using the indicated antibodies. Arrow heads indicate the position of ADP-ribosylated H2B protein in $h2b^{wt}$ but not $h2b^{E18AE19A}$ cells

8. Transfect the AX2 cells by electroporation with 4–8 μg of the targeting construct in H50 buffer. Following the transfection prepare serial dilutions with 1×10^5, 1×10^4, and 1×10^3 cells/mL, and transfer them into 96-well plates. The aim is to achieve a dilution that following selection in blasticidin will result in a single colony in a well, ensuring a clonal strain. The next day, add 10 μg/mL blasticidin and select cultures for approximately 2 weeks until colonies are apparent.
9. Transfer blasticidin-resistant clones into a 24-well plate, and expand by culturing in selection for another 2 days.
10. Verify gene replacement using PCR to amplify the segment of the gene of interest spanning the mutated site and confirming mutation by DNA sequencing. If a novel restriction enzyme site was introduced in the mutation strategy, the PCR fragment can be initially screened by assessing whether the PCR fragments are cleaved by the appropriate restriction enzyme.
11. Once a strain containing the desired mutations has been identified, re-clone the strain by spreading the cells on a KA lawn on SM Agar [34]. After 3–4 days, pick single colonies and expand under standard growth conditions.
12. Repeat **step 13** to confirm that the selected clonal strains contain the desired mutation.

4 Notes

1. The reaction setup of all ART ADP-ribosylation assays should be carried out on ice to avoid initiation of the catalytic activity of the ART enzymes prior to the experiment. Also, sheared salmon sperm DNA and NAD should be gently defrosted on ice.
2. Quenched samples of the ART ADP-ribosylation assays can be stored at −20 °C before analyzing them by SDS-PAGE.
3. Before adding the X-ray film in the dark in Subheading 3.1.1, **step 2**, check the radioactive signal of the gel, especially at around the size of the ART, using a Geiger counter to confirm ADP-ribosylation using ^{32}P-NAD. Incubate the cassette containing the film and the gel for 24–48 h, and store it in a −80 °C freezer to increase the signal. Incubation with the X-ray film can be repeated and shortened or extended depending of the signal intensity/radioactivity.
4. Blocking, as well as incubation with the primary antibody, can be carried out at room temperature for 1 h or overnight at 4 °C during Western blotting, as determined during antibody optimization or according to the manufacturers' instructions.
5. Coverslips should never be allowed to dry out. To ensure this, coverslips can be processed in small batches. Furthermore,

during antibody incubation on Parafilm, the coverslips should be covered with the base of the 24-well plate so that each coverslip is contained within one well. Blocking and antibody solutions should cover the entire coverslip during the incubation steps while being restricted to the coverslip by surface tension (40–60 μL is sufficient for a 10 mm diameter coverslip). At all stages, coverslips should be treated gently when adding or removing liquid to minimize loss of cells.

6. The pellets that occur during chromatin extraction and histone extraction can be very small and might appear transparent. Be careful when removing the supernatant after spinning to avoid losing the pellet and to minimize loss of cells.

Acknowledgments

We thank members of the Lakin and Pears laboratories for constructive comments during the preparation of this manuscript. N.L.'s laboratory is supported by Cancer Research UK (www.cancerresearch.org.uk; grant C1521/A12353), Medical Research Council (www.mrc.ac.uk; MR/L000164/1), and NC3Rs (www.nc3rs.org.uk; NC/K00137X/1). AR was supported by a Clarendon Award (University of Oxford). C.P.'s laboratory is supported by NC3Rs (www.nc3rs.org.uk; NC/M000834/1).

References

1. Citarelli M, Teotia S, Lamb RS (2010) Evolutionary history of the poly(ADP-ribose) polymerase gene family in eukaryotes. BMC Evol Biol 10:308. https://doi.org/10.1186/1471-2148-10-308
2. Otto H, Reche PA, Bazan F, Dittmar K, Haag F, Koch-Nolte F (2005) In silico characterization of the family of PARP-like poly(ADP-ribosyl) transferases (pARTs). BMC Genomics 6:139. https://doi.org/10.1186/1471-2164-6-139
3. Vyas S, Matic I, Uchima L, Rood J, Zaja R, Hay RT, Ahel I, Chang P (2014) Family-wide analysis of poly(ADP-ribose) polymerase activity. Nat Commun 5:4426. https://doi.org/10.1038/ncomms5426
4. Block WD, Lees-Miller SP (2005) Putative homologues of the DNA-dependent protein kinase catalytic subunit (DNA-PKcs) and other components of the non-homologous end joining machinery in Dictyostelium discoideum. DNA Repair (Amst) 4(10):1061–1065. https://doi.org/10.1016/j.dnarep.2005.06.008
5. Hsu DW, Gaudet P, Hudson JJ, Pears CJ, Lakin ND (2006) DNA damage signaling and repair in Dictyostelium discoideum. Cell Cycle 5(7):702–708
6. Hudson JJ, Hsu DW, Guo K, Zhukovskaya N, Liu PH, Williams JG, Pears CJ, Lakin ND (2005) DNA-PKcs-dependent signaling of DNA damage in Dictyostelium discoideum. Curr Biol 15(20):1880–1885. https://doi.org/10.1016/j.cub.2005.09.039
7. Muramoto T, Chubb JR (2008) Live imaging of the Dictyostelium cell cycle reveals widespread S phase during development, a G2 bias in spore differentiation and a premitotic checkpoint. Development 135(9):1647–1657. https://doi.org/10.1242/dev.020115
8. Zhang XY, Langenick J, Traynor D, Babu MM, Kay RR, Patel KJ (2009) Xpf and not the Fanconi anaemia proteins or Rev3 accounts for the extreme resistance to cisplatin in Dictyostelium discoideum. PLoS Genet 5(9):e1000645. https://doi.org/10.1371/journal.pgen.1000645
9. Pontel LB, Langenick J, Rosado IV, Zhang XY, Traynor D, Kay RR, Patel KJ (2016) Xpf suppresses the mutagenic consequences

of phagocytosis in Dictyostelium. J Cell Sci 129(24):4449–4454. https://doi.org/10.1242/jcs.196337
10. Pears CJ, Couto CA, Wang HY, Borer C, Kiely R, Lakin ND (2012) The role of ADP-ribosylation in regulating DNA double-strand break repair. Cell Cycle 11(1):48–56. https://doi.org/10.4161/cc.11.1.18793
11. Couto CA, Hsu DW, Teo R, Rakhimova A, Lempidaki S, Pears CJ, Lakin ND (2013) Nonhomologous end-joining promotes resistance to DNA damage in the absence of an ADP-ribosyltransferase that signals DNA single strand breaks. J Cell Sci 126(Pt 15):3452–3461. https://doi.org/10.1242/jcs.128769
12. Couto CA, Wang HY, Green JC, Kiely R, Siddaway R, Borer C, Pears CJ, Lakin ND (2011) PARP regulates nonhomologous end joining through retention of Ku at double-strand breaks. J Cell Biol 194(3):367–375. https://doi.org/10.1083/jcb.201012132
13. Rakhimova A, Ura S, Hsu DW, Wang HY, Pears CJ, Lakin ND (2017) Site-specific ADP-ribosylation of histone H2B in response to DNA double strand breaks. Sci Rep 7:43750. https://doi.org/10.1038/srep43750
14. Ahel I, Ahel D, Matsusaka T, Clark AJ, Pines J, Boulton SJ, West SC (2008) Poly(ADP-ribose)-binding zinc finger motifs in DNA repair/checkpoint proteins. Nature 451(7174):81–85. https://doi.org/10.1038/nature06420
15. Gunn AR, Banos-Pinero B, Paschke P, Sanchez-Pulido L, Ariza A, Day J, Emrich M, Leys D, Ponting CP, Ahel I, Lakin ND (2016) The role of ADP-ribosylation in regulating DNA interstrand crosslink repair. J Cell Sci 129(20):3845–3858. https://doi.org/10.1242/jcs.193375
16. Strmecki L, Greene DM, Pears CJ (2005) Developmental decisions in Dictyostelium discoideum. Dev Biol 284(1):25–36
17. Robinson DN, Spudich JA (2000) Dynacortin, a genetic link between equatorial contractility and global shape control discovered by library complementation of a Dictyostelium discoideum cytokinesis mutant. J Cell Biol 150(4):823–838
18. Chen L, Iijima M, Tang M, Landree MA, Huang YE, Xiong Y, Iglesias PA, Devreotes PN (2007) PLA2 and PI3K/PTEN pathways act in parallel to mediate chemotaxis. Dev Cell 12(4):603–614. https://doi.org/10.1016/j.devcel.2007.03.005
19. Li G, Alexander H, Schneider N, Alexander S (2000) Molecular basis for resistance to the anticancer drug cisplatin in Dictyostelium. Microbiology 146(Pt 9):2219–2227
20. Van Driessche N, Alexander H, Min J, Kuspa A, Alexander S, Shaulsky G (2007) Global transcriptional responses to cisplatin in Dictyostelium discoideum identify potential drug targets. Proc Natl Acad Sci U S A 104(39):15406–15411. https://doi.org/10.1073/pnas.0705996104
21. Williams RS, Eames M, Ryves WJ, Viggars J, Harwood AJ (1999) Loss of a prolyl oligopeptidase confers resistance to lithium by elevation of inositol (1,4,5) trisphosphate. EMBO J 18(10):2734–2745. https://doi.org/10.1093/emboj/18.10.2734
22. Hsu DW, Chubb JR, Muramoto T, Pears CJ, Mahadevan LC (2012) Dynamic acetylation of lysine-4-trimethylated histone H3 and H3 variant biology in a simple multicellular eukaryote. Nucleic Acids Res 40(15):7247–7256. https://doi.org/10.1093/nar/gks367
23. Stevense M, Chubb JR, Muramoto T (2011) Nuclear organization and transcriptional dynamics in Dictyostelium. Develop Growth Differ 53(4):576–586. https://doi.org/10.1111/j.1440-169X.2011.01271.x
24. Leiting B, Noegel A (1988) Construction of an extrachromosomally replicating transformation vector for Dictyostelium discoideum. Plasmid 20(3):241–248
25. Manstein DJ, Schuster HP, Morandini P, Hunt DM (1995) Cloning vectors for the production of proteins in Dictyostelium discoideum. Gene 162(1):129–134
26. D'Amours D, Desnoyers S, D'Silva I, Poirier GG (1999) Poly(ADP-ribosyl)ation reactions in the regulation of nuclear functions. Biochem J 342(Pt2):249–268
27. Gibson BA, Kraus WL (2012) New insights into the molecular and cellular functions of poly(ADP-ribose) and PARPs. Nat Rev Mol Cell Biol 13(7):411–424. https://doi.org/10.1038/nrm3376
28. Slade D, Dunstan MS, Barkauskaite E, Weston R, Lafite P, Dixon N, Ahel M, Leys D, Ahel I (2011) The structure and catalytic mechanism of a poly(ADP-ribose) glycohydrolase. Nature 477(7366):616–620. https://doi.org/10.1038/nature10404
29. Barkauskaite E, Brassington A, Tan ES, Warwicker J, Dunstan MS, Banos B, Lafite P, Ahel M, Mitchison TJ, Ahel I, Leys D (2013) Visualization of poly(ADP-ribose) bound to PARG reveals inherent balance between exo- and endo-glycohydrolase activities. Nat Commun 4:2164. https://doi.org/10.1038/ncomms3164
30. Sharifi R, Morra R, Denise Appel C, Tallis M, Chioza B, Jankevicius G, Simpson MA, Matic I, Ozkan E, Golia B, Schellenberg MJ, Weston

R, Williams JG, Rossi MN, Galehdari H, Krahn J, Wan A, Trembath RC, Crosby AH, Ahel D, Hay R, Ladurner AG, Timinszky G, Williams RS, Ahel I (2013) Deficiency of terminal ADP-ribose protein glycohydrolase TARG1/C6orf130 in neurodegenerative disease. EMBO J 32:1225. https://doi.org/10.1038/emboj.2013.51

31. Rosenthal F, Feijs KL, Frugier E, Bonalli M, Forst AH, Imhof R, Winkler HC, Fischer D, Caflisch A, Hassa PO, Luscher B, Hottiger MO (2013) Macrodomain-containing proteins are new mono-ADP-ribosylhydrolases. Nat Struct Mol Biol 20(4):502–507. https://doi.org/10.1038/nsmb.2521
32. Jankevicius G, Hassler M, Golia B, Rybin V, Zacharias M, Timinszky G, Ladurner AG (2013) A family of macrodomain proteins reverses cellular mono-ADP-ribosylation. Nat Struct Mol Biol 20(4):508–514. https://doi.org/10.1038/nsmb.2523
33. Messner S, Hottiger MO (2011) Histone ADP-ribosylation in DNA repair, replication and transcription. Trends Cell Biol 21(9):534–542. https://doi.org/10.1016/j.tcb.2011.06.001
34. Couto AM, Lakin ND, Pears CJ (2013) Investigation of DNA repair pathway activity. Methods Mol Biol 983:295–310. https://doi.org/10.1007/978-1-62703-302-2_16
35. Faix J, Linkner J, Nordholz B, Platt JL, Liao XH, Kimmel AR (2013) The application of the Cre-loxP system for generating multiple knock-out and knock-in targeted loci. Methods Mol Biol 983:249–267. https://doi.org/10.1007/978-1-62703-302-2_13
36. Daniels CM, Ong SE, Leung AK (2014) Phosphoproteomic approach to characterize protein mono- and poly(ADP-ribosyl)ation sites from cells. J Proteome Res 13(8):3510–3522. https://doi.org/10.1021/pr401032q
37. Daniels CM, Ong SE, Leung AK (2015) The promise of proteomics for the study of ADP-ribosylation. Mol Cell 58(6):911–924. https://doi.org/10.1016/j.molcel.2015.06.012
38. Gagne JP, Isabelle M, Lo KS, Bourassa S, Hendzel MJ, Dawson VL, Dawson TM, Poirier GG (2008) Proteome-wide identification of poly(ADP-ribose) binding proteins and poly(ADP-ribose)-associated protein complexes. Nucleic Acids Res 36(22):6959–6976. https://doi.org/10.1093/nar/gkn771
39. Isabelle M, Gagne JP, Gallouzi IE, Poirier GG (2012) Quantitative proteomics and dynamic imaging reveal that G3BP-mediated stress granule assembly is poly(ADP-ribose)-dependent following exposure to MNNG-induced DNA alkylation. J Cell Sci 125(Pt 19):4555–4566. https://doi.org/10.1242/jcs.106963
40. Jungmichel S, Rosenthal F, Altmeyer M, Lukas J, Hottiger MO, Nielsen ML (2013) Proteome-wide identification of poly(ADP-ribosyl)ation targets in different genotoxic stress responses. Mol Cell 52(2):272–285. https://doi.org/10.1016/j.molcel.2013.08.026
41. Zhang Y, Wang J, Ding M, Yu Y (2013) Site-specific characterization of the Asp- and Glu-ADP-ribosylated proteome. Nat Methods 10(10):981–984. https://doi.org/10.1038/nmeth.2603
42. Bonfiglio JJ, Fontana P, Zhang Q, Colby T, Gibbs-Seymour I, Atanassov I, Bartlett E, Zaja R, Ahel I, Matic I (2017) Serine ADP-ribosylation depends on HPF1. Mol Cell 65(5):932–940 e936. https://doi.org/10.1016/j.molcel.2017.01.003
43. Leidecker O, Bonfiglio JJ, Colby T, Zhang Q, Atanassov I, Zaja R, Palazzo L, Stockum A, Ahel I, Matic I (2016) Serine is a new target residue for endogenous ADP-ribosylation on histones. Nat Chem Biol 12(12):998–1000. https://doi.org/10.1038/nchembio.2180
44. Faix J, Kreppel L, Shaulsky G, Schleicher M, Kimmel AR (2004) A rapid and efficient method to generate multiple gene disruptions in Dictyostelium discoideum using a single selectable marker and the Cre-loxP system. Nucleic Acids Res 32(19):e143. https://doi.org/10.1093/nar/gnh136
45. Garcia MX, Alexander H, Mahadeo D, Cotter DA, Alexander S (2003) The Dictyostelium discoideum prespore-specific catalase B functions to control late development and to protect spore viability. Biochim Biophys Acta 1641(1):55–64
46. Gaudet P, Tsang A (1999) Regulation of the ribonucleotide reductase small subunit gene by DNA-damaging agents in Dictyostelium discoideum. Nucleic Acids Res 27(15):3042–3048
47. Podgorski G, Deering RA (1980) Effect of methyl methanesulfonate on survival of radiation-sensitive strains of Dictyostelium discoideum. Mutat Res 73(2):415–418
48. Bronner CE, Welker DL, Deering RA (1992) Mutations affecting sensitivity of the cellular slime mold Dictyostelium discoideum to DNA-damaging agents. Mutat Res 274(3):187–200
49. Freeland TM, Guyer RB, Ling AZ, Deering RA (1996) Apurinic/apyrimidinic (AP) endonuclease from Dictyostelium discoideum: cloning, nucleotide sequence and induction by sublethal levels of DNA damaging agents. Nucleic Acids Res 24(10):1950–1953
50. Ling AZ, Guyer RB, Deering RA (2001) Dictyostelium discoideum plasmid containing an AP-endonuclease upstream sequence:

bleomycin induction of a luciferase reporter. Environ Mol Mutagen 38(2–3):244–247

51. Deering RA, Guyer RB, Stevens L, Watson-Thais TE (1996) Some repair-deficient mutants of Dictyostelium discoideum display enhanced susceptibilities to bleomycin. Antimicrob Agents Chemother 40(2):464–467
52. Hsu DW, Kiely R, Couto CA, Wang HY, Hudson JJ, Borer C, Pears CJ, Lakin ND (2011) DNA double-strand break repair pathway choice in Dictyostelium. J Cell Sci 124(Pt 10):1655–1663. https://doi.org/10.1242/jcs.081471
53. Lee SK, Yu SL, Alexander H, Alexander S (1998) A mutation in repB, the dictyostelium homolog of the human xeroderma pigmentosum B gene, has increased sensitivity to UV-light but normal morphogenesis. Biochim Biophys Acta 1399(2–3):161–172
54. Hurley DL, Skantar AM, Deering RA (1989) Nuclear DNA synthesis is blocked by UV irradiation in Dictyostelium discoideum. Mutat Res 217(1):25–32

Chapter 10

Mono-ADP-Ribosylation Catalyzed by Arginine-Specific ADP-Ribosyltransferases

Linda A. Stevens and Joel Moss

Abstract

Methods are described for determination of arginine-specific mono-ADP-ribosyltransferase activity of purified proteins and intact cells by monitoring the transfer of ADP-ribose from NAD^+ to a model substrate, e.g., arginine, agmatine, and peptide (human neutrophil peptide-1 [HNP1]), and for the nonenzymatic hydrolysis of ADP-ribose-arginine to ornithine, a noncoded amino acid. In addition, preparation of purified ADP-ribosylarginine is included as a control substrate for ADP-ribosylation reactions.

Key words ADP-ribosylarginine, HNP-1, NAD^+, Mono-ADP-ribosylation, ADP-ribose, Defensins

1 Introduction

Arginine-specific mono-ADP-ribosyltransferases catalyze the transfer of the ADP-ribose moiety of NAD^+ to arginine residues in proteins [1]. The best described arginine-specific transferases are the bacterial toxins, cholera toxin (CT), and heat-labile *Escherichia coli* enterotoxin (LT), which ADP-ribosylate arginine in Gαs, the guanine nucleotide-binding protein involved in the regulation of adenylyl cyclase activity [2, 3]. The hypothesis that eukaryotic transferases existed that catalyze similar reactions lead to the purification of an arginine-specific ADP-ribosyltransferase from turkey erythrocytes [4]. In addition to arginine in proteins, the bacterial toxins and the turkey transferase ADP-ribosylate free arginine and some guanidino compounds [5], consistent with the finding that it is the guanidino group in arginine that is modified by ADP-ribose.

Mammalian 36-kD arginine-specific ADP-ribosyltransferases (ART1) were purified from human and rabbit skeletal muscle, characterized, and cloned [6]. Based on similarities in sequence, five ART genes have been identified (ART1–5) [7]; ART1 was found in the heart and skeletal muscle [8, 9]. ART2 is found in rat and mouse T cells but not in human, where it appears to be a

Paul Chang (ed.), *ADP-ribosylation and NAD+ Utilizing Enzymes: Methods and Protocols*, Methods in Molecular Biology, vol. 1813,
https://doi.org/10.1007/978-1-4939-8588-3_10,

pseudogene [10]. Human ART3 is expressed in the testis and cardiac muscle and ART4 in the heart, lung, liver, and spleen [7, 11, 12]. ART1–4 are extracellular proteins attached to the membrane by a glycosylphosphatidylinositol (GPI) [13, 14] anchor, whereas ART5, which was cloned from mouse lymphoma cells and expressed in mouse skeletal muscle and mouse and human testis, is membrane associated and secreted [15].

The crystal structures of ADP-ribosyltransferases revealed six beta strands [16]. The glutamic acid in loop beta 5 is part of a conserved R(β1)-S(β2)-EXE(β5) motif, and the glutamic acid two-residue upstream (E-X-E) determines the substrate specificity of ART1, ART2, and ART5 arginine-specific ADP-ribosyltransferases [17]. Changes in the motif with E-E-E (ART1) replaced by E-R-I (hART3) or K-K-E (hART4) are believed to be responsible for a lack of transferase activity [7, 18]. ART3 is expressed in untreated monocytes [19]. ART4, which is identical to the GPI-linked, glycosylated, polymorphic Dombrock blood group antigen, is expressed on the surface of erythrocytes [20]. Several unidentified cell surface proteins on human monocytes were ADP-ribosylated after treatment with radiolabeled NAD^+, suggesting the presence of active ADP-ribosyltransferases on the cell surface [21]. Lipopolysaccharide stimulated ART4 expression, resulting in a 31-kD ADP-ribosylated protein [19]. The modification was sensitive to $HgCl_2$, consistent with an ADP-ribose-cysteine linkage and the presence of cysteine-specific transferases.

ART1, expressed in cells of the immune system, e.g., neutrophils lymphocytes, may regulate the immune response by targeting extracellular and cell surface proteins [14, 22]. An arginine-specific ADP-ribosyltransferase was identified on the surface of human polymorphonuclear leukocytes (PMNs), where it was linked to the cell surface by a GPI anchor, consistent with its identification as ART1 [23, 24]. Several cell membrane substrates were identified after incubation of the PMNs with radiolabeled NAD^+ [23]. Proliferation of cytotoxic T cells by stimulator cells and the ability to lyse targets were suppressed by incubation with NAD^+ [25]. Cell surface proteins were ADP-ribosylated on arginine, consistent with the hypothesis that an arginine-specific ADP-ribosyltransferase is responsible for the effects [26]. The T-cell lymphoma cell line (EL-4), stably transformed with ART1, exhibited ADP-ribosylation of T-cell receptors with disruption of the TCR complex [27]. CD38, a transmembrane glycoprotein, catalyzes, in the presence of NAD^+, the dual activities of formation and hydrolysis of cyclic adenosine diphosphate ribose (cADPr) [28]. It is expressed primarily on hematopoietic cells and can be induced on T lymphocytes with antibodies against specific antigen receptors. After incubation of the cells with NAD^+, CD38 was ADP-ribosylated on arginine and cysteine, presumably by ecto-transferases [29]. Both CD38 activities were inhibited by ADP-ribosylation on arginine, whereas

ADP-ribosylation of cysteine only inhibited hydrolase activity. ADP-ribosylation on arginine resulted in apoptosis of the activated T cells.

Additional extracellular proteins that may interact with arginine-specific ADP-ribosyltransferases include growth factors, e.g., basic fibroblast growth factor (FGF-2), platelet-derived growth factor (PDGF), involved in angiogenesis [30]. Adult bovine aortic arch endothelial (AEBE) cells and human hepatoma cells (SK-Hep1) that express ART1 incubated with NAD^+, ADP-ribosylated FGF-2 [31]. The arginine target for ADP-ribosylation in FGF-2 appeared to be in the putative receptor-binding domain; the addition of heparin blocked the modification [32]. Chinese hamster lung fibroblasts stably transfected with ART1 and rat skeletal muscle membranes that express endogenous ART1 when incubated with NAD^+, ADP-ribosylated PDGF-BB but not PDGF-AA. Purified ADP-ribosylated PDGF-BB, when compared to the unmodified protein, was unable to stimulate a mitogenic or chemotactic response from human pulmonary smooth muscle cells and showed reduced binding to its receptor [33].

Human neutrophil peptides (HNP), part of the large family of antimicrobial peptides (defensins), are released from activated neutrophils in the lung as part of the innate immune response [34, 35]. Its arginine-rich sequence suggested that HNP-1 might serve as a substrate for the arginine-specific ADP-ribosyltransferase (ART1) expressed on the apical surface of airway epithelial cells [36] as well as by inflammatory cells recruited to the lung that are in contact with extracellular NAD^+ [22, 25, 26, 37]. HNP-1 was ADP-ribosylated on arginine 14 in vitro with a loss of antimicrobial and cytotoxic activity [38]. The modified peptide released interleukin-8 (IL-8) from A549 cells and enhanced T-cell chemotaxis. C2C12 myotubes endogenously express ART1, ADP-ribosylated HNP-1 on arginine 14 with a secondary site on arginine 24 [39]. Surprisingly, in vitro, ADP-ribosylated HNP-1 with ADP-ribose on arginine 24 exhibited nonenzymatic conversion of ADP-ribose arginine 14 to ornithine, a noncoded amino acid [40]. HNP1 with arginine 14 and 24 replaced with ornithine lacked the epithelial cell cytotoxicity of HNP1 [41]. Most notably, mono- and di-ADP-ribosylated HNP-1 and ADP-ribosylated-HNP-ornithine were found in the bronchoalveolar lavage of patients with idiopathic pulmonary fibrosis, consistent with replacement of ADP-ribosylarginine with ornithine occurring in human airways in vivo [39, 40]. In addition, mono- and di-ADP-ribosylated HNP-1 were found in the BALF of smokers, asthmatics, and a patient with idiopathic pulmonary fibrosis, confirming that HNP-1 is an endogenous human substrate of ART1 and suggesting that ADP-ribosylation is involved in the immune inflammatory response in disease [38, 39].

Increase in expression of ART1 in C2C12 muscle cells correlated with the differentiation of myoblasts to myotubes presented an in vitro model for identifying ART1 substrates [42]. Incubation of myotubes with exogenous NAD^+ resulted in ADP-ribosylation of integrin $\alpha7$ on arginine. Integrin $\alpha7\beta1$ dimer is the major laminin-binding receptor on skeletal myoblasts [43]. Altered function or expression of integrin alpha 7 is seen in multiple myopathies [44]).

ART1 is a GPI-linked ecto-protein [42]. Intracellular rabbit ART1, however, was expressed in rabbit skeletal muscle sarcoplasmic reticulum [45]. ART1, which is expressed on the apical surface of polymorphonuclear leukocytes [23], was translocated to the cell surface from an intracellular pool, in response to chemotaxins [46]. GRP/BIP is a molecular chaperone that binds newly synthesized proteins and facilitates their translocation to the ER [47]. As a response to nutritional stress, GRP78/BIP was inactivated by arginine mono-ADP-ribosylation in its binding domain during its trafficking in the ER [48, 49]. In HeLa cells, transfected hART1 was localized to the endoplasmic reticulum and appears to be responsible for ADP-ribosylarginines in GRP78/BIP [50]. An endogenous ADP-ribosyltransferase was reported to modify arginine on the beta subunit of Gs, the heterotrimeric G protein responsible for activating adenylyl cyclase activity. The association of Gαs and $G_{\beta\gamma}$ is, in part, regulated by the exchange of GTP for GDP on Gαs [51]. Free $G_{\beta\gamma}$ modulates multiple cellular functions that are altered by ADP-ribosylation of the G_β subunit [52, 53]. An unidentified arginine-specific ADP-ribosyl-acceptor hydrolase cleaves the modified G_β, suggesting that an ADP-ribosylation cycle regulates the function of the $\beta\gamma$ heterodimer.

The link between arginine-specific ADP-ribosylation and tumorigenicity was established in the ARH1-deficient mouse model (ARH1$^{-/-}$) [54]. ADP-ribosylarginine hydrolase (ARH1), a component of the ADP-ribosylation cycle, is no longer available to cleave the ADP-ribose from arginine in the ARH1$^{-/-}$ mouse model and thus regulate the effect of arginine modification on proteins. Mouse embryonic fibroblasts (MEFs) generated from ARH1-deficient mice proliferated faster and formed more and larger colonies in soft agar when compared to WT-derived MEFs. Tumorigenicity was assessed by subcutaneously injecting ARH1$^{+/+}$ and ARH1$^{-/-}$ cells into athymic mice. Tumors were observed in mice injected with ARH$^{-/-}$ cells but not ARH$^{+/+}$ cells. Consistent with the MEF results, the number of lymphomas, carcinomas, and sarcomas were increased in ARH1$^{-/-}$ mice compared to WT [55]. Thus, an increase in ADP-ribosylation of arginine and disruption of the ADP-ribosylation cycle were associated with tumorigenesis.

Purified GPI-linked ART1 can modify free arginine and arginine in several proteins which may or may not be in vivo substrates.

New tools were needed as ADP-ribosylated targets in intact cells have been difficult to detect due to a lack of specific anti-ADP-ribosylarginine antibodies. The observation that macro-domain modules in proteins selectively bound ADP-ribose and poly-ADP-ribose suggested that modified proteins could be isolated using macro-domain binding [56]. Macro-domain, a 190 amino acid module identified in the Af1521 protein from *Archaeoglobus fulgidus,* binds ADP-ribose and poly-ADP-ribose, in pull-down assays from cells ADP-ribosylated by bacterial toxins and endogenous transferases [57]. MS analysis identified 12 mono- and poly-ADP-ribosylated substrates. In addition, a protocol was developed to identify endogenous ADP-ribosylation sites in both cells and tissues [58]. Proteins isolated from cells were digested to peptides, incubated with poly-ADP-ribose glycohydrolase (PARG) to convert poly- to mono-ADP-ribosylated peptides, enriched for ADP-ribose by Af1521 pull down and analyzed by LC-MS. During oxidative stress, Lys, Glu, Asp, and Arg were ADP-ribosylated. Interestingly, Arg was the predominant modified residue in normal liver tissue [58]. Proteomic studies with mass spectroscopy have identified many more ADP-ribosylated substrates. A data base of ADP-ribosylated proteins has been created to catalog modified proteins with details from published data. To date there are 2389 unique proteins, 93% human [59].

The techniques for determining arginine-specific ADP-ribosyltransferase activity described in Methods are based on measuring and collecting ADP-ribosylated substrates. The addition of ADP-ribose (two negative charges) to arginine (positive charge) on proteins or the ADP-ribosylation of agmatine (decarboxylated arginine) as a model substrate results in an increase of net negative charge of the product. Agmatine was determined to be a better substrate than arginine due to the absence of the negatively charged carboxyl group [60]. Purification of the products of arginine-specific ADP-ribosyltransferase reactions is best achieved by two types of chromatography, hydrophobic and ion exchange, to separate, identify, and quantify the reaction products. The column or resin selection is based on the components of the reaction and the requirements of the study. High-performance liquid chromatography (HPLC) separations are ideal for collecting and measuring the amount of ADP-ribosylated peptide and protein as well as ADP-ribosylarginine [61]. A more precise quantification of the amount of substrate and product can be achieved by using radioactive NAD^+ in the reaction mix. The activity of cholera toxin, a bacterial arginine-specific ADP-ribosyltransferase, historically has been the standard for measuring the transfer of ADP-ribose to arginine and the synthesis of reaction products such as ADP-ribosylagmatine and ADP-ribosylarginine [62, 63].

2 Materials

HPLC (high-performance liquid chromatography instrument with/auto sampler/diode-array detector/florescence detector) and columns, scintillation counter, X-ray film processor.

Cell lysis buffer: 3% SDS, 0.1M sodium acetate pH 6.8, 5 mM EDTA. NAD^+, DTT (dithiothreitol), agmatine, ovalbumin, CTA, AG1-X2 resin, scintillation fluid, phosphatidylinositol-specific phospholipase C (PIPLC), NAD^+ [*adenine*-14C] (0.05 μCi), C2C12 cells, DMEM cell media with 10% fetal bovine serum (FBS), trichloroacetic acid (TCA), RIPA buffer: 10 mM Tris–HCl, pH 8.0, 1 mM EDTA, 1% Triton X-100, 0.1% deoxycholate, 0.1% SDS, 140 mM NaCl, 1 mM PMSF, Af1521 macro-domain-GST resin, [^{32}P]-NAD^+, poly ADP-ribose glycohydrolase (PARG), LysC, trypsin, hydroxylamine.

3 Methods

3.1 A Radioassay to Measure the Arginine-Specific ADP-Ribosyltransferase Activity of Cholera Toxin

Historically cholera toxin is considered the standard arginine-specific ADP-ribosyltransferase. The toxin consists of two subunits, A which has two units 1 and 2 and B. The B subunit binds to the ganglioside GM1 receptor on the cell surface permitting A subunits to enter the cell. The A1 subunit, the catalytically active subunit, is released from the A2 and then transfers ADP-ribose from NAD^+ to arginine on Gαs. In addition, CTA can ADP-ribosylate free arginine [62, 64].

Cholera toxin A is reduced by DTT to generate the A1 subunit which enzymatically transfers radioactive ADP-ribose from [*adenine*-^{14}C] NAD^+ to agmatine. Radioactive ADP-ribose-agmatine is separated from radioactive NAD^+ that binds to a strong anion exchange resin.

1. Incubate cholera toxin A subunit (CTA) 60 μg with 30 mM dithiothreitol (DTT), 0.1 mM NAD^+ [*adenine*-^{14}C] (0.05 μCi), 20 mM agmatine as an ADP-ribose acceptor, 30 μg ovalbumin in 20 mM potassium phosphate, pH 7.5, final volume of 300 μL, 3 h at 30 °C. (Include negative control without CTA.)
2. Solubilize dry AG1-X2, a strong anion exchange resin with water. Prepare columns (0.4 × 4 cm Pasteur pipettes) by adding resin, and equilibrate with water. Apply samples (100 μL) of the reaction mixture. Elute the columns with 5 1-mL aliquots of water, and collect the eluate in scintillation vials. Add scintillation fluid to measure the amount of ADP-ribose [^{14}C]-agmatine in a scintillation counter. The transferase activity is determined by the amount of ADP-ribose (nmol)/h attached to the acceptor [63, 65]. The suggested activity is that 1 μg of CTA generates 2 pmole ADP-ribose in 30 min at 30 °C.

3.2 An Assay to Prepare ADP-Ribosylarginine, Using the Reaction Mixture Described Above, Substituting Arginine for Agmatine

Cholera toxin A1 enzymatically transfers radioactive ADP-ribose to arginine. ADP-ribose arginine is purified by HPLC [63].

1. Prepare radioactive ADP-ribose-arginine as described in Subheading 3.1, **step 1**, substituting 20 mM arginine for 20 mM agmatine.
2. Separate the reaction products on an HPLC strong anion exchange column equilibrated in solution A by gradient elution. The HPLC gradient elution program is as follows: Solution A (20 mM sodium phosphate, pH 4.5), for 30 min, followed by a linear gradient for 10 min to 100% B (solution A + 1.0 M NaCl), followed by 100% B, 10 min, flow = 1.0 mL/min. Monitor the separation at absorbance 258 nm (ADP-ribose). Two ADP-ribosylarginine peaks will elute that represent the alpha and beta isomers of ADP-ribosylarginine (*see* **Note 1**).
3. Collect 1 mL fractions. Locate the fraction containing radioactive ADP-ribosylarginine, which is quantified with a scintillation counter [66].

3.3 A Radioassay to Measure the Activity of ART1, an Arginine-Specific ADP-Ribosyltransferase, Expressed in Cells by Quantifying the Transfer of ADP-Ribose from NAD+ to Agmatine

ART1, attached to the cell membrane by a glycosylphosphatidylinositol (GPI) anchor, can be released from cells by phosphatidylinositol-specific phospholipase C (PIPLC). The released ART1 is assayed by incubation with radioactive NAD^+ and agmatine as an ADP-ribose acceptor. The number of cells assayed will have to be determined experimentally.

1. To determine if cells express ART1, collect a minimum of one million cells from tissue culture plates at 37 °C; centrifuge briefly to generate a cell pellet in 1.5 mL microfuge tubes. Wash cell pellet with PBS three times to remove cell media, and incubate with 0.05 U PIPLC in 350 μL PBS at 37 °C, for 1 h with periodic vortexing. ART1 will be released into the supernatant. Centrifuge to remove the cells.
2. Incubate the ART1 sample with 20 mM agmatine as an ADP-ribose acceptor with 0.1 mM NAD^+ [*adenine*-14C] (0.05 μCi) in phosphate-buffered saline (PBS) in 150 μL for 1 h at 30 °C.
3. Apply samples (50 μL) of the reaction mixture to AG1-X2 columns, as described in Subheading 3.1, **step 1**, to collect radioactive ADP-ribosylagmatine [^{14}C] [60].

3.4 A Radioassay of GPI-Anchored Arginine-Specific ADP-Ribosylation in Intact Cells

Differentiated C2C12 myoblasts endogenously express ART1 and thus can act as a source of ART1 enzyme. PIPLC treatment of C2C12 cells removes GPI-linked ART1 transferase from the cells. A reaction mix containing radioactive NAD^+ is added to the intact cells to activate the ART1 transferase activity that was previously determined in Subheading 3.2. The cells are centrifuged to pellet the cells and to remove the reaction mixture. The cells are lysed and the lysate is analyzed for arginine ADP-ribosylated proteins [42]. This method can be applied to other cell types after determination of expressed ART1 activity; *see* Subheading 3.2.

1. Grow the indicated cells as well as C2C12 mouse skeletal muscle cells on 60 mm plates in DMEM with 10% fetal bovine serum (FBS) in 5% CO_2, 37 °C as a positive control. After 3 days observe the C2C12 cells and note a change in morphology from round to fiber-like, indicating the cells are differentiated and express ART1. *See* Subheading 3.2 for preparation of ART1 from cells and identifying ADP-ribosylated protein substrates in C2C12 cells [42].
2. Collect a minimum of one million C2C12 cells or the indicated cells from tissue culture plates at 37 °C. Wash the cell pellets with PBS. Collect the cells in 1.5 mL microfuge tubes. To one half the cell pellets, add 0.05 U PIPLC in 350 μL PBS at 37 °C for 1 h to cleave the GPI-anchored transferase, and collect the supernatant for ART1 activity assay (*see* Subheading 3.2, **step 1**). Wash the PIPLC incubated cells. To both cell pellets (+/− PIPLC), add PBS (350 μL), 5 μM [*adenylate*-^{32}P] NAD^+ (4 Ci/mmol), and 1 mM ADP-ribose (to block nonenzymatic ADP-ribose transferase activity) (*see* **Note 2**) for 1 h at 37 °C with periodic vortexing. PIPLC-treated cells will be a negative control. Centrifuge the cells to remove the reaction mixture. Wash the cells twice with PBS before lysis in 0.5 mL lysis buffer (3% SDS, 01.M sodium acetate pH 6.8, 5 mM EDTA). Collect the lysed cells and boil for 10 min to stop the reaction.
3. Determine the protein concentration. Precipitate the protein (100 μg) with 10% trichloroacetic acid (TCA), collect the pellet, and neutralize with buffer. Suspend the sample in SDS-PAGE loading buffer before separation on 10–20% SDS-PAGE gradient gels. Stain the gels with Coomassie Blue. Dry the gels and expose the dried gels to Kodak X-AR film for 24 h to determine the molecular weight of proteins that are ADP-ribosylated. A second SDS-PAGE separation can be utilized for Western blot analysis with specific antibodies.
4. Before drying, proteins can be extracted from the gels for "in gel digestion" and identified by mass spectrometry [67]. In addition, ADP-ribosylated proteins from cell lysates can be pulled down with Af1521-GST resin; *see* Subheading 3.6.

3.5 An Assay Using a Peptide Substrate for the Determination of Arginine-Specific ADP-Ribosyltransferase (ART1) Activity

Human neutrophil peptide-1 (HNP-1) is arginine-rich and an in vivo substrate for ART1. HNP-1 is di-(arginine 14 and 24) and mono-ADP-ribosylated on arginine 14 by ART1 [39]. ART1 can be collected from NMU cells transfected with ART1 subcloned into a pMAMneo expression vector (*see* **Note 3**) or generated from C2C12 cells (*see* Subheadings 3.2 and 3.3).

HNP-1 is incubated with ART1 and NAD^+. The products are separated by HPLC, the peaks are collected, and the identity is confirmed by MS [39].

1. Incubate HNP-1, 3 nmol with 5 mM NAD^+, ART1, 2 nmol/hr transferase activity (*see* Subheading 3.2) in 50 mM potassium phosphate pH 7.5 in 150 μL at 30 °C overnight. Terminate the reaction by addition of guanidine HCL (final concentration 6 M).
2. The products, the amount of ADP-ribosylated-HNP-1, and the ADP-ribosylation sites are determined by a reverse-phase HPLC separation equipped with a diode-array detector DAD, which takes advantage of the hydrophobicity of HNP and the change in charge by the addition of ADP-ribose. Apply the reaction mixture to a Discovery BIO Wide Pore reverse-phase C18 column equilibrated with HPLC-grade water containing 0.1% trifluoroacetic acid (TFA), mobile phase (A).
3. The products are separated by gradient elution: solution A, 20 min followed by a linear gradient from 0 to 60% solution B (100% isopropanol, 0.2% TFA), followed by 95% solution B for 25 min, flow = 0.8 mL/min. Monitor the separation at absorbance 210 nm (peptide backbone) and 258 nm (ADP-ribose). The collected purified products can be confirmed by mass spectroscopic analysis [39]. To determine the amount of mass in an HPLC peak: mol = Area (under the peak at 280 nm) × HPLC flow rate ÷ Molar extinction (ξ_M) × path length (HPLC cell) [68] (*see* Fig. 1).

3.6 Production of ADP-Ribosylated-HNP-1-Ornithine and Conversion of ADP-Ribosylated Arginine to Ornithine

The conversion of ADP-ribosyl-HNP-1 to ADP-ribosyl-HNP-1-ornithine is nonenzymatic. ART1 ADP-ribosylates HNP-1 on arginine 14 and 24. ADP-ribosylarginine 14 is replaced by ornithine [41]. Ornithine is a noncoded amino acid and is identified by amino acid analysis.

HNP-1 is incubated with ART1 and NAD^+ to ADP-ribosylarginines. The products are separated by HPLC. The HPLC peaks are collected, and ADP-ribosyl-HNP-1-ornithine is confirmed by amino acid analysis.

1. Incubate HNP-1 (6 nmol) with 5 mM NAD^+ and ART1 (5.8 nmol/h) in 150μL of 50 mM potassium phosphate (pH 7.5) at 30 °C for 24 h. The amount of ADP-ribosylated-HNP-ornithine can be optimized by longer incubation time. Terminate the reaction by addition of guanidine HCl (final concentration 6 M). *See* Fig. 2.
2. Separate the products by HPLC under the conditions described in Subheading 3.4. ADP-ribosyl-HNP-1-ornithine is identified by elution time (second eluted peak), and ornithine is con-

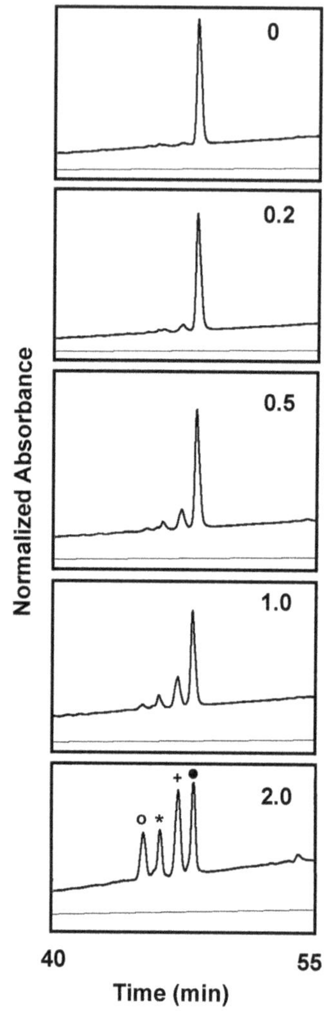

Fig. 1 HPLC separation of ADP-ribosly-HNP-1. The reaction products from the incubation of ART1 at 0, 0.2, 0.5, 1.0, and 2.0 nmol/h activity and HNP-1 separated by HPLC at absorbance 210 nm as described (Subheading 3.4). Di-ADP-ribosylated-HNP-1 (open circle), ADP-ribosyl-HNP-1-ornithine (asterisk), mono-ADP-ribosylated HNP-1 (+) and HNP-1 (filled circle) [40]

firmed by HPLC (with autosampler, florescence detector) amino acid analysis (*see* Fig. 1).

3. To analyze for ornithine in the sample by amino acid analysis, collect the HPLC-eluted peak, dry under vacuum, and then dissolve in 200 μL 6 N HCl, 5 μL 40 mM DTT before acid hydrolysis at 155 °C for 45 min under nitrogen. Vacuum dry the hydrolysate; then solubilize in 25 μL water, 0.05% TFA before ortho-phthalaldehyde (OPA, Agilent Technologies) HPLC precolumn automated derivatization. The OPA derivatives are separated on an Eclipse-AAA (4.6 × 150 mm, 5 μm particle size) HPLC column (Agilent) equilibrated with mobile

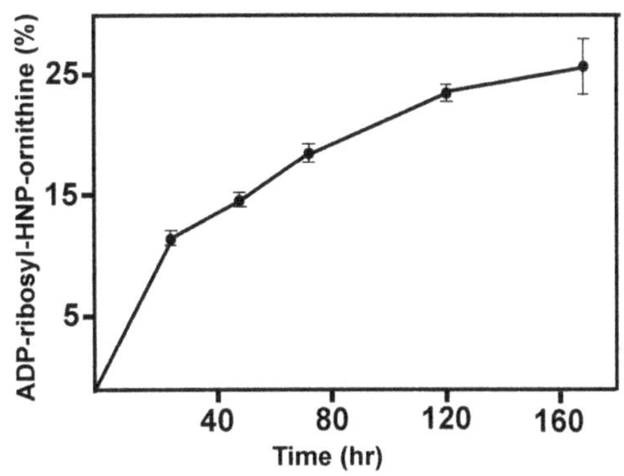

Fig. 2 Amount of ADP-ribosyl-HNP-1-ornithine as a function of time. HNP-1 is incubated with ART1 at the indicated times as described (Subheading 3.5). ADP-ribosyl-HNP-1-ornithine is separated by HPLC, and the amount is quantified by calculating the number of picomoles from the area under the peak (mAu) at 280 nm. The percentage can be calculated by the total amount of reaction products in the separation [40]

phase A, 40 mM sodium dibasic phosphate buffer (pH 7.8). The amino acids are eluted with a linear gradient of 0–40% solvent B, acetonitrile/MeOH/water (45:45:10) for 1.9–15 min to 57% B, and 18.1–18.6 min gradient to 100% B. (*See* Fig. 3.) The conditions for derivatization of the sample and the standard amino acids are described in Agilent Technologies Technical Note (publication no 5980-1193) [40] (*see* **Note 4**).

3.7 Detection of Endogenous Substrates ADP-Ribosylated on Arginine by Macro-domain Pull Down

Af1521 from *Archaeoglobus fulgidus* is a macro-domain module that binds ADP-ribose and poly-ADP-ribose [56, 69] and has been used to identify mono-ADP-ribosylated proteins in CHO cells. ADP-ribosylated proteins in cell lysates can be detected by Af1521 macro-domain pull down. The ADP-ribose arginine linkage can be detected by sensitivity to cleavage by hydroxylamine [70, 71] (*see* **Note 5**).

Lysed cells or tissues are incubated with [^{32}P]-NAD$^+$. Mono- or poly-ADP-ribosylated proteins are pulled down by Af1521 macro-domain. Proteins that are ADP-ribosylated on arginine are incubated with hydroxylamine to release the ADP-ribose.

1. Solubilize the tissue or cells in RIPA buffer (10 mM Tris–HCl, pH 8.0, 1 mM EDTA, 1% Triton-X-100, 0.1% deoxycholate, 0.1% SDS, 140 mM NaCl, 1 mM PMSF). To the solubilized tissue or lysed cells, incubate with 4.5 μCi [^{32}P]-NAD$^+$ for 1 h at 37 °C or determine the time required for maximum ADP-ribosylation [57]. Add 20 μg Af1521 macro-domain-GST resin to the lysate and rotate overnight at 4 °C. Wash the resin three times with PBS; then elute the radioactive mono- and poly-ADP-ribosylated proteins with 20 mM ADP-ribose. To identify proteins that are ADP-ribosylated on arginine, incu-

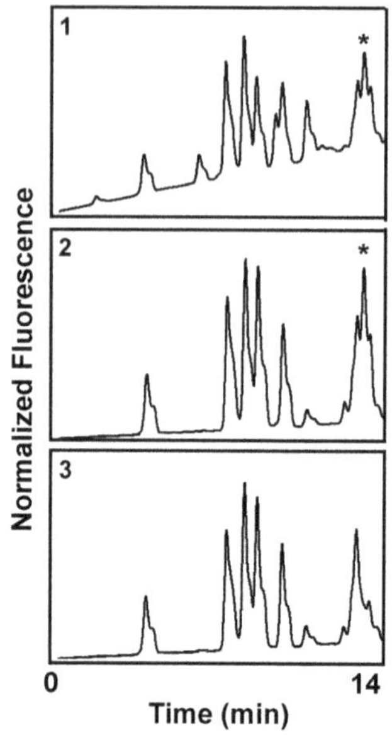

Fig. 3 Amino acid analysis of purified ADP-ribosyl-HNP-1-ornithine. ADP-ribosyl-HNP-1-ornithine purified from the HPLC separation in described in Fig. 1 and prepared for amino acid analysis as described (Subheading 3.5). After OPA derivatization, the amino acids are separated by HPLC with monitoring at fluorescence excitation 340 nm and emission 450 nm. Peptides analyzed were (*1*) ADP-ribosyl-HNP-1-ornithine (asterisk), (*2*) ADP-ribosly-HNP-1 plus ornithine (asterisk), and (*3*) HNP-1 [40]

bate the Af1521 pull down with or without 1 M hydroxylamine (pH 7.0) for 2 h at 37 °C to release ADP-ribose [72]. Add SDS-PAGE loading buffer to the resin or the treated eluate, boil and separate the supernatant on SDS-polyacrylamide gels, stain with Coomassie Blue, dry the gel, and expose to X-AR film as described in Subheading 3.3, **step 4**, to separate all the ADP-ribosylated proteins with or without hydroxylamine treatment. The assay can be repeated, separated on SDS-polyacrylamide gels, and analyzed by Western blotting. ADP-ribosylated proteins can be identified by protein-specific antibodies to probe the immunoblot. For suggestions on applying Af1521 macro-domain resin-GST and other similar products from Tulip Biolabs, see http://www.tulipbiolabs.com/.

2. To identify mono-ADP-ribosylated proteins, poly-ADP-ribosylated proteins can be hydrolyzed to mono-modified proteins by incubation with PAR glycohydrolase (PARG) and then compared to **step 1**. Briefly, the proteins are digested to peptides with LysC or trypsin and incubated with PARG. The mono-ADP-ribosylated peptides are pulled down with Af1521 and identified by MS [58].

4 Notes

1. ADP-ribosylarginine product is in the alpha configuration but readily converts to the beta form with time. Only the alpha form is a substrate for ADP-ribosylarginine hydrolase (ARH1) [73, 74].
2. ADP-ribose can form nonenzymatic linkages to cysteine and lysine. The nonenzymatic transfer of free ADP-ribose to arginine has not been reported. ADP-ribose can be released from a modified arginine substrate by chemical incubation with hydroxylamine or by enzymatic cleavage by ADP-ribosylarginine hydrolase (ARH1) [75–78].
3. ART1 can be collected from NMU cells transfected with ART1 cDNA subcloned into pMAMneo expression vector. Expression is induced by 1 μM dexamethasone sodium phosphate for 24 hours. Incubation of the cells with PIPLC releases ART1. The amount is difficult to detect by SDS-PAGE; the activity is measured as in Subheading 3.2, **step 2** [79].
4. Amino acids are identified by comparison to the elution times of primary amino acids. To be separated by HPLC chromatographic analysis, the amino acids must undergo a derivatization reaction with ortho-phthalaldehyde (OPA) for primary amino acids, e.g., arginine, aspartate, fluorenylmethyl chloroformate (FMOC) for secondary amino acids, e.g., lysine, proline. Briefly, the reactions are carried out prior to separation with an HPLC equipped with an autosampler, a fluorescence detector, and a diode-array (DAD) UV detector. Vials containing the reagents for the derivatization are positioned in the autosampler and added according to a detailed HPLC program. Amino acid standards can be purchased from Agilent. Up-to-date information and support is available at www.agilent.com/chem.
5. The macro-domain Af1521 is a mono-ADP-ribosyl-acceptor hydrolase that can cleave ADP-ribose from ARTD10 [80]. ARTD10 is a mono-ADP-ribosyltransferase that appears to be auto-modified on aspartates or glutamates but not on arginines [81].

Acknowledgment

Funding: The study was funded by the Intramural Research Program, NIH, NHLBI.

References

1. Moss J, Balducci E, Cavanaugh E, Kim HJ, Konczalik P, Lesma EA, Okazaki IJ, Park M, Shoemaker M, Stevens LA, Zolkiewska A (1999) Characterization of NAD:arginine ADP-ribosyltransferases. Mol Cell Biochem 193(1–2):109–113
2. Spangler BD (1992) Structure and function of cholera toxin and the related Escherichia coli heat-labile enterotoxin. Microbiol Rev 56(4):622–647
3. Freissmuth M, Gilman AG (1989) Mutations of GS alpha designed to alter the reactivity of the protein with bacterial toxins. Substitutions at ARG187 result in loss of GTPase activity. J Biol Chem 264(36):21907–21914
4. Moss J, Vaughan M (1978) Isolation of an avian erythrocyte protein possessing ADP-ribosyltransferase activity and capable of activating adenylate cyclase. Proc Natl Acad Sci U S A 75(8):3621–3624
5. Osborne JC Jr, Stanley SJ, Moss J (1985) Kinetic mechanisms of two NAD:arginine ADP-ribosyltransferases: the soluble, salt-stimulated transferase from Turkey erythrocytes and choleragen, a toxin from Vibrio cholerae. Biochemistry 24(19):5235–5240
6. Zolkiewska A, Nightingale MS, Moss J (1992) Molecular characterization of NAD:arginine ADP-ribosyltransferase from rabbit skeletal muscle. Proc Natl Acad Sci U S A 89(23):11352–11356
7. Glowacki G, Braren R, Firner K, Nissen M, Kuhl M, Reche P, Bazan F, Cetkovic-Cvrlje M, Leiter E, Haag F, Koch-Nolte F (2002) The family of toxin-related ecto-ADP-ribosyltransferases in humans and the mouse. Protein Sci 11(7):1657–1670. https://doi.org/10.1110/ps.0200602
8. Peterson JE, Larew JS, Graves DJ (1990) Purification and partial characterization of arginine-specific ADP-ribosyltransferase from skeletal muscle microsomal membranes. J Biol Chem 265(28):17062–17069
9. Okazaki IJ, Zolkiewska A, Nightingale MS, Moss J (1994) Immunological and structural conservation of mammalian skeletal muscle glycosylphosphatidylinositol-linked ADP-ribosyltransferases. Biochemistry 33(43):12828–12836
10. Koch F, Haag F, Thiele HG (1990) Nucleotide and deduced amino acid sequence for the mouse homologue of the rat T-cell differentiation marker RT6. Nucleic Acids Res 18(12):3636
11. Lindskog C, Linne J, Fagerberg L, Hallstrom BM, Sundberg CJ, Lindholm M, Huss M, Kampf C, Choi H, Liem DA, Ping P, Varemo L, Mardinoglu A, Nielsen J, Larsson E, Ponten F, Uhlen M (2015) The human cardiac and skeletal muscle proteomes defined by transcriptomics and antibody-based profiling. BMC Genomics 16:475. https://doi.org/10.1186/s12864-015-1686-y
12. Koch-Nolte F, Haag F, Braren R, Kuhl M, Hoovers J, Balasubramanian S, Bazan F, Thiele HG (1997) Two novel human members of an emerging mammalian gene family related to mono-ADP-ribosylating bacterial toxins. Genomics 39(3):370–376. https://doi.org/10.1006/geno.1996.4520
13. Corda D, Di Girolamo M (2002) Mono-ADP-ribosylation: a tool for modulating immune response and cell signaling. Sci STKE 2002(163):PE53
14. Zolkiewska A (2005) Ecto-ADP-ribose transferases: cell-surface response to local tissue injury. Physiology (Bethesda) 20:374–381. https://doi.org/10.1152/physiol.00028.2005
15. Okazaki IJ, Kim HJ, Moss J (1996a) Cloning and characterization of a novel membrane-associated lymphocyte NAD:arginine ADP-ribosyltransferase. J Biol Chem 271(36):22052–22057
16. Domenighini M, Rappuoli R (1996) Three conserved consensus sequences identify the NAD-binding site of ADP-ribosylating enzymes, expressed by eukaryotes, bacteria and T-even bacteriophages. Mol Microbiol 21(4):667–674
17. Han S, Tainer JA (2002) The ARTT motif and a unified structural understanding of substrate recognition in ADP-ribosylating bacterial toxins and eukaryotic ADP-ribosyltransferases. Int J Med Microbiol 291(6–7):523–529
18. Hottiger MO, Hassa PO, Luscher B, Schuler H, Koch-Nolte F (2010) Toward a unified nomenclature for mammalian ADP-ribosyltransferases. Trends Biochem Sci 35(4):208–219. https://doi.org/10.1016/j.tibs.2009.12.003
19. Grahnert A, Friedrich M, Pfister M, Haag F, Koch-Nolte F, Hauschildt S (2002) Mono-ADP-ribosyltransferases in human monocytes: regulation by lipopolysaccharide. Biochem J 362(Pt 3):717–723
20. Parusel I, Kahl S, Braasch F, Glowacki G, Halverson GR, Reid ME, Schawalder A,

Ortolan E, Funaro A, Malavasi F, Hardie D, Halder S, Buckley CD, Haag F, Koch-Nolte F (2005) A panel of monoclonal antibodies recognizing GPI-anchored ADP-ribosyltransferase ART4, the carrier of the Dombrock blood group antigens. Cell Immunol 236(1–2):59–65. https://doi.org/10.1016/j.cellimm.2005.08.008

21. Heine H, Ulmer AJ, Flad HD, Hauschildt S (1995) Lipopolysaccharide-induced change of phosphorylation of two cytosolic proteins in human monocytes is prevented by inhibitors of ADP-ribosylation. J Immunol 155(10):4899–4908
22. Saxty BA, Kefalas P, Yadollahi-Farsani M, MacDermot J (1998) Arginine-specific ADP-ribosyltransferases in leukocytes. J Leukoc Biol 63(1):15–21
23. Donnelly LE, Rendell NB, Murray S, Allport JR, Lo G, Kefalas P, Taylor GW, MacDermot J (1996) Arginine-specific mono(ADP-ribosyl) transferase activity on the surface of human polymorphonuclear neutrophil leucocytes. Biochem J 315(Pt 2):635–641
24. Kefalas P, Allport JR, Donnelly LE, Rendell NB, Murray S, Taylor GW, Lo G, Yadollahi-Farsani M, MacDermot J (1997) Arginine-specific mono(ADP-ribosyl)transferase activity in human neutrophil polymorphs. A possible link with the assembly of filamentous actin and chemotaxis. Adv Exp Med Biol 419:241–244
25. Wang J, Nemoto E, Dennert G (1997) Regulation of cytotoxic T cell functions by a GPI-anchored ecto-ADP-ribosyltransferase. Adv Exp Med Biol 419:191–201
26. Wang J, Nemoto E, Kots AY, Kaslow HR, Dennert G (1994) Regulation of cytotoxic T cells by ecto-nicotinamide adenine dinucleotide (NAD) correlates with cell surface GPI-anchored/arginine ADP-ribosyltransferase. J Immunol 153(9):4048–4058
27. Liu ZX, Yu Y, Dennert G (1999) A cell surface ADP-ribosyltransferase modulates T cell receptor association and signaling. J Biol Chem 274(25):17399–17401
28. Mehta K, Shahid U, Malavasi F (1996) Human CD38, a cell-surface protein with multiple functions. FASEB J 10(12):1408–1417
29. Han MK, Cho YS, Kim YS, Yim CY, Kim UH (2000) Interaction of two classes of ADP-ribose transfer reactions in immune signaling. J Biol Chem 275(27):20799–20805. https://doi.org/10.1074/jbc.M001189200
30. Tonnesen MG, Feng X, Clark RA (2000) Angiogenesis in wound healing. J Investig Dermatol Symp Proc 5(1):40–46. https://doi.org/10.1046/j.1087-0024.2000.00014.x
31. Boulle N, Jones EM, Auguste P, Baird A (1995) Adenosine diphosphate ribosylation of fibroblast growth factor-2. Mol Endocrinol 9(6):767–775. https://doi.org/10.1210/mend.9.6.8592522
32. Jones EM, Baird A (1997) Cell-surface ADP-ribosylation of fibroblast growth factor-2 by an arginine-specific ADP-ribosyltransferase. Biochem J 323(Pt 1):173–177
33. Saxty BA, Yadollahi-Farsani M, Upton PD, Johnstone SR, MacDermot J (2001) Inactivation of platelet-derived growth factor-BB following modification by ADP-ribosyltransferase. Br J Pharmacol 133(8):1219–1226. https://doi.org/10.1038/sj.bjp.0704187
34. van Wetering S, Tjabringa GS, Hiemstra PS (2005) Interactions between neutrophil-derived antimicrobial peptides and airway epithelial cells. J Leukoc Biol 77(4):444–450
35. Lehrer RI, Ganz T (1992) Defensins: endogenous antibiotic peptides from human leukocytes. Ciba Found Symp 171:276–290; discussion 290–3
36. Balducci E, Horiba K, Usuki J, Park M, Ferrans VJ, Moss J (1999) Selective expression of RT6 superfamily in human bronchial epithelial cells. Am J Respir Cell Mol Biol 21(3):337–346
37. Okazaki IJ, Moss J (1999) Characterization of glycosylphosphatidylinositiol-anchored, secreted, and intracellular vertebrate mono-ADP-ribosyltransferases. Annu Rev Nutr 19:485–509
38. Paone G, Wada A, Stevens LA, Matin A, Hirayama T, Levine RL, Moss J (2002) ADP ribosylation of human neutrophil peptide-1 regulates its biological properties. Proc Natl Acad Sci U S A 99(12):8231–8235. https://doi.org/10.1073/pnas.122238899
39. Paone G, Stevens LA, Levine RL, Bourgeois C, Steagall WK, Gochuico BR, Moss J (2006a) ADP-ribosyltransferase-specific modification of human neutrophil peptide-1. J Biol Chem 281(25):17054–17060
40. Stevens LA, Levine RL, Gochuico BR, Moss J (2009) ADP-ribosylation of human defensin HNP-1 results in the replacement of the modified arginine with the noncoded amino acid ornithine. Proc Natl Acad Sci U S A 106(47):19796–19800. https://doi.org/10.1073/pnas.0910633106
41. Stevens LA, Barbieri JT, Piszczek G, Otuonye AN, Levine RL, Zheng G, Moss J (2014) Nonenzymatic conversion of ADP-ribosylated arginines to ornithine alters the biological activities of human neutrophil peptide-1.

J Immunol 193(12):6144–6151. https://doi.org/10.4049/jimmunol.1303068
42. Zolkiewska A, Moss J (1993) Integrin alpha 7 as substrate for a glycosylphosphatidylinositol-anchored ADP-ribosyltransferase on the surface of skeletal muscle cells. J Biol Chem 268(34):25273–25276
43. Zolkiewska A, Thompson WC, Moss J (1998) Interaction of integrin alpha 7 beta 1 in C2C12 myotubes and in solution with laminin. Exp Cell Res 240(1):86–94. https://doi.org/10.1006/excr.1998.4002
44. Burkin DJ, Kaufman SJ (1999) The alpha7beta1 integrin in muscle development and disease. Cell Tissue Res 296(1):183–190
45. Soman G, Mickelson JR, Louis CF, Graves DJ (1984) NAD: guanidino group specific mono ADP-ribosyltransferase activity in skeletal muscle. Biochem Biophys Res Commun 120(3):973–980
46. Kefalas P, Saxty B, Yadollahi-Farsani M, MacDermot J (1999) Chemotaxin-dependent translocation of immunoreactive ADP-ribosyltransferase-1 to the surface of human neutrophil polymorphs. Eur J Biochem 259(3):866–871
47. Little E, Ramakrishnan M, Roy B, Gazit G, Lee AS (1994) The glucose-regulated proteins (GRP78 and GRP94): functions, gene regulation, and applications. Crit Rev Eukaryot Gene Expr 4(1):1–18
48. Chambers JE, Petrova K, Tomba G, Vendruscolo M, Ron D (2012) ADP ribosylation adapts an ER chaperone response to short-term fluctuations in unfolded protein load. J Cell Biol 198(3):371–385. https://doi.org/10.1083/jcb.201202005
49. Leno GH, Ledford BE (1989) ADP-ribosylation of the 78-kDa glucose-regulated protein during nutritional stress. Eur J Biochem 186(1–2):205–211
50. Fabrizio G, Di Paola S, Stilla A, Giannotta M, Ruggiero C, Menzel S, Koch-Nolte F, Sallese M, Di Girolamo M (2015) ARTC1-mediated ADP-ribosylation of GRP78/BiP: a new player in endoplasmic-reticulum stress responses. Cell Mol Life Sci 72(6):1209–1225. https://doi.org/10.1007/s00018-014-1745-6
51. Cassel D, Selinger Z (1977) Mechanism of adenylate cyclase activation by cholera toxin: inhibition of GTP hydrolysis at the regulatory site. Proc Natl Acad Sci U S A 74(8):3307–3311
52. Lupi R, Corda D, Di Girolamo M (2000) Endogenous ADP-ribosylation of the G protein beta subunit prevents the inhibition of type 1 adenylyl cyclase. J Biol Chem 275(13):9418–9424
53. Lupi R, Dani N, Dietrich A, Marchegiani A, Turacchio S, Berrie CP, Moss J, Gierschik P, Corda D, Di Girolamo M (2002) Endogenous mono-ADP-ribosylation of the free Gbetagamma prevents stimulation of phosphoinositide 3-kinase-gamma and phospholipase C-beta2 and is activated by G-protein-coupled receptors. Biochem J 367(Pt 3):825–832. https://doi.org/10.1042/BJ20020660
54. Kato J, Vekhter D, Heath J, Zhu J, Barbieri JT, Moss J (2015) Mutations of the functional ARH1 allele in tumors from ARH1 heterozygous mice and cells affect ARH1 catalytic activity, cell proliferation and tumorigenesis. Oncogene 4:e151. https://doi.org/10.1038/oncsis.2015.5
55. Kato J, Zhu J, Liu C, Stylianou M, Hoffmann V, Lizak MJ, Glasgow CG, Moss J (2011) ADP-ribosylarginine hydrolase regulates cell proliferation and tumorigenesis. Cancer Res 71(15):5327–5335. https://doi.org/10.1158/0008-5472.CAN-10-0733
56. Karras GI, Kustatscher G, Buhecha HR, Allen MD, Pugieux C, Sait F, Bycroft M, Ladurner AG (2005) The macro domain is an ADP-ribose binding module. EMBO J 24(11):1911–1920. https://doi.org/10.1038/sj.emboj.7600664
57. Dani N, Stilla A, Marchegiani A, Tamburro A, Till S, Ladurner AG, Corda D, Di Girolamo M (2009) Combining affinity purification by ADP-ribose-binding macro domains with mass spectrometry to define the mammalian ADP-ribosyl proteome. Proc Natl Acad Sci U S A 106(11):4243–4248. https://doi.org/10.1073/pnas.0900066106
58. Martello R, Leutert M, Jungmichel S, Bilan V, Larsen SC, Young C, Hottiger MO, Nielsen ML (2016) Proteome-wide identification of the endogenous ADP-ribosylome of mammalian cells and tissue. Nat Commun 7:12917. https://doi.org/10.1038/ncomms12917
59. Vivelo CA, Wat R, Agrawal C, Tee HY, Leung AK (2017) ADPriboDB: the database of ADP-ribosylated proteins. Nucleic Acids Res 45(D1):D204–D209. https://doi.org/10.1093/nar/gkw706
60. Moss J, Stevens LA, Cavanaugh E, Okazaki IJ, Bortell R, Kanaitsuka T, Mordes JP, Greiner DL, Rossini AA (1997) Characterization of mouse Rt6.1 NAD:arginine ADP-ribosyltransferase. J Biol Chem 272(7):4342–4346
61. Oka S, Kato J, Moss J (2006) Identification and characterization of a mammalian 39-kDa poly(ADP-ribose) glycohydrolase. J Biol Chem 281(2):705–713. https://doi.org/10.1074/jbc.M510290200

62. Moss J, Vaughan M (1977) Mechanism of action of choleragen. Evidence for ADP-ribosyltransferase activity with arginine as an acceptor. J Biol Chem 252(7):2455–2457
63. Kato J, Zhu J, Liu C, Moss J (2007) Enhanced sensitivity to cholera toxin in ADP-ribosylarginine hydrolase-deficient mice. Mol Cell Biol 27(15):5534–5543
64. Wernick NL, Chinnapen DJ, Cho JA, Lencer WI (2010) Cholera toxin: an intracellular journey into the cytosol by way of the endoplasmic reticulum. Toxins (Basel) 2(3):310–325. https://doi.org/10.3390/toxins2030310
65. Stevens LA, Bourgeois C, Bortell R, Moss J (2003) Regulatory role of arginine 204 in the catalytic activity of rat alloantigens ART2a and ART2b. J Biol Chem 278(22):19591–19596. https://doi.org/10.1074/jbc.M210364200
66. Konczalik P, Moss J (1999) Identification of critical, conserved vicinal aspartate residues in mammalian and bacterial ADP-ribosylarginine hydrolases. J Biol Chem 274(24):16736–16740
67. Wisniewski JR, Zougman A, Nagaraj N, Mann M (2009) Universal sample preparation method for proteome analysis. Nat Methods 6(5):359–362. https://doi.org/10.1038/nmeth.1322
68. Levine RL, Williams JA, Stadtman ER, Shacter E (1994) Carbonyl assays for determination of oxidatively modified proteins. Methods Enzymol 233:346–357
69. Allen MD, Buckle AM, Cordell SC, Lowe J, Bycroft M (2003) The crystal structure of AF1521 a protein from Archaeoglobus fulgidus with homology to the non-histone domain of macroH2A. J Mol Biol 330(3):503–511
70. Moss J, Yost DA, Stanley SJ (1983) Amino acid-specific ADP-ribosylation. J Biol Chem 258(10):6466–6470
71. Payne DM, Jacobson EL, Moss J, Jacobson MK (1985) Modification of proteins by mono(ADP-ribosylation) in vivo. Biochemistry 24(26):7540–7549
72. Weng B, Thompson WC, Kim HJ, Levine RL, Moss J (1999) Modification of the ADP-ribosyltransferase and NAD glycohydrolase activities of a mammalian transferase (ADP-ribosyltransferase 5) by auto-ADP-ribosylation. J Biol Chem 274(45):31797–31803
73. Tsuchiya M, Tanigawa Y, Mishima K, Shimoyama M (1986) Determination of ADP-ribosyl arginine anomers by reverse-phase high-performance liquid chromatography. Anal Biochem 157(2):381–384
74. Moss J, Oppenheimer NJ, West RE Jr, Stanley SJ (1986) Amino acid specific ADP-ribosylation: substrate specificity of an ADP-ribosylarginine hydrolase from Turkey erythrocytes. Biochemistry 25(19):5408–5414
75. Moss J, Stanley SJ, Nightingale MS, Murtagh JJ Jr, Monaco L, Mishima K, Chen HC, Williamson KC, Tsai SC (1992) Molecular and immunological characterization of ADP-ribosylarginine hydrolases. J Biol Chem 267(15):10481–10488
76. McDonald LJ, Moss J (1994) Enzymatic and nonenzymatic ADP-ribosylation of cysteine. Mol Cell Biochem 138(1–2):221–226
77. Cervantes-Laurean D, Minter DE, Jacobson EL, Jacobson MK (1993) Protein glycation by ADP-ribose: studies of model conjugates. Biochemistry 32(6):1528–1534
78. Hsia JA, Tsai SC, Adamik R, Yost DA, Hewlett EL, Moss J (1985) Amino acid-specific ADP-ribosylation. Sensitivity to hydroxylamine of [cysteine(ADP-ribose)]protein and [arginine(ADP-ribose)]protein linkages. J Biol Chem 260(30):16187–16191
79. Okazaki IJ, Kim HJ, McElvaney NG, Lesma E, Moss J (1996b) Molecular characterization of a glycosylphosphatidylinositol-linked ADP-ribosyltransferase from lymphocytes. Blood 88(3):915–921
80. Rosenthal F, Feijs KL, Frugier E, Bonalli M, Forst AH, Imhof R, Winkler HC, Fischer D, Caflisch A, Hassa PO, Luscher B, Hottiger MO (2013) Macrodomain-containing proteins are new mono-ADP-ribosylhydrolases. Nat Struct Mol Biol 20(4):502–507. https://doi.org/10.1038/nsmb.2521
81. Kleine H, Poreba E, Lesniewicz K, Hassa PO, Hottiger MO, Litchfield DW, Shilton BH, Luscher B (2008) Substrate-assisted catalysis by PARP10 limits its activity to mono-ADP-ribosylation. Mol Cell 32(1):57–69. https://doi.org/10.1016/j.molcel.2008.08.009

Chapter 11

Monitoring Expression and Enzyme Activity of Ecto-ARTCs

Stephan Menzel, Sahil Adriouch, Peter Bannas, Friedrich Haag, and Friedrich Koch-Nolte

Abstract

Mammalian ARTCs are expressed as glycosylphosphatidylinositol (GPI)-anchored ectoenzymes (ARTC1–ARTC4) or secretory proteins (ARTC5) by different cell types. The ARTC2 enzymes catalyze mono-ADP-ribosylation of arginine residues in the extracellular domain of membrane proteins or secretory proteins. In this chapter we provide protocols to monitor the expression and activity of ARTCs on the cell membrane of living cells and in soluble form in biological fluids.

Key words ADP-ribosylation, Ecto-ADP-ribosyltransferases, ARTC, SDS-PAGE autoradiography, Flow cytometry, Near-infrared in vivo imaging

1 Introduction

ADP-ribosyltransferases of the clostridial toxin-like NAD$^+$-ADP-ribosyl-acceptor ADP- ribosyltransferase type C (ARTCs) catalyze mono-ADP-ribosylation of substrate proteins by transferring the ADP-ribose moiety from NAD$^+$ to the guanidino group of accessible arginine residues of the target [1, 2]. In mammals, the ARTC family consists of four glycosylphosphatidylinositol (GPI)-anchored membrane proteins (ART1 through ART4) and a secretory protein (ART5) [3, 4]. These ecto-ARTCs—in contrast to their intracellular ARTD counterparts—are active in the extracellular space and ADP-ribosylate membrane proteins or proteins present in extracellular body fluids [4, 5, 6]. The best characterized members of the ARTC family are ARTC2.1 and ARTC2.2 on murine immune cells and ARTC1 on human and murine muscle cells [7–10]. Activation of murine T cells by T cell receptor engagement or by activation of the purinergic P2X7 receptor leads to shedding of CD62L [11] and the ecto-domain of ARTC2.2, releasing a catalytically active soluble form environment [12–14].

Paul Chang (ed.), *ADP-ribosylation and NAD+ Utilizing Enzymes: Methods and Protocols*, Methods in Molecular Biology, vol. 1813, https://doi.org/10.1007/978-1-4939-8588-3_11,

The different cellular localization and target specificity of ARTCs vs. ARTDs require a distinct set of methods for monitoring their expression and activity. Membrane-anchored extracellular ARTCs are accessible for antibodies on living cells. Thus, cell surface expression of ARTCs can be monitored by flow cytometry using fluorochrome-conjugated monoclonal antibodies without the need for cell permeabilization [15–17] (Subheading 3.1). Similarly, soluble ARTCs that have been released into extracellular fluids can be detected by sandwich ELISA using a pair of monoclonal antibodies recognizing two non-overlapping epitopes on the same ARTC [13] (Subheading 3.2).

Two approaches have been developed to monitor the enzymatic activity of ARTCs and to detect their ADP-ribosylated target proteins: using either ^{32}P-radiolabeled NAD$^+$ (Subheading 3.3) or etheno-NAD$^+$ (Subheading 3.4) as substrate (Fig. 1) [7, 18]. Advantages of ^{32}P-NAD$^+$ are the high sensitivity and the possibility to quantify incorporation of the radiolabel by autoradiography. Limitations of this method are the special handling requirements for radioactive materials and the difficulties to use such materials for in vivo experiments [7]. The use of etheno-NAD$^+$ leads to etheno-ADP-ribosylation of arginine residue on the target proteins. The exposed moiety can conveniently be detected using ethenoadenosine-specific monoclonal antibody 1G4 [18]. This approach thus can be used in classical antibody-based assays including flow cytometry, immunoprecipitation, and Western blotting (Subheading 3.4) and is also applicable for in vivo studies.

The following materials and methods have been developed for measuring expression and activity of mouse ARTC2.2. The protocols can readily be adapted to monitor other ecto-ARTCs (*see* **Note 1**).

2 Materials

1. ARTC-transfected cells: purchase a synthetic gene block encompassing the full length open reading frame of the desired ARTC flanked by suitable restriction enzyme sites, and clone the gene block into a eukaryotic expression vector, e.g., pCDNA6 (*see* **Note 2**). Transfect HEK-293T cells or lymphoma cells using a suitable transfection system, e.g., jetPEI for HEK-293T cells or electroporation for lymphoma cells (*see* **Note 3**). Optional, in order to facilitate visualization of transfected vs. non-transfected cells, co-transfect cells with a cDNA expression vector encoding GFP. Harvest cells 24–48 h post transfection. For stable selection, replace medium 24–48 h after transfection with cell culture medium containing a suitable selection drug (e.g., blasticidin in case of pCDNA6).

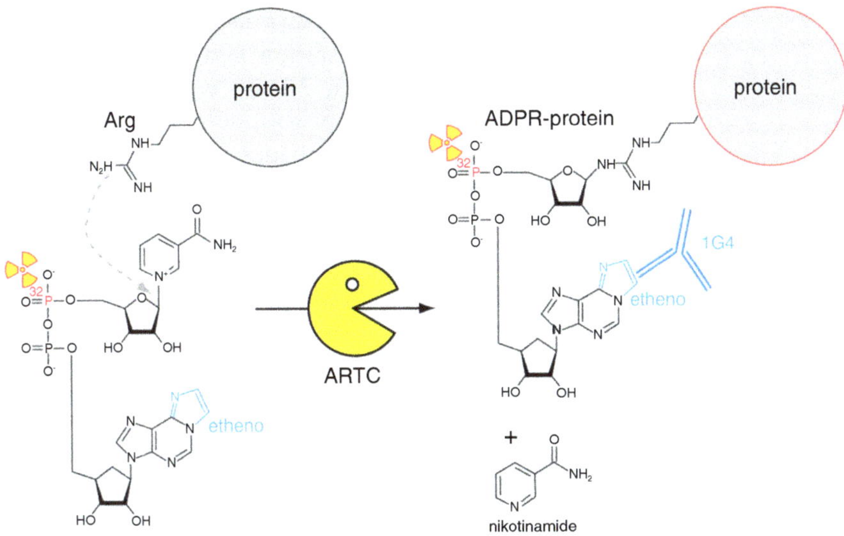

Fig. 1 Monitoring ARTC-catalyzed ADP-ribosylation of target proteins. ^{32}P-NAD$^+$ or etheno-NAD$^+$ can be used as substrates to track ARTC-mediated ADP-ribosylation. ^{32}P-ADP-ribosylated proteins can be detected by SDS-PAGE autoradiography; etheno-ADP-ribosylated proteins can be detected by flow cytometry or Western blotting using ethenoadenosine-specific monoclonal antibody 1G4

Monitor expression and enzyme activity of the respective ARTC as described below.

2. ARTC-specific antibodies: for flow cytometry, immunoprecipitation, and ELISA, use validated monoclonal antibodies that recognize ARTC in native conformation (*see* **Note 1**). For Western blot analyses, use polyclonal anti-peptide antibodies.

2.1 Monitoring the Expression of Ecto-ARTCs by Flow Cytometry

1. Antibodies: titrate each lot of fluorochrome-conjugated antibodies (final concentrations typically are in the range of 0.1–1 μg/mL), anti-ARTC2.2 (Nika102), anti-B220 (RA3-6B2), anti-CD3 (145-2C11), anti-CD4 (RM4-5), anti-CD8 (RPA-T8), and anti-CD25 (PC61). Dynabead-immobilized goat anti-mouse IgG for B cell depletion of splenocytes.
2. FACS buffer: PBS pH 7.4, 1 mM EDTA, 0.5% bovine serum albumin (BSA).
3. ARTC2.2-expression vector: standard eukaryotic expression vector, e.g., pCDNA6-blasticidin (addgene #V22120) containing the full length open reading frame of ARTC2.2 (NCBI Gene ID, 11872), jetPEI transfection reagent (Q-Biogen), and blasticidin (15 μg/mL).
4. Cells: ARTC2.2-expressing lymphoma cell lines Yac-1 (ATCC, TIB160) and ARTC2.2-transfected HEK-293T cells (ATCC, CRL11268), RPMI or DMEM tissue culture medium containing 10% fetal calf serum (FCS), and incubator. Primary

cells from murine blood, thymus, spleen, and lymph nodes: scissors, forceps, a preparation board, 70% ethanol, 15 and 50 mL centrifuge tubes, 70 μm cell strainers, 10 cm petri dishes, 10 mL syringe piston, serological pipettes, pipette boy, cell centrifuge, Neubauer chamber, cell microscope, Dynabead-immobilized goat anti-mouse IgG (Dynal Biotech), and a suitable magnet for depletion of B cells.

5. Propidium iodide: 100× stock solution at 20 μg/mL for discrimination of dead cells.

2.2 Monitoring Soluble ARTCs in Biological Fluids and Cell Culture Supernatants by ELISA

1. Antibodies: catcher, Nika109 (100 ng/well); detector, biotinylated Nika102 (100 ng/well); and peroxidase-conjugated anti-rat IgG.
2. Plates: common ELISA plates like Nunc MaxiSorp F96 plates.
3. *Buffers: antibody and sample dilution buffer, PBS, 1% BSA; coating buffer, bicarbonate buffer,* pH 9.6, or PBS pH 7.4; blocking buffer, PBS, 5% BSA; washing buffer, PBS, 0.05% Tween 20.
4. Substrate: 3,3′,5,5′-tetramethylbenzidin (use 100 μL per well).
5. Stop solution: sulfuric acid, 1 M (use 100 μL per well).
6. Biological fluids: serum, collect mouse blood into 1.5 mL centrifuge tube, and store at room temperature for 1–2 h; remove the blood clot, centrifuge the sample for 5 min at 16,000 *g* at 4 °C, and harvest the supernatant. Cell culture supernatants, for inducing ecto-domain shedding of ARTC2.2, incubate cells for 30 min at 37 °C with PI-PLC (1 IU/ml), ATP (1 mM), or PMA (10 mM). Spin down cells for 10 min at 1.200 rpm and harvest supernatant.
7. Standard: recombinant ecto-domain of ARTC2 (serial log-10 dilution of ARTC2.2 in blocking buffer from 100 ng/mL to 100 pg/mL, i.e., from 10 ng to 10 pg/well).

2.3 Detecting ADP-Ribosylated Target Proteins with ^{32}P-NAD+ as Substrate

1. NAD^+: radiolabeled NAD^+, ^{32}P-NAD^+. Use 5–50 μCi/mL, in a buffer containing unlabeled NAD^+ at a final concentration of 1–10 μM NAD^+ (*see* **Note 4**). Unlabeled NAD^+, 1 mM stock solution of β-NAD^+ in PBS, store in aliquots at −80 °C. ADP-ribose 10 mM stock solution of ADP-ribose (Sigma) in PBS, store in aliquots at −80 °C.
2. 10× labeling solution: PBS containing 10 μM NAD^+, 5 mM ADP-ribose.
3. Isotope lab with appropriate shielding (1 cm plastic shields for low-dose beta-emitting isotopes), radiation detectors, and personal dosimeter.
4. SDS-PAGE: common SDS-PAGE apparatus and precast 10% acrylamide gels, e.g., Novex/NuPAGE System. Use MES

(50 mM MES, 50 mM Tris base, 0.1% SDS, 1 mM EDTA, pH 7.3) as gel running buffer for the detection of small- to medium-sized proteins. Gels can be dried using the DryEase System (Life Technologies).

5. Autoradiography: X-ray films suitable for autoradiography, e.g., Hyperfilm MP (GE Healthcare). X-ray cassettes. Dark room. −80 °C freezer. Standard film developer, e.g., AGFA Curix60.

2.3.1 ^{32}P-ADP-Ribosylation of Membrane Proteins

1. Cells: *see* Subheading 2.
2. Reaction buffer: PBS or cell culture medium without FCS.
3. Antibodies: conjugate monoclonal antibodies to a suitable matrix, e.g., covalently to AminoLink agarose beads (Thermo Fisher) or non-covalently to protein G Sepharose beads (Pharmacia), anti-LFA-1 (M17/4), anti-CD8 (RPA-T8), anti-CD25 (PC61), and anti-P2X7 (Hano44) [19] (*see* **Note 5**).
4. Lysis and wash buffer: PBS containing 1% Triton X-100 and protease inhibitors such as AEBSF (1 mM).
5. ACK erythrocyte lysis buffer containing 155 mM NH_4Cl + 10 mM $KHCO_3$ + 0.1 mM EDTA at pH 7.2.

2.3.2 ^{32}P-ADP-Ribosylation of Soluble Proteins

1. Target proteins: cell lysates, serum, 50–500 ng purified recombinant proteins.
2. Recombinant ARTCs: 10–50 ng. ARTC-containing cell supernatant or other ARTC-containing biological fluid.
3. Reaction buffer: PBS, optionally add BSA as carrier protein (10 μg/mL). BSA is a weak target for ARTC-mediated ADP-ribosylation.

2.4 Detecting Etheno-ADP-Ribosylated Target Proteins

2.4.1 Monitoring Etheno-ADP-Ribosylation of Membrane Proteins by Flow Cytometry

1. Cells: *see* Subheading 2.
2. Etheno-NAD$^+$: 100 μM stock solution of β-NAD$^+$ in PBS, store in aliquots at −80 °C.
3. Buffers: RPMI medium; FACS buffer, PBS pH 7.4, 1 mM EDTA, and 0.5% bovine serum albumin (BSA).
4. Antibodies: ethenoadenosine-specific monoclonal antibody 1G4 (mouse IgG1) and fluorochrome-conjugated secondary antibody, e.g., FITC-labeled anti-mouse IgG.

2.4.2 Detecting Etheno-ADP-Ribosylated Proteins by Western Blot Analysis

1. PVDF membranes.
2. Buffers: transfer buffer, Tris-glycine 20% methanol; blocking buffer, TBS and 5% milk powder; antibody dilution buffer, TBS, 1% milk powder, and 0.05% Tween 20.
3. Antibodies: primary antibody, 1G4 (0.5 μg/mL); secondary antibody, peroxidase-conjugated goat anti-mouse IgG.

4. Detection system: enhanced chemiluminescence (ECL) system.
5. Film: ECL Hyperfilm (GE Healthcare).

2.5 In Vivo Imaging of ARTC-Expressing Cells

2.5.1 Near-Infrared In Vivo Imaging of ARTC2.2-Expressing T Cells

1. Antibodies: AF680-conjugated nanobody s+16a and AF680-conjugated mAb Nika102.
2. Mice: BALBc wild type, ARTC2.2 knockout, and ARTC2.2-overexpressing transgenic mice. To reduce autofluorescence of the intestine, keep mice on an alfalfa-free diet for 1 week prior to in vivo imaging.
3. 1 mL syringe, lamp, or water bath.
4. Whole-body in vivo small animal near-infrared fluorescence (NIRF) imaging system (e.g., IVIS-Spectrum or IVIS-200, Caliper Life Sciences intravital imaging system), Living Image 4.2 software (Caliper Life Sciences).

2.5.2 Near-Infrared In Vivo Imaging of ARTC2.2-Expressing Lymphoma Cells

1. Antibodies: AF680-conjugated nanobody s+16a and AF680-conjugated mAb Nika102.
2. Cells: untransfected DC27.10 lymphoma cells and DC27.10 cells stably transfected with ARTC2.2.
3. Mice: athymic nude mice (NMRI-Foxn1^{nu}). To reduce autofluorescence of the intestine, keep mice on an alfalfa-free diet for 1 week prior to in vivo imaging.
4. Basement matrix (Matrigel), 1 mL syringes, 30G needles.
5. 1 mL syringe, lamp, or water bath.
6. In vivo NIRF imaging system. IVIS-200.

3 Methods

3.1 Monitoring Expression of Ecto-ARTCs by Flow Cytometry

To detect ARTCs on the surface of living or fixed cells, stain ecto-ARTCs by incubation of cells with fluorochrome-labeled ARTC-specific antibodies. Wash cells to remove unbound antibody and detect cell-bound antibodies by flow cytometry.

1. Conjugate ARTC2.2-specific monoclonal antibody to a suitable fluorochrome, e.g., Alexa488 according to the manufacturer's instructions.
2. Resuspend ARTC2.2-expressing lymphoma cells or primary lymphocytes from peripheral blood, thymus, spleen, lymph nodes, etc. in FACS buffer at 2 × 10^{7} cells/mL (*see* **Note 6**).
3. Transfer 50 μL aliquots of cells (1 × 10^{6} cells) into 5 mL polystyrol tubes.
4. Prepare twofold concentrated antibody cocktails in RPMI medium, e.g., anti-ARTC2.2, anti-CD3, anti-CD4, anti-CD8, and anti-CD25 in FACS buffer. Add 50 μL of antibody solution per tube; incubate in the dark for 30 min at 4 °C.

5. Wash samples twice with 2 mL FACS buffer by centrifugation at 300 *g* for 10 min at 4 °C
6. Resuspend cells in 250 μL RPMI, optional: add 10 μL propidium iodide (20 μg/mL) for dead cell discrimination.
7. Analyze stained cells by flow cytometry, e.g., with a FACS Canto II (BD) and appropriate data analysis software, e.g., FlowJo (Tristar) (Fig. 2).

3.2 Detecting and Quantifying Soluble ARTCs by ELISA

To detect soluble ARTCs by sandwich ELISA, coat the wells of an ELISA plate with a catcher antibody, block free binding sites with BSA, add samples, and allow ARTCs to bind to the catcher antibody. After washing, detect bound ARTs by sequential incubation with a biotinylated secondary antibody, peroxidase-conjugated streptavidin, and TMB substrate. To stop the reaction, add sulfuric acid and measure absorbance in a plate reader.

1. Coat an ELISA plate with the ARTC2.2-specific catcher antibody Nika109 (200 ng/100 μL per well) over night at 4 °C.
2. Remove antibody solution, and block free binding sites by incubating each well with 200 μL blocking buffer (PBS, 5% BSA) for 60 min at RT.
3. Add 100 μL of sample per well, e.g., mouse serum (*see* **Note 7**) or cell culture supernatants (*see* **Note 8**). As a standard, use a serial dilution of recombinant ARTC2.2 in blocking buffer.
4. Incubate plates for 2 h at RT.
5. Wash wells 4× with 200 μL washing buffer (PBS, 0.05% Tween 20).
6. Add biotinylated Nika102 (100 ng in 100 μL sample buffer per well) as detector antibody, and incubate at RT for 1 h (*see* **Note 9**).
7. Wash wells 4× with 200 μL washing buffer (PBS, 0.05% Tween 20).
8. Detect bound Nika102 with peroxidase-conjugated streptavidin.
9. Wash wells 4× with 200 μL washing buffer (PBS, 0.05% Tween 20).
10. Add 100 μL of TMB-based reaction reagent, and wait until a blue staining is visible.
11. Stop reaction by adding stop solution (0.5 M sulfuric acid).
12. Read absorbance at 450 nm in a plate reader (Fig. 3).

3.3 Detecting ^{32}P-ADP-Ribosylated Target Proteins

3.3.1 ^{32}P-ADP-Ribosylation of Specific Membrane Proteins

To detect ^{32}P-ADP-ribosylated membrane proteins, incubate cells with ^{32}P-NAD, and then wash cells to remove unbound radiolabel. Lyse the cell membrane by incubation in a nonionic detergent, and remove nuclei and insoluble material by centrifugation. Keep an aliquot of the cell lysate to monitor total radiolabeled proteins. Immunoprecipitate specific target proteins of interest using

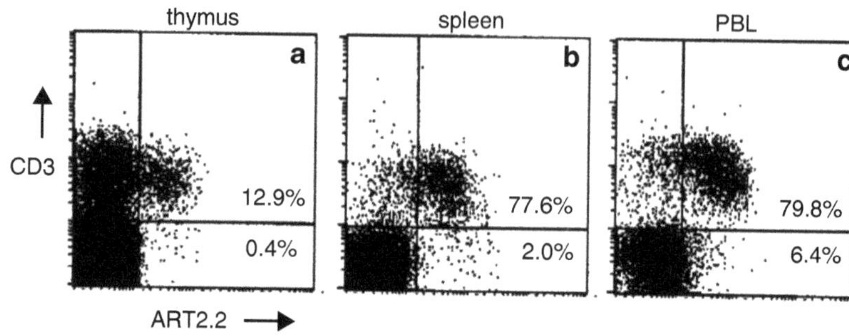

Fig. 2 Monitoring cell surface expression of ARTC2.2 by flow cytometry. Thymocytes (**a**), splenocytes (**b**), and peripheral blood leukocytes (**c**) from a C57BL/6J mouse were stained with FITC-conjugated anti-ARTC2.2 (Nika102) and with PE-conjugated anti-CD3. Numbers indicate the percentage of ARTC2.2+ cells within subpopulations of $CD3^+$ and $CD3^-$. In the thymus, only a subset of $CD3^{int}$ (mature) T cells express ARTC2.2. In the spleen and peripheral blood, most T cells express high levels of ARTC2.2. Only a few, if any, $CD3^-$ cells express ARTC2.2. Reproduced by permission from the American Association of Immunologists ©1999 (Koch-Nolte et al. 1999, J Immunol 163:6014–6022, Fig. 3)

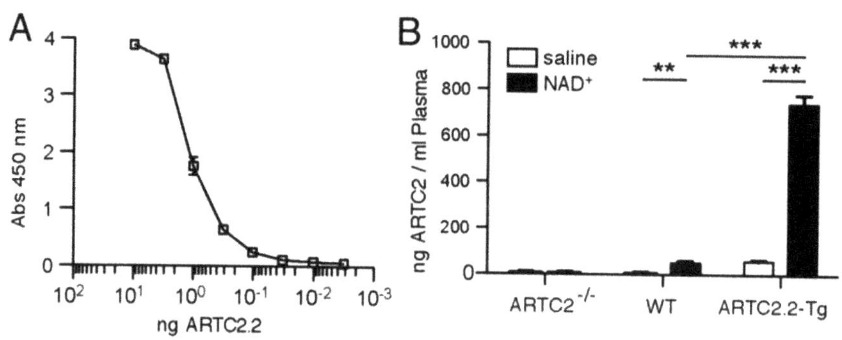

Fig. 3 Detecting soluble ARTC2.2 by ELISA. ELISA plates coated with ARTC2.2-specific mAb Nika109 were incubated with serial dilution of purified shed ARTC2.2 (**a**) or with mouse serum (1:10) (**b**). Bound ARTC2.2 was detected with biotinylated Nika102 and peroxidase-conjugated streptavidin. (**b**) C57BL/6J mice received intravenous injections of saline (200 μL) or NAD (2 mg in 200 μL saline) 20 min before sacrifice. ARTC2.2−/−, ARTC2.2-deficient mice; WT, wild-type mice; ARTC2.2-Tg, ARTC2.2-transgenic mice. Reproduced by permission from the American Association of Immunologists ©2015 (Menzel et al. J Immunology 195:2057–2066, Fig. 7)

immobilized antigen-specific monoclonal antibodies. Detect radiolabeled target proteins by SDS-PAGE autoradiography.

1. Resuspend ARTC-expressing lymphoma cells in RPMI medium at 5×10^6 cells/mL.
2. Resuspend T cells (*see* **Note 10**) in RPMI medium at 1×10^7 cells/mL. Gently perfuse the spleen of a sacrificed mouse with 10 mL ice-cold medium using a 1.0 gauge needle and 10 mL syringe. Pass cell suspension through a 70 μm cell strainer. Lyse erythrocytes by resuspending splenocytes in ACK erythrocyte lysis buffer. Save an aliquot (1/10th) of sple-

nocytes for comparative flow cytometry. Deplete B cells by adding Dynabead-immobilized goat anti-mouse IgG (Dynal Biotech) at 4 °C. Gently rock the solution for 3 min at 4 °C. Beads will clump as a result of binding to B cells. Insert the test tube into a magnetic rack, and wait for 2–3 min until the clumped beads have gathered at the side of the tube. Transfer the cell suspension to a fresh tube and repeat the procedure. Confirm the purity of T cells by flow cytometry, e.g., using APC-conjugated anti-B220 and PE-conjugated anti-CD3.

3. In the isotope lab, prepare the 10× labeling solution of in RPMI medium (10 μM unlabeled NAD^+ and 5 mM ADP-ribose).
4. Behind a 1 cm plastic shield, add ^{32}P-labeled NAD^+ to 50 μCi/mL (for a final concentration of 5 μCi/mL).
5. To each 90 μL aliquot of cells, add 10 μL of the 10× labeling solution.
6. Incubate the cells for 30 min at 4 °C or at 37 ° C (*see* **Note 11**). Every few minutes, gently shake the tubes to keep cells in suspension.
7. Wash cells 4 × in PBS, 1% BSA, 1 mM ADP-ribose.
8. Lyse cells in 250 μL PBS, 1% IGEPAL, 1 mM AEBSF at 4 °C or at 37 °C for 30 min (*see* **Note 12**).
9. Remove nuclei and insoluble material by centrifugation (1000 *g* at 4 °C for 10 min).
10. Transfer supernatant to a fresh tube, and preclear cell lysates by high-speed centrifugation (14,000 *g* at 4 °C for 30 min). Remove an aliquot of the cell lysate (1/10th) for direct analysis of total radiolabeled proteins by SDS-PAGE autoradiography.
11. Incubate cell lysates with 20 μL protein G Sepharose for 30 min at 4 °C.
12. Immunoprecipitate target proteins using validated, target antigen-specific monoclonal antibodies immobilized on AminoLink agarose or protein G Sepharose (*see* **Note 1**).
13. Size fractionate proteins by SDS-PAGE (e.g., on 10% precast NuPAGE gels, Invitrogen).
14. Detect radiolabeled proteins by autoradiography by exposing the membrane to an X-ray film (e.g., Hyperfilm MP (GE Healthcare)) at −80 °C (*see* **Note 13**) (Fig. 4).

3.3.2 ^{32}P-ADP-Ribosylation of Soluble Target Proteins

To monitor ^{32}P-ADP-ribosylation of soluble target proteins, incubate solutions containing target proteins and ARTC enzyme with ^{32}P-NAD. Stop reaction by adding SDS-PAGE sample buffer, and detect radiolabeled proteins by SDS-PAGE autoradiography.

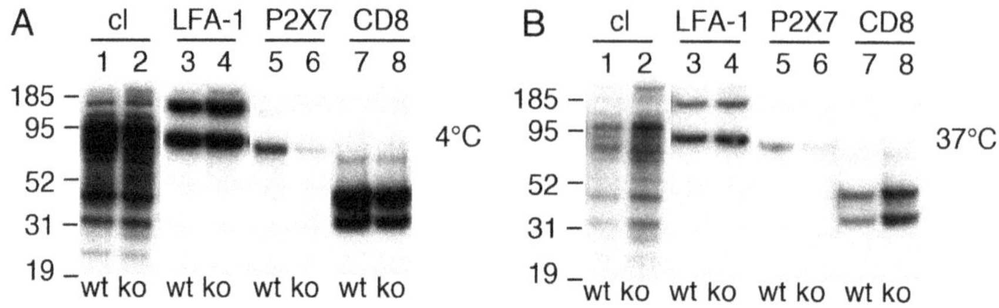

Fig. 4 Detecting ^{32}P-ADP-ribosylated membrane proteins by SDS-PAGE autoradiography. Purified lymph node T cells of BALB/c mice were incubated for 20 min at 4 °C (**a**) or at 37 °C (**b**) in the presence of exogenous ^{32}P-NAD$^+$ (1 μM). Cells were washed and lysed in 1% Triton X-100. Cell lysates (cl) were clarified by high-speed centrifugation and subjected to immunoprecipitation with protein G immobilized monoclonal antibodies directed against LFA-1, P2X7, or CD8. Radiolabeled proteins were detected by SDS-PAGE autoradiography. *WT* wild-type mice, *ko* CD38-deficient mice. Incubation of cells at 37 °C results in lower radiolabeling of proteins than incubation at 4 °C, consistent with reversion of ADP-ribosylation at 37 °C (*see* **Note 11**). At 37 °C, overall radiolabeling of proteins is lower in wild type than in CD38-deficient cells, consistent with CD38-mediated hydrolysis of ^{32}P-NAD by WT cells (*see* **Note 10**). Reproduced by permission from the American Association of Immunologists ©2009 (Scheuplein et al., J Immunology 182:2898–2908, Fig. 9)

1. Prepare a solution of target proteins in 17 μL PBS, e.g., 100–500 ng recombinant protein, 0.5 μL serum, or cell lysate containing 1–20 μg of protein.
2. Add recombinant ARTC2.2 (5–100 ng in 1 μl PBS).
3. In the isotope lab, prepare the 10× labeling solution in PBS (10 μM unlabeled NAD$^+$ and 5 mM ADP-ribose).
4. Behind a 1 cm plastic shield, add ^{32}P-labeled NAD to 50 μCi/ml (for a final concentration of 5 μCi/mL).
5. To each aliquot of target protein solution, add 2 μL of the 10× labeling solution.
6. Incubate for 15 min at 37 °C.
7. Add 6.6 μL of 4× sample buffer and heat samples for 10 min at 70 °C.
8. Size fractionate proteins by SDS-PAGE, e.g., on a 10% precast NuPAGE gel.
9. Stain the gel with Coomassie blue overnight; destain and dry the gel (*see* **Note 14**).
10. Detect radiolabeled proteins by autoradiography by exposing the dried gel to an X-ray film, e.g., Hyperfilm MP (GE Healthcare), at −80 °C .

3.4 Detecting Etheno ADP-Ribosylated Target Proteins

3.4.1 Monitoring Etheno-ADP-Ribosylation of Membrane Proteins by Flow Cytometry

To monitor etheno-ADP-ribosylation of membrane proteins by flow cytometry, incubate primary or transfected cells with etheno-NAD. Wash cells, and stain etheno-ADP-ribosylated cell surface proteins by incubation of cells with fluorochrome-labeled ethenoadenosine-specific monoclonal antibody 1G4. Wash cells to remove unbound antibody, and detect cell-bound antibodies by flow cytometry.

1. Conjugate ethenoadenosine-specific monoclonal antibody 1G4 to a suitable fluorochrome, e.g., Alexa488 according to the manufacturer's (Molecular Probes, Thermo Fisher) instructions.
2. Resuspend ARTC-expressing lymphoma cells or T cells in RPMI medium at 1×10^7 cells/mL.
3. Add 1/10 volume of 100 μM etheno-NAD^+, 5 mM ADP-ribose, and incubate cells for 15 min at 4–37 °C (*see* **Note 15**).
4. Dilute cells 20-fold in medium lacking NAD^+ and wash cells twice.
5. Incubate cells with fluorochrome-conjugated ethenoadenosine-specific mAb 1G4 (1 μg/mL) and other fluorochrome-conjugated antibodies for 30 min at 4 °C.
6. Wash cells twice in FACS buffer.
7. Resuspend cells in 250 μL RPMI, optional: add 10 μL propidium iodide (20 μg/mL) for dead cell discrimination.
8. Analyze stained cells by flow cytometry, e.g., with a FACS Canto II (BD) and appropriate data analysis software, e.g., FlowJo (Tristar) (Fig. 5).

3.4.2 Detecting Specific Etheno-ADP-Ribosylated Membrane Proteins by Immunoprecipitation and Western Blotting

To monitor etheno-ADP-ribosylation of membrane proteins by Western blotting, incubate cells with etheno-NAD. Wash cells to remove excess substrate. Lyse the cell membrane by incubation in a nonionic detergent, and remove nuclei and insoluble material by centrifugation. Keep an aliquot of the cell lysate to monitor total etheno-ADP-ribosylated proteins. Immunoprecipitate specific target proteins of interest using immobilized antigen-specific monoclonal antibodies. Size fractionate proteins by SDS-PAGE and blot proteins onto a PVDF membrane. Block free binding sites with BSA, and detect etheno-ADP-ribosylated proteins by sequential incubation with monoclonal antibody 1G4, peroxidase-labeled anti-mouse IgG, and ECL reagent. Expose the membrane to ECL film in the dark room.

1. Incubate cells with etheno-NAD^+ as described in the previous section (Subheading 3.4.1, **steps 1–5**).
2. Lyse cells, immunoprecipitate specific etheno-ADP-ribosylated proteins, and separate proteins by SDS-PAGE as described above for ^{32}P-ADP-ribosylated proteins (Subheading 3.3.1, **steps 8–11**).
3. Blot proteins onto a PVDF membrane.
4. Block membrane using blocking buffer for 1 h at RT.
5. Incubate membrane with monoclonal antibody 1G4 (0,5 μg/mL in PBS, 1% BSA, 0.05% Tween 20) for 1 h at RT.

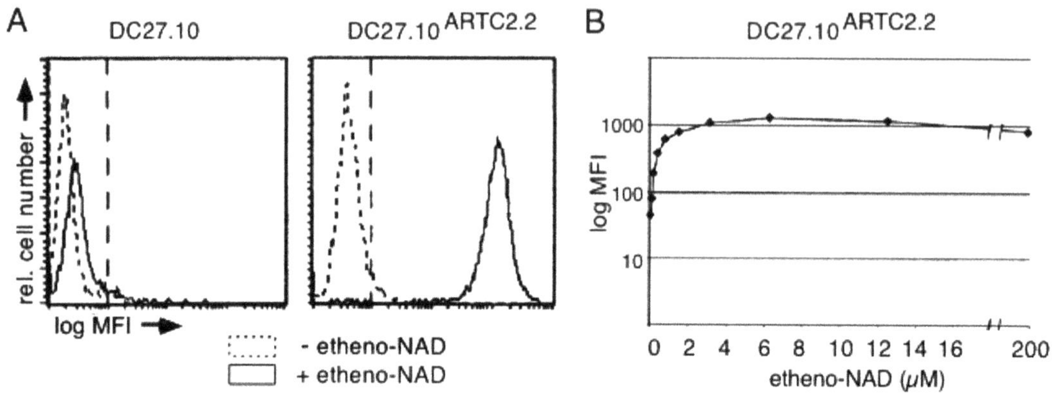

Fig. 5 Monitoring etheno-ADP-ribosylation of membrane proteins by flow cytometry. (**a**) Untransfected and DC27.10 lymphoma cells stably transfected with ARTC2.2 were incubated for 10 min in the absence (dashed lines) or presence (solid lines) of 50 μM etheno-NAD$^+$. Cells were washed, stained with Alexa488-conjugated mAb 1G4, and subjected to flow cytometry. Numbers indicate mean fluorescence intensity (MFI) of cells incubated with etheno-NAD$^+$. (**b**) Dose-response analysis of etheno-ADP-ribosylation. ARTC2.2-transfected DC27.10 cells were incubated for 30 s in the presence of the indicated concentrations of etheno-NAD. Cells were washed, stained with 1G4, and analyzed by flow cytometry. Reprinted from *Anal Biochem* 314, Krebs, et al. Flow cytometric and immunoblot assays for cell surface ADP-ribosylation using a monoclonal antibody specific for ethenoadenosine. 108–115, ©2003, with permission from Elsevier

6. Wash membrane 4× with TBS, 0.05% Tween 20.
7. Incubate membrane with peroxidase-labeled anti-mouse IgG in PBS, 1% BSA, 0.05% Tween 20) for 1 h at RT.
8. Wash membrane 4× with TBS, 0.05% Tween 20.
9. In the dark room, place membrane on a sheet of cellophane, and add ECL reagent.
10. Wrap the membrane with telephone, and expose the membrane to ECL film for 2 min (*see* **Note 16**) (Fig. 6).

3.5 In Vivo Imaging of ARTC-Expressing Cells

To detect ARTC-expressing cells by near-infrared in vivo imaging, use ARTC-specific nanobody or monoclonal antibody conjugated to an appropriate fluorescent dye, e.g., AF680 [20, 21]. The day prior to in vivo imaging, shave and depilate the ventral and lateral throat, thorax, and abdomen of the mice. Anesthetize mice and position mice on heated imaging stage with the ventral side directed toward the camera. Intravenously inject fluorochrome-conjugated nanobody or antibody. Perform NIRF imaging before and at various time points after injection of fluorochrome-conjugates. Estimate radiant efficiency using manually drawn regions of interest (ROIs) around cervical and axillary lymph node regions.

3.5.1 Near-Infrared In Vivo Imaging of ARTC2.2-Expressing T Cells

1. Conjugate ARTC2.2-specific nanobody s+16a [22] or monoclonal antibody Nika102 by random reaction of primary amines to the succinimidyl ester moiety of an appropriate

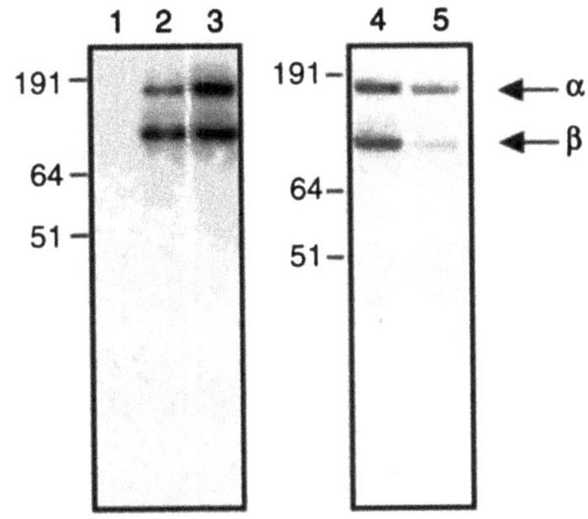

Fig. 6 Detecting etheno-ADP-ribosylated membrane proteins by Western blotting. Untransfected and DC27.10 cells stably transfected with ARTCs were incubated for 30 min in the presence of 20 μM etheno-NAD^+. Cells were washed and lysed for 30 min at 4 °C in 1% Triton X-100. Proteins in cleared lysates were subjected to immunoprecipitation with immobilized mAb M17/4 (anti-LFA-1). Immunoprecipitates were size fractionated by SDS-PAGE followed by immunoblot analyses with mAb 1G4. The upper and lower bands correspond to the alpha and the beta chain of LFA-1, respectively. Lane 1, untransfected cells; lane 2, mouse ARTC2.1-transfected cells; lane 3, mouse ARTC2.2-transfected cells; lane 4, mouse ARTC1-transfected cells; lane 5, human ARTC1-transfected cells. Reprinted from *Anal Biochem* 314, Krebs, et al. Flow cytometric and immunoblot assays for cell surface ADP-ribosylation using a monoclonal antibody specific for ethenoadenosine. 108–115, ©2003, with permission from Elsevier

fluorescent dye, e.g., AF680, according to the manufacturer's instructions. Remove unbound fluorochrome by gel filtration (*see* **Note 17**).

2. Monitor purity of antibodies by SDS-PAGE and Coomassie staining. Monitor incorporation of the dye by capturing a fluorescent image of the gel with the NIRF imaging system.
3. Monitor expression of ARTC2.2 on lymph node cells by flow cytometry as described above in Subheading 3.1, **steps 1–7**.
4. The day prior to in vivo imaging, shave and depilate the ventral and lateral throat, thorax, and abdomen of the mice.
5. Anesthetize mice with 1–2% isofluorane, and position mice in the imaging chamber, allowing for maintenance of gaseous anesthesia. Maintain 1–2% isoflurane during the imaging procedure using the isoflurane manifold housed in the imaging chamber. Protect animal's corneas from drying with an eye ointment while under anesthesia. Position mice on heated imaging stage with the ventral side directed toward the camera. Use fluorescent filter sets of 615–665 nm for excitation, 695–770 nm for emission, and 580–610 nm excitation for background subtraction.

6. Place mice into a restrainer, and warm the tail region with a lamp or water bath in order to expand the tail vein. Intravenously inject 200 μL saline containing 50 μg s+16a^{AF680} or 10 μg of Nika102^{AF680} (corresponding to approximately 3000 and 60 pmol, respectively) (*see* **Note 18**).
7. Perform NIRF imaging before and at various time points after injection of fluorochrome-conjugates (*see* **Note 19**).
8. Estimate radiant efficiency using manually drawn ROIs around cervical and axillary lymph node regions (Fig. 7a, b).

3.5.2 Near-Infrared In Vivo Imaging of ARTC2.2-Expressing Lymphoma Cells

To detect ARTC-expressing lymphoma cells by near-infrared in vivo imaging, use ARTC-specific nanobody or monoclonal antibody conjugated to an appropriate fluorescent dye, e.g., AF680. Subcutaneously inject an aliquot of lymphoma cells in Matrigel basement matrix, and monitor tumor growth daily [23]. Intravenously inject fluorochrome-conjugates and perform NIRF imaging after positioning of mice on a heated imaging stage with the tumors directed toward the camera. Estimate radiant efficiency using manually drawn ROIs around the tumor regions.

1. Conjugate ARTC2.2-specific nanobody s+16a or monoclonal antibody Nika102 to a suitable fluorochrome as described in Subheading 3.5.1, **steps 1–2** (*see* **Note 17**).
2. Monitor expression of ARTC2.2 on lymphoma cells by flow cytometry as described in Subheading 3.1, **steps 1**–7.
3. Aliquot lymphoma cells (0.5–2 × 10^6 cells) in 100 μL RPMI medium in 1.5 mL microcentrifuge tubes, and mix carefully with 100 μL of ice-cold Matrigel basement matrix (*see* **Note 18**).
4. Draw cell suspension into precooled syringes and put on ice until injection.
5. Anesthetize mice with 2% isoflurane in an induction chamber, and inject lymphoma cells subcutaneously into the shoulder flanks. Use thumb and index finger to pull the skin away from the body of the mouse. Inject cells slowly evenly into the pouch created by the fingers. Matrigel helps keeping injected cells in place.
6. Monitor tumor growth daily. Prepare mice for imaging when tumors have reached a size of diameter of 5–10 mm. Position mice on heated imaging stage with the tumors (dorsal side) directed toward the camera (*see* **Note 19, 20**).
7. Intravenously inject fluorochrome-conjugates, and perform NIRF imaging as described in Subheading 3.5.1, **steps 6–7** (*see* **Note 21, 22, 23**).

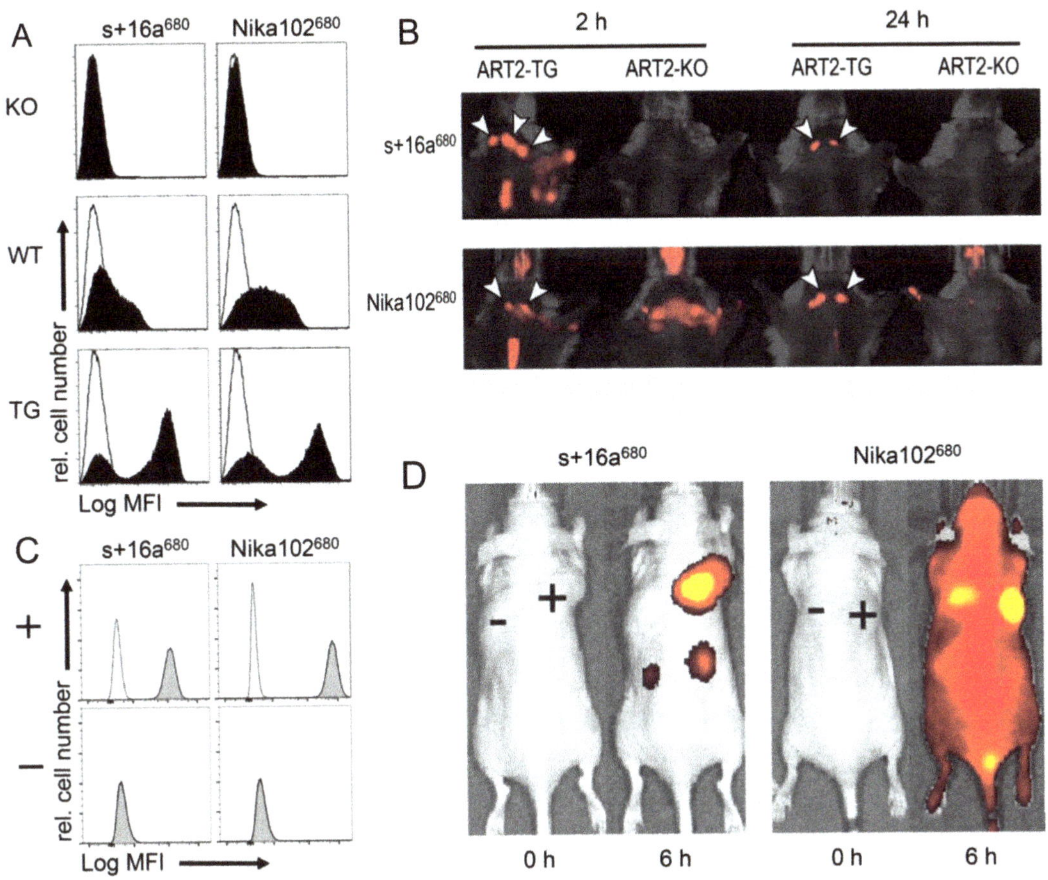

Fig. 7 In vivo imaging of ARTC2.2-expressing cells. (**a**) Lymph node cells from BALBc mice were co-stained with anti-CD3^{PE} and ARTC2.2-specific nanobody s+16a^{AF680} or mAb Nika102^{AF680} and subjected to flow cytometry. Mean fluorescence intensities (MFI) of CD3^{+} T cells are shown (solid histograms). Open histograms show isotype controls. *KO* ARTC2.2-knockout mice, *WT* wild-type mice, *TG* transgenic mice overexpressing ARTC2.2. (**b**) ARTC2.2-TG mice and ARTC2.2-KO mice received i.v. injections of 200 μL saline containing either 50 μg s+16a^{AF680} or 10 μg Nika102^{AF680}. Two and 24 h after injection, mice were shaved and subjected to near-infrared in vivo imaging. Arrowheads indicate cervical and axillary lymph nodes. (**c**) Untransfected (−) and ARTC2.2-transfected DC27.10 lymphoma cells (+) were stained with s+16a^{AF680} or Nika102^{AF680} and subjected to flow cytometry (shaded histograms). Open histograms show isotype controls. MFI of ARTC2.2 expression is plotted. (**d**) Athymic nude mice were injected s.c. with untransfected (−) and ARTC2.2-transfected (+) DC27.10 lymphoma cells. Seven days later, mice received i.v. injections of 200 μL saline containing either 50 μg s+16a^{AF680} or 10 μg Nika102^{AF680}. Mice were subjected to near-infrared in vivo imaging before (0 h) and 6 h after injection. (**a**, **b**) Reprinted from *Contrast Media Mol Imaging 9*, Bannas, et al., In vivo near-infrared fluorescence targeting of T cells: comparison of nanobodies and conventional monoclonal antibodies. *135–142*, ©2014, with permission from Wiley. (**a**, **b**) Reprinted from *Contrast Media Mol Imaging 10*, Bannas, et al., Molecular imaging of tumors with nanobodies and antibodies: Timing and dosage are crucial factors for improved in vivo detection. *367–378*, ©2015, with permission from Wiley

8. Estimate radiant efficiency using manually drawn ROIs around the tumor regions (Fig. 7c, d).

4 Notes

1. For monitoring expression of ecto-ARTCs, use appropriate, validated mAbs, e.g., antihuman ARTC1 (GUGU1-03), anti-mouse ARTC2.1 (GUGU2-32), anti-mouse ARTC2.2 (Nika102), anti-mouse ART3 (GUGU3-43), antihuman ART3 (GUGU3-51), anti-mouse ARTC4 (GUGU3-34), and antihuman ARTC4 (NONI-04). ARTC1-ARTC4 all are expressed as GPI-anchored membrane proteins, but ARTC5 is expressed as a secretory protein. For monitoring the activity of other ecto-ARTCs, note that ARTC1 and ARTC5 show similar activity as ARTC2.2, but ARTC3 and ARTC4 show little if any detectable enzyme activity, and mouse ARTC2.1 requires DTT (i.e., add 1 mM DTT to all buffers). ARTC2 is not expressed in human or other primates (the corresponding genes are inactivated by premature stop codons) [24]. Most mouse strains carry two functional, tandem ARTC2 genes. Noteworthy exceptions are C57Bl/6 mice which carry a knockout allele of ARTC2.1 [13] and NZW mice, which carry a deletion of the ARTC2.2 gene [25].
2. The ARTC open reading frame typically encompasses 700–900 nucleotides, with a start codon followed by the coding region for the N-terminal signal sequence and the ARTC catalytic domain. Some ARTCs further encode a stalk region of varying length and a C-terminal GPI-signal sequence. It may be useful to add a consensus Kozak sequence immediately upstream of the ART start codon.
3. Transfection of HEK-293T cells using jetPEI typically results in 40–80% of cells expressing ARTC by 24–48 h post transfection. Transfection of lymphoma cells by jetPEI or electroporation typically results in much lower numbers of transfected cells (<1% of cells). For such cells, selection of stable transfectants by cultivation in an appropriate selection medium is recommended. The efficiency of stable transfection can be improved by linearizing the expression vector using a restriction enzyme that cuts outside of the expression cassette, e.g., in the antibiotic resistance gene used for propagating the plasmid in *E. coli* (usually ampR).
4. The half-life of ^{32}P is 14.3 days. The actual radiation has to be calculated according to the reference date. Always use 1 cm acrylic shielding, and wear personal dosimeter when handling

^{32}P-NAD$^+$. Lead shielding is less effective due to the bremsstrahlung produced by the beta particle and the shielding.

5. In some cases, ADP-ribosylation of a target protein can affect recognition by a target-specific antibody. ADP-ribosylation has been reported to inhibit binding of the following monoclonal antibodies: Hano43 (anti-P2X7), 7D4 (anti-CD25) [26], and YTS156.7.7 (anti-CD8) [27].
6. It is important to prepare cell suspensions from mouse lymphatic organs at 4 °C because tissue disruption leads to the release of NAD$^+$. On cells co-expressing ARTC2.2 and the P2X7 ion channel, this will lead to ADP-ribosylation of P2X7. ADP-ribosylation causes gating of P2X7 at 20–37 °C, leading to ecto-domain shedding of several cell surface proteins, including ARTC2.2. Prolonged incubation of cells with ADP-ribosylated P2X7 at 20–37 °C will also result in cell death [28]. ADP-ribosylation of P2X7 and other target proteins during cell preparation can be blocked by systemic (i.p. or i.v.) injection of the ARTC2-blocking nanobody s+16a 20–60 min before sacrifice [29]. *See also the chapter by Rissiek et al.* "Monitoring the Sensitivity of T cell Populations Toward NAD$^+$ Released During Cell Preparation."
7. The concentration of ARTC2.2 in murine serum samples can vary depending on the strain mouse strain. Treatment of mice with pro-inflammatory agents, e.g., LPS or NAD$^+$, leads to increased serum levels of ARTC2.2 [13].
8. Treatment of ARTC2.2-expressing T cells or lymphoma cells with PI-PLC, NAD$^+$, ATP, PMA, or anti-CD3 can induce shedding of ARTC2.2. Even trypsinization or mechanical manipulation of cells can induce shedding of ARTC2.2.
9. It is important to use a pair of antibodies that recognize non-overlapping epitopes on the target antigen, e.g., for ARTC2.2 Nika109 and Nika102. Since both antibodies share a common isotype (rat IgG2a kappa), a secondary antibody cannot distinguish between catcher and detector. Therefore, it is necessary to biotinylate the detector, e.g., using the EZ-Link NHS-PEG solid phase biotinylation kit from Thermo Fisher.
10. Murine B cells express high levels of CD38, a potent NADase. Presence of B cells and other CD38-expressing cells will lead to rapid degradation of extracellular NAD+ and, consequentially, to reduced NAD-dependent ADP-ribosylation of membrane proteins catalyzed by ARTCs. In the presence of CD38, it may be necessary to increase the concentration of exogenous ^{32}P-NAD or etheno-NAD (e.g., to 10–50 μM) in order to detect ARTC-catalyzed ADP-ribosylation.

11. ARTCs effectively ADP-ribosylate their targets at 4 °C and at 37 °C. ADP-ribosyl-arginine can be cleaved by phosphodiesterases and ADP-ribosylhydrolases at 37 °C; ADP-ribose inhibits these reactions. When using cells that co-express ARTC2.2 and P2X7, incubation at 37 °C can cause ectodomain shedding of ARTC2.2.
12. Exposure time for optimal detection depends on various factors (*see* **Note 10**). Develop film for 1 day at −80 °C, and adjust exposure time according to signal and background intensities. Consider the half-life of ^{32}P (14 days). Films for autoradiography are light sensitive. Handle in a dark room, using only a suitable red light source.
13. The gel can be Coomassie stained and dried before autoradiography. Prolonged incubation in destaining solution will reduce background signals at the bottom of the gel due to unbound ^{32}P-NAD.
14. Phosphodiesterases and ADP-ribosylhydrolases may hydrolyze the etheno-ADP-ribosylated targets (*see* **Note 11**). ARTC2.2 is active at 4–37 °C. Titrate the concentration of extracellular etheno-NAD, e.g., from 0.1 to 100 μM in order to determine the EC50 for etheno-ADP-ribosylation of membrane proteins on the cell type under investigation.
15. Exposure time for optimal detection depends on various factors (*see* **Note 10**). Develop the film after 2 min at RT, and adjust exposure time according to signal and background intensity. ECL films are light sensitive. Handle in a dark room, using only a suitable red light source.
16. Calculate the number of dyes per nanobody s + 16a and mAb Nika102 using molar extinction coefficients of 15,720 mol^{-1} cm^{-1} and 203,000 mol^{-1} cm^{-1}, respectively. Assess the purity of antibody constructs by SDS-PAGE and Coomassie staining.
17. The small (15 kD) nanobody penetrates tissues much faster than the large (150 kD) monoclonal antibody. The monovalent nanobody has a faster off rate than the bivalent antibody. Unbound nanobody is rapidly eliminated by renal filtration, while unbound monoclonal antibody remains in circulation. Dose titration and kinetic experiments are recommended to optimize imaging results.
18. The optimal amount of fluorochrome-conjugated nanobodies/antibodies depends on their size, in vivo half-life, labeling efficacy, and expression level of the target antigen. The optimal amount should be determined empirically for each construct and target.

19. To monitor unbound circulating nanobody or monoclonal antibody, prepare serum from small samples of blood. Estimate concentration of fluorochrome-conjugated nanobodies/antibodies in serum by comparative flow cytometry analyses of ARTC2.2-expressing cells with serial dilutions of serum vs. fluorochrome-conjugated standards of known concentration.
20. Shorter exposure times enable faster frame rates; longer exposure times provide greater sensitivity. Binning determines the pixel size of the CCD camera; higher pixel size increases sensitivity and frame rate but reduces spatial resolution.
21. To validate in vivo data, perform NIRF imaging on lymph nodes and other organs harvested from sacrificed mice at different time points after injection of fluorochrome-conjugates.
22. Untransfected DC27.10 cells grow slightly faster than ARTC2.2-transfected cells in vivo. To obtain tumors of similar size after 7 days, adjust the numbers of injected cells, e.g., 0.5×10^6 untransfected and 1.5×10^6 ARTC2.2 transfected cells.
23. To validate in vivo data, perform NIRF imaging on tumors and other organs harvested from sacrificed mice at different time points after injection of fluorochrome-conjugates.

References

1. Hottiger MO, Hassa PO, Lüscher B, Schüler H, Koch-Nolte F (2010) Toward a unified nomenclature for mammalian ADP-ribosyltransferases. Trends Biochem Sci 35:208–219
2. Laing S, Unger M, Koch-Nolte F, Haag F (2010) ADP-ribosylation of arginine. Amino Acids 41:257–269
3. Koch-Nolte F, Adriouch S, Bannas P, Krebs C, Scheuplein F, Seman M, Haag F (2006) ADP-ribosylation of membrane proteins: unveiling the secrets of a crucial regulatory mechanism in mammalian cells. Ann Med 38:188–199
4. Glowacki G, Braren R, Firner K, Nissen M, Kühl M, Reche P, Bazan F, Cetkovic-Cvrlje M, Leiter E, Haag F, Koch-Nolte F (2002) The family of toxin-related ecto-ADP-ribosyltransferases in humans and the mouse. Protein Sci 11:1657–1670
5. Otto H, Reche PA, Bazan F, Dittmar K, Haag F, Koch-Nolte F (2005) In silico characterization of the family of PARP-like poly(ADP-ribosyl)transferases (pARTs). BMC Genomics 6:139
6. Koch-Nolte F, Petersen D, Balasubramanian S, Haag F, Kahlke D, Willer T, Kastelein R, Bazan F, Thiele HG (1996) Mouse T cell membrane proteins Rt6-1 and Rt6-2 are arginine/protein mono(ADPribosyl)transferases and share secondary structure motifs with ADP-ribosylating bacterial toxins. J Biol Chem 271:7686–7693
7. Zolkiewska A, Moss J (1993) Integrin alpha 7 as substrate for a glycosylphosphatidylinositol-anchored ADP-ribosyltransferase on the surface of skeletal muscle cells. J Biol Chem 268:25273–25276
8. Moss J, Stanley SJ, Watkins PA (1980) Isolation and properties of an NAD- and guanidine-dependent ADP-ribosyltransferase from Turkey erythrocytes. J Biol Chem 255:5838–5840
9. Koch-Nolte F, Kernstock S, Mueller-Dieckmann C, Weiss MS, Haag F (2008) Mammalian ADP-ribosyltransferases and ADP-ribosylhydrolases. Front Biosci 13:6716–6729
10. Kahl S, Nissen M, Girisch R, Duffy T, Leiter EH, Haag F, Koch-Nolte F (2000) Metalloprotease-mediated shedding of enzymatically active mouse ecto-ADP-ribosyltransferase ART2.2 upon T cell activation. J Immunol 165:4463–4469
11. Menzel S, Rissiek B, Bannas P, Jakoby T, Miksiewicz M, Schwarz N, Nissen M, Haag F, Tholey A, Koch-Nolte F (2015) Nucleotide-induced membrane-proximal proteolysis con-

trols the substrate specificity of T cell ecto-ADP-ribosyltransferase ARTC2.2. J Immunol 195:2057–2066
12. Adriouch S, Bannas P, Schwarz N, Fliegert R, Guse AH, Seman M, Haag F, Koch-Nolte F (2008) ADP-ribosylation at R125 gates the P2X7 ion channel by presenting a covalent ligand to its nucleotide binding site. FASEB J 22:861–869
13. Koch-Nolte F, Duffy T, Nissen M, Kahl S, Killeen N, Ablamunits V, Haag F, Leiter EH (1999) A new monoclonal antibody detects a developmentally regulated mouse ecto-ADP-ribosyltransferase on T cells: subset distribution, inbred strain variation, and modulation upon T cell activation. J Immunol 163:6014–6022
14. Koch-Nolte F, Glowacki G, Bannas P, Braasch F, Dubberke G, Ortolan E, Funaro A, Malavasi F, Haag F (2005) Use of genetic immunization to raise antibodies recognizing toxin-related cell surface ADP-ribosyltransferases in native conformation. Cell Immunol 236:66–71
15. Parusel I, Kahl S, Braasch F, Glowacki G, Halverson GR, Reid ME, Schawalder A, Ortolan E, Funaro A, Malavasi F, Hardie D, Halder S, Buckley CD, Haag F, Koch-Nolte F (2005) A panel of monoclonal antibodies recognizing GPI-anchored ADP-ribosyltransferase ART4, the carrier of the Dombrock blood group antigens. Cell Immunol 236:59–65
16. Krebs C, Koestner W, Nissen M, Welge V, Parusel I, Malavasi F, Leiter EH, Santella RM, Haag F, Koch-Nolte F (2003) Flow cytometric and immunoblot assays for cell surface ADP-ribosylation using a monoclonal antibody specific for ethenoadenosine. Anal Biochem 314:108–115
17. Möller S, Jung C, Adriouch S, Dubberke G, Seyfried F, Seman M, Haag F, Koch-Nolte F (2007) Monitoring the expression of purinoceptors and nucleotide-metabolizing ecto-enzymes with antibodies directed against proteins in native conformation. Purinergic Signal 3:359–366
18. Koch-Nolte F, Reyelt J, Schossow B, Schwarz N, Scheuplein F, Rothenburg S, Haag F, Alzogaray V, Cauerhff A, Goldbaum FA (2007) Single domain antibodies from llama effectively and specifically block T cell ecto-ADP-ribosyltransferase ART2.2 in vivo. FASEB J 21:3490–3498
19. Haag F, Koch-Nolte F, Kühl M, Lorenzen S, Thiele HG (1994) Premature stop codons inactivate the RT6 genes of the human and chimpanzee species. J Mol Biol 243:537–546
20. Koch-Nolte F, Klein J, Hollmann C, Kühl M, Haag F, Gaskins HR, Leiter E, Thiele HG (1995) Defects in the structure and expression of the genes for the T cell marker Rt6 in NZW and (NZB x NZW)F1 mice. Int Immunol 7:883–890
21. Teege S, Hann A, Miksiewicz M, MacMillan C, Rissiek B, Buck F, Menzel S, Nissen M, Bannas P, Haag F, Boyer O, Seman M, Adriouch S, Koch-Nolte F (2015) Tuning IL-2 signaling by ADP-ribosylation of CD25. Sci Rep 5:8959
22. Lischke T, Schumacher V, Wesolowski J, Hurwitz R, Haag F, Koch-Nolte F, Mittrücker H-W (2013) CD8-β ADP-ribosylation affects CD8(+) T-cell function. Eur J Immunol 43:1828–1838
23. Seman M, Adriouch S, Scheuplein F, Krebs C, Freese D, Glowacki G, Deterre P, Haag F, Koch-Nolte F (2003) NAD-induced T cell death: ADP-ribosylation of cell surface proteins by ART2 activates the cytolytic P2X7 purinoceptor. Immunity 19:571–582
24. Rissiek B, Danquah W, Haag F, Koch-Nolte F (2014) Technical advance: a new cell preparation strategy that greatly improves the yield of vital and functional Tregs and NKT cells. J Leukoc Biol 95:543–549
25. Scheuplein F, Schwarz N, Adriouch S, Krebs C, Bannas P, Rissiek B, Seman M, Haag F, Koch-Nolte F (2009) NAD+ and ATP released from injured cells induce P2X7-dependent shedding of CD62L and externalization of phosphatidylserine by murine T cells. J Immunol 182:2898–2908
26. Bannas P, Adriouch S, Kahl S, Braasch F, Haag F, Koch-Nolte F (2005) Activity and specificity of toxin-related mouse T cell ecto-ADP-ribosyltransferase ART2.2 depends on its association with lipid rafts. Blood 105:3663–3670
27. Bannas P, Well L, Lenz A, Rissiek B, Haag F, Schmid J, Hochgräfe K, Trepel M, Adam G, Ittrich H, Koch-Nolte F (2014) In vivo near-infrared fluorescence targeting of T cells: comparison of nanobodies and conventional monoclonal antibodies. Contrast Media Mol Imaging 9:135–142
28. Bannas P, Lenz A, Kunick V, Fumey W, Rissiek B, Schmid J, Haag F, Leingärtner A, Trepel M, Adam G, Koch-Nolte F (2015) Validation of nanobody and antibody based in vivo tumor xenograft NIRF-imaging experiments in mice using ex vivo flow cytometry and microscopy. J Vis Exp 98:52462
29. Bannas P, Lenz A, Kunick V, Well L, Fumey W, Rissiek B, Haag F, Schmid J, Schütze K, Eichhoff A, Trepel M, Adam G, Ittrich H, Koch-Nolte F (2015) Molecular imaging of tumors with nanobodies and antibodies: timing and dosage are crucial factors for improved in vivo detection. Contrast Media Mol Imaging 10:367–378

Chapter 12

ADP-Ribosyl-Acceptor Hydrolase Activities Catalyzed by the ARH Family of Proteins

Masato Mashimo and Joel Moss

Abstract

The ARH family of ADP-ribosyl-acceptor hydrolases is composed of three 39-kDa proteins (ARH1, 2, and 3), which hydrolyze specific ADP-ribosylated substrates. ARH1 hydrolyzes mono(ADP-ribosyl)ated arginine, which results from actions of cholera toxin and other nicotinamide adenine dinucleotide (NAD^+):arginine ADP-ribosyl-transferases, while ARH3 hydrolyzes poly(ADP-ribose) and *O*-acetyl-ADP-ribose, resulting from the action of poly(ADP-ribose) polymerases and sirtuins, respectively. ARH2 has not been reported to have enzymatic activity, because of differences in the catalytic domain. Thus, the substrate specificities of ARH1 and ARH3 proteins result in unique cellular functions. In this chapter, we introduce several methods to monitor the activities of the ARH family members.

Key words ADP-ribosylation, ADP-ribosyl-acceptor hydrolase (ARH), ADP-ribosylated arginine, Poly(ADP-ribose) polymerase (PARP) 1, *O*-acetyl-ADP-ribose (*O*AADPr), Cholera toxin

1 Introduction

ADP-ribosylation involves the transfer of the ADP-ribose moiety of nicotinamide adenine dinucleotide (NAD^+) to specific acceptors, forming ADP-ribosylated acceptor [1, 2]. ADP-ribosylation can be subclassified based on the number of linked ADP-ribose moieties attached to acceptors, resulting in mono(ADP-ribosyl)ation and poly(ADP-ribosyl)ation [1, 2]. Mono(ADP-ribosyl)ation reactions are initiated by attaching ADP-ribose to specific amino acid residues (e.g., arginine, cysteine, aspartate, glutamate, diphthamide) [3–5]. In poly(ADP-ribosyl)ation reactions, the initial ADP-ribose is attached at lysine, glutamate, aspartate, and serine residues, which serve as an acceptor for other ADP-ribose moieties, forming poly(ADP-ribose) (PAR) polymers via *O*-glycosidic linkages [6–9]. Some bacterial toxins utilize the mono(ADP-ribosyl)ation reaction to target specific amino acid residues of protein acceptors and thereby to disrupt host cell biosynthetic, regulatory, and metabolic pathways [10–13].

Paul Chang (ed.), *ADP-ribosylation and NAD+ Utilizing Enzymes: Methods and Protocols*, Methods in Molecular Biology, vol. 1813, https://doi.org/10.1007/978-1-4939-8588-3_12, © Springer Science+Business Media, LLC, part of Springer Nature 2018

In mammalian cells, NAD^+ arginine ADP-ribosyltransferases (ART) and poly(ADP-ribose) polymerases (PARP) catalyze mono(ADP-ribosyl)ation and poly(ADP-ribosyl)ation, respectively, of acceptor proteins, which alters their activity and thus critical cellular functions [4, 5, 14–16]. ADP-ribosylation also results from the Sir2-catalyzed NAD^+-dependent histone/protein deacetylase (SIRT) reaction in which ADP-ribose is transferred to an acetate group of acetylated histones, generating *O*-acetyl-ADP-ribose (*O*AADPr) [17, 18]. *O*AADPr synthesized by this reaction might serve as a signal molecule to regulate diverse biological processes [17–19].

The ADP-ribosyl-acceptor hydrolase (ARH) family consists of three 39-kDa proteins (ARH1–3) [20–22]. ARH1 and ARH3 use unique ADP-ribosylated substrates, while ARH2 has not been shown to be an active hydrolase, possibly due to differences at the active site [20, 21, 23, 24]. ARH1, a cytoplasmic protein, appears to be the only enzyme in mammalian cells to catalyze the hydrolysis of the *N*-glycosidic bond between ADP-ribose and the guanidino group of arginine, releasing ADP-ribose and generating unmodified arginine [20, 22, 25]. Mono(ADP-ribosyl)ated arginine, an ARH1 substrate, is synthesized by catalytic action of some bacterial toxins and ARTs expressed in mammalian cells [26]. Cholera toxin secreted by *Vibrio cholerae* catalyzes mono(ADP-ribosyl)ation of an arginine residue of the α-subunit of the stimulatory guanine nucleotide-binding (Gαs) protein of the adenylyl cyclase system, resulting in loss of its GTPase activity [12, 27]. Sustained activation of Gαs by inhibition of its GTPase activity results in fluid and electrolyte accumulation in the intestine, leading to the severe diarrhea characteristic of cholera [28–30]. *ARH1*$^{-/-}$ mice exhibited enhanced mono(ADP-ribosyl)ation of Gαs proteins and fluid accumulation in intestinal loops, compared to their WT counterparts [31]. Thus, ARH1 reduces bacterial toxin-induced mono(ADP-ribosyl)ation and subsequent symptoms, by hydrolyzing the mono(ADP-ribosyl)ated arginine bond. In addition, ARH1 serves as a tumor suppressor [32, 33]. *ARH1*$^{-/-}$ and *ARH1*$^{+/-}$ mice developed tumors more frequently than WT mice. Tumors arising in *ARH1*$^{+/-}$ mice and some tumors in humans exhibit loss of heterozygosity of the *ARH1* gene and/or mutations that alter ARH1 enzymatic activity [33]. Thus, effects on the *ARH1* gene results in enhanced tumorigenesis in mice and perhaps in human.

Despite significant amino acid similarity to ARH1, ARH3 hydrolyzes an *O*-glycosidic bond between an acetate and ADP-ribose in *O*AADPr and between ADP-ribose and ADP-ribose in PAR [21, 22, 24, 34]. ARH3 is localized mainly in the cytoplasm, followed by mitochondria and nucleus [35]. Because of different substrate preferences and cellular localization, ARH3 has different roles from ARH1 in cellular functions [36]. PAR hydrolysis by ARH3 serves to suppress parthanatos, which is PAR-dependent cell death [35, 37]. This type of cell death is seen in neuronal cells after brain ischemic/reperfusion injury and in Parkinson's disease [38–40]. Parthanatos is initiated by PARP1 activation after DNA damage, which results in PAR synthesis in the nucleus. PAR translocation to the cytoplasm is a critical step for the induction of parthanatos [38]. Because most PAR is attached to PARP1 itself via its automodification domain [1, 2], endoglycosidase activity by the 110 kDa nuclear PARG is needed to generate protein-free PAR, which can be translocated to the cytoplasm and then bind apoptosis-inducing factor (AIF) anchored on mitochondrial membranes [35, 37]. The binding of PAR causes AIF release from mitochondria. As AIF contains a nuclear localization sequence, AIF is translocated to the nucleus and then causes large-scale DNA fragmentation by nucleases [41]. PAR also binds hexokinase, a primary enzyme in glycolysis, inhibiting that process [42]. ARH3 appears to hydrolyze protein-free PAR in the nucleus and cytoplasm, preventing PAR translocation to the cytoplasm and in turn parthanatos [35]. In addition, ARH3 might be primarily responsible for PAR degradation in the mitochondria [43]. When PAR was synthesized by expression of the catalytic domain of PARP1 with a mitochondrial-targeting sequence (mito-PARP1cd), overexpression of ARH3 reduced PAR content in mitochondria, while *ARH3*$^{-/-}$ mouse embryonic fibroblasts exhibited excessive and prolonged PAR accumulation in mitochondria compared to WT MEFs [43]. As enzymes responsible for PAR synthesis in mitochondria have not been identified, functions of ARH3 in mitochondria are still unclear. However, SIRT3 and SIRT5 are present in mitochondria and possess deacetylase activity, generating *O*AADPr [44, 45]. In mitochondria, ARH3 might participate in the regulation of *O*AADPr-induced cellular functions.

To characterize the substrate specificities and biological properties of the ARH family, we have prepared several [^{14}C] or [^{32}P]-labeled ADP-ribosylated substrates using bacterial toxins and recombinant ARTs (Table 1). [^{14}C]-labeled ADP-ribosylated substrates allow for the quantitative analysis of ARH activity by measuring the formation of [^{14}C]-labeled products, while [^{32}P]-labeled ADP-ribosylated substrates allow visualization of the hydrolysis of ADP-ribosylated substrates by the ARH family. In this chapter, we discuss the preparation of specific ADP-ribosylated substrates and their use in determining ARH substrates and measuring ARH activity.

2 Materials

2.1 Preparation of Model Substrates

2.1.1 Preparation of Mono(ADP-Ribosyl)ated [^{14}C]Arginine

1. [U-^{14}C]L-arginine (50 mCi/mmol, 0.1 mCi/mL).
2. 2 mM NAD$^+$ (*see* **Note 1**).
3. Cholera toxin A subunit.
4. Cholera toxin reaction buffer: 30 μg ovalbumin, 20 mM dithiothreitol (DTT), 50 mM potassium phosphate, pH 7.5.

Table 1
Substrates of ADP-ribosylation by bacterial toxins, SIRT1, and PARP1

Products		Enzymes	Acceptors
ADP-ribosylated arginine	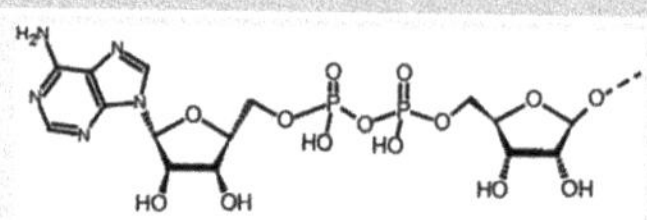 ADP-ribose	Cholera toxin A subunit	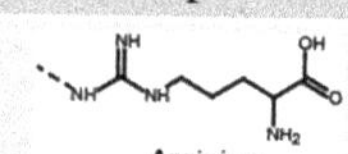 Arginine
ADP-ribosylated cysteine		Pertussis toxin	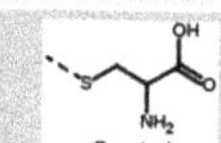 Cysteine
ADP-ribosylated diphthamide		*Pseudomonas* exotoxin A	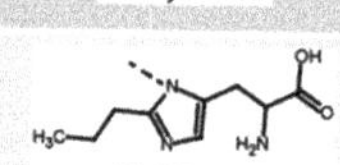 Diphthamide
ADP-ribosylated asparagine		*Clostridium botulinum* C3	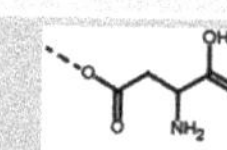 Asparagine
PAR		PARP1	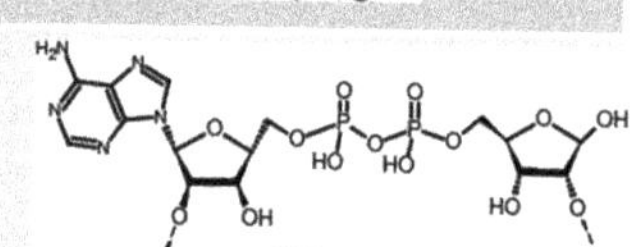 ADP-ribose
OAADPR		SIRT1	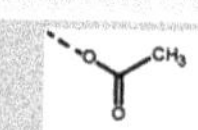 Acetyl group

Bacterial toxins catalyze amino acid-specific mono(ADP-ribosyl)ation [10–13]. PARP1 transfers ADP-ribose to glutamate, aspartate, and lysine residues, followed by elongation of the chain by further addition of ADP-ribose residues. SIRT1 transfers an acetate from acetyl histone to ADP-ribose, generating *OAADPr* [17]

5. High-performance liquid chromatography (HPLC) system with a diode-array spectrophotometric detector.
6. Strong anion exchange (SAX) column (4.6 × 250 mm, 300 Å pore size).
7. SAX buffer A: 20 mM potassium phosphate (pH 4.5).
8. SAX buffer B: 1 M NaCl.
9. Fraction collector.

2.1.2 Preparation of [^{32}P] or [^{14}C]-Labeled Poly(ADP-Ribosyl)ated PARP1

1. [Adenine-U-^{32}P]NAD^+ (800 Ci/mmol, 5 mCi/mL), [adenine-U-^{14}C]NAD^+ (270 mCi/mmol, 0.025 mCi/mL).
2. Recombinant human PARP1 full-length protein (no tag).
3. Deoxyribonucleic acid sodium salt from calf thymus.
4. PARP1 reaction buffer: 5 μM NAD^+, 10 mM $MgCl_2$, 5 mM DTT, 100 mM Tris–HCl, pH 8.0.
5. Ovalbumin: 4 mg/mL in water.

6. Trichloroacetic acid (TCA): 20% (w/v) solution in water.
7. Ice-cold acetone: store at 4 °C.
8. Application buffer: 6 M guanidine HCl, 50 mM MOPS, 0.5 M NH_2OH, pH 8.5.

2.1.3 Preparation of Dihydroboronyl-Bio Rex Affinity Resin

1. Bio-Rex™ 70 cation exchange resin, analytical grade, 100–200 mesh, sodium form.
2. 3-Aminophenylboronic acid.
3. *N*-(3-dimethylaminopropyl)-*N'*-ethylcarbodiimide.
4. Gravity flow column.
5. Storage buffer: 6 M guanidine HCl, 50 mM MOPS, pH 6.0.
6. Application buffer: 6 M guanidine HCl, 50 mM MOPS, 0.5 M NH_2OH, pH 8.5.
7. Wash buffer: 1 M NH_4HCO_3, pH 9.0.

2.1.4 Preparation of OAADPr

1. [Adenine-U-^{32}P]NAD^+ (800 Ci/mmol, 5 mCi/mL), [adenine-U-^{14}C]NAD^+ (270 mCi/mmol, 0.025 mCi/mL).
2. Recombinant human SIRT1 full-length protein (N-terminal His-tag).
3. Acetyl-histone H3 (Lys14) peptide.
4. SIRT1 reaction buffer: 50 mM Tris–HCl, 2.7 mM KCl, 1 mM $MgCl_2$, 0.2 mg bovine serum albumin (BSA), pH 7.0.
5. HPLC system with a diode-array spectrophotometric detector.
6. C18 polymeric reversed-phase column (4.6 × 250 mm, 300 Å pore size).
7. C18 buffer A: 0.05% (v/v) trifluoroacetic acid (TFA) in water.
8. C18 buffer B: 0.05% (v/v) TFA in acetonitrile.
9. Liquid scintillation analyzer.

2.2 Preparation of Substrates for Mono(ADP-Ribosyl)ation from Mouse Brain Tissue

1. Homogenization buffer: 1 mM EDTA, 1 mM NaN_3, 1 mM DTT, 250 mM sucrose, 20 mM Tris–HCl, pH 8.0, containing protease inhibitor cocktails.
2. Dounce tissue grinder.
3. Sodium cholate.
4. Cholera toxin A subunit, pertussis toxin, *Pseudomonas aeruginosa* exotoxin A, *Clostridium botulinum* C3 enzyme.
5. Mono(ADP-ribosyl)ation reaction buffer: 10 mM $MgCl_2$, 5 mM DTT, 50 mM potassium phosphate, pH 7.5.
6. TCA: 20% (w/v) solution in water.

7. Ice-cold acetone: store at 4 °C.
8. Application buffer: 6 M guanidine HCl, 50 mM MOPS, 0.5 M NH_2OH, pH 8.5.
9. 50 mM potassium phosphate, pH 7.5.

2.3 Assay of ARH Hydrolase Activity Using Mono(ADP-Ribosyl)ated [^{14}C] Arginine as a Substrate

1. Mono(ADP-ribosyl)ated [^{14}C]arginine (*see* Subheadings 2.1.1 and 3.1.1).
2. ARH reaction buffer: 50 mM potassium phosphate, 10 mM $MgCl_2$, 5 mM DTT, pH 7.0.
3. Gravity flow column.
4. Affi-Gel boronate gel.
5. Elution buffer: 0.1 M glycine, pH 9.0, 0.1 M NaCl, 10 mM $MgCl_2$.
6. Liquid scintillation analyzer.

2.4 Assay of ARH Activity with Mono(ADP-Ribosyl)ated Proteins, PAR, and OAADPr Using Sodium Dodecyl Sulfate-Polyacrylamide Gel Electrophoresis (SDS-PAGE) Autoradiography

1. [^{32}P]-labeled mono(ADP-ribosyl)ated proteins, PAR, *O*AADPr (*see* Subheadings 2.1.2, 2.1.4, 2.2, 3.1.3, 3.1.4, and 3.2).
2. ARH reaction buffer: 50 mM potassium phosphate, 10 mM $MgCl_2$, 5 mM DTT, pH 7.0.
3. 5× Laemmli buffer.
4. Polyacrylamide electrophoresis (PAGE) and Western blot equipment for mini-sized and large-sized gels.
5. 12 or 4–20% precast polyacrylamide mini-sized gel (8 cm × 8 cm × 0.15 cm).
6. 20% polyacrylamide large-sized gel (20 cm × 20 cm × 0.15 cm).
7. Nitrocellulose membrane.
8. X-ray film.

2.5 Assay of ARH Activity with Poly(ADP-Ribosyl)ated PARP1 and OAADPr by HPLC

1. [^{14}C]-labeled poly(ADP-ribosyl)ated PARP1, *O*AADPr (*see* Subheadings 2.1.2, 2.1.4, 3.1.2 and 3.1.4).
2. ARH reaction buffer: 50 mM potassium phosphate, 10 mM $MgCl_2$, 5 mM DTT, pH 7.0.
3. HPLC system with a diode-array spectrophotometric detector.
4. C18 polymeric reversed-phase column (4.6 × 250 mm, 300 Å Pore Size).
5. C18 buffer A: 0.05% (v/v) TFA in water.
6. C18 buffer B: 0.05% (v/v) TFA in acetonitrile.
7. Liquid scintillation analyzer.

3 Methods

3.1 Preparation of Model Substrates

3.1.1 Mono(ADP-Ribosyl)ated [^{14}C]Arginine

For quantitative analysis of ARH hydrolysis of mono(ADP-ribosyl)ated arginine, mono(ADP-ribosyl)ated [^{14}C]arginine is synthesized as a product in the reaction of cholera toxin with NAD^+ and [^{14}C]arginine. After the reaction of cholera toxin, mono(ADP-ribosyl)ated [^{14}C]arginine and substrates are separated by HPLC.

1. Mono(ADP-ribosyl)ated [^{14}C]arginine is synthesized by incubation of 50 μg of cholera toxin A subunit (*see* **Note 2**) with 2 mM [^{14}C]arginine (50 μCi) and 2 mM NAD^+ in 300 μL of cholera toxin reaction buffer (*see* Subheading 2.1.1, **step 4**) at 30 °C overnight.
2. Mono(ADP-ribosyl)ated [^{14}C]arginine is separated by HPLC with a SAX column. Isocratic elution (1 mL/min) with SAX buffer A for 30 min is followed by a liner gradient of SAX buffer B from 30 to 40 min and then SAX buffer B in the same buffer from 40 to 50 min (*see* Subheadings 2.1.1, **steps 7** and **8** and *see* **Note 3**). Mono(ADP-ribosyl)ated [^{14}C]arginine is eluted at 11–12 min.
3. Collect eluate fractions containing mono(ADP-ribosyl)ated [^{14}C]arginine with a fraction collector.
4. Lyophilize mono(ADP-ribosyl)ated [^{14}C]arginine, and dissolve in 50 μL of water.

3.1.2 [^{32}P]- or [^{14}C]-Labeled Poly(ADP-Ribosyl)ated PARP1

For analysis of ARH hydrolysis of PAR, [^{32}P]- or [^{14}C]-labeled PAR polymers are synthesized by poly(ADP-ribosyl)ated recombinant PARP1 proteins with either [^{32}P] or [^{14}C]NAD^+. The addition of DNA is required for optimal PARP1 activation.

1. [^{32}P]- or [^{14}C]-labeled poly(ADP-ribosyl)ated PARP1 is synthesized by incubation of 6 μg of recombinant PARP1 protein in the presence of 12 μg of calf thymus DNA (*see* **Note 4**) and 10 μM [adenine-U-^{32}P]NAD^+ (10 μCi/reaction) or 10 μM [adenine-U-^{14}C]NAD^+ (0.5 μCi/reaction) (*see* **Note 5**) in 300 μL of PARP1 reaction buffer (*see* Subheading 2.1.2, **step 4**) at 37 °C for 10 min.
2. Add 25 μL of 4 mg/mL ovalbumin to enhance the precipitation of poly(ADP-ribosyl)ated PARP1 before the addition of 300 μL of 20% TCA. Mix thoroughly and place at −20 °C for >1 h.
3. Collect the precipitated protein by centrifugation at 16,000 × *g* for 30 min at 4 °C.
4. Wash three times with ice-cold acetone.
5. Resuspend the pellet with 500 μL of application buffer (*see* Subheading 2.1.2, **step 8**).

3.1.3 Protein-Free [^{32}P]-Labeled PAR

High-resolution SDS-PAGE can separate and determine the size of the [^{32}P]-labeled PAR degradation products of ARH hydrolysis. Affinity chromatography on dihydroxyboryl-Bio-Rex affinity resin is useful to isolate protein-free [^{32}P]-labeled PAR. The boronate group of dihydroboronyl-Bio-Rex efficiently captures PAR and removes PARP1 and other substrates (*see* **Note 6**).

1. Suspend Bio-Rex™ 70 cation exchange resin (50 g) in water, and incubate with rotation for 18 h at room temperature.
2. Remove the supernatant.
3. Dissolve 5 g of 3-aminophenylboronic acid in 50 mL of water, and add to the Bio-Rex slurry. Adjust pH of the solution to 5.0.
4. Add 5 g of *N*-(3-dimethylaminopropyl)-*N'*-ethylcarbodiimide to the mixture, and incubate with rotation for 18 h at room temperature.
5. Wash with water three times and with storage buffer (*see* Subheading 2.1.3, **step 5**) once.
6. Resuspend in storage buffer (*see* Subheading 2.1.3, **step 5**), and store at 4 °C.
7. Transfer 200 μL of dihydroboronyl-Bio Rex Affinity resin slurry to a gravity flow column (*see* **Note 6**).
8. Equilibrate the slurry with application buffer (*see* Subheading 2.1.3, **step 6**) until pH of the solution is 8.6.
9. Apply 500 μL of [^{32}P]-labeled poly(ADP-ribosyl)ated PARP1 to gravity flow column, and incubate with rotation at 25 °C, overnight.
10. Wash with 5 mL of application buffer (*see* Subheading 2.1.3, **step 6**) once, and then with wash buffer (*see* Subheading 2.1.3, **step 7**) twice.
11. Elute protein-free [^{32}P]-labeled PAR with 3 ml of hot water (40 °C).
12. Concentrate the eluate by evaporation to approximately 500 μL.
13. Measure its radioactivity with a liquid scintillation counter.
14. Store at −20 °C.

3.1.4 [^{32}P] and [^{14}C]-Labeled *O*AADPr

*O*AADPr is generated by NAD$^+$-dependent deacetylation of acetylated histone by SIRT1. The *O*AADPr product and substrates are separated by HPLC. ARH catalyzes the hydrolysis of [^{14}C]*O*AADPr to [^{14}C]ADP-ribose. To quantify ARH activity, the amount of radioactive [^{14}C]ADP-ribose is measured.

1. [^{32}P]- or [^{14}C]-labeled *O*AADPr is synthesized by incubation of 6.1 μg of recombinant human SIRT1 protein in the pres-

ence of 10 μM [^{32}P]NAD$^+$ (10 μCi/reaction) or 100 μM [^{14}C]NAD$^+$ (5 μCi/reaction) (*see* **Note 5**) and 100 μg of acetyl-histone peptide H3 in 200 μL of SIRT1 reaction buffer (*see* Subheading 2.1.4, **step 4**) at 30 °C for 4 h.

2. [^{14}C]-labeled *O*AADPr and substrates are separated by HPLC equipped with a C18 polymeric reversed-phase column (4.6 × 250 mm). Isocratic elution (1 mL/min) with C18 buffer A for 5 min is followed by a linear gradient to 60% buffer A and 40% C18 buffer B from 5 to 45 min (*see* Subheading 2.1.4, **steps** 7 and **8** and *see* **Note** 7). [^{14}C]-labeled *O*AADPr is eluted at 7 min.
3. Collect eluate fractions containing *O*AADPr with a fraction collector.
4. Measure the radioactivity of [^{14}C]-labeled *O*AADPr with a liquid scintillation counter.

3.2 Isolation of [^{32}P]-Labeled Mono(ADP-Ribosyl)ated Gαs, Gαi/Gαo, EF-II, or Rho from Mouse Brain Tissues Synthesized by Bacterial Toxins

Preparation of [^{32}P]-labeled mono(ADP-ribosyl)ation at specific amino acid residues is essential to assess substrate specificity of ARH enzymes. [^{32}P]-labeled mono(ADP-ribosyl)ated substrate proteins are synthesized by cholera toxin A subunit (mono(ADP-ribosyl)ated arginine of Gαs protein), pertussis toxin (mono(ADP-ribosyl)ated cysteine of Gαi/Gαo proteins), *Pseudomonas aeruginosa* exotoxin A (mono(ADP-ribosyl)ated diphthamide of EF-II), or *Clostridium botulinum* C3 enzyme (mono(ADP-ribosyl)ated asparagine of Rho). These substrates can be obtained by fractionation of mouse brain tissue. Gαs and Gαi/Gαo proteins are present in the membrane fraction, while EF-II and Rho proteins are in the cytoplasm [21]. Note that ARH1 but not ARH3 hydrolyzes mono(ADP-ribosyl)ated arginine of Gαs.

3.2.1 Preparation of Gαs, Gαi/Gαo, EF-II, and Rho Proteins from Mouse Brain Tissues

1. Mouse brain tissues are homogenized by Dounce tissue grinder in homogenization buffer (1 g of brain/5 mL of buffer) (*see* Subheading 2.2, **step 1** and *see* **Note 8**). The following procedure should be performed on ice.
2. Centrifuge at 1000 × *g* for 10 min to remove cell debris and nucleus.
3. Collect the postnuclear supernatant, and centrifuge at 100,000 × *g* for 90 min to separate cytosol and membrane fractions. The cytoplasmic fraction contains EF-II and Rho.
4. Stir the membrane fraction (50 mg) in 4 mL of homogenization buffer (*see* Subheading 2.2, **step 1**) plus 1% sodium cholate on ice for 1 h.
5. Centrifuge the membrane fraction at 100,000 × *g* for 1 h at 4 °C. The supernatant contains G-proteins.

3.2.2 [^{32}P]-Labeled Mono(ADP-Ribosyl)ated Arginine of Gαs Subunit

1. Gαs protein exacted from brain membrane fraction (1 mg of protein) is incubated with activated cholera toxin A subunit (*see* **Note 2**) and 10 μM [^{32}P]NAD$^+$ (10 μCi/reaction), in 1 mL of mono(ADP-ribosyl)ation reaction buffer reaction buffer (*see* Subheading 2.2, **step 5**) for 2 h at 37 °C.
2. Add 1 mL of 20% TCA to terminate mono(ADP-ribosyl)ation reaction by bacterial toxins. Mix thoroughly and place at −20 °C for >1 h.
3. Collect the precipitated protein by centrifugation at 16000 × *g* for 30 min at 4 °C.
4. Wash three times with ice-cold acetone.
5. Resuspend the pellet with 50 mM potassium phosphate, pH 7.5.

3.2.3 [^{32}P]-Labeled Mono(ADP-Ribosyl)ated Cysteine of Gαi/Gαo Subunits

Gαi/Gαo proteins exacted from brain membrane fraction (1 mg of protein) are incubated with activated pertussis toxin (50 μg) (*see* **Note 2**) and 10 μM [^{32}P]NAD$^+$ (10 μCi/reaction) in 1 mL of mono(ADP-ribosyl)ation reaction buffer (*see* Subheading 2.2, **step 5**) for 2 h at 37 °C. The reaction is terminated by addition of 20% TCA as described above (*see* Subheading 3.2.2).

3.2.4 [^{32}P]-Labeled Mono(ADP-Ribosyl)ated Diphthamide of EF-II

Mouse brain cytosol (1 mg of protein) is incubated with activated *Pseudomonas aeruginosa* exotoxin A (400 μg) (*see* **Note 2**) and 10 μM [^{32}P] NAD$^+$ (10 μCi/reaction) in 1 mL of mono(ADP-ribosyl)ation reaction buffer (*see* Subheading 2.2, **step 5**) for 2 h at 37 °C. The reaction is terminated by addition of 20% TCA as described above (*see* Subheading 3.2.2).

3.2.5 [^{32}P]-Labeled Mono(ADP-Ribosyl)ated Asparagine of Rho

Mouse brain cytosol containing Rho protein (1 mg of protein) is incubated with activated *Clostridium botulinum* C3 enzyme (10 μg) (*see* **Note 2**) and 10 μM [^{32}P]NAD$^+$ (10 μCi/reaction) in 1 mL of mono(ADP-ribosyl)ation reaction buffer (*see* Subheading 2.2, **step 5**) for 2 h at 37 °C. The reaction is terminated by addition of 20% TCA as described above (*see* Subheading 3.2.2).

3.3 ARH Hydrolase Assays

3.3.1 Assay of ARH Hydrolysis of Mono(ADP-Ribosyl)ated [^{14}C]Arginine

Affi-Gel boronate affinity gel chromatography can separate mono(ADP-ribosyl)ated [^{14}C]arginine from cleaved [^{14}C]arginine, the product of ARH hydrolysis. The boronate group of Affi-Gel boronate affinity gel forms a complex with coplanar *cis*-diol groups of ADP-ribose. ARH1 hydrolyzes *N*-glycosidic bond between ADP-ribose and arginine [20]; [^{14}C]arginine is eluted, while mono(ADP-ribosyl)ated [^{14}C]arginine and ADP-ribose are retained. ARH activity is assessed by measuring the radioactivity of the eluate containing [^{14}C]arginine (Fig. 1).

1. Mono(ADP-ribosyl)ated [^{14}C]arginine (50 μM, 6000 cpm) is incubated with recombinant ARH proteins (50 ng) in 100 μL of ARH reaction buffer (*see* Subheading 2.3, **step 2**) at

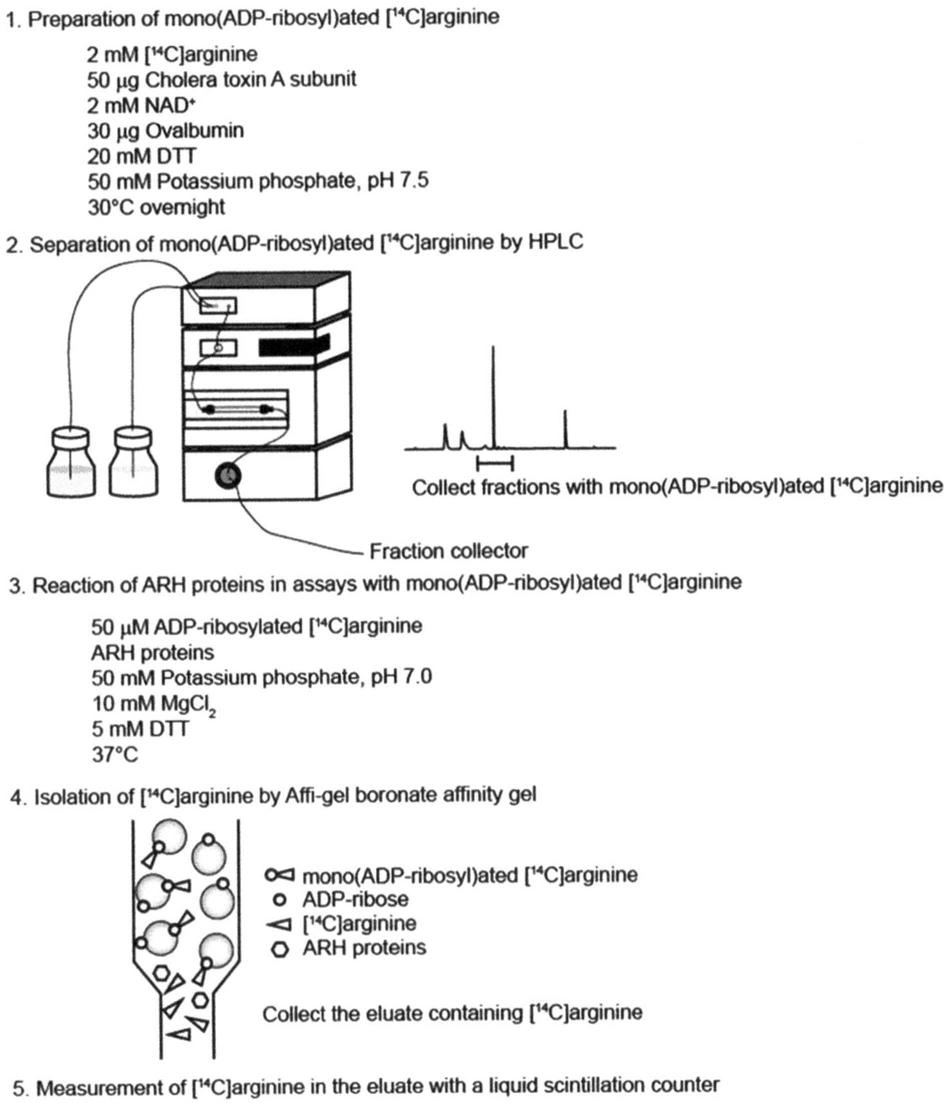

Fig. 1 Scheme for assay of ARH hydrolysis of mono(ADP-ribosyl)ated [^{14}C]arginine. Details are described in Subheadings 3.1.1 and 3.3.1

37 °C. Recombinant ARH1 proteins can be prepared and expressed GST tagged [32].

2. The reaction mixture is applied to a Poly-Prep chromatography column containing Affi-Gel boronate affinity gel (100 mg) (*see* **Note 6**). Affi-Gel boronate affinity gel is equilibrated with elution buffer (*see* Subheading 2.3, **step 5**).
3. Elute with 5 mL of elution buffer (*see* Subheading 2.3, **step 5**).
4. Measure the radioactive of [^{14}C]arginine in the eluate by a liquid scintillation counter.

3.3.2 Autoradiographic Assay of ARH Activity with [^{32}P]-Labeled Mono(ADP-Ribosyl)ated Proteins, PAR, and *O*AADPr

Hydrolysis of [^{32}P]-labeled mono(ADP-ribosyl)ated substrates is utilized to assess the enzymatic activity of ARH enzymes. High-resolution SDS-PAGE in 20% polyacrylamide large-sized gels separates and determines the size of the PAR degradation products.

1. [^{32}P]-labeled mono(ADP-ribosyl)ated proteins (50 μg), PAR (300 nM ADP-ribose), or *O*AADPr (2.5 μM) is incubated with recombinant human ARH proteins (1.5 pmol) in 25 μL of ARH reaction buffer (*see* Subheading 2.4, **step 2** and **Note 9**) at 37 °C.
2. Add 6 μL of 5× Laemmli buffer to terminate the reaction.
3. Samples (20 μg) are separated on SDS-PAGE in 12 or 4–20% polyacrylamide precast mini-sized gels or high-resolution SDS-PAGE in 20% polyacrylamide large-sized gels (*see* **Note 10**) and transferred to nitrocellulose membranes, according to standard protocols.
4. Detect the radioactivity by exposing the membrane to X-ray films for 10 h at −80 °C. Exposure time is varied as required for optimal detection (*see* Fig. 2).

3.3.3 Assay of ARH Activity with [^{14}C]-Labeled Poly(ADP-Ribosyl)ated PARP1 and *O*AADPr by HPLC

ARH3 hydrolyzes PAR and *O*AADPr, releasing ADP-ribose. Reaction products of ARH3 incubated with poly(ADP-ribosyl)ated PARP1 or *O*AADPr are separated by HPLC. By collecting [^{14}C]ADP-ribose fraction and measuring its radioactivity by liquid scintillation counter, ARH3 hydrolysis of PAR and *O*AADPr can be estimated (*see* Figs. 3, 4, and 5).

1. [^{14}C]-labeled poly(ADP-ribosyl)ated PARP1 (250 ng) or *O*AADPr (2.5 μM) is incubated with recombinant ARH protein (1.5 pmol) in 200 μL of ARH reaction buffer (*see* Subheading 2.5, **step 2** and **Note 9**) at 37 °C.
2. [^{14}C]ADP-ribose, a product of ARH activity, and substrate are separated by HPLC equipped with C18 column and collected with a fraction collector, as described above (*see* Subheading 2.5, **steps 5** and **6** and **Note 7**). ADP-ribose is eluted at 4 min.
3. Measure the radioactivity of [^{14}C]ADP-ribose by a liquid scintillation counter.

Fig. 2 (continued) generating ADP-ribose. Data are from [21]. (**c**) ADP-ribosyl-acceptor hydrolase activity of ARH3 with [^{32}P]-labeled *O*AADPr. [^{32}P]-labeled *O*AADPr (2.5 μM) was incubated with recombinant mouse ARH3 protein (2 pmol) for 2 h at 30 °C. Samples were subjected to high-resolution PAGE in 20% polyacrylamide large-sized gels (20 cm × 20 cm × 0.15 cm) and transferred to nitrocellulose membranes, which were exposed to X-ray films. [^{32}P]-labeled standards: NAD$^+$, AMP, ADP-ribose, and *O*AADPr. ARH3 hydrolyzes *O*AADPr, releasing ADP-ribose. Data are from [24]

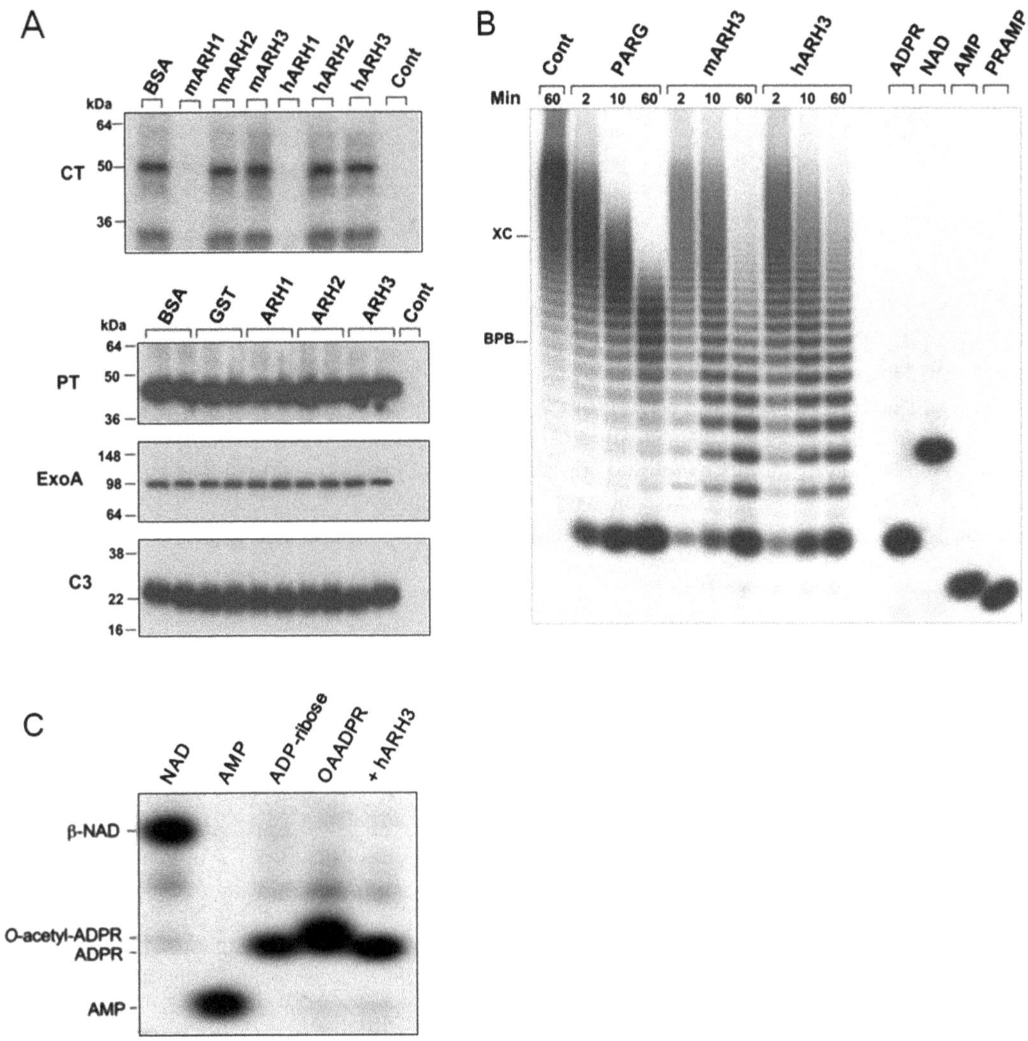

Fig. 2 Assay by autoradiography of ADP-ribosyl-acceptor hydrolase activity of ARH1–3 with substrates; mono(ADP-ribosyl)ated arginine, cysteine, diphthamide, and asparagine, *O*AADPr, and PAR. (**a**) ADP-ribosyl-acceptor hydrolase activity of ARH1, 2, and 3 with [^{32}P]-labeled mono(ADP-ribosyl)ated arginine, cysteine, diphthamide, and asparagine by bacterial toxins. [^{32}P]-labeled mono(ADP-ribosyl)ated amino acids in acceptor proteins (20–100 μg) including Gs, Gi, EF-2, and Rho catalyzed by bacterial toxins (CT, cholera toxin; PT, pertussis toxin; ExoA, *Pseudomonas aeruginosa* exotoxin A; C3, *Clostridium botulinum* C3 enzyme) were incubated with recombinant mouse or human ARH1, 2, and 3 proteins (5 μM), BSA, or GST at 30 °C overnight. *Cont* indicates brain samples had been incubated with [^{32}P] NAD$^+$ in the absence of bacterial toxin. Samples (30 μg) were subjected to SDS-PAGE in 12 or 4–20% polyacrylamide gels and transferred to nitrocellulose membrane, followed by exposure to X-ray films. ARH1, but not ARH2 and 3, hydrolyzes mono(ADP-ribosyl)ated arginine of Gs proteins by cholera toxin. Data are from [21]. (**b**) ADP-ribosyl-acceptor hydrolase activity of PARG and ARH3 with [^{32}P]-labeled PAR substrate. [^{32}P]-labeled PAR (5.5 × 10^5 cpm, ~300 nM ADP-ribose) was incubated with human or mouse ARH3 (1 μM) or PARG (1.5 nM) for indicated times. Samples were subjected to high-resolution PAGE in 20% polyacrylamide large-sized gel (20 cm × 20 cm × 0.15 cm) and transferred to nitrocellulose membranes, which were exposed to X-ray films. Control (Cont), reaction without enzyme incubated for 60 min. [^{32}P]-labeled standards: ADP-ribose (ADPR), NAD$^+$, AMP, and phosphoribosyl-AMP (PRAMP). Positions of BPB and XC correspond to (ADP-ribose)$_8$ and (ADP-ribose)$_{18}$, respectively. PARG and ARH3 degrade PAR polymers,

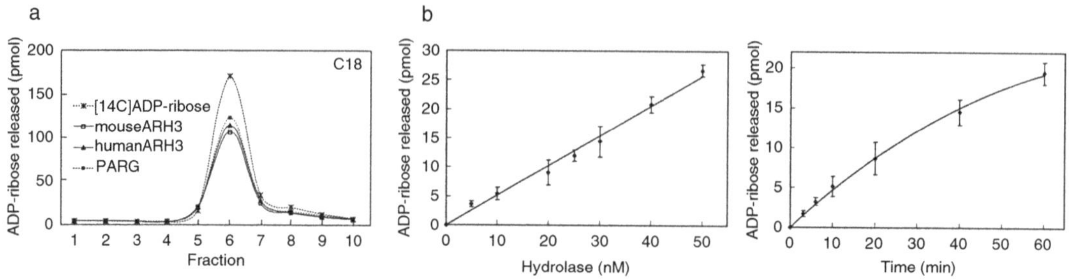

Fig. 3 Assay of ADP-ribosyl-acceptor hydrolase activity of ARH3 with PAR by HPLC. (**a**) PAR hydrolysis by ARH3 and PARG. [^{14}C]-labeled poly(ADP-ribosyl)ated PARP1 (900 ng, ~2.5 μM ADP-ribose) was incubated with recombinant human or mouse ARH3 and calf thymus PARG proteins at indicated concentrations at 37 °C for 60 min. [^{14}C]-labeled *O*AADPr (peak *2*) and ADP-ribose (peak *1*) were separated by HPLC, and radioactivity was quantified by a liquid scintillation counter. ARH3 catalyzes the hydrolysis of PAR [21]. (**b**) Effect of time and concentration on PAR hydrolysis by ARH3. [^{14}C]-labeled poly(ADP-ribosyl)ated PARP1 (250 ng, ~850 nM ADP-ribose) was incubated with recombinant human ARH3 proteins at 30 °C as indicated before separation of substrate and products by HPLC. Data are means ±1/2 the range of values from duplicate assays. All data presented are from [21]

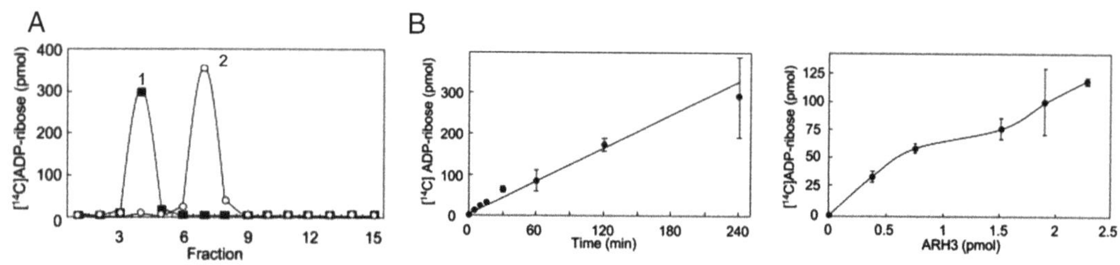

Fig. 4 Assay of ADP-ribosyl-acceptor hydrolase activity of ARH3 with *O*AADPr by HPLC. (**a**) *O*AADPr hydrolysis by ARH3. [^{14}C]-labeled *O*AADPr (2.5 μM) is incubated with (closed rectangle) or without (open circle) recombinant human ARH3 (1.5 pmol) for 2 h at 30 °C. [^{14}C]-labeled *O*AADPr (peak *2*) and ADP-ribose (peak *1*) were separated by HPLC, and radioactivity was quantified by a liquid scintillation counter. ARH3 catalyzes the hydrolysis of *O*AADPr, generating ADP-ribose. (**b**) Effect of time and concentration on *O*AADPr hydrolysis by ARH3. [^{14}C]-labeled *O*AADPr (2.5 μM) is incubated with recombinant human ARH3 proteins at 30 °C as indicated before separation of substrate and products by HPLC. Data are means ±1/2 the range of values from duplicate assays. All data presented are from [24]

4 Notes

1. 100 mM aliquots of NAD^+ are stored at −20 °C. It should be diluted immediately prior to use.
2. Bacterial toxins including cholera toxin A subunit, pertussis toxin, *Pseudomonas aeruginosa* exotoxin A, and *Clostridium botulinum* C3 enzyme should be fully activated by incubation with 100 mM DTT for 30 °C, 15 min before use.
3. Mono(ADP-ribosyl)ated arginine is eluted at 11–12 min. Arginine and nicotinamide are eluted between 3 and

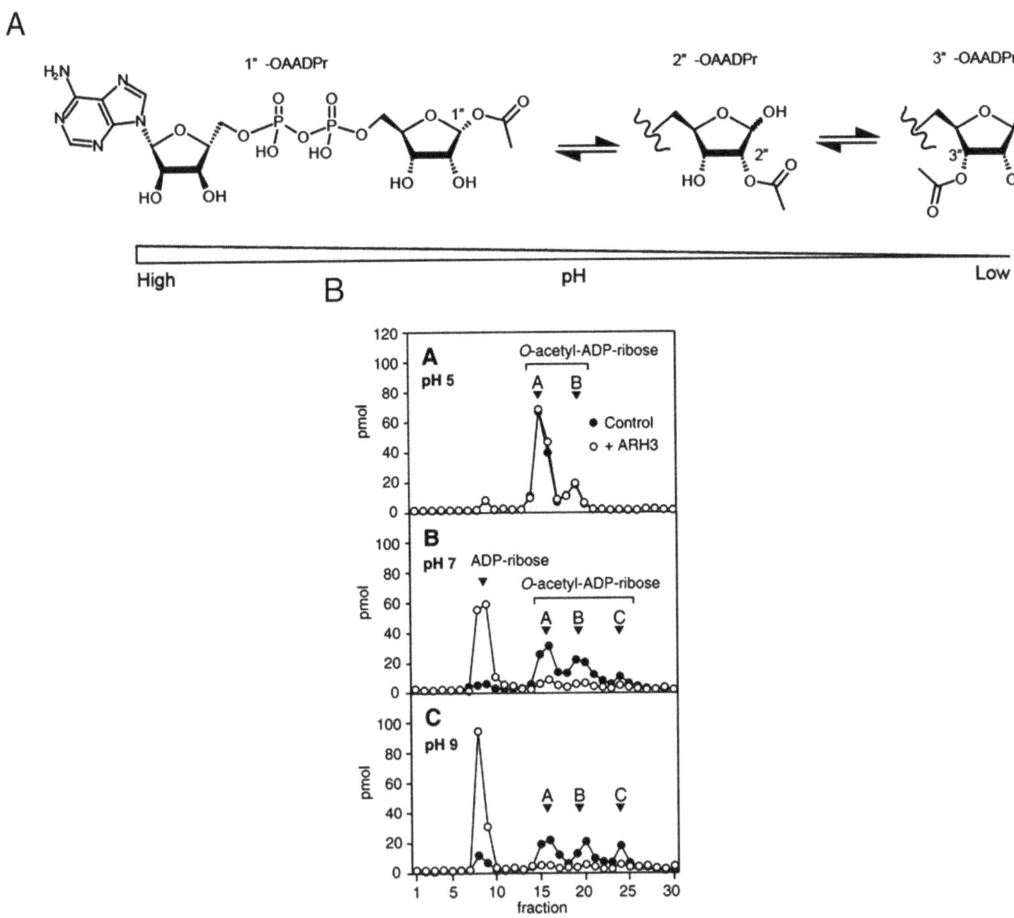

Fig. 5 Interconversion of an acetyl group of *O*AADPr in a pH-dependent manner affects ADP-ribosyl hydrolase activity of ARH3. (**a**) pH-dependent interconversion of an acetate moiety among 1″-3″ hydroxyl group of ADP-ribose. (**b**) ADP-ribose hydrolase activity of ARH3 with *O*AADPr occurs in a pH-dependent manner. [^{14}C]-labeled *O*AADPr (1 μM, 200 μL) was incubated at pH 5 (*A*), 7 (*B*), or 9 (*C*) in the presence (open circles) or absence (closed circles) of recombinant human ARH3 protein (1.5 pmol) for 1 h at 30 °C. [^{14}C]-labeled *O*AADPr and ADP-ribose were separated by HPLC, and the radioactivity was measured by a liquid scintillation counter. 1"-*O*AADPr (peak *C*) is present at low level at pH 7 and high abundance at pH 9, while 2″ and 3″-*O*AADPr (peaks *B* and *A*, respectively) appear at pH 5. All peaks appeared to be in rapid equilibrium at pH 9. ARH3 activity at pH 9 was significantly faster than that at pH 7. Thus, ARH3 hydrolyzes 1"-*O*AADPr preferentially rather than 2″- and 3″-*O*AADPr. These data are from [34]

6 min. NAD^+ is eluted at 17–19 min and ADP-ribose at 38–39 min.

4. The addition of calf thymus DNA allows activation of recombinant PARP1 proteins. For optimum synthesis of PAR by PARP1, we recommend formation of DNA strand breaks by UV irradiation or treatment with restriction enzymes such as EcoRI before use.

5. [^{14}C]NAD$^+$ should be purified before use with HPLC equipped with a SAX column (*see* Subheading 3.1.1, **step 2**). NAD$^+$ is eluted at 17 to 19 min. Eluate fractions containing [^{14}C] NAD$^+$ are collected with the fraction collector.
6. Dihydroboronyl-Bio Rex affinity resin and Affi-Gel boronate affinity resin separate ADP-ribosylated proteins. Immobilized boronate forms a complex with sugars that possess coplanar *cis*-diol groups [46]. Binding of the solution is pH dependent; pH higher than pH 7.5 is required.
7. Nicotinamide is eluted at 3 min, ADP-ribose at 4 min, *O*AADPr at 7 min, NAD$^+$ at 9 min, histone at 17 min, and acetyl-histone peptide H3 at 20 min.
8. Recombinant proteins of Gαs, Gαi, and Gαo subunits, EF-II, and Rho as mono(ADP-ribosyl)ated substrates are now commercially available. To quantify ARH activity by measuring [^{14}C]-labeled ADP-ribose released from these proteins, recombinant G proteins should be used, instead of brain extracts.
9. The pH of the reaction buffer must be accurate. Acetic moiety is interconverted among 1″-3″ hydroxyl group of ADP-ribose in a pH-dependent manner. The interconversion affects ARH activity (*see* Fig. 5).
10. High-resolution SDS-PAGE allows identification of the size of PAR polymers. Electrophoresis at 400 V is stopped when bromophenol blue (BPB) moves 9 cm from the origin. Positions of BPB and xylene cyanol (XC) in the gel correspond to $(\text{ADP-ribose})_8$ and $(\text{ADP-ribose})_{18}$, respectively.

Acknowledgment

Funding: The study was funded by the Intramural Research Program, NIH, NHLBI.

References

1. Verheugd P, Butepage M, Eckei L, Luscher B (2016) Players in ADP-ribosylation: readers and erasers. Curr Protein Pept Sci 17(7):654–667
2. Palazzo L, Mikoc A, Ahel I (2017) ADP-ribosylation: new facets of an ancient modification. FEBS J 284:2932. https://doi.org/10.1111/febs.14078
3. Okazaki IJ, Moss J (1996) Mono-ADP-ribosylation: a reversible posttranslational modification of proteins. Adv Pharmacol 35:247–280
4. Corda D, Di Girolamo M (2003) Functional aspects of protein mono-ADP-ribosylation. EMBO J 22(9):1953–1958. https://doi.org/10.1093/emboj/cdg209
5. Bütepage M, Eckei L, Verheugd P, Luscher B (2015) Intracellular mono-ADP-ribosylation in signaling and disease. Cell 4(4):569–595. https://doi.org/10.3390/cells4040569
6. Ogata N, Ueda K, Kagamiyama H, Hayaishi O (1980) ADP-ribosylation of histone H1. Identification of glutamic acid residues 2, 14, and the COOH-terminal lysine residue as

modification sites. J Biol Chem 255(16): 7616–7620
7. Altmeyer M, Messner S, Hassa PO, Fey M, Hottiger MO (2009) Molecular mechanism of poly(ADP-ribosyl)ation by PARP1 and identification of lysine residues as ADP-ribose acceptor sites. Nucleic Acids Res 37(11):3723–3738. https://doi.org/10.1093/nar/gkp229
8. Zhang Y, Wang J, Ding M, Yu Y (2013) Site-specific characterization of the Asp- and Glu-ADP-ribosylated proteome. Nat Methods 10(10):981–984. https://doi.org/10.1038/nmeth.2603
9. Leidecker O, Bonfiglio JJ, Colby T, Zhang Q, Atanassov I, Zaja R, Palazzo L, Stockum A, Ahel I, Matic I (2016) Serine is a new target residue for endogenous ADP-ribosylation on histones. Nat Chem Biol 12(12):998–1000. https://doi.org/10.1038/nchembio.2180
10. Iglewski BH, Kabat D (1975) NAD-dependent inhibition of protein synthesis by Pseudomonas aeruginosa toxin. Proc Natl Acad Sci U S A 72(6):2284–2288
11. Pappenheimer AM Jr (1977) Diphtheria toxin. Annu Rev Biochem 46:69–94. https://doi.org/10.1146/annurev.bi.46.070177.000441
12. Cassel D, Pfeuffer T (1978) Mechanism of cholera toxin action: covalent modification of the guanyl nucleotide-binding protein of the adenylate cyclase system. Proc Natl Acad Sci U S A 75(6):2669–2673
13. Tamura M, Nogimori K, Murai S, Yajima M, Ito K, Katada T, Ui M, Ishii S (1982) Subunit structure of islet-activating protein, pertussis toxin, in conformity with the A-B model. Biochemistry 21(22):5516–5522
14. Vyas S, Chesarone-Cataldo M, Todorova T, Huang YH, Chang P (2013) A systematic analysis of the PARP protein family identifies new functions critical for cell physiology. Nat Commun 4:2240. https://doi.org/10.1038/ncomms3240
15. Bock FJ, Chang P (2016) New directions in poly(ADP-ribose) polymerase biology. FEBS J 283(22):4017–4031. https://doi.org/10.1111/febs.13737
16. Gupte R, Liu Z, Kraus WL (2017) PARPs and ADP-ribosylation: recent advances linking molecular functions to biological outcomes. Genes Dev 31(2):101–126. https://doi.org/10.1101/gad.291518.116
17. Borra MT, O'Neill FJ, Jackson MD, Marshall B, Verdin E, Foltz KR, Denu JM (2002) Conserved enzymatic production and biological effect of O-acetyl-ADP-ribose by silent information regulator 2-like NAD+-dependent deacetylases. J Biol Chem 277(15):12632–12641. https://doi.org/10.1074/jbc.M111830200
18. Liou GG, Tanny JC, Kruger RG, Walz T, Moazed D (2005) Assembly of the SIR complex and its regulation by O-acetyl-ADP-ribose, a product of NAD-dependent histone deacetylation. Cell 121(4):515–527. https://doi.org/10.1016/j.cell.2005.03.035
19. Grubisha O, Rafty LA, Takanishi CL, Xu X, Tong L, Perraud AL, Scharenberg AM, Denu JM (2006) Metabolite of SIR2 reaction modulates TRPM2 ion channel. J Biol Chem 281(20):14057–14065. https://doi.org/10.1074/jbc.M513741200
20. Moss J, Jacobson MK, Stanley SJ (1985) Reversibility of arginine-specific mono(ADP-ribosyl)ation: identification in erythrocytes of an ADP-ribose-L-arginine cleavage enzyme. Proc Natl Acad Sci U S A 82(17):5603–5607
21. Oka S, Kato J, Moss J (2006) Identification and characterization of a mammalian 39-kDa poly(ADP-ribose) glycohydrolase. J Biol Chem 281(2):705–713. https://doi.org/10.1074/jbc.M510290200
22. Mashimo M, Kato J, Moss J (2014) Structure and function of the ARH family of ADP-ribosyl-acceptor hydrolases. DNA Repair (Amst) 23:88–94. https://doi.org/10.1016/j.dnarep.2014.03.005
23. Konczalik P, Moss J (1999) Identification of critical, conserved vicinal aspartate residues in mammalian and bacterial ADP-ribosylarginine hydrolases. J Biol Chem 274(24):16736–16740
24. Ono T, Kasamatsu A, Oka S, Moss J (2006) The 39-kDa poly(ADP-ribose) glycohydrolase ARH3 hydrolyzes O-acetyl-ADP-ribose, a product of the Sir2 family of acetyl-histone deacetylases. Proc Natl Acad Sci U S A 103(45):16687–16691. https://doi.org/10.1073/pnas.0607911103
25. Moss J, Tsai SC, Adamik R, Chen HC, Stanley SJ (1988) Purification and characterization of ADP-ribosylarginine hydrolase from Turkey erythrocytes. Biochemistry 27(15): 5819–5823
26. Moss J, Balducci E, Cavanaugh E, Kim HJ, Konczalik P, Lesma EA, Okazaki IJ, Park M, Shoemaker M, Stevens LA, Zolkiewska A (1999) Characterization of NAD:arginine ADP-ribosyltransferases. Mol Cell Biochem 193(1–2):109–113
27. Freissmuth M, Gilman AG (1989) Mutations of GS alpha designed to alter the reactivity of the protein with bacterial toxins. Substitutions at ARG187 result in loss of GTPase activity. J Biol Chem 264(36):21907–21914

28. Gill DM, Meren R (1978) ADP-ribosylation of membrane proteins catalyzed by cholera toxin: basis of the activation of adenylate cyclase. Proc Natl Acad Sci U S A 75(7):3050–3054
29. Kahn RA, Gilman AG (1984) Purification of a protein cofactor required for ADP-ribosylation of the stimulatory regulatory component of adenylate cyclase by cholera toxin. J Biol Chem 259(10):6228–6234
30. Kahn RA, Gilman AG (1986) The protein cofactor necessary for ADP-ribosylation of Gs by cholera toxin is itself a GTP binding protein. J Biol Chem 261(17):7906–7911
31. Kato J, Zhu J, Liu C, Moss J (2007) Enhanced sensitivity to cholera toxin in ADP-ribosylarginine hydrolase-deficient mice. Mol Cell Biol 27(15):5534–5543. https://doi.org/10.1128/MCB.00302-07
32. Kato J, Zhu J, Liu C, Stylianou M, Hoffmann V, Lizak MJ, Glasgow CG, Moss J (2011) ADP-ribosylarginine hydrolase regulates cell proliferation and tumorigenesis. Cancer Res 71(15):5327–5335. https://doi.org/10.1158/0008-5472.CAN-10-0733
33. Kato J, Vekhter D, Heath J, Zhu J, Barbieri JT, Moss J (2015) Mutations of the functional ARH1 allele in tumors from ARH1 heterozygous mice and cells affect ARH1 catalytic activity, cell proliferation and tumorigenesis. Oncogene 4:e151. https://doi.org/10.1038/oncsis.2015.5
34. Kasamatsu A, Nakao M, Smith BC, Comstock LR, Ono T, Kato J, Denu JM, Moss J (2011) Hydrolysis of O-acetyl-ADP-ribose isomers by ADP-ribosylhydrolase 3. J Biol Chem 286(24):21110–21117. https://doi.org/10.1074/jbc.M111.237636
35. Mashimo M, Kato J, Moss J (2013) ADP-ribosyl-acceptor hydrolase 3 regulates poly (ADP-ribose) degradation and cell death during oxidative stress. Proc Natl Acad Sci U S A 110(47):18964–18969. https://doi.org/10.1073/pnas.1312783110
36. Fontana P, Bonfiglio JJ, Palazzo L, Bartlett E, Matic I, Ahel I (2017) Serine ADP-ribosylation reversal by the hydrolase ARH3. elife 6. https://doi.org/10.7554/eLife.28533
37. Mashimo M, Moss J (2016) Functional role of ADP-Ribosyl-acceptor hydrolase 3 in poly(ADP-ribose) polymerase-1 response to oxidative stress. Curr Protein Pept Sci 17(7):633–640
38. Yu SW, Wang H, Poitras MF, Coombs C, Bowers WJ, Federoff HJ, Poirier GG, Dawson TM, Dawson VL (2002) Mediation of poly(ADP-ribose) polymerase-1-dependent cell death by apoptosis-inducing factor. Science 297(5579):259–263. https://doi.org/10.1126/science.1072221
39. Andrabi SA, Kang HC, Haince JF, Lee YI, Zhang J, Chi Z, West AB, Koehler RC, Poirier GG, Dawson TM, Dawson VL (2011) Iduna protects the brain from glutamate excitotoxicity and stroke by interfering with poly(ADP-ribose) polymer-induced cell death. Nat Med 17(6):692–699. https://doi.org/10.1038/nm.2387
40. Lee Y, Karuppagounder SS, Shin JH, Lee YI, Ko HS, Swing D, Jiang H, Kang SU, Lee BD, Kang HC, Kim D, Tessarollo L, Dawson VL, Dawson TM (2013) Parthanatos mediates AIMP2-activated age-dependent dopaminergic neuronal loss. Nat Neurosci 16(10):1392–1400. https://doi.org/10.1038/nn.3500
41. Wang Y, An R, Umanah GK, Park H, Nambiar K, Eacker SM, Kim B, Bao L, Harraz MM, Chang C, Chen R, Wang JE, Kam TI, Jeong JS, Xie Z, Neifert S, Qian J, Andrabi SA, Blackshaw S, Zhu H, Song H, Ming GL, Dawson VL, Dawson TM (2016) A nuclease that mediates cell death induced by DNA damage and poly(ADP-ribose) polymerase-1. Science 354(6308):aad6872. https://doi.org/10.1126/science.aad6872
42. Andrabi SA, Umanah GK, Chang C, Stevens DA, Karuppagounder SS, Gagne JP, Poirier GG, Dawson VL, Dawson TM (2014) Poly(ADP-ribose) polymerase-dependent energy depletion occurs through inhibition of glycolysis. Proc Natl Acad Sci U S A 111(28):10209–10214. https://doi.org/10.1073/pnas.1405158111
43. Niere M, Mashimo M, Agledal L, Dolle C, Kasamatsu A, Kato J, Moss J, Ziegler M (2012) ADP-ribosylhydrolase 3 (ARH3), not poly(ADP-ribose) glycohydrolase (PARG) isoforms, is responsible for degradation of mitochondrial matrix-associated poly(ADP-ribose). J Biol Chem 287(20):16088–16102. https://doi.org/10.1074/jbc.M112.349183
44. Parihar P, Solanki I, Mansuri ML, Parihar MS (2015) Mitochondrial sirtuins: emerging roles in metabolic regulations, energy homeostasis and diseases. Exp Gerontol 61:130–141. https://doi.org/10.1016/j.exger.2014.12.004
45. Osborne B, Bentley NL, Montgomery MK, Turner N (2016) The role of mitochondrial sirtuins in health and disease. Free Radic Biol Med 100:164–174. https://doi.org/10.1016/j.freeradbiomed.2016.04.197
46. Alvarez-Gonzalez R, Juarez-Salinas H, Jacobson EL, Jacobson MK (1983) Evaluation of immobilized boronates for studies of adenine and pyridine nucleotide metabolism. Anal Biochem 135(1):69–77

Chapter 13

Mono-ADP-Ribosylhydrolase Assays

Jeannette Abplanalp, Ann-Katrin Hopp, and Michael O. Hottiger

Abstract

Despite substantial progress in ADP-ribosylation research in recent years, the identification of ADP-ribosylated proteins, their ADP-ribose acceptors sites, and the respective writers and erasers remains challenging. The use of recently developed mass spectrometric methods helps to further characterize the ADP-ribosylome and its regulatory enzymes under different conditions and in different cell types. Validation of these findings may be achieved by in vitro assays for the respective enzymes. In the below method, we describe how recombinant ADP-ribosylated proteins are demodified in vitro with mono-ADP-ribosylhydrolases of choice to elucidate substrate and potentially also site specificity of these enzymes.

Key words Mono-ADP-ribosylhydrolases, Macrodomain, De-ADP-ribosylation assay, PARG, MACROD2, MACROD1, ARH1, ARH3, TARG, OARD1, C6ORF130

1 Introduction

ADP-ribosylation is an evolutionary conserved covalent posttranslational modification (PTM) mainly catalyzed by ADP-ribosyltransferases (ARTs) [1, 2]. These enzymes use nicotinamide adenine dinucleotide (NAD^+) as a substrate to transfer ADP-ribose moieties onto specific amino acid residues of target proteins [3], leading to mono-ADP-ribosylation (MARylation), or to extend ADP-ribosylated sites to linear and branched poly-ADP-ribose chains (PARylation) [1]. While the role of PARylation is well established and has extensively been studied within the last decades, especially in the context of DNA damage and cell death [2, 4], the physiological relevance of MARylation remains poorly understood. Nonetheless, an increasing number of studies suggest MARylation to be implicated in various cellular processes, including immunomodulation, endoplasmatic reticulum (ER) stress, cytoskeleton rearrangement, cell metabolism, and host-pathogen interactions

Jeannette Abplanalp and Ann-Katrin Hopp contributed equally to this work.

Paul Chang (ed.), *ADP-ribosylation and NAD+ Utilizing Enzymes: Methods and Protocols*, Methods in Molecular Biology, vol. 1813, https://doi.org/10.1007/978-1-4939-8588-3_13, © Springer Science+Business Media, LLC, part of Springer Nature 2018

[5]. To date, at least 16 different enzymes are described to catalyze MARylation in mammals. The cholera toxin-like ARTs (ARTCs) ARTC1, 2, and 5 have been shown to specifically mono-ADP-ribosylate their targets at arginines [6]. Moreover, the majority of diphtheria toxin-like ARTs (ARTDs), as well as two sirtuins, namely, Sirt4 and 6, have been identified to possess MARylation activity [7, 8]. The target amino acids modified by specific mono-ARTDs remain unknown.

ADP-ribosylation of proteins is reversed by enzymes capable of hydrolyzing both MAR- and PARylation. In mammals, two enzymes, poly-(ADP-ribose)-glycohydrolase (PARG) and ADP-ribosylhydrolase 3 (ARH3), are known to degrade PAR chains [9, 10]. The first enzyme described to hydrolyze MARylation was ADP-ribosylarginine hydrolase 1 (ARH1) [11], which releases ADP-ribose from arginine residues. ARH1-deficient mice show spontaneous development of various tumors, suggesting a role for MARylation in cancer development [12]. Recently, three studies identified the mammalian proteins MACROD1, MACROD2, and TARG (also referred to as OARD1 or C6ORF130) as novel glutamate- and aspartate-specific MAR hydrolases, respectively [13–15]. The physiological role of each of these three hydrolases as well as their precise site specificity, however, is still poorly understood. Mutations within the gene encoding for TARG have been linked to the development of severe neurological dysfunction in humans [15]. MACROD1 and MACROD2 were previously described to be mutated or differentially expressed in the context of different cancers [16, 17]. MACROD1 has also been proposed to be a regulator of adipogenesis and insulin secretion [18] as well as a transcriptional co-factor for estrogen and androgen receptor signaling [19, 20]. Recent studies suggest an implication of MACROD1 in NFκB-mediated gene expression [21]. Whether the reported phenotypes are dependent on the ability of the three enzymes to hydrolyze MAR has yet to be elucidated. While no hydrolases capable of releasing ADP-ribose from lysine are published to date, we and others reported very recently that serine mono-ADP-ribosylation is reversed by ARH3 [22, 23].

Substrate specificities of MAR hydrolases are thus far difficult to assess due to technical limitations in identifying ADP-ribosylated proteins and the respective acceptor amino acids. Recent advances in the field of proteomics helped to resolve this important question. Improved enrichment protocols and the adaptation of new fragmentation techniques allowed the identification of the cellular ADP-ribosylome in untreated and hydrogen peroxide treated cells as well as in mouse organs [24, 25]. Serine was recently identified as the major ADP-ribose acceptor site besides glutamate, aspartate, lysine, and arginine [25, 26]. Whether additional sites are modified is currently still under investigation. Despite the technical progresses described above, mass spectrometry alone is not always sensitive enough to investigate the contribution of a specific enzyme, e.g., a specific MAR hydrolase, to the ADP-ribosylation status of a given protein. Low abundant proteins might be below the detection limit. In addition to this limitation, different hydrolases might exhibit partially redundant functions in vivo. Thus, the knockout phenotype of one hydrolase might be masked due to another hydrolase with the same catalytic specificity. In vitro hydrolase activity assays on targets premodified with radioactively labeled NAD^+ allow circumventing the problems discussed above. Currently, these hydrolase activity assays represent one of the best possibilities to assess substrate specificity of a hydrolase. Furthermore, they allow the validation of hits identified by either ADP-ribosylome or interactome analyses performed by mass spectrometry.

2 Materials

2.1 In Vitro MARylation

1. GST- or HIS-tagged recombinant mono-ARTD of interest (e.g., N-terminally GST-tagged hARTD10 (residues 818–1025) expressed and purified from *E. coli*) (*see* **Note 1**).
2. GST- or HIS-tagged substrate proteins that are subsequently modified (Subheading 3.2) and serve as MARylated substrate for the de-MARylation experiments in Subheading 3.3 (e.g., GST-tagged histone tails purified from bacteria). Alternatively the de-MARylation of the mono-ARTD can be tested in Subheading 3.3 (omit **step 2** of Subheading 3.2).
3. $[^{32}P]$-NAD^+ (800 Ci/mmol, 5 mCi/mL) (*see* **Note 2**).
4. 6.25 μM NAD^+ (β-nicotinamide adenine dinucleotide hydrate, ≥99%) (*see* **Note 3**).
5. 5× ADP-ribosylation reaction buffer (RB); always prepare freshly: 250 mM Tris–HCl pH 7.4, 20 mM $MgCl_2$, 1.25 mM DTT.
6. 10 mM PARP inhibitor PJ-34 (>98%, hydrochloride hydrate); the 10 mM stock solution has to be stored at −20 °C.
7. Illustra MicroSpin G50 columns (GE Healthcare).

2.2 In Vitro De-MARylation

1. GST- or HIS-tagged recombinant hydrolase of interest (e.g., N-terminally GST-tagged MACROD1 and MACROD2, expressed and purified in *E. coli*). Recombinant protein(s) without MAR hydrolase activity as negative control (e.g., C-terminally HIS-tagged PARG purified from insect cells or enzymatically inactive MAR hydrolase mutants purified from *E. coli*).

2. MARylated substrates (e.g., auto-modified ARTDs or premodified target proteins).
3. 6× SDS loading buffer: 300 mM Tris–HCl pH 6.8, 30% glycerol (87%), 0.05% bromophenol blue, 12% sodium dodecyl sulfate (SDS, >97%, Calbiochem), 12% β-mercaptoethanol (≥99.0%).

2.3 SDS-PAGE and Autoradiography

1. Resolving gel: 375 mM Tris–HCl pH 8.8, 0.1% SDS, 7.5–15% acrylamide (Serva Electrophoresis GmbH) (*see* **Note 4**), 0.1% ammonium persulfate (≥99.99%), and 0.1% *N*,*N*,*N*′,*N*′-tetramethylethylenediamine (TEMED, ~99%).
2. Stacking gel: 125 mM Tris–HCl pH 6.8, 0.1% SDS, 0.05% bromophenol blue, 4.8% acrylamide, 0.1% ammonium persulfate, and 0.1% TEMED.
3. Coomassie blue staining solution: 45% methanol (technical grade), 10% acetic acid (80%, both Thommen-Furler AG), and 0.25% Coomassie Brilliant Blue R250.
4. Destaining solution: 45% methanol (technical grade) and 10% acetic acid (80%, both Thommen-Furler AG).
5. Storage Phosphor Screen BAS-IP MS 2040 E multipurpose standard, 20 × 40 cm (or similar, GE Healthcare).
6. Phosphor imager Typhoon FLA 7000 (or similar, GE Healthcare) (*see* **Note 5**).

3 Methods

3.1 Overview of the Protocol

The ADP-ribosylhydrolase assay presented in this chapter allows addressing the substrate—as well as ADP-ribose acceptor site preference of a given MAR hydrolase in vitro. In brief, first, a potential target protein is recombinantly expressed, purified, and modified with the help of a MARylating enzyme. Depending on the enzymes used, reaction times and temperatures as well as NAD^+ concentrations might need optimization and will deviate from the presented protocol. For simplification, the catalytic domain of human ARTD10 (residues 818–1025) is used as an example in this protocol. After either auto- or transmodification of the enzyme and/or target substrate, the reaction is stopped either by adding of 10 μM PJ34 (or another PARP inhibitor) or by removing residual NAD^+ using a G50 column. The ADP-ribosylated protein is subsequently incubated with the MAR hydrolase of choice. The reaction is terminated by adding SDS loading buffer. The products are analyzed by SDS-PAGE, staining and destaining of the gel, exposure on a phosphor screen, and subsequent visualization using a phosphor imaging system. The quantification of the demodification allows determining the activity of the tested MAR hydrolases.

3.2 In Vitro MARylation

To generate MARylated substrates for assaying MAR hydrolase activity, mono-ADP-ribosyltransferases are either automodified or used to transmodify other potential target proteins. To this end, the MARylating enzyme alone, or together with a target protein, is incubated in the presence of radioactively labeled NAD^+.

1. For in vitro auto-ADP-ribosylation of hARTD10, 17 pmol of the recombinant catalytic domain of hARTD10 are used. This is important to reach a final amount of 10 pmol per demodification reaction (*see* **Note 6**). hARTD10 (818–1025) is incubated in 1× RB with 100 nM radioactively labeled $[^{32}P]$-NAD^+ (0.3 μCi/μL end concentration reached by adding cold 6.25 μM NAD^+) in a final volume of 25 μL for 15 min at 37 °C (pipet reaction on ice, NAD^+ added last) (*see* **Note 7**). Depending on the number of planned demodification reactions (Subheading 3.3), the reaction mix might need to be scaled up.
2. For the generation of MARylated target proteins in trans, 30–50 pmol of the respective protein are added to the reaction mix described above (i.e., point 1).
3. After incubation, the MARylation reaction is stopped by either:
   (a) Adding 10 μM PJ34.
   (b) Filtering the reaction using a prepacked illustra MicroSpin G50 column. To this end, vortex the column briefly, remove the cap, centrifuge for 2 min at 700 *g* at room temperature, place the column in a new microcentrifuge tube, immediately add the reaction mix, and centrifuge again. The eluate is further used for demodification reactions (*see* Subsection 3.3).

3.3 In Vitro De-MARylation

To test the activity of a given MAR hydrolase toward a specific substrate, the hydrolase is incubated together with MARylated target proteins.

1. For each demodification sample, 15 μL of the MARylation reaction mix (from Subheading 3.2) are incubated with 10 pmol of the hydrolase of interest in 1× RB in a total volume of 25 μL for 15 min at 37 °C (*see* **Note 8**). As a control, always use one reaction where no hydrolase is added, and one reaction using, e.g., recombinant PARG, or an enzymatically inactive mutant of the MAR hydrolase tested which should not lead to a decrease in signal intensity (*see* Fig. 1a).
2. The reaction is stopped by addition of SDS-loading buffer and subsequent boiling at 95 °C for 5 min. Samples can be either stored at −20 °C or directly analyzed by SDS-PAGE.

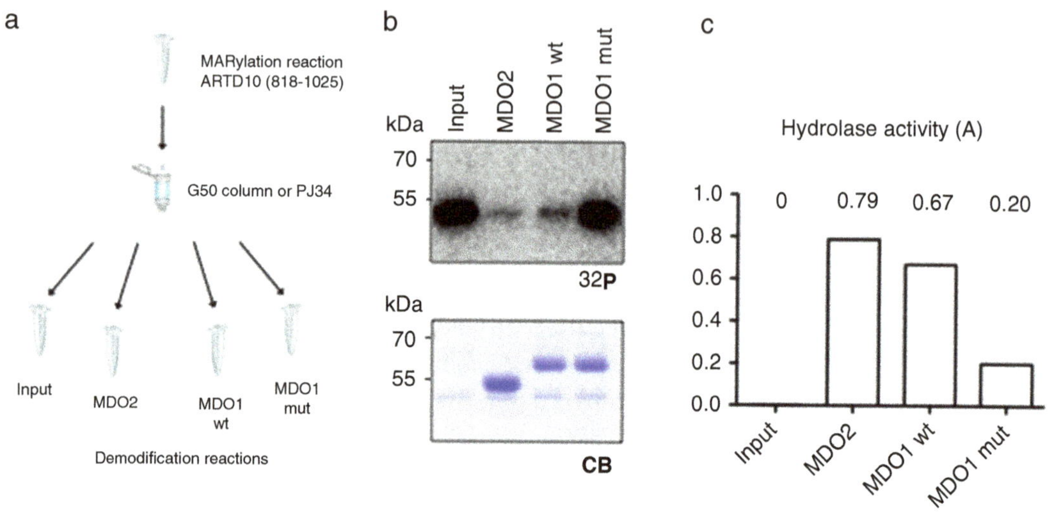

Fig. 1 (**a**) Scheme of a typical hydrolase assay setup. (**b**) Radiography and Coomassie pictures of exemplary reactions. (**c**) Quantification of hydrolase activities. The radioactivity signal (^{32}P) normalized over Coomassie blue (CB) signal is expressed as a fraction of the input. *MDO* macrodomain, *wt* wild type, *mut* enzymatically inactive mutant

3.4 SDS-PAGE and Autoradiography

To visualize and allow for subsequent quantification of the MAR hydrolase activity, the reactions are loaded on an SDS-PAGE gel and exposed on a phosphor screen.

1. Samples are loaded on an SDS-PAGE gel and run at 120 V for 1–2 h.
2. Due to potential interference with the ADP-ribosylation signal of the modified protein, the running front containing free [^{32}P]-NAD$^+$ as well as the stacking gel are cut away.
3. The gel is stained with Coomassie blue for 30 min and subsequently destained with destaining solution for 30 min – 2 h until sharp protein bands are visible. The Coomassie stained gel is photographed, and the bands are quantified using an image processing software for subsequent normalization of the radioactive signals (*see* Subheading 3.5).
4. To protect the sensitive phosphor screen, the gel can be either dried or is directly packed in a sealable plastic bag or with plastic wrap, before exposure for 1 h or up to several days (*see* **Note 9**).
5. Using a phosphor imager, the radioactivity signals are visualized and can be further analyzed and quantified using image processing software (*see* Fig. 1b).

3.5 Analysis

For the quantification of the MAR hydrolase activity, the signal intensity (radioactivity and Coomassie) for every stained band is measured using an image processing software such as ImageJ. Normalization of the radioactive signal intensity with the

Coomassie protein signal intensity for every tested target protein provides a measure of the MARylation status. The MARylation status of each sample is correlated to the input sample (modified, but not demodified), which is arbitrarily set to 1, to allow cross-comparison among different samples and to obtain a measure of MAR hydrolase activity.

The radioactivity signal R normalized to the respective Coomassie signal C results in the specific modification signal S.

$$S = \frac{R}{C}$$

The specific activity A of the demodifying enzyme is given by

$$A = 1 - \frac{S_X}{S_i}$$

where S_X is the specific modification signal of a given condition (enzyme X) and S_i is the specific modification signal of the input sample (modified, but not demodified) (*see* Fig. 1c).

4 Notes

1. For long-term storage, recombinant proteins should be stored in liquid nitrogen. Once thawed, proteins can be stored at −20 °C for several weeks.
2. For nonradioactive detection, alternatively biotin-NAD^+ or etheno-NAD^+ can be used in combination with streptavidin and the corresponding antibody, respectively.
3. The 6.25 μM NAD^+ stock needs to be stored at −80 °C.
4. Choose a low percentage separating gel for large proteins (>100 kDa) and a high percentage separating gel for small proteins.
5. Alternatively, radioactive signals can be also visualized using exposure on film.
6. For other enzymes or other ADP-ribosylated substrate proteins, the optimal concentrations might slightly differ from the concentrations indicated above and thus, might have to be adjusted.
7. For the in vitro auto-ADP-ribosylation, we found 15 min incubation at 37 °C to be the optimal condition for hARTD10cat. However, reaction time, as well as temperature, might need to be optimized for other mono-ADP-ribosyltranferases.
8. Depending on the hydrolase, reaction time as well as temperature of the demodification reaction might need to be adjusted.

9. Exposure time strongly depends on the MARylating activity of the corresponding ARTD used. The higher the enzymatic activity, the lower the exposure time.

Acknowledgments

The authors would like to thank Tobias Suter (University of Zurich) for providing editorial assistance and critical input during manuscript writing. Work on ADP-ribosyltransferases and hydrolases in the laboratory of MOH is supported by the Kanton of Zurich and the Swiss National Science Foundation (SNF 310030_157019 and 31003A_176177).

References

1. Hottiger MO (2015) Nuclear ADP-ribosylation and its role in chromatin plasticity, cell differentiation, and epigenetics. Annu Rev Biochem 84:227–263
2. Luo X, Kraus WL (2012) On PAR with PARP: cellular stress signaling through poly(ADP-ribose) and PARP-1. Genes Dev 26(5):417–432
3. Barkauskaite E, Jankevicius G, Ahel I (2015) Structures and mechanisms of enzymes employed in the synthesis and degradation of PARP-dependent protein ADP-ribosylation. Mol Cell 58(6):935–946
4. Caldecott KW (2014) Protein ADP-ribosylation and the cellular response to DNA strand breaks. DNA Repair 19:108–113
5. Butepage M et al (2015) Intracellular mono-ADP-ribosylation in signaling and disease. Cell 4(4):569–595
6. Laing S et al (2011) ADP-ribosylation of arginine. Amino Acids 41(2):257–269
7. Du J, Jiang H, Lin H (2009) Investigating the ADP-ribosyltransferase activity of sirtuins with NAD analogues and 32P-NAD. Biochemistry 48(13):2878–2890
8. Pan PW et al (2011) Structure and biochemical functions of SIRT6. J Biol Chem 286(16):14575–14587
9. Ueda K et al (1972) Poly ADP-ribose glycohydrolase from rat liver nuclei, a novel enzyme degrading the polymer. Biochem Biophys Res Commun 46(2):516–523
10. Oka S, Kato J, Moss J (2006) Identification and characterization of a mammalian 39-kDa poly(ADP-ribose) glycohydrolase. J Biol Chem 281(2):705–713
11. Moss J, Jacobson MK, Stanley SJ (1985) Reversibility of arginine-specific mono(ADP-ribosyl)ation: identification in erythrocytes of an ADP-ribose-L-arginine cleavage enzyme. Proc Natl Acad Sci U S A 82(17):5603–5607
12. Kato J et al (2011) ADP-ribosylarginine hydrolase regulates cell proliferation and tumorigenesis. Cancer Res 71(15):5327–5335
13. Jankevicius G et al (2013) A family of macrodomain proteins reverses cellular mono-ADP-ribosylation. Nat Struct Mol Biol 20(4):508
14. Rosenthal F et al (2013) Macrodomain-containing proteins are new mono-ADP-ribosylhydrolases. Nat Struct Mol Biol 20(4):502–507
15. Sharifi R et al (2013) Deficiency of terminal ADP-ribose protein glycohydrolase TARG1/C6orf130 in neurodegenerative disease. EMBO J 32(9):1225–1237
16. Xi HQ, Zhao P, Han WD (2010) Clinicopathological significance and prognostic value of LRP16 expression in colorectal carcinoma. World J Gastroenterol 16(13):1644–1648
17. Mohseni M et al (2014) MACROD2 overexpression mediates estrogen independent growth and tamoxifen resistance in breast cancers. Proc Natl Acad Sci U S A 111(49):17606–17611
18. Zang L et al (2013) Identification of LRP16 as a negative regulator of insulin action and adipogenesis in 3T3-L1 adipocytes. Horm Metab Res 45(5):349–358

19. Han WD et al (2007) Estrogenically regulated LRP16 interacts with estrogen receptor alpha and enhances the receptor's transcriptional activity. Endocr Relat Cancer 14(3):741–753
20. Yang J et al (2009) The single-macro domain protein LRP16 is an essential cofactor of androgen receptor. Endocr Relat Cancer 16(1):139–153
21. Wu Z et al (2011) LRP16 integrates into NF-kappaB transcriptional complex and is required for its functional activation. PLoS One 6(3):e18157
22. Abplanalp J et al (2017) Proteomic analyses identify ARH3 as a serine mono-ADP-ribosylhydrolase. Nat Commun 8(1):2055
23. Fontana P et al (2017) Serine ADP-ribosylation reversal by the hydrolase ARH3. elife 6:e28533
24. Martello R et al (2016) Proteome-wide identification of the endogenous ADP-ribosylome of mammalian cells and tissue. Nat Commun 7:12917
25. Bilan V et al (2017) Combining higher-energy collision dissociation and electron-transfer/higher-energy collision dissociation fragmentation in a product-dependent manner confidently assigns proteomewide ADP-ribose acceptor sites. Anal Chem 89(3):1523–1530
26. Leidecker O et al (2016) Serine is a new target residue for endogenous ADP-ribosylation on histones. Nat Chem Biol 12(12):998

Chapter 14

Hydrolysis of ADP-Ribosylation by Macrodomains

Melanija Posavec Marjanović, Gytis Jankevicius, and Ivan Ahel

Abstract

ADP-ribosylation is the process of transferring the ADP-ribose moiety from NAD^+ to a substrate. While a number of proteins represent well described substrates accepting ADP-ribose modification, a recent report demonstrated biological role for DNA ADP-ribosylation as well. The conserved macrodomain fold of several known hydrolyses was found to possess de-ADP-ribosylating activity and the ability to hydrolyze (reverse) ADP-ribosylation. Here we summarize the methods that can be employed to study mono-ADP-ribosylation hydrolysis by macrodomains.

Key words Macrodomain fold, ADP-ribosylation, Hydrolysis, Toxin-antitoxin system, NAD^+

1 Introduction

Protein ADP-ribosylation is a highly dynamic process [1] that affects diverse cellular processes in eukaryotes [2–4]. It is therefore priority for the cell to balance the synthesis of ADP-ribosylation and its removal. Recent reports point toward a complex network of de-ADP-ribosylating enzymes consisting of: PAR glycohydrolase (PARG) [5, 6], ADP-ribosylhydrolase 1 and 3 (ARH1 and ARH3, [7–9]), terminal ADP-ribose hydrolase (TARG1) [10], MACROD1 and MACROD2 [11, 12], nucleoside diphosphate linked to another moiety X (NUDIX) proteins [13], and ectonucleotide pyrophosphate/phosphodiesterases (ENPP/NPP) [14]. PARG and ARH3 are capable of degrading PAR chains, while ARH1, ARH3, MACROD1, MACROD2, TARG1, NUDIX, and ENPP/NPP remove/cleave the protein-proximal unit of ADP-ribose [9, 15]. The enzymes PARG, TARG1, MACROD1 and MACROD2 share in common the macrodomain fold. Although these proteins belong to different phylogenetic macrodomain branches and use distinct catalytic mechanisms to remove ADP-ribose [16], they all adopt a globular macrodomain fold which forms a deep cleft where free ADP-ribose or substrate-linked ADP-ribose is accommodated. The macrodomain fold is present also in other proteins that are not

Paul Chang (ed.), *ADP-ribosylation and NAD+ Utilizing Enzymes: Methods and Protocols*, Methods in Molecular Biology, vol. 1813, https://doi.org/10.1007/978-1-4939-8588-3_14,

catalytically active but can bind ADP-ribose and PAR [17]. ADP-ribose forms part of the crucial cellular metabolite NAD^+, and therefore it is not surprising that the evolution of macrodomain follows the evolution of NAD^+ signaling and NAD^+-consuming proteins and processes, including PARP and sirtuins. Hence, macrodomains play a key role in the regulation of protein ADP-ribosylation reversal.

Macrodomain homologues are not restricted only to eukaryotic organism but can be found in bacteria, archaea, and some viruses, where they also exhibit specific roles [18]. Additionally, a recent finding demonstrated that macrodomain-containing protein DarG reverses toxic DNA ADP-ribosylation set by DarT enzyme in bacteria [19]. Altogether, macrodomains emerge as adaptive folds that act as "readers, erasers, and interpreters" [16] in the sense that they (1) can bind their ligand alone or when attached to a substrate, (2) degrade long PAR chains and substrate-proximal unit of ADP-ribose, and (3) impact many crucial processes in organisms of all kingdoms of life.

In this chapter we describe protocols to study the hydrolytic activity of different macrodomain proteins on protein and DNA mono-ADP-ribosylation using gel-based assays.

2 Materials

2.1 Hydrolysis of PARP1 E988Q Mutant Auto-ADP-Ribosylation by Macrodomain Proteins

1. Human PARP1 E988Q mutant protein, expressed from pET28 PARP1 E988Q vector containing 6× His tagged full length PARP1 with E988Q mutation in *E. coli* expression strain Rosetta™(DE3)pLysS with 0.2 mM IPTG added for protein production induction. PARP1 E988Q protein was purified using Ni-NTA resin (e.g., from Qiagen) (*see* **Note 1**).
2. 20 mM NAD^+ prepared in nuclease-free water.
3. NAD^+, [^{32}P]-800 Ci/mmol 5 mCi/mL from PerkinElmer.
4. 10× activated DNA from Trevigen (*see* **Note 2**).
5. 10× PARP1 buffer: 500 mM Tris pH 8, 500 mM NaCl, 40 mM $MgCl_2$, 2 mM DTT (*see* **Note 3**).
6. Recombinant human Macro1.1 domain, expressed from pET-MCN vector containing macroH2A1.1 amino acids 155–369 with 6× His-TEV-V5 tag in *E. coli* expression strain Rosetta™(DE3)pLysS with 0.2 mM IPTG for protein production induction. Recombinant Macro1.1 protein was purified using Ni-NTA resin (e.g., from Qiagen) (*see* **Note 4**).
7. Recombinant human MACROD2 macrodomain, expressed from pET-MCN vector containing MACROD2 amino acids 7–243 with 6× His-TEV-V5 tag in *E. coli* expression strain Rosetta™(DE3)pLysS with 0.2 mM IPTG for protein

production induction. MACROD2 protein was purified using Ni-NTA resin (e.g., from Qiagen).

8. Full length human TARG1 protein, expressed from pET-MCN vector with 6× His-TEV-V5 tag in *E. coli* expression strain Rosetta™(DE3)pLysS with 0.1 mM IPTG added for protein production induction. TARG1 protein was purified using Ni-NTA resin (e.g., from Qiagen).
9. 4× NuPAGE LDS sample buffer (from Life Technologies).
10. β-mercaptoethanol.
11. 4× NuPAGE LDS sample buffer with β-mercaptoethanol.
12. Thermoblock.
13. 12% Bis-Tris Gel (e.g., NuPAGE Novax from Thermo Fisher Scientific).
14. 20× NuPAGE MOPS SDS running buffer.
15. Gel running apparatus.
16. Pre-stained protein ladder.
17. Ready to use Coomassie staining solution.
18. Fixing solution: 10% (v/v) acetic acid, 10% (v/v) methanol.
19. Whatman™ 3MM Chr Chromatography paper.
20. Gel dryer.
21. BIMAX Maximum Sensitivity Film (Kodak) or Amersham ECL HyperFilm.

2.2 Hydrolysis of DNA ADP-Ribosylation by Macrodomain Proteins

2.2.1 DNA ADP-Ribosylated Substrate Preparation

1. ADP-ribosylation buffer: 50 mM Tris–HCl pH 8, 150 mM NaCl.
2. 20 mM NAD^+ prepared in nuclease-free water.
3. Substrate oligonucleotides for ADP-ribosylation.

Oligonucleotide	Sequence
F	GAGCTGTACAAGTCAGATCTCGAGCTC
R	GAGCTCGAGATCTGACTTGTACAGCTC

4. Full length *Taq*DarT ADP-ribosyl-transferase, expressed from pBAD33 vector tagged with 6× His-TEV-V5, in *E. coli* BL21 strain using 0.8% arabinose for protein production induction. Protein was purified using TALON affinity resin from Clontech.
5. 1× TE buffer: 10 mM Tris pH 8, 1 mM EDTA.
6. Denaturing polyacrylamide gels (7–8 M urea, 15–20% polyacrylamide (29:1), 1× TBE), also available commercially: e.g., Novex® TBE-Urea Gels from Life Technologies.
7. Gel running apparatus.

8. TLC plate with indicator green 254 nm.
9. Centrifuge tube filters with 0.22 μm filter (e.g., Costar SPIN-X centrifuge filter).
10. Single-use microspin column for clean-up (e.g., PD SpinTrap G-25).

2.2.2 Hydrolysis of DNA ADP-Ribosylated Substrate by TaqDarG Enzyme

1. 4× Macro buffer: 200 mM Tris pH 8, 200 mM NaCl.
2. Full length *Taq*DarG, expressed from pET28a vector with 6× His tag in *E. coli* BL21 bacterial strain with the addition of 0.2 mM IPTG. *Taq*DarG protein was purified using Ni-NTA resin (Qiagen).
3. ADP-ribosylated substrate oligonucleotide prepared as in Subheading 3.2.1.
4. Thermoblock.
5. 2× urea loading buffer: 8 M urea, 20 mM EDTA, 2mM Tris pH 7.5
6. 1× TBE: 90 mM Tris-borate pH 8, 2 mM EDTA.
7. Denaturing polyacrylamide gels (7–8 M urea, 15–20% polyacrylamide (29:1), 1× TBE), also available commercially: Novex® TBE-Urea Gels (Life Technologies).
8. Gel running apparatus (e.g., Life Technologies).
9. Ethidium bromide.
10. UV documentation system.

3 Methods

3.1 Hydrolysis of PARP1 E988Q Mutant Auto-ADP-Ribosylation by Macrodomain Proteins (See Fig. 1)

Fresh 10× PARP1 buffer is prepared and used to mix with other components of the master mix: activated DNA, PARP1 E988Q protein, NAD^+ and [^{32}P]- NAD^+ and nuclease-free water. The master mix is divided into four reaction tubes. After 20 min of incubation at room temperature, different macrodomain proteins are added to the reaction tubes and incubated for another 15 min. Macrodomains with hydrolytic activity will remove the mono-ADP-ribosylation from auto-modified PARP1 E988Q [21]. The reaction is stopped by adding 4× Sample buffer containing β-mercaptoethanol; samples are boiled and loaded in duplicate onto two SDS-PAGE gels. One gel is stained with ready-to-use Coomassie stain, while the other is washed, fixed, dried, and finally exposed using ECL or high-sensitivity X-ray films.

1. Prepare fresh 10× PARP1 buffer.
2. Calculate the volumes of activated DNA to 1× final, PARP1 E988Q to 0.5 μM, 10× buffer to 1×, NAD^+ to 2 μM, 0.5 μCi [^{32}P]- NAD^+ and nuclease-free water needed for reactions with

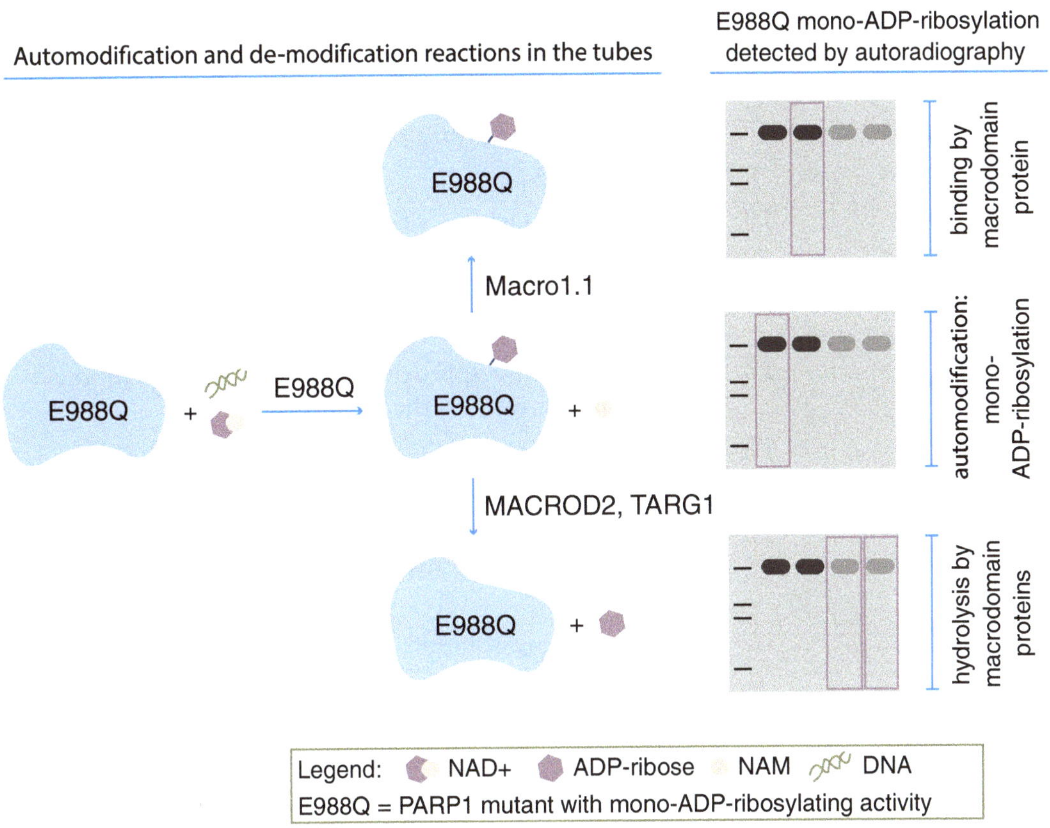

Fig. 1 Schematic representation of analysis of macrodomain hydrolytic activity of different proteins on automodified PARP1 E988Q protein using autoradiography. PARP1 E988Q is activated by DNA and modified in the presence of [^{32}P]- NAD$^+$. MACROD2 and TARG1 macrodomains hydrolyze the ADP-ribose bound to PARP1 E988Q, while Macro1.1 macrodomain has no hydrolytic activity. After SDS-PAGE electrophoresis, the signal was detected by autoradiography

three different macrodomain proteins (three reactions), one control reaction, and an extra reaction for pipetting error. In total, prepare the mix for five reactions but prepare four reaction tubes. Set the total volume of one reaction to 10 μL.

3. Calculate the volumes of macrodomain proteins according to their concentrations needed to take 5 μM of each macrodomain protein.
4. Prepare master mix containing everything but the macrodomains—activated DNA, PARP1 E988Q protein, 1× buffer, NAD$^+$, [^{32}P]- NAD$^+$ and nuclease-free water—mix gently and distribute into four clean reaction tubes.
5. Incubate the reaction 20 min at room temperature.
6. Then add the macrodomain proteins in the following order: (1) no macrodomain protein, (2) Macro1.1, (3) MACROD2, (4) PARG.

7. Incubate 15 additional minutes at room temperature.
8. Add 5 μL of 4× Sample buffer with β-mercaptoethanol to stop the reaction. Add it to all tubes in the same order as you added all the other components.
9. Boil 2 min at 95 °C.
10. Centrifuge the samples prior to loading at maximal speed (18,000 × *g*) for 30 s.
11. Load in duplicate two SDS-PAGE gels in the same order as you prepared the samples.
12. Run the samples for approximately 1 h at 175 V until the dye front has almost reached the gel border.
13. Stain one gel directly with Coomassie stain (*see* **Note 5**).
14. Wash the other gel twice in water for 5 min at room temperature while gently agitating.
15. Remove the water and add fixing solution. Incubate at least 30 min.
16. Discard the fixing solution and continue washing the gel in water for at least an hour.
17. Lay the gel on 3MM Chromatography paper and dry it using the gel dryer.
18. Finally expose the gel using ECL or high-sensitivity X-ray films.

3.2 Hydrolysis of DNA ADP-Ribosylation by Macrodomain Proteins

*Taq*DarT protein is mixed with substrate oligonucleotides in 1× ADP-ribosylation buffer and NAD^+ and incubated in order to ADP-ribosylate the oligos. Then, the modified oligonucleotides are gel purified using 1× TE buffer and desalted using single-use microspin column.

3.2.1 DNA ADP-Ribosylated Substrate Preparation

1. Calculate the quantities of components necessary to modify 20–50 μM of the DNA substrate oligonucleotide using 2 μM *Taq*DarT in the presence of 40–100 μM NAD^+ in 1× ADP-ribosylation buffer, in a total reaction volume of 50 μL (*see* **Note 6**).
2. Mix the components and incubate 1 h at 37 °C.
3. To gel purify the modified oligonucleotide substrate, run the reaction mix on denaturing polyacrylamide gel for approximately 1 h at 5 W.
4. When the run is over, use a TLC plate with fluorescent indicator green (254 nm) to put it underneath the gel.
5. Then visualize the modified substrate as UV shadow.
6. Cut out the upper band (the modified substrate band) and transfer the gel piece to a clean microcentrifuge tube.

7. Crush the gel piece with a pipette tip.
8. Add 1× TE buffer and incubate 2 h at 37 °C (or at room temperature overnight).
9. Elute the modified substrate by passing the solubilized substrate through 0.22 μm filter (Costar SPIN-X centrifuge filter or similar).
10. Desalt into water using single-use microspin column (PD SpinTrap G-25 or similar) (*see* **Note 7**).

3.2.2 Hydrolysis of DNA ADP-Ribosylated Substrate by TaqDarG Enzyme

*Taq*DarG protein and control reaction (no protein) are mixed with 1× Macro buffer in the way that ADP-ribosylated oligonucleotide is added to final concentration of 2 μM. Reactions are incubated at 37 °C and afterward stopped by adding 2× Urea Loading buffer. Reactions are boiled and then loaded onto denaturing polyacrylamide gel. The gel is stained with ethidium bromide and visualized under UV light.

1. Calculate the required volumes of reagents to achieve 1× Macro buffer and 100 nM *Taq*DarG, include a reaction with no protein to serve as a control and a marker.
2. Make up the volume to 18 μL with deionized water.
3. Start the reaction by adding 2 μL of ADP-ribosylated oligonucleotide (2 μM final).
4. Incubate at 37 °C for 15 min (for kinetic analyses take samples at different time points).
5. Add 20 μL 2× Urea Loading buffer.
6. Heat at 95 °C for 2 min.
7. Load 5 μL on denaturing polyacrylamide gels in 1× TBE.
8. Run the samples for approximately 1 h at 5 W.
9. Disassemble the gel cast and rinse the gel briefly in 1× TBE.
10. Stain with 0.25 μg/mL ethidium bromide in 1× TBE for 5 min.
11. Rinse the gel briefly in 1× TBE.
12. Visualize under UV and capture image using documentation system.

4 Notes

1. PARP1 E988Q mutant (expressed and purified as previously described for wild type PARP1 [20]) lacks polymerase activity and has only mono-ADP-ribosylating activity. PARP10 can also be used as mono-ADP-ribosylated protein since PARP10 automodifies itself. GST-fused PARP10 (fusion protein

containing PARP10 amino acids 818–1025) can be expressed in *E. coli* BL21 (DE3) strain [21].

2. Activated DNA can be also prepared in the lab according to the instructions in the paper by Clark et al. [22]. Still, the activation potential of this homemade activated DNA should be tested and concentrations optimized for the reaction.
3. 10× PARP1 buffers may contain 10–100 mM $MgCl_2$. This concentration varies depending on the complex you want to form with PARP1 enzyme [22].
4. Macro1.1 domain binds ADP-ribose and PAR but does not possess hydrolytic activity, so unlike MACROD2 and TARG1, it cannot remove the ADP-ribose from PARP1 E988Q protein.
5. Ready to use Coomassie stains that leave low background are very convenient. Proteins are stained in 30–60 min and stand out as the background staining remains minimal.
6. Since the following step is de-modification, prepare the substrate in excess.
7. Evaporate water if higher concentrations are needed.

Acknowledgments

The work in Ahel lab is supported by Wellcome Trust (grant number 101794), the European Research Council (grant number 281739) and by Cancer Research UK (grant number C35050/A22284). MPM is financed by Croatian National Centre of Research Excellence in Personalized Healthcare grant.

References

1. Alvarez-Gonzalez R, Althaus FR (1989) Poly(ADP-ribose) catabolism in mammalian cells exposed to DNA-damaging agents. Mutat Res 218(2):67–74
2. Barkauskaite E, Jankevicius G, Ahel I (2015) Structures and mechanisms of enzymes employed in the synthesis and degradation of PARP-dependent protein ADP-ribosylation. Mol Cell 58(6):935–946. https://doi.org/10.1016/j.molcel.2015.05.007
3. Canto C, Menzies KJ, Auwerx J (2015) NAD(+) metabolism and the control of energy homeostasis: a balancing act between mitochondria and the nucleus. Cell Metab 22(1):31–53. https://doi.org/10.1016/j.cmet.2015.05.023
4. Hottiger MO (2015) Nuclear ADP-ribosylation and its role in chromatin plasticity, cell differentiation, and epigenetics. Annu Rev Biochem 84:227–263. https://doi.org/10.1146/annurev-biochem-060614-034506
5. Lin W, Ame JC, Aboul-Ela N, Jacobson EL, Jacobson MK (1997) Isolation and characterization of the cDNA encoding bovine poly(ADP-ribose) glycohydrolase. J Biol Chem 272(18):11895–11901
6. Slade D, Dunstan MS, Barkauskaite E, Weston R, Lafite P, Dixon N, Ahel M, Leys D, Ahel I (2011) The structure and catalytic mechanism of a poly(ADP-ribose) glycohydrolase. Nature 477(7366):616–620. https://doi.org/10.1038/nature10404

7. Moss J, Stanley SJ, Nightingale MS, Murtagh JJ Jr, Monaco L, Mishima K, Chen HC, Williamson KC, Tsai SC (1992) Molecular and immunological characterization of ADP-ribosylarginine hydrolases. J Biol Chem 267(15):10481–10488
8. Oka S, Kato J, Moss J (2006) Identification and characterization of a mammalian 39-kDa poly(ADP-ribose) glycohydrolase. J Biol Chem 281(2):705–713. https://doi.org/10.1074/jbc.M510290200
9. Fontana P, Bonfiglio JJ, Palazzo L, Bartlett E, Matic I, Ahel I (2017) Serine ADP-ribosylation reversal by the hydrolase ARH3. Elife 6. pii: e28533. https://doi.org/10.7554/eLife.28533
10. Sharifi R, Morra R, Appel CD, Tallis M, Chioza B, Jankevicius G, Simpson MA, Matic I, Ozkan E, Golia B, Schellenberg MJ, Weston R, Williams JG, Rossi MN, Galehdari H, Krahn J, Wan A, Trembath RC, Crosby AH, Ahel D, Hay R, Ladurner AG, Timinszky G, Williams RS, Ahel I (2013) Deficiency of terminal ADP-ribose protein glycohydrolase TARG1/C6orf130 in neurodegenerative disease. EMBO J 32(9):1225–1237. https://doi.org/10.1038/emboj.2013.51
11. Jankevicius G, Hassler M, Golia B, Rybin V, Zacharias M, Timinszky G, Ladurner AG (2013) A family of macrodomain proteins reverses cellular mono-ADP-ribosylation. Nat Struct Mol Biol 20(4):508–514. https://doi.org/10.1038/nsmb.2523
12. Rosenthal F, Feijs KL, Frugier E, Bonalli M, Forst AH, Imhof R, Winkler HC, Fischer D, Caflisch A, Hassa PO, Luscher B, Hottiger MO (2013) Macrodomain-containing proteins are new mono-ADP-ribosylhydrolases. Nat Struct Mol Biol 20(4):502–507. https://doi.org/10.1038/nsmb.2521
13. Palazzo L, Thomas B, Jemth AS, Colby T, Leidecker O, Feijs KL, Zaja R, Loseva O, Puigvert JC, Matic I, Helleday T, Ahel I (2015) Processing of protein ADP-ribosylation by Nudix hydrolases. Biochem J 468(2):293–301. https://doi.org/10.1042/BJ20141554
14. Palazzo L, Daniels CM, Nettleship JE, Rahman N, McPherson RL, Ong SE, Kato K, Nureki O, Leung AK, Ahel I (2016) ENPP1 processes protein ADP-ribosylation in vitro. FEBS J 283(18):3371–3388. https://doi.org/10.1111/febs.13811
15. Palazzo L, Mikoc A, Ahel I (2017) ADP-ribosylation: new facets of an ancient modification. FEBS J 84(18):2932–2946. https://doi.org/10.1111/febs.14078
16. Rack JG, Perina D, Ahel I (2016) Macrodomains: structure, function, evolution, and catalytic activities. Annu Rev Biochem 85:431–454. https://doi.org/10.1146/annurev-biochem-060815-014935
17. Kustatscher G, Hothorn M, Pugieux C, Scheffzek K, Ladurner AG (2005) Splicing regulates NAD metabolite binding to histone macroH2A. Nat Struct Mol Biol 12(7):624–625
18. de Souza RF, Aravind L (2012) Identification of novel components of NAD-utilizing metabolic pathways and prediction of their biochemical functions. Mol Biosyst 8(6):1661–1677. https://doi.org/10.1039/c2mb05487f
19. Jankevicius G, Ariza A, Ahel M, Ahel I (2016) The toxin-antitoxin system DarTG catalyzes reversible ADP-ribosylation of DNA. Mol Cell 64(6):1109–1116. https://doi.org/10.1016/j.molcel.2016.11.014
20. Langelier MF, Planck JL, Roy S, Pascal JM (2012) Structural basis for DNA damage-dependent poly(ADP-ribosyl)ation by human PARP-1. Science 336(6082): 728–732. https://doi.org/10.1126/science.1216338
21. Kleine H, Poreba E, Lesniewicz K, Hassa PO, Hottiger MO, Litchfield DW, Shilton BH, Luscher B (2008) Substrate-assisted catalysis by PARP10 limits its activity to mono-ADP-ribosylation. Mol Cell 32(1):57–69. https://doi.org/10.1016/j.molcel.2008.08.009
22. Clark NJ, Kramer M, Muthurajan UM, Luger K (2012) Alternative modes of binding of poly(ADP-ribose) polymerase 1 to free DNA and nucleosomes. J Biol Chem 287(39):32430–32439. https://doi.org/10.1074/jbc.M112.397067

Chapter 15

HPLC-Based Enzyme Assays for Sirtuins

Jun Young Hong, Xiaoyu Zhang, and Hening Lin

Abstract

Sirtuins are a class of enzymes that utilize nicotinamide adenine dinucleotide, NAD^+, to remove various acyl groups from protein lysine residues. They have important biological functions and regulate numerous biological pathways. Small molecules that can modulate sirtuin enzymatic activities are potential therapeutic candidates to treat various human diseases. This protocol describes a high-performance liquid chromatography (HPLC)-based method to measure the enzyme kinetics for SIRT2 and SIRT6's demyristoylase activities and SIRT5's desuccinylase activity. This method uses peptide substrates that resemble physiological substrates and thus can give more reliable kinetic parameters (K_m and k_{cat} values) for these enzymes. The data obtained are useful for understanding the biological function of sirtuins and developing sirtuin modulators.

Key words Enzymatic activity assay, High-performance liquid chromatography assay, SIRT2, SIRT5, SIRT6, Demyristoylase, Desuccinylase

1 Introduction

Sirtuins are an interesting class of enzymes that regulate aging and transcription among numerous other biological processes [1, 2]. Initially, they were thought to be NAD^+-dependent protein lysine deacetylases and exert their biological functions by deacetylating various substrate proteins [3, 4]. However, several sirtuins do not have efficient deacetylase activities in vitro and were later discovered to remove acyl groups other than acetyl. For example, SIRT5 can efficiently remove negatively charged succinyl and malonyl groups, while SIRT6 can remove long chain fatty acyl groups [5, 6]. The physiological functions of these activities have also been demonstrated. Because of the important biological functions of sirtuins, many laboratories around the world are interested in developing small molecules that can modulate the activities of sirtuins. Enzyme activities assays are important for both understanding the physiological function of enzymes and developing small molecule modulators. Here, we describe a HPLC-based method

Paul Chang (ed.), *ADP-ribosylation and NAD+ Utilizing Enzymes: Methods and Protocols*, Methods in Molecular Biology, vol. 1813, https://doi.org/10.1007/978-1-4939-8588-3_15,

for measuring SIRT2 and SIRT6's demyristoylase activities and SIRT5's desuccinylase activity.

Primary data from the assay can be fitted to Michaelis-Menten equation to derive K_m and k_{cat} values. K_m represents the substrate concentration at which the reaction rate is half-maximal. K_m also reflects the binding affinity between the substrate and the enzyme. A lower K_m value indicates a stronger interaction between the enzyme and the substrate. The k_{cat} value represents the turnover number, which characterizes the number of substrates converted to a product in a given time. Combined, k_{cat}/K_m represents the catalytic efficiency of the enzyme [7, 8]. The k_{cat}/K_m value is also referred to the specificity constant as substrates with higher k_{cat}/K_m values are more preferred substrates than those with lower k_{cat}/K_m values. Therefore, measuring the k_{cat}/K_m values can help to understand the physiological function of enzymes. Furthermore, developing small molecule inhibitors of sirtuins typically require the measurement of IC_{50} values (50% inhibition concentration). IC_{50} measurement should be carried out with substrate concentrations around K_m values. Substrate concentrations, being too low or too high, will alter IC_{50} values, making it harder to differentiate effective inhibitors among the candidates.

In these assays, we used histone H3-based peptide sequences (amino acids 4–13) with different acyl groups: myristoylated H3K9 for SIRT2 and SIRT6 and succinylated H3K9 for SIRT5 (Fig. 1) [5, 6, 9, 10]. Both peptide substrates are loaded with two tryptophans at the C-terminals, which facilitate the detection and quantification by UV absorption at 280 nm in HPLC. Then, the area of the product peak divided by the sum of both substrate and product peak areas can tell us how much substrate is converted to product, which can in turn yield the initial reaction rate of the enzymatic reaction.

$$\frac{\text{Peak area of product}}{\text{Peak area of product} + \text{peak area ofsubstrate}} = \text{conversion}\%$$

2 Materials

2.1 General Materials for High-Performance Liquid Chromatography UV Assay

1. High-performance liquid chromatography system with a UV-Vis detector and an autosampler that can take 96-well plates.
2. C18 HPLC Column (100 A, 75 mm × 4.6 mm, 2.6 μm) (*see* **Note 1**).
3. Buffer A: Millipore purified water or HPLC grade water with 0.1% trifluoroacetic acid (*see* **Note 2**).
4. Buffer B: HPLC grade acetonitrile with 0.1% trifluoroacetic acid.
5. High-speed centrifuge: Centrifuge that can reach to 15,000 × *g*.

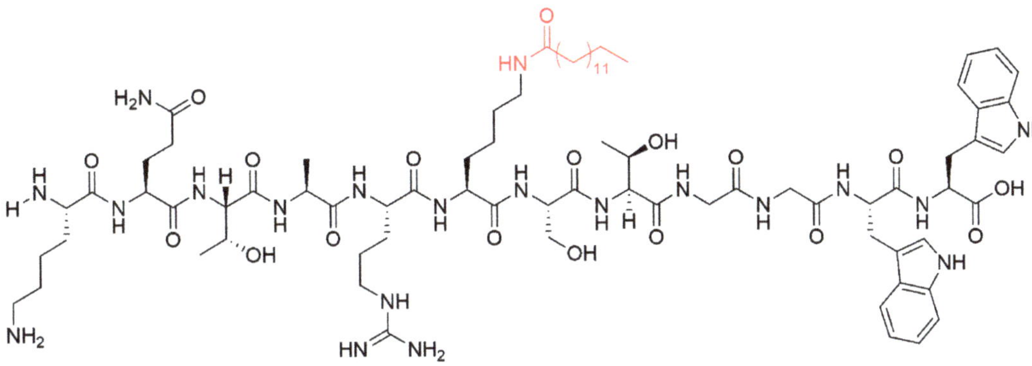

Myristoyl H3K9

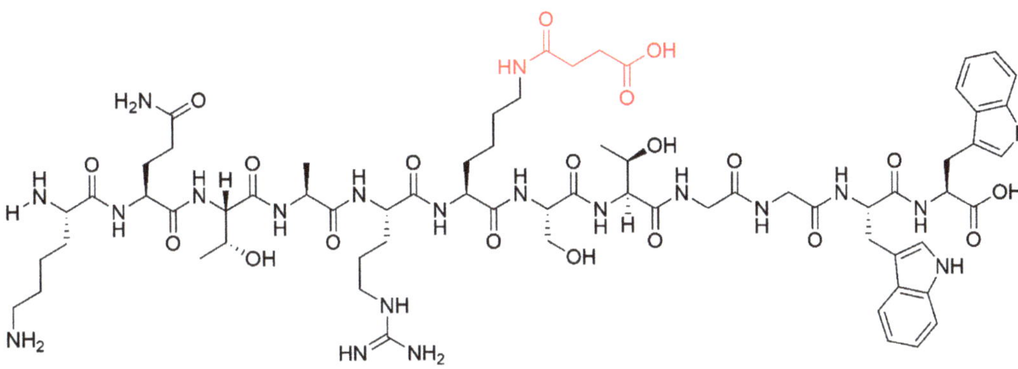

Succinyl H3K9

Fig. 1 Two sirtuins peptide substrates used in the HPLC assay

2.2 SIRT2 Demyristoylase Activity Assay

1. Reaction buffer: 20 mM Tris–HCl (pH 8.0), 1 mM DTT, 1 mM NAD^+ (*see* **Note 3**). To 50 μL of water, add 1.2 μL of 1 M Tris–HCl (pH 8.0), 2.4 μL of 25 mM DTT, and 2.4 μL of 25 mM NAD^+.
2. Quench buffer: 200 mM HCl and 320 mM acetic acid in methanol.
3. SIRT2 enzyme: Prepare 6 μM of purified SIRT2 (*see* **Note 4**) [11]. The enzyme is dissolved in 20 mM Tris–HCl (pH 8.0), 500 mM NaCl, and 10% glycerol (*see* **Note 5**).
4. Myristoyl H3K9 peptide: Dissolve the peptide in water at concentrations 7.5–960 μM (*see* **Note 6**).
5. Ice.

2.3 SIRT5 Desuccinylase Activity Assay

1. Reaction Buffer: 20 mM Tris–HCl (pH 8.0), 1 mM DTT, 1 mM NAD^+ (*see* **Note 3**). To 50 μL of water, add 1.2 μL of 1 M Tris–Hcl (pH 8.0), 2.4 μL of 25 mM DTT, and 2.4 μL of 25 mM NAD^+.
2. Quench buffer: Water with 10% trifluoroacetic acid.
3. SIRT5 Enzyme: Prepare 30 μM of purified enzyme (*see* **Note 7**) [5]. The enzyme is dissolved in 20 mM Tris–HCl (pH 8.0), 500 mM NaCl, and 10% glycerol (*see* **Note 5**).
4. Succinyl H3K9 Peptide: Dissolve the peptide in water at concentrations 7.5–750 μM (*see* **Note 6**).
5. Ice.

2.4 SIRT6 Demyristoylase Activity Assay

1. Reaction buffer: 20 mM Tris–HCl (pH 8.0), 1 mM DTT, 1 mM NAD^+ (*see* **Note 3**). To 50 μL of water, add 1.2 μL of 1 M Tris–HCl (pH 8.0), 2.4 μL of 25 mM DTT, and 2.4 μL of 25 mM NAD^+.
2. Quench buffer: HPLC-grade acetonitrile.
3. SIRT6 Enzyme: Prepare 3 μM of purified SIRT6 enzyme (*see* **Note 8**) [6]. The enzyme is dissolved in 20 mM Tris–HCl (pH 7.2), 500 mM NaCl, and 10% glycerol (*see* **Note 4**).
4. Myristoyl H3K9 Peptide: Dissolve the peptide in water at concentrations 10–1280 μM (*see* **Note 6**).
5. Ice.

3 Methods

Carry out all procedures at room temperature unless otherwise specified.

3.1 SIRT2 Demyristoylase Activity Assay

This method involves incubating SIRT2 with varying concentrations of myristoyl H3K9, its substrate, for a certain period to ensure that the assumption for the Michaelis-Menten equation, $[S] \approx [S_o]$, is met [12]. The reaction is then stopped with the quench buffer and centrifuged vigorously to remove SIRT2 precipitate. The purpose of removing SIRT2 precipitate is to avoid blockade in the HPLC system, which will be used to detect and quantify myristoyl and free H3K9 peptide.

1. Thaw the purified SIRT2 on ice.
2. Prepare the reaction buffer in a 1.5 mL centrifuge tube.
3. Add 2 μL of purified SIRT2 to the reaction buffer.
4. As soon as the timer starts, add 2 μL of the myristoyl H3K9 peptide with the given concentration.
5. Vortex briefly (~2 s) to mix.

6. Centrifuge the tube for less than 3 s, using a table top centrifuge (*see* **Note 9**).
7. Place the tube at 37 °C using dry block incubator for 3 min (*see* **Note 10**).
8. After 3 min, remove the tube from the incubator and add 60 μL of the quench buffer.
9. Vortex the tube thoroughly for 5 s (*see* **Note 11**).
10. Place the tube on a tube rack.
11. Repeat from #1 for different concentrations of the peptide (*see* **Note 12**).
12. Centrifuge all the tubes at 15,000 × *g* for 10 min to remove the precipitate SIRT2. The supernatant will be used for HPLC analysis (*see* **Note 13**).

3.2 SIRT5 Desuccinylase Activity Assay

This method involves incubating SIRT5 with varying concentrations of succinyl H3K9, its substrate, for a certain period to ensure that the assumption for the Michaelis-Menten equation, $[S] \approx [S_o]$, is met [12]. The reaction is then stopped with the quench buffer and centrifuged vigorously to remove SIRT5 precipitate. The purpose of removing SIRT5 precipitate is to avoid blockade in the HPLC system, which will be used to detect and quantify succinyl and free H3K9 peptide.

1. Thaw the purified SIRT5 on ice.
2. Prepare the reaction buffer in a 1.5 mL centrifuge tube.
3. Add 2 μL of purified SIRT5 to the reaction buffer.
4. As soon as the timer starts, add 2 μL of the succinyl H3K9 peptide with the given concentration.
5. Vortex briefly (~2 s) to mix.
6. Centrifuge the tube for less than 3 s, using a table top centrifuge (*see* **Note 9**).
7. Place the tube at 37 °C incubator for 4 min (*see* **Note 10**).
8. After 4 min, remove the tube from the incubator, and add 60 μL of the quench buffer.
9. Vortex the tube thoroughly for 5 s (*see* **Note 11**).
10. Place the tube on a tube rack.
11. Repeat from #1 for different concentrations of the peptide (*see* **Note 12**).
12. Centrifuge all the tubes at 15,000 × *g* for 10 min to remove the precipitated SIRT5. The supernatant will be used for HPLC analysis (*see* **Note 13**).

3.3 SIRT6 Demyristoylase Activity Assay

This method involves incubating SIRT6 with varying concentrations of myristoyl H3K9, its substrate, for a certain period to ensure that assumption for the Michaelis-Menten equation, $[S] \approx [S_o]$, is met [12]. The reaction is then stopped with the quench buffer and centrifuged vigorously to remove SIRT6 precipitate. The purpose of removing SIRT6 precipitate is to avoid blockade in the HPLC system, which will be used to detect and quantify myristoyl and free H3K9 peptide.

1. Thaw the purified SIRT6 in ice.
2. Prepare the reaction buffer in a 1.5 mL centrifuge tube.
3. Add 2 μL of purified SIRT6 to the reaction buffer.
4. As soon as the timer starts, add 2 μL of the myristoyl H3K9 peptide with the given concentration.
5. Vortex briefly (~2 s) to mix.
6. Centrifuge the tube for less than 3 seconds, using a table top centrifuge (*see* **Note 9**).
7. Place the tube at 37 °C incubator for 10 min (*see* **Note 10**).
8. After 10 min, remove the tube from the incubator, and add 40 μL of the quench buffer.
9. Vortex the tube thoroughly for 5 s (*see* **Note 11**)
10. Place the tube on a tube rack.
11. Repeat from #1 for different concentrations of the peptide (*see* **Note 12**).
12. Centrifuge all the tubes at 15,000 × *g* for 10 min to remove the precipitate SIRT6. The supernatant will be used for HPLC analysis (*see* **Note 13**).

3.4 HPLC UV Detection and Data Analysis

After the reaction and removing protein precipitate, the reaction mixture is loaded onto the HPLC system with a C18 column to detect myristoyl/succinyl and free lysine peptide using the 280 nm UV absorption. Then, by identifying the peak area of corresponding peptides, the conversion rate can be determined, which can be used to calculate the K_m and k_{cat} values.

1. Add 70 μL of centrifuged reaction sample to a well of a 96-well plate (*see* **Note 14**).
2. Program the HPLC software to inject 60 μL of the sample to the system (*see* **Note 15**).
3. For SIRT2 and SIRT6, the chromatography gradient was 0% buffer B for 2 min, 0–20% buffer B for 2 min, 20–40% buffer B for 13 min, 40–100% buffer B for 2 min, and then 100% buffer B for 5 min. The flow rate for SIRT2 and SIRT6 was 0.5 mL/min. For SIRT5, the gradient was 0% to 50% buffer B for 20 min with a flow rate of 1 mL/min (*see* **Note 16**).

4. UV-Vis detector was set to measure at wavelength 280 nm.
5. To locate the product or substrate peptide peaks, run a sample only with the standard product or substrate peptide under the same chromatography gradient condition.
6. Integrate the unmodified and myristoyl or succinyl H3K9 peptide peaks (*see* **Note 17**).
7. Divide the peak area of unmodified to the total peak area of unmodified and myristoyl or succinyl H3K9 peptide to obtain the conversion rate for each substrate concentration.
8. Multiply the conversion rate with the corresponding substrate concentration, and divide by the reaction time to obtain the initial reaction rate V_i.
9. Plot Vi versus substrate concentration $[S_0]$.
10. Use software such as KaleidaGraph or SigmaPlot to fit the plot to the Michaelis-Menten equation $V_i = V_{max}*[S_0]/(K_m + [S_0])$ to obtain V_{max} and K_m. V_{max} divide by enzyme concentration $[E_0]$ will give k_{cat}.

4 Notes

1. Place a guard column right before the start of the C18 column to avoid precipitated protein from entering the C18 column.
2. Recommend using trifluoroacetic acid (≥99.0%) for HPLC.
3. Prepare DTT and NAD^+ fresh every time.
4. SIRT2 is purified in truncated length (aa38–256). The Sirt2 gene is inserted into pET28a vector with N-terminal His6-SUMO tag. The vector is transformed into *E. coli* BL21 cells. The cells are cultured in 2 L luria broth with 50 μg/mL kanamycin (50 mg/mL stock in H_2O, 1:1000 dilution) and 20 μg/mL chloramphenicol (20 mg/mL stock in ethanol, 1:1000 dilution). At OD600 of 0.6, 20 μM of isopropyl β-D-1-thiogalactopyranoside (IPTG) is added to induce expression (1 M stock in H_2O, 40 μL of IPTG to 2 L of luria broth). The cells are grown at 15 °C, 200 rpm for 16 h. The following day, the cells are collected by centrifuging at 4 °C, 11305 × *g* for 5 min, and resuspended in 20 mL of 20 mM Tris–HCl (pH 8.0), 500 mM NaCl, 10% glycerol, and 1 mM PMSF. The collected cells are passed through a cell disruptor at least three times to lyse the cell. After removing cellular debris by centrifugation at 4 °C, 48298 × *g* for 30 min, the supernatant is loaded onto a nickel column, pre-equilibrated with 20 mM Tris–HCl (pH 8.0) and 500 mM NaCl. Solution of imidazole is added in increasing concentration, from 0 to 500 mM, to elute the desired protein. 1.5 mL of each imidazole solution

was used to elute the protein. Each fraction is analyzed by SDS-PAGE to determine whether the desired protein is present. Fractions comprising the desired protein are collected and added into a dialysis tubing. Then 0.1 mL of ULP1, a SUMO protease purchased from commercial source, was added to remove the SUMO tag on SIRT2. The dialysis tubing is incubated at 4 °C overnight with 20 mM Tris–HCl (pH 8.0) and 500 mM NaCl. The protein is additionally purified through a Superdex 75 column (Biorad), using 20 mM Tris-HCl (pH 8.0) and 500 mM NaCl. The desired fraction is then concentrated, and 50% glycerol is added accordingly to make the final glycerol concentration to be 10%. After the purification, the protein is stored at −80 °C for future use [11].

5. Avoid repetitive freeze-thaw. Aliquot the purified enzyme into smaller volumes, so that the leftover enzyme can be discarded after the experiment. The sirtuin enzyme loses activity significantly after several freeze-thaw actions. Store all the enzyme at -80 °C before the experiment.
6. The peptides used in this protocol can either be synthesized by a solid-phase peptide synthesizer or purchased from custom peptide synthesis companies. The sequence for myristoylated H3K9 peptide used for this assay is KQTAR-(Myr-K)-STGGWW. The sequence for succinylated H3K9 peptide is KQTAR-(Succ-K)-STGGWW. To validate the concentration, NanoDrop UV-Vis spectrophotometer can be used. The substrate concentration is recommended to be approximately between 0.2 and 5 equivalent of K_m value. In between the interval, more samples with different concentrations are preferred for more accurate measurement. All experiments should be done in triplicate [11].
7. SIRT5 is purified in truncated form (aa34–302). It is inserted into pET28a vector with N-terminal His6 tag and expressed in *E. coli* BL21 pRARE2 strain. The cells are cultured in 2 L luria broth media at 37 °C until the OD600 reaches approximately 0.5. Then, 0.2 mM isopropyl β-D-1-thiogalactopyranoside (IPTG) is added to induce protein expression at 15 °C for 16 h. The cells are harvested by centrifuging at 4 °C, 11305 × *g* for 5 min, and resuspended in 20 mL of 20 mM Tris–HCl (pH 8.0), 500 mM NaCl, 10% glycerol, and 1 mM PMSF. Then, the cells are lysed by a cell disruptor three times. The cellular debris is removed by centrifuging at 4 °C, 48298 × *g* for 30 min. The supernatant from the lysate is loaded onto a nickel column, pre-equilibrated with 20 mM Tris–HCl (pH 8.0), 500 mM NaCl, and 10% glycerol. Solution of imidazole is added in increasing

concentration, from 0 to 500 mM, to elute out the desired protein. 1.5 mL of each imidazole solution was used to elute different protein fractions. To check the quality of purified protein, SDS-PAGE gel is used. The protein is further purified by Superdex 75 column (Biorad), using 20 mM Tris–HCl (pH 8.0), 500 mM NaCl, and 10% glycerol. The desired fraction is then combined and concentrated. The protein is stored at −80 °C for future use [5].

8. SIRT6 is purified in full-length form. It is inserted into pET28a vector with N-terminal His6 tag and expressed in *E. coli* BL21 pRARE2 strain. The cell culture and protein purification use the same method as SIRT5 purification described above in **Note** 7.
9. Spin down very briefly. Spending too much time spinning down will cause loss of activity of the enzyme.
10. The enzyme concentration, peptide concentration, and reaction time can vary based on the activity of the purified enzyme. The target conversion is less than 10% at all substrate concentrations. This is to make sure that the assumption for the Michaelis-Menten equation, $[S] \approx [S_o]$, is met [12].
11. This is to precipitate out all the enzyme in the tube.
12. Peptide concentrations should vary depending on the activity of the sirtuin. The concentrations should be 0.2–5 K_m values [12].
13. Avoid dropping or any physical collision, which could cause any resuspension of the enzyme.
14. This is to ensure that 60 μL of the sample is injected into the system. If less volume is placed in the well, the system will take in air with the sample, causing pressure fluctuation.
15. It is recommended to only run 1–2 samples at a time. The fatty acyl peptide can precipitate out, being absorbed by the plastic after long period of time and making the measurement inaccurate.
16. For every run, keep an eye on the pressure level. The expected pressure for the experiment is less than 1250 psi. If it is over 1250 psi or much lower than 50 psi, there could be either blockade or leakage in the system. Also, under pump failures, the pressure could fluctuate widely. These would cause inaccurate readings of the UV trace.
17. Typically, the unmodified H3K9 peptide elutes at retention time of 10.0 min, succinyl H3K9 peptide at 10.9 min, myristoyl H3K9 peptide at 21.3 min.

References

1. Jing H, Lin H (2015) Sirtuins in epigenetic regulation. Chem Rev 115:2350–2375
2. Imai S, Guarente L (2010) Ten years of NAD-dependent SIR2 family deacetylases: implications for metabolic diseases. Trends Pharmacol Sci 31:212–220
3. Imai S, Armstrong CM, Kaeberlein M, Guarente L (2000) Transcriptional silencing and longevity protein Sir2 is an NAD-dependent histone deacetylase. Nature 403:795–800
4. Sauve AA, Wolberger C, Schramm VL et al (2006) The biochemistry of sirtuins. Annu Rev Biochem 75:435–465
5. Du J, Zhou Y, Su X et al (2011) Sirt5 is a NAD-dependent protein lysine demalonylase and desuccinylase. Science 334:806–809
6. Zhang X, Khan S, Jiang H et al (2016) Identifying the functional contribution of the defatty-acylase activity of SIRT6. Nat Chem Biol 12:614–620
7. Marangoni AG (2002) Enzyme kinetics. John Wiley & Sons, Inc., Hoboken, NJ
8. Brooks HB, Geeganage S, Kahl SD et al (2004) Basics of enzymatic assays for HTS. In: Sittampalam GS, Coussens NP, Brimacombe K et al (eds) Assay guidance manual. Eli Lilly & Company and the National Center for Advancing Translational Sciences, Bethesda, MD
9. Teng Y-B, Jing H, Aramsangtienchai P et al (2015) Efficient demyristoylase activity of SIRT2 revealed by kinetic and structural studies. Sci Rep 5:8529
10. Chalkiadaki A, Guarente L (2015) The multifaceted functions of sirtuins in cancer. Nat Rev Cancer 15:608–624
11. Jing H, Hu J, He B et al (2016) A SIRT2-selective inhibitor promotes c-Myc oncoprotein degradation and exhibits broad anticancer activity. Cancer Cell 29:297–310
12. Berg JM, Tymoczko JL, Stryer L et al (2002) Biochemistry. W.H. Freeman, New York, NY

Part IV

Small Molecule Screening Assays of NAD+ Utilizing Enzymes

Chapter 16

Small-Molecule Screening Assay for Mono-ADP-Ribosyltransferases

Teemu Haikarainen, Sudarshan Murthy, Mirko M. Maksimainen, and Lari Lehtiö

Abstract

Mono-ADP-ribosyltransferases of the PARP/ARTD enzyme family are enzymes catalyzing the transfer of a single ADP-ribose unit to target proteins. The enzymes have various roles in vital cellular processes such as DNA repair and transcription, and many of the enzymes are linked to cancer-relevant functions. Thus inhibition of the enzymes is a potential way to discover and develop new drugs against cancer. Here we describe an activity-based screening assay for mono-ADP-ribosyltransferases. The assay utilizes the natural substrate of the enzymes, NAD^+, and it is based on chemically converting the leftover substrate to a fluorophore and measuring its relative concentration after the enzymatic reaction. The assay is homogenous, robust, and cost-effective and, most importantly, applicable to mono-ADP-ribosyltransferases as well as poly-ADP-ribosyltransferases for screening of small-molecule inhibitors against the enzymes.

Key words PARP, ARTD, Mono-ADP-ribosyltransferase, Screening assay, Inhibitor

1 Introduction

The transfer of an ADP-ribose from NAD^+ to a target protein is a posttranslational modification catalyzed mainly by proteins belonging to the PARP/ARTD family. These enzymes can be further divided to mono-ADP-ribosyltransferases and poly-ADP-ribosyltransferases. Mono-ADP-ribosyltransferases can catalyze the transfer of a single ADP-ribose unit to target proteins, whereas poly-ADP-ribosyltransferases can catalyze the polymerization reaction resulting in long chains of ADP-ribose units attached to target proteins. Most of the enzymes are also able to modify themselves through automodification. There are 18 members in the human ARTD protein family [1, 2]. ARTD1–ARTD6 (PARP1–PARP5) are poly-ADP-ribosyltransferases, whereas ARTD7–ARTD18 (PARP7–PARP17, TPT1, potentially also PARP3) are mono-ADP-ribosyltransferases, with the exception of ARTD13, which is an inactive member of the ARTD family [3].

Paul Chang (ed.), *ADP-ribosylation and NAD+ Utilizing Enzymes: Methods and Protocols*, Methods in Molecular Biology, vol. 1813, https://doi.org/10.1007/978-1-4939-8588-3_16, © Springer Science+Business Media, LLC, part of Springer Nature 2018

Poly-ADP-ribosyltransferases of the ARTD/PARP family are the most extensively studied in the context of drug discovery especially due to their roles in cancer-related functions, such as DNA repair and Wnt-signaling [4–6]. Recently, the mono-ADP-ribosyltransferases have also gained more attention due to discovery of their roles in various cellular functions such as cell death and transcriptional regulation [7].

Assays based on biophysical methods and on enzyme activity have been described in the literature. Biophysical assays such as differential scanning fluorimetry [8] and microarray [9] are based on measuring binding of the compound to the target protein. Activity-based ELISA methods utilize a biotinylated NAD^+ as a substrate, and the signal is detected via luminescence produced after reaction with streptavidin-conjugated HRP [10]. Similar ELISA-based assays utilizing biotinylated NAD^+ and streptavidin-conjugated HRP are available commercially.

We describe here a robust, time- and cost-effective assay, which can be used for screening of small-molecule inhibitors for all active members of the ARTD family [11, 12]. The assay was initially adapted and optimized from an assay developed by Putt & Hergenrother [13] for ARTD1/PARP1, and it is based on quantifying the leftover of NAD^+ after enzymatic hydrolysis by ARTDs. In the reaction, an ARTD enzyme uses NAD^+ to attach a single ADP-ribose to a substrate protein such as SRPK2. It should be noted that the assay does not differentiate between trans- and automodification (of ARTD itself) or NAD^+ hydrolysis. Leftover NAD^+ is converted to a fluorophore via a chemical reaction with acetophenone and formic acid, and relative activity can be measured when compared to control without the enzyme. This screening method can be directly applied for all mono-ADP-ribosyltransferases taken that conditions can be found where these enzymes show robust activity. Assay can also be applied for other enzyme families carrying out similar cleavage of NAD^+. The assay conditions optimized for the catalytic domains of human ARTD7/PARP15, ARTD8/PARP14, ARTD10/PARP10, ARTD12/PARP12, and ARTD15/PARP16 using a 96-well plate are described in Table 1 and Fig. 1. In the following we will describe the assay in detail and address specific points to take into account when applying the assay for these enzymes.

2 Materials

Prepare all solutions in ultrapure water. Store all reagents at room temperature unless otherwise stated. Follow local regulations when disposing hazardous waste.

Table 1

Optimized assay conditions for mono-ADP-ribosyltransferases. Incubation is carried out at room temperature with shaking

PARP/ARTD	Proteins	Substrate	Buffer	Incubation time
PARP10/ ARTD10	100 nM PARP10 2 μM SRPK2	500 nM NAD^+	50 mM Tris–HCl pH 7.0	13 h
PARP12/ ARTD12	500 nM PARP12	500 nM NAD^+	50 mM NaH_2PO_4/Na_2HPO_4 pH 7.0	20 h
PARP14/ ARTD8	500 nM PARP14	500 nM NAD^+	50 mM NaH_2PO_4/Na_2HPO_4 pH 7.0	18 h
PARP15/ ARTD7	200 nM PARP15 500 nM SRPK2	500 nM NAD^+	50 mM NaH_2PO_4/Na_2HPO_4 pH 7.0	3 h
PARP16/ ARTD15	2 μM PARP16.	500 nM NAD^+	50 mM HEPES pH 7.0, 2 mM $NiCl_2$	24 h

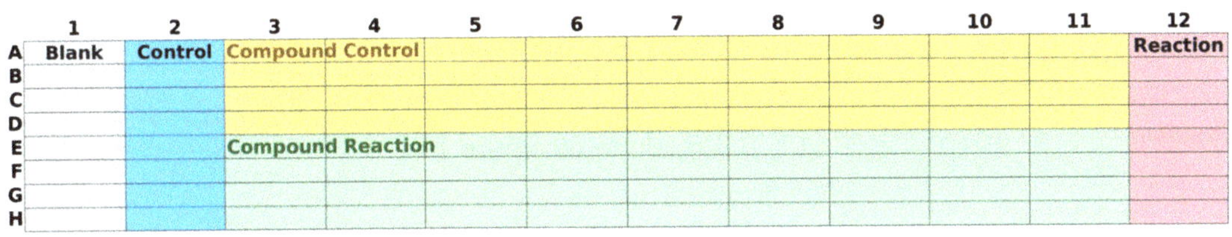

Fig. 1 Assay layout for a 96-well plate

2.1 Enzymatic Reaction

1. Active, purified recombinant mono-ADP-ribosyltransferase and substrate proteins (*see* Tables 1 and 2). The recombinant proteins can be purchased, or the catalytic fragments can be expressed in *E. coli* host using expression vectors listed in Table 2. Recombinant proteins can be purified using Ni-affinity chromatography with optional removal of tag and a polishing size exclusion chromatography. Purified proteins should be flash frozen in liquid nitrogen in small aliquots (e.g. in PCR tubes). Only use freshly thawed protein samples for the assays.
2. Assay buffer depends on the enzyme and the buffers we have tested are listed in Table 1.
3. 1 mM NAD^+ in water (*see* **Note 1**). Store in −20 °C and dilute before use in assay buffer to appropriate concentration. Assay pH will be set by the buffer used in dilution of all the reagents (Table 1).
4. Assay plates: Black polypropylene 96-well plates with U-shaped or flat bottom.
5. Compound dilution plates: Transparent polypropylene V-shaped 96-well plates.

Table 2
Expression constructs for recombinant proteins used in the assay

Protein	Residues	Vector
PARP10/ARTD10	809–1017	pNIC-CH
PARP12/ARTD12	489–684	pNIC28-Bsa4
PARP14/ARTD8	1535–1801	pNIC28-Bsa4
PARP15/ARTD7	460–656	pNIC28-Bsa4
PARP16/ARTD15	1–280	pNIC28-Bsa4
SRPK2	51–688	pNIC28-Bsa4

6. Plate shaker: Optional, but recommended due to long incubation times to achieve efficient mixing and temperature control.
7. Adhesive PCR foil seals for plates. Other seals can be used, but PCR seals have performed best for us.

2.2 Chemical Reaction

1. 20% acetophenone in ethanol (*see* **Note 2**).
2. 2 M KOH made in ultrapure H_2O.
3. 100% formic acid.
4. Plastic plate covers.
5. Plate reader based on filters or monochromators with fluorescence intensity capability (ex. 372 nm; em. 444 nm).

3 Methods

3.1 Testing Enzymes

Before screening of small-molecule inhibitors, it is advisable to test the enzymatic activity using the assay and to carry this out by varying both the time and the enzyme concentration [11]. This should be repeated for each new batch of protein in order to confirm the suitable assay parameters. The activity of each protein batch should be evaluated using the optimized assay conditions (Table 1). A dilution series of the enzyme using twofold serial dilution (e.g., 50, 100, 200, 400, 800, 1600, 3200 nM) should be assayed using different time points usually ranging from few hours up to 24 h (*see* Table 1). The conversion-% (*see* Subheading 3.5) for a robust screening assay should be 60–70%. It is advisable to select lowest possible enzyme concentration, which still gives 60–70% conversion within 24 h. For convenience, one should also ensure that there is enough protein to carry out a screening campaign with the same protein batch.

3.2 Before Screening

When screening small focused libraries, we recommend using quadruplicate wells for both the enzymatic reactions and compound control wells (Fig. 1). If planning screening of large compound libraries, we recommend the researcher to get familiar with the assay optimization and validation for high-throughput screening and to confirm that, e.g., the screening window coefficient (Z′) indicates that the assay would be suitable for screening of compounds in singlets (two wells per compound: control and reaction). A good source of information would be the assay guidance manual from the National Institutes of Health [14].

Assay conditions can be optimized for buffer components, incubations time, and NAD^+ concentration [11]. The same assay protocol described below can be used for these optimizations. We have observed that many enzymes of the ARTD family are very sensitive to DMSO and this requires special attention. It may be necessary to add equal amounts of DMSO to all wells taking into account the amount transferred with the compound, which are commonly stored as DMSO stocks. Same is true for the dose-response measurements where the compound concentration varies.

3.3 Compound Screening

Compound screening is carried out by mixing the appropriate concentration of the compound, proteins used, and NAD^+ followed by incubation for the enzymatic reaction to occur. Blank and control wells are needed to quantify the inhibition.

1. Dilute the compounds to be screened (stock usually 10 mM in 100% DMSO) to e.g. 100 μM in assay buffer (Table 1) or in assay buffer supplemented with 1% DMSO in compound dilution plates (*see* **Note 3**). Use always fresh compound dilutions in screening. As the volume of the enzymatic reaction is 50 μL, only 5 μL of the diluted compound solution is needed for each well. Dilution can be made directly from the compound stock to a microtube or microplate and the solution should be thoroughly mixed.
2. Transfer 5 μL of the compounds from the dilution plates to assay plates in Compound Control and Compound Reaction wells. Add 5 μL of buffer with 1% DMSO to Control wells to match the DMSO concentration in the Compound wells. Screening is usually done in singlets (one control + one reaction) when assay reproducibility is sufficient [14] (*see* Fig. 1 for layout).
3. Add 25 μL NAD^+ to all wells except Blank wells (*see* Table 1 for concentration and Fig. 1 for locations).
4. Add 10 μL of substrate (SRPK2) (or buffer) to all wells except Blank wells (*see* Table 1 for concentration and Fig. 1 for layout) (*see* **Note 4**).

5. Add 10 μL of ADP-ribosyltransferase to the Reaction and Compound Reaction wells (*see* Fig. 1 for layout).
6. Add 10 μL assay buffer to Control and Compound Control wells (*see* Fig. 1 for layout) (*see* **Note 5**).
7. Add 50 μL assay buffer to Blank wells (*see* Fig. 1 for layout).
8. Seal the plates with adhesive PCR foil seals. Incubate in plate shaker at 300 rpm at room temperature until 50–70% conversion is reached (*see* Table 1 for examples of incubation times). Please note that the incubation time needs to be optimized for each protein batch. Relatively long incubation times, efficient mixing, and temperature control are often required in order to obtain robust and stable signal.
9. Some compounds can produce fluorescence signal with used wavelengths and therefore interfere with the assay. This is more typical than quenching of the signal (*see* **Note 6**). Fluorescence signal caused by the compounds can be observed by comparing the Compound Control and Control wells (Fig. 1).

3.4 Chemical Reaction

After the enzymatic reaction, a chemical reaction is carried out in order to convert NAD^+ to a fluorophore.

1. Move the plate to the chemical fume hood (*see* **Note 7**). The chemicals used in the conversion of NAD^+ to a fluorophore are hazardous. Acetophenone also has a distinctive and strong odor.
2. Add 20 μL of 20% acetophenone in ethanol to all the wells. Add 20 μL of 2 M KOH to all the wells in the plate. Insert plastic plate cover on the plate. Incubate for 10 min. This will stop the enzymatic reaction and convert leftover NAD^+ to a fluorophore after an incubation in excess of acid in step 3.
3. Add 90 μL of 100% formic acid to all the wells in the plate. Insert plastic plate cover on the plate. Incubate for 20 min (*see* **Note 8**).
4. Measure the fluorescence intensity with a plate reader. Use 372 nm excitation and 444 nm emission wavelengths. It is advisable to optimize parameters, such as slits and *Z*-focus for the plate reader and detector gain for the NAD^+ concentration used.

3.5 Data Analysis

The flow for the simple evaluation of the screening result plate is described below.

1. Average the Blank wells and subtract the blank from all the other wells.
2. Activity of an enzyme can be calculated by dividing the difference between the mean of the Control wells from the mean of the Reaction wells and dividing this by the mean of the Control

wells. Typically this is expressed as % of conversion of NAD^+ (for robust screening assay ~60%) and can be set to 100% activity.

3. Repeat the above calculation, but by taking the difference between the Compound Control and the Compound Reaction, and divide this by the mean of the Control wells. This procedure assumes that compound contribution to the fluorescence is the same for both the Compound Control and Reaction wells (*see* **Note 9**).
4. The conversion-% can be compared between the Reaction wells and plotted as activity-% using, e.g. Graphpad Prism.
5. Set the hit limit according to the purpose [14].
6. Proceed to dose-response measurements of the hit compounds and further validation using an orthogonal assay.

3.6 Dose-Response Measurements

For dose-response measurement, the same procedure can be used, but instead of using different compounds, varying concentration of a single compound (e.g. half-log dilution in quadruplicates) is used. Pay attention to DMSO concentration (*see* **Note 3**) and use assay conditions resulting in approximately 30% conversion of NAD^+. This is a compromise between the effect of substrate consumption during the enzymatic assay and the robustness of the signal [11]. Fluorescence signal can be used directly, and is recommended, for fitting the dose response (e.g., in Graphpad Prism). The mean fluorescence of Control and Reaction wells can be added 2 log units away from the lowest and highest concentration, respectively to improve fitting. In the case of compound interference, conversion-% (or activity-%) must be used for fitting.

4 Notes

1. Store the NAD^+ stock solution in 20 μL aliquots at −20 °C. Do not reuse the thawed aliquots.
2. Prepare and store 20% acetophenone in ethanol in fume hood. Protect the solution from light.
3. The assay has low DMSO tolerance, and control wells have to include same DMSO concentration as reaction wells [11]. Use of 1% DMSO has shown to give enough precision to the DMSO concentration in dose-response measurements, and this shortcut simplifies the pipetting procedure.
4. In order to simplify the pipetting procedure, it is possible to premix SRPK2 into the NAD^+-containing buffer.
5. If using a protein concentration higher than 1 μM, 10 μL of protein (the same concentration as used in the reaction wells)

should be added to the Compound Controls instead of buffer just before stopping the reaction by acetophenone and KOH.

6. One way to overcome the interference is to increase the NAD^{+} concentration and therefore the signal, but this may not be possible due to, e.g. limited number of ADP-ribosylation sites available and inhibition through automodification.
7. Carry out all the pipetting and incubation steps in Subheading 3.4 inside the fume hood.
8. The fluorophore is stable in the dark at room temperature for 2 h. We have observed that high protein concentration can affect the fluorescence signal.
9. This procedure does not work if the compounds quench the fuorescence signal, which has not been very typical in our experiments. In this case the calculation should be carried by taking the difference between the Compound Control and the Compound Reaction and by dividing this by the mean of the Compound Control wells.

References

1. Schreiber V, Dantzer F, Ame J-C, de Murcia G (2006) Poly(ADP-ribose): novel functions for an old molecule. Nat Rev Mol Cell Biol 7:517–528
2. Hottiger MO, Hassa PO, Lüscher B, Schüler H, Koch-Nolte F (2010) Toward a unified nomenclature for mammalian ADP-ribosyltransferases. Trends Biochem Sci 35:208–219
3. Kleine H et al (2008) Substrate-assisted catalysis by PARP10 limits its activity to mono-ADP-ribosylation. Mol Cell 32:57–69
4. Gupte R, Liu Z, Kraus WL (2017) PARPs and ADP-ribosylation: recent advances linking molecular functions to biological outcomes. Genes Dev 31:101–126
5. Vyas S, Chang P (2014) New PARP targets for cancer therapy. Nat Rev Cancer 14:502–509
6. Haikarainen T, Krauss S, Lehtiö L (2014) Tankyrases: structure, function and therapeutic implications in cancer. Curr Pharm Des 20:6472–6488
7. Daniels CM, Ong S-E, Leung AKL (2015) The promise of proteomics for the study of ADP-ribosylation. Mol Cell 58:911–924
8. Wahlberg E et al (2012) Family-wide chemical profiling and structural analysis of PARP and tankyrase inhibitors. Nat Biotechnol 30:283–288
9. Peng B, Thorsell A-G, Karlberg T, Schüler H, Yao SQ (2017) Small molecule microarray based discovery of PARP14 inhibitors. Angew Chem Int Ed Engl 56:248–253
10. Ekblad T et al (2015) Towards small molecule inhibitors of mono-ADP-ribosyltransferases. Eur J Med Chem 95:546–551
11. Venkannagari H, Fallarero A, Feijs KLH, Lüscher B, Lehtiö L (2013) Activity-based assay for human mono-ADP-ribosyltransferases ARTD7/PARP15 and ARTD10/PARP10 aimed at screening and profiling inhibitors. Eur J Pharm Sci 49:148–156
12. Narwal M, Fallarero A, Vuorela P, Lehtiö L (2012) Homogeneous screening assay for human tankyrase. J Biomol Screen 17:593–604
13. Putt KS, Hergenrother PJ (2004) An enzymatic assay for poly(ADP-ribose) polymerase-1 (PARP-1) via the chemical quantitation of NAD(+): application to the high-throughput screening of small molecules as potential inhibitors. Anal Biochem 326:78–86
14. Sittampalam GS, Coussens NP, Brimacombe K, Grossman A, Arkin M, Auld D, Austin C, Baell J, Bejcek B, TDY C, Dahlin JL, Devanaryan V, Foley TL, Glicksman M, Hall MD, Hass JV, Inglese J, Iversen PW, Kahl SD, Kales SC, Lal-Nag M, Li Z, McGee J, McManus O, Riss T, Trask OJ Jr, Weidner JR, Xia M, Xu X (eds) (2004) Assay guidance manual. Eli Lilly & Company and the National Center for Advancing Translational Sciences, Bethesda, MD

Chapter 17

A Simple, Sensitive, and Generalizable Plate Assay for Screening PARP Inhibitors

Ilsa T. Kirby, Rory K. Morgan, and Michael S. Cohen

Abstract

Poly-ADP-ribose polymerases (also known as ADP-ribosyltransferases or ARTDs) are a family of 17 enzymes in humans that catalyze the reversible posttranslational modification known as ADP-ribosylation. PARPs are implicated in diverse cellular processes, from DNA repair to the unfolded protein response. Small-molecule inhibitors of PARPs have improved our understanding of PARP-mediated biology and, in some cases, have emerged as promising treatments for cancers and other human diseases. However these advancements are hindered, in part, by a poor understanding of inhibitor selectivity across the PARP family. Here, we describe a simple, sensitive, and generalizable plate assay to test the potency and selectivity of small molecules against several PARP enzymes in vitro. In principle, this assay can be extended to all active PARPs, providing a convenient and direct comparison of inhibitors across the entire PARP enzyme family.

Key words ADP-ribosylation, PARPs, ARTDs, Small-molecule inhibitor, Screen, Click chemistry, 6-a-NAD^+

1 Introduction

Poly-ADP-ribose polymerases (PARPs1-16, also known as ADP-ribosyltransferases or ARTDs) catalyze the reversible posttranslational modification known as ADP-ribosylation, which involves the transfer of ADP-ribose from nicotinamide adenine dinucleotide (NAD^+) to amino acids in target proteins. PARPs have emerged as key regulators of diverse cellular processes, and their dysregulation has been linked to a wide range of pathologies, such as allergic asthma [1] and cancer [2]. As such, there has been increasing interest in the development of PARP inhibitors as potential therapeutics as well as tools to further explore PARP biology. While potent small-molecule inhibitors exist for some PARPs (e.g., PARP1, 2, 3, 5a, and 5b), potent and selective inhibitors for the majority of PARP family members are scarce [3]. Moreover, there are few studies that explore the selectivity of existing PARP

Paul Chang (ed.), *ADP-ribosylation and NAD+ Utilizing Enzymes: Methods and Protocols*, Methods in Molecular Biology, vol. 1813, https://doi.org/10.1007/978-1-4939-8588-3_17,

inhibitors [4, 5]. An understanding of inhibitor selectivity across the PARP family is critical for their use as tool compounds and as potential clinical treatments. To facilitate the development of selective PARP inhibitors, we have devised a plate assay for screening PARP inhibitors against several recombinant PARPs in parallel.

Our lab has explored several methods for measuring PARP activity, including clickable NAD^+ analogues [6] and a clickable aminooxy alkyne (AO-alkyne) probe with in-gel readouts [7]. These methods are useful for cellular and biochemical analysis of PARP-mediated ADP-ribosylation, but are not practical for high-throughput inhibitor screening. To address this challenge, we adapted our PARP activity assay to a 96-well plate format. In this method SRSF protein kinase 2 (SRPK2), a known PARP substrate [8], is bound to Ni-NTA plates (Pierce). SRPK2 serves as a pan-ADP-ribosylation target for PARPs using physiologically relevant concentrations [9] of a clickable NAD^+ analogue, 6-alkyne-NAD^+ (6-a-NAD^+) [6]. Following incubation with 6-a-NAD^+ and a given recombinant PARP, samples are subjected to click conjugation with biotin-azide, followed by incubation with streptavidin-HRP (Strep-HRP), and the extent of ADP-ribosylation of SRPK2 is detected with QuantaRed™ enhanced chemifluorescent HRP substrate kit (Thermo Scientific) (Fig. 1).

Using this method we tested two FDA-approved PARP inhibitors, olaparib [10] and rucaparib [11], and a recently published PARP10 inhibitor, OUL35 [4], against $PARP1_{FL}$, $PARP2_{FL}$, $PARP3_{FL}$, $PARP5b_{cat}$, $PARP10_{cat}$, $PARP11_{FL}$, $PARP14_{wwe\text{-}cat}$, and $PARP15_{cat}$ (Fig. 2, Table 1).

2 Materials

1. Ni-NTA plates: Thermo Scientific Pierce nickel coated white 96-well plates, store at 4 °C.
2. Human PARP buffer (hB): 50 mM HEPES pH 7.5, 100 mM NaCl, 4 mM $MgCl_2$, 0.2 mM TCEP (*see* **Note 1**).
3. Inhibitor stock solutions: stock solution in DMSO (800×), store at −20 °C (*see* **Note 2**).
4. 4× inhibitor solutions: 0.5% DMSO in hB, make fresh for each use (*see* **Note 3**).
5. 4× 6-a-NAD^+ solution: 400 μM in hB, make fresh for each use (*see* **Note 4**). 6-a-NAD^+ used in assay development was synthesized in house by the Cohen lab. This reagent is also available through Biology Life Science Institute (Cat # N 051–01).
6. Activated DNA: Sigma-Aldrich deoxyribonucleic acid from calf thymus, type XV, activated, lyophilized powder. 0.1 mg/mL stock in DI H_2O, store at −20 °C.

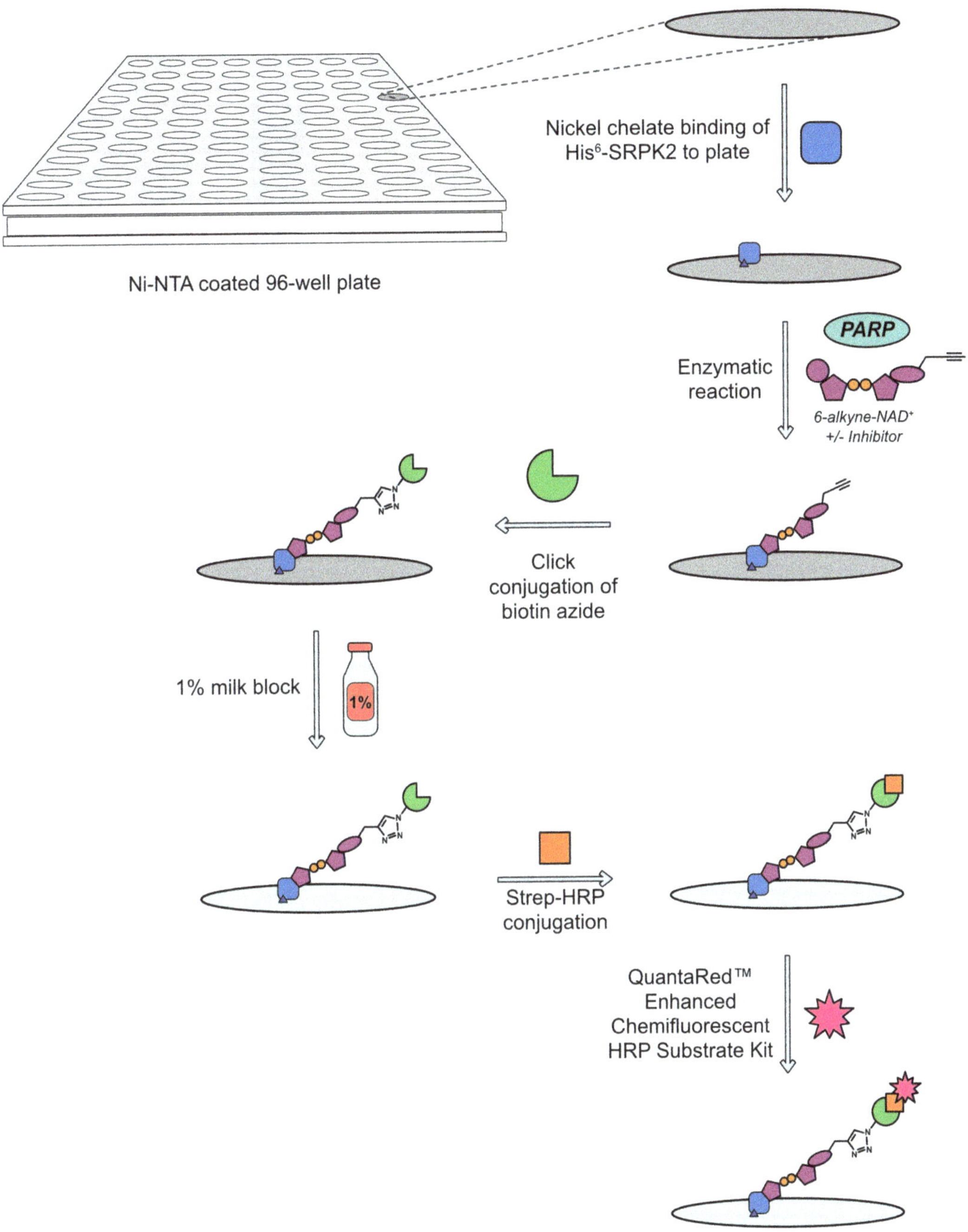

Fig. 1 Schematic showing the SRPK2 96-well plate assay

7. Dumbbell nick-5'phosphorylated DNA hairpin (dNick-5'P): hairpin formed by slow annealing step in annealing buffer [12], 100 μM stock stored at −20 °C.
8. 2× inhibitor/2× 6-a-NAD$^+$ solution: equal volumes of 4× inhibitor solution and 4× 6-a-NAD$^+$ solution, 0.25% DMSO.

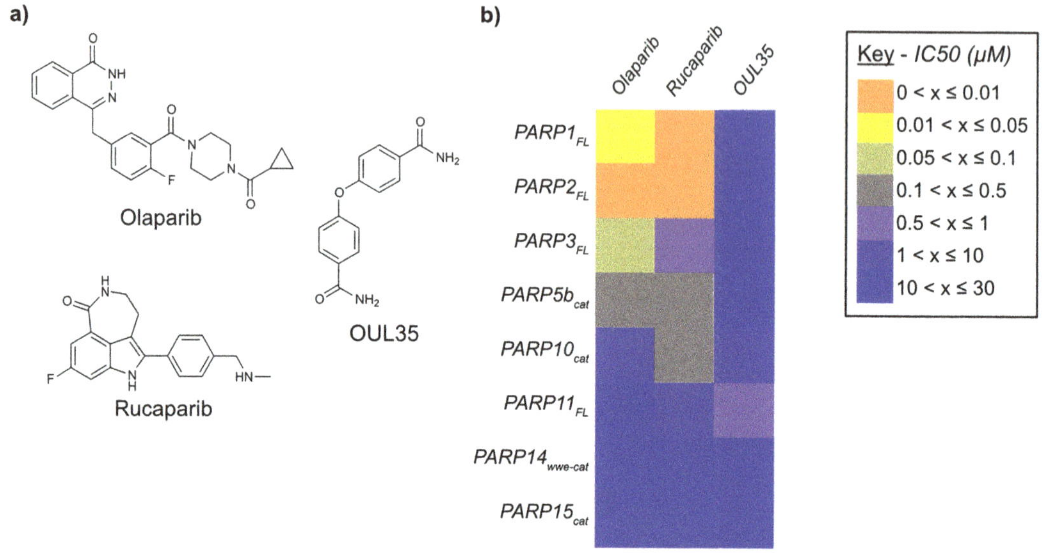

Fig. 2 (**a**) Structures of olaparib and rucaparib, two recently FDA approved PARP inhibitors, as well as OUL35, a recently described PARP10 inhibitor; (**b**) heat map showing IC50 values (μM) obtained using the SRPK2 plate assay

Table 1
In vitro IC_{50} values for olaparib, rucaparib, and OUL35 (SEM from a minimum of three dose-response experiments)

	Olaparib (IC_{50}, μM)	Rucaparib (IC_{50}, μM)	OUL35 (IC_{50}, μM)
$PARP1_{FL}$	0.025 ± 0.002	0.0075 ± 0.002	>10
$PARP2_{FL}$	0.0066 ± 0.0003	0.0049 ± 0.001	>10
$PARP3_{FL}$	0.083 ± 0.04	0.72 ± 0.3	>10
$PARP5b_{cat}$	0.21 ± 0.08	3.6 ± 0.3	>10
$PARP10_{cat}$	4.5 ± 0.3	2.5 ± 0.2	4.9 ± 0.9
$PARP11_{FL}$	>10	3.4 ± 0.5	0.93 ± 0.1
$PARP14_{wwe\text{-}cat}$	>10	>10	>10
$PARP15_{cat}$	>10	>10	>10

IC_{50} values for olaparib and rucaparib against several PARPs were consistent with previous studies, whereas the IC_{50} value for OUL35 against $PARP10_{cat}$ was higher than determined previously [4, 5]

9. $PARP1_{FL}$ 2× solution: 10 nM in hB with 0.1 mg/mL activated DNA.
10. $PARP2_{FL}$ 2× solution: 10 nM in hB with 0.1 mg/mL activated DNA.
11. $PARP3_{FL}$ 2× solution: 600 nM in hB with 2.5 μM dNick 5'P [12].
12. $PARP5b_{cat}$ 2× solution: 200 nM in hB.

13. $PARP10_{cat}$ 2× solution: 2 nM in hB.
14. $PARP11_{FL}$ 2× solution: 4 nM in hB.
15. $PARP14_{wwe\text{-}cat}$ 2× solution: 5 nM in hB.
16. $PARP15_{cat}$ 2× solution: 10 nM in hB.
17. 1× PBST: 0.02% Tween-20 in 1× PBS, store at room temperature.
18. Click chemistry:

    Combine the following components in the following order for 1 mL click buffer (CB)

    (a) 20 μL Tris[(1-benzyl-1*H*-1,2,3,-triazol-4-yl)methyl]amine (TBTA) (Click Chemistry Tools): 5 mM stock solution in DMSO (50×). Store at −20 °C. Final concentration: 100 μM.

    (b) 20 μL copper(II) sulfate ($CuSO_4$): 50 mM stock solution in water (50×). Store at room temperature. Final concentration: 1 mM.

    (c) 20 μL Biotin-PEG3-Azide (Click Chemistry Tools): 5 mM stock solution in DMSO (50×). Store at −20 °C. Final concentration: 100 μM.

    (d) 20 μL Tris(2-carboxyethyl)phosphine hydrochloride (TCEP): 50 mM stock solution in water (50×). Final concentration: 1 mM (*see* **Note 5**).

    (e) 100 μL 10× PBS (10×).

    (f) 820 μL DI H_2O.

19. 1% milk solution: 1% w/v nonfat dry milk in 1× PBST.
20. Strep-HRP solution: 0.05 ng/μL in 1× PBS.
21. QuantaRed Detection Solution: Thermo Scientific QuantaRed Enhanced Chemifluorescent HRP Substrate.
22. QuantaRed Working Solution: 49.5% QuantaRed Enhancer Solution, 49.5% QuantaRed Enhancer Solution, 1% ADHP Concentrate Solution (*see* **Note 6**).
23. Molecular Devices SpectraMax i3 multimode platform plate reader. Settings: fluorescence intensity top mode, monochromator, endpoint reading; landscape height 14.6 mm; excitation wavelength 570 nm, bandwidth 9 nm; emission wavelength 600 nm, bandwidth 15 nm; 6 flashes per read; read height 1 mm.

3 Methods

This procedure provides a simply, sensitive, and generalizable assay for screening PARP inhibitors in parallel. The assay allows for head-to-head comparison of inhibitors across the PARP family with accurate and precise results over many independent trials. For optimal results all incubations should be performed at the specified temperature with gentle agitation. Recombinant proteins are stored in 10% glycerol at −80 °C and undergo no more than two freeze thaw cycles.

3.1 Ni-NTA Plate Preparation with SRPK2: Binding of His_6-SRPK to the Ni-NTA Plates Provide Immobilized Targets for ADP-Ribosylation by Various PARPs in the Enzymatic Reaction (Subheading 3.2)

1. Thaw SRPK2 on ice and dilute into cold hB buffer to a concentration of 0.3 μM.
2. Add 50 μL of SRPK2 solution to each well of Ni-NTA plate and incubate for 1 h.
3. Wash each well with 1× PBST (3×, 100 μL), 1× PBS (1×, 100 μL), and hB (1×, 100 μL), and incubate each wash for 5 min.

3.2 ADP-Ribosylation Reaction with 6-a-NAD^+: PARP Enzymes Transfer 6-Alkyne-ADP-Ribose from 6-a-NAD^+ onto His_6-SRPK2 Bound to the Plate in Subheading 3.1

Addition of PARP inhibitors will reduce 6-alkyne-ADP-ribosylation levels according to their potency. The clickable 6-alkyne-ADP-ribose is conjugated with biotin-azide via click chemistry. Strep-HRP and a chemifluorescent HRP substrate provide a readout of ADP-ribosylation levels.

1. Add 25 μL 2× enzyme to each well followed by 25 μL 2× inhibitor/2× 6-a-NAD^+ to each well. Incubate at 30 °C for 1 h.
2. Remove reaction mixture and wash each well with 1× PBST (3×, 100 μL), 1× PBS (1×, 100 μL).
3. Add 50 μL 1× click buffer to each well and incubate for 30 min at RT.
4. Remove click buffer and wash each well with 1× PBST (3×, 100 μL), 1× PBS (1×, 100 μL).
5. Add 50 μL 1% milk solution and incubate for 30 min at RT.
6. Remove milk solution and wash each well with 1× PBST (3×, 100 μL), 1× PBS (1×, 100 μL).
7. Add 50 μL Strep-HRP solution to each well and incubate for 30 min at RT.
8. Remove Strep-HRP solution and wash each well with 1× PBST (3×, 100 μL), 1× PBS (1×, 100 μL).
9. Add 100 μL QuantaRed Working Solution to each well and incubate for 30–45 s to develop plate at RT.
10. Add 10 μL QuantaRed stop solution to quench reaction. Image within 5 min of quenching.
11. Image plate within 5 min of quenching (*see* **Note 7**).

4 Notes

1. Make 100 mM TCEP stock solution in water fresh for every use.
2. Storage and shelf life depends on individual compound stability.
3. Some inhibitors may be insoluble at higher concentrations; this can be solved with agitation and gentle heating within the bounds of confirmed stability for each inhibitor on a case-by-case basis.
4. 6-a-NAD^+ should be stored at −80 °C for up to 1 year; multiple freeze thaws should be avoided to maintain optimal activity.
5. Prepare fresh immediately before use, addition of TCEP should cause a color change from light blue to green.
6. Let components warm to room temperature before combining. Use solution within 5 min of making.
7. Background signal increases significantly within 1 h but then remain stable up to at least 24 h.

Acknowledgments

We thank members of the Cohen lab for many helpful discussions. We thank H. Schuler for the construct for $PARP14_{wwe\text{-}cat}$. We thank J. Pascal (Université de Montréal) for helpful discussions regarding PARP3 enzyme activity. This work was funded by the NIH (NIH 1R01NS088629) and a grant from the Pew Charitable Trust (M.S.C.).

References

1. Mehrotra P, Hollenbeck A, Riley JP et al (2013) Poly (ADP-ribose) polymerase 14 and its enzyme activity regulates TH2 differentiation and allergic airway disease. J Allergy Clin Immunol 131:521–531.e12. https://doi.org/10.1016/j.jaci.2012.06.015
2. Barbarulo A, Iansante V, Chaidos A et al (2013) Poly(ADP-ribose) polymerase family member 14 (PARP14) is a novel effector of the JNK2-dependent pro-survival signal in multiple myeloma. Oncogene 32:4231–4242. https://doi.org/10.1038/onc.2012.448
3. Wahlberg E, Karlberg T, Kouznetsova E et al (2012) Family-wide chemical profiling and structural analysis of PARP and tankyrase inhibitors. Nat Biotechnol 30:283–288. https://doi.org/10.1038/nbt.2121
4. Venkannagari H, Verheugd P, Koivunen J et al (2016) Small-molecule chemical probe rescues cells from mono-ADP-ribosyltransferase ARTD10/PARP10-induced apoptosis and sensitizes cancer cells to DNA damage. Cell Chem Biol 23:1251–1260. https://doi.org/10.1016/j.chembiol.2016.08.012
5. Thorsell A-G, Ekblad T, Karlberg T et al (2017) Structural basis for potency and promiscuity in poly(ADP-ribose) polymerase (PARP) and tankyrase inhibitors. J Med Chem 60:1262–1271. https://doi.org/10.1021/acs.jmedchem.6b00990
6. Carter-OConnell I, Jin H, Morgan RK et al (2014) Engineering the substrate specificity of ADP-ribosyltransferases for identifying direct protein targets. J Am Chem Soc 136:5201–5204. https://doi.org/10.1021/ja412897a

7. Morgan RK, Cohen MS (2015) A clickable aminooxy probe for monitoring cellular ADP-ribosylation. ACS Chem Biol 10:1778–1784. https://doi.org/10.1021/acschembio.5b00213
8. Venkannagari H, Fallarero A, Feijs KLH et al (2013) Activity-based assay for human mono-ADP-ribosyltransferases ARTD7/PARP15 and ARTD10/PARP10 aimed at screening and profiling inhibitors. Eur J Pharm Sci 49:148–156. https://doi.org/10.1016/j.ejps.2013.02.012
9. Cambronne XA, Stewart ML, Kim D et al (2016) Biosensor reveals multiple sources for mitochondrial NAD. Science 352:1474–1477. https://doi.org/10.1126/science.aad5168
10. Menear KA, Adcock C, Boulter R et al (2008) 4-[3-(4-cyclopropanecarbonylpiperazine-1-carbonyl)-4-fluorobenzyl]-2H-phthalazin-1-one: a novel bioavailable inhibitor of poly(ADP-ribose) polymerase-1. J Med Chem 51:6581–6591. https://doi.org/10.1021/jm8001263
11. Thomas HD, Calabrese CR, Batey MA et al (2007) Preclinical selection of a novel poly(ADP-ribose) polymerase inhibitor for clinical trial. Mol Cancer Ther 6:945–956. https://doi.org/10.1158/1535-7163.MCT-06-0552
12. Langelier M-F, Riccio AA, Pascal JM (2014) PARP-2 and PARP-3 are selectively activated by 5′ phosphorylated DNA breaks through an allosteric regulatory mechanism shared with PARP-1. Nucleic Acids Res 42:7762–7775. https://doi.org/10.1093/nar/gku474

Part V

Mass Spectrometry Techniques for Detection of Mono-ADP-Ribosylation

Chapter 18

Nonlocalized Searching of HCD Data for Fast and Sensitive Identification of ADP-Ribosylated Peptides

Thomas Colby, Juan José Bonfiglio, and Ivan Matic

Abstract

ADP-ribosylation is a technically challenging PTM which has just emerged into the field of PTM-specific proteomics. But this fragile modifier requires special treatment on both a data acquisition and data processing level: it is highly labile under higher-energy collisional dissociation (HCD), and the degree of lability can depend on the site it modifies. Its behavior thus violates some assumptions on which proteomics algorithms are based. Here we present **nonlocalized ADPr searching**: a simple principle for maximizing sensitivity toward ADP-ribosylation when searching conventional HCD data. By scoring the strong fragment ions generally observed in ADPr spectra rather than the weak and often absent localization-dependent ions, nonlocalized searches are more sensitive. They also run significantly faster, due to reduced search space, and require no assumptions about which amino acids can be modified. We illustrate implementation in three search systems: Morpheus, MaxQuant, and MASCOT, and we also present a means of rapidly finding and extracting ADP-ribosylated peptide spectra from large datasets for more focused searching. This approach both improves identification of ADP-ribosylated peptides and avoids mis-localization of the modification sites.

Key words ADP-ribosylation, Serine ADPr, HCD fragmentation, Localization, Lability, Nonlocalized

1 Introduction

The principle behind identifying and localizing posttranslational modifications by mass spectrometry is essentially the same as that of identifying unmodified peptides, except that the amino acid "alphabet" used to predict peptide ion masses is extended to include the modified forms of specified amino acids. The theoretical mass of a precursor peptide with a given sequence is calculated by summing up the masses of the constituent "letters" in its sequence. To model a modification on a given residue, the residue's mass is replaced by its modified mass when calculating masses for any peptide or peptide fragment containing it. One assumption implicit in these calculations is that the modifier remains intact when the peptide is fragmented.

Paul Chang (ed.), *ADP-ribosylation and NAD+ Utilizing Enzymes: Methods and Protocols*, Methods in Molecular Biology, vol. 1813, https://doi.org/10.1007/978-1-4939-8588-3_18,

Typically peptides are fragmented by collisions with gas molecules which cause breaks at random positions in the peptide backbone, yielding a fragment ion series characteristic of the peptide's sequence. Two common variants of this type of fragmentation are CID (collision-induced decay), which causes only one break per molecule, and HCD (higher-energy collisional dissociation), which can cause multiple breakages. Many common modifiers (methylation, acetylation, etc.) remain intact when peptides are fragmented by CID or HCD. For these modifiers, the modified peptide's mass equals the unmodified peptide mass plus the modification mass, and in the fragmentation spectrum, a peak-to-peak distance corresponding to the modified residue mass can often be observed, in which case the modifier is localized by the data. Another default assumption in peptide matching is that where the unmodified mass difference between peaks is observed, the residue is not modified.

But under CID/HCD some modifiers are "labile"—fragile under these measurement conditions. The best-known example of this phenomenon is phosphorylation (+79.966 Dalton), which is partially labile. The modifier can break off, taking the hydroxyl group of the modified residue with it, a so-called neutral loss of phosphoric acid (97.977 Dalton), a behavior commonly used to recognize phosphopeptides during acquisition [1]. When analyzing HCD fragmentation spectra of phosphopeptides, both the modified residue mass and the mass after the loss can appear in the ion series.

One should keep lability and neutral losses in mind when examining an ADP-ribosylated peptide spectrum. In the following example spectrum (Fig. 1) from a recently published dataset [2] (*see* **Note 1a**), the most prominent features are the diagnostic peaks corresponding to fragments of ADP-ribose. The next most intense peak (marked "?") is presumably related to the modified peptide.

These labeled diagnostic peaks can be used as markers for ADP-ribosylated peptides, either for manual assessment (Subheading 3.1) or for automated post-acquisition enrichment (Subheading 3.2). The strength of these signals already indicates the lability of the modification.

A closer look at the peptide fragment ions in this spectrum reveals that it is of sufficient quality for de novo sequencing (Fig. 2). One can annotate both the complete b- and the y-ion series predicted for the sequence, SMMN(I or L) QTK.

These two ion series, which constitute the strongest peaks in the spectrum, bear **no trace of the modification**. The strong peak previously observed can now be annotated as the unmodified peptide (now marked M + H). Since the precursor mass is equal to that of the sequenced peptide plus ADP-ribose, one might assume a non-covalent modification. But a closer examination (Fig. 3) reveals ions corresponding to sequence ions plus fragments of

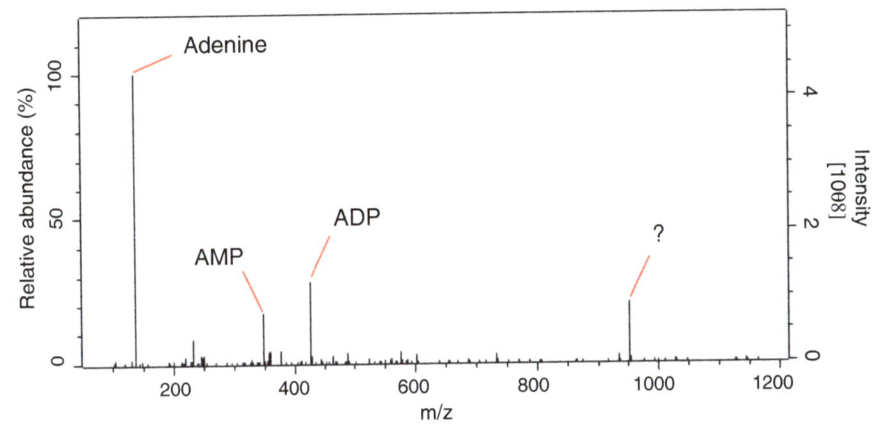

Fig. 1 An example ADP-ribosylated peptide spectrum from a publicly available dataset, specifically scan 8276 from the file 20140515_QE6_UPLC5_SCL_SA_Hela_PAR_PD_Rep1.raw [2]. Diagnostic peaks corresponding to adenine, AMP, and ADP fragments of ADP-ribose are labeled. View created with the expert annotation system [3]

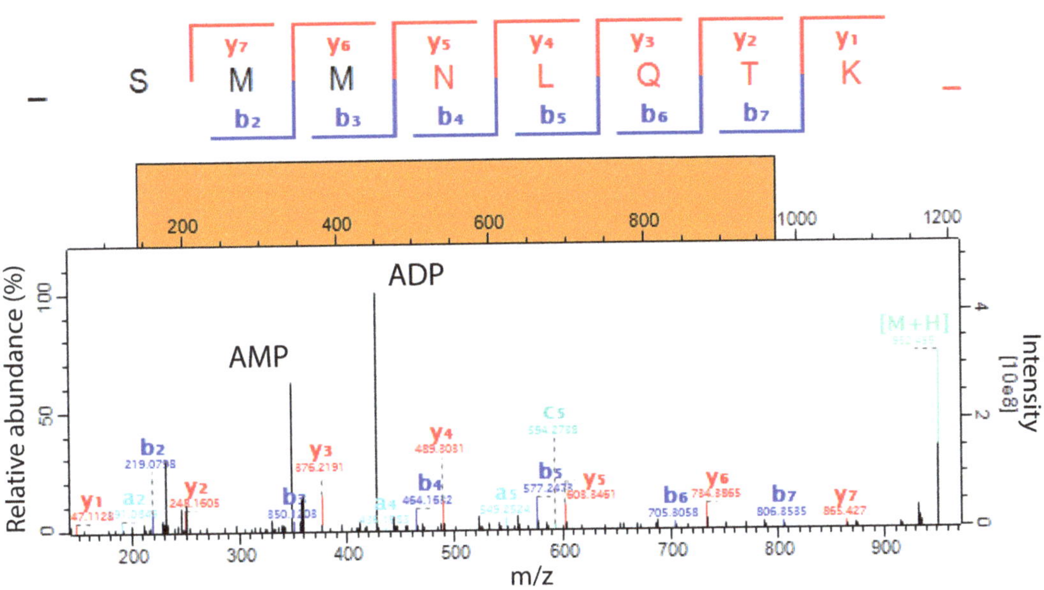

Fig. 2 A zoomed view of the same spectrum with annotations for the y- and b- ion series for the sequenced peptide. The peak corresponding to the unmodified peptide is marked "M + H"

ADPr (annotated with "*"), though they are among the weakest peaks.

Thus the more likely explanation is a covalent linkage that is extremely labile and most often breaks off without leaving a trace—a complete neutral loss, as is observed with modifiers like O-GlcNAc [4] (*see* **Note 1b**). Despite the intensity of this example spectrum, the fragment ions bearing traces of the modifier constitute only a

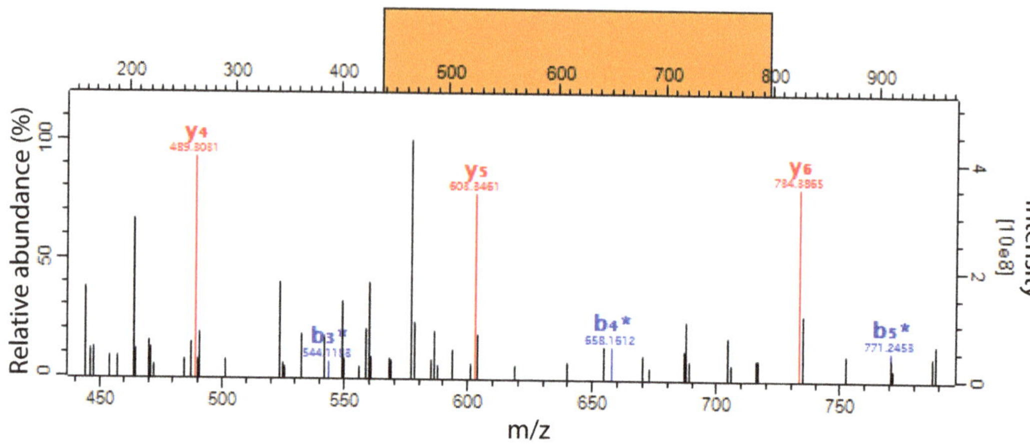

Fig. 3 A further zoom of the same spectrum with annotations for the y- and b*- ion series. The b* fragments include part of the ADP-ribosylation left after the loss of AMP. These are the most prominent fragment ions which still contain part of the labile modification

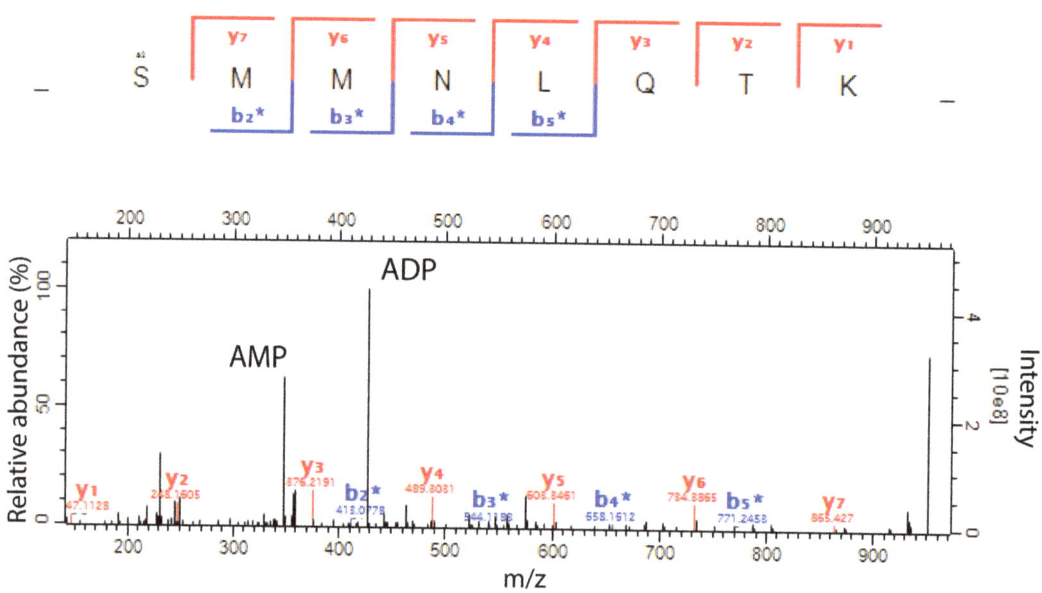

Fig. 4 A zoomed view similar to Fig. 2 but with annotations for the y- and b*- ion series

partial series (Fig. 4), shorter than the native b-ion series initially annotated.

This particular dataset was acquired with parameterization for exceptionally high-quality MS/MS spectra (*see* **Note 1c**). If these spectra had been acquired with more conventional acquisition parameters, even more of these weaker localization-specific signals would have been missing. Any peptide match requiring these ions will score poorly, which puts ADP-ribosylated peptides at a disadvantage relative to unmodified peptides regarding identification. Likewise, localized searches will favor peptide models in which these intense peaks are fitted, localizing the modification wherever peaks are weakest—generating arbitrary localizations (*see* **Note 1d**). **We can counter both this loss of search sensitivity toward ADPr peptides and this mis-localization by scoring the stronger, unmodified sequence ions instead,** by considering the modifier to undergo a **complete neutral loss.**

Searching mass spectrometry data with a complete neutral loss may appear counterproductive at first, but it has distinct advantages. Though this approach yields *no ADPr site mapping within the peptide sequence*, it delivers *no localization misinformation* and requires *no assumptions* about the amino acid specificity of the modifier. Treating ADPr as a completely labile modification on the peptide terminus rather than specifying particular amino acids also reduces the "search space" significantly—making analyses faster. It also appears to model the modifier's behavior better and may more accurately reflect the real information content of the data. If one uses this complete neutral loss model and an AMP neutral loss model to analyze the largest publicly available HCD dataset for ADPr peptides [2], one can compare the performance statistically.

2 Materials

Since this is a post-acquisition methodology, it can (but need not) be used in combination with biochemical enrichment of ADP-ribosylated peptides, like that described in the Leung et al. of this volume (CITE). It is intended to get the most information possible about ADP-ribosylated peptides out of HCD data. It is *not* intended for use with targeted or triggered ETD fragmentation data, which contain much more localization information. While a software system for protein identification from MS data is required, the user may choose from several options, depending on the hardware and software resources they have available (*see* Subheading 2.2). **Supplementary materials**, where mentioned, are available for download at https://datashare.mpcdf.mpg.de/s/MVkv5zXHWj5npUp.

2.1 Data

Use high-resolution HCD data collected on samples containing ADP-ribosylated peptides. Though the observed fragmentation behavior occurs in lower-resolution instruments as well, the advantage of these methodologies is most clearly seen using high-resolution (>15,000) fragmentation data (obtainable from Q-Exactive, Orbitrap Fusion, or Lumos instruments: *see* **Note 2**). For illustration purposes, we use the most extensive publicly available HCD dataset on ADP-ribosylated peptides [2]. A list of some suitable published datasets can be found in **Note 2a**.

2.2 Analysis System

2.2.1 Optional

Xcalibur (Thermo)—the ability to view and analyze raw data is highly valuable for manual quality control and initial qualitative analysis of data on ADP-ribosylated peptide samples, but it is not strictly necessary in order to apply the search methods described.

MSConvert from the ProteoWizard package (http://proteowizard.sourceforge.net/)—this utility is necessary if one uses MASCOT or any other search system that cannot use .raw files directly or if one wants to perform "post-acquisition enrichment" (Subheading 3.2).

Expert system for computer-assisted annotation of MS/MS spectra [3]—this stand-alone tool allows the user to annotate spectra directly, testing various modification hypotheses. It is available at no cost from http://www.biochem.mpg.de/mann/tools/—use the modification and configuration files supplied in supplementary materials. See included notes on file conversion.

Perl—for using conversion and data filtering scripts included in the supplementary materials, the ActivePerl community edition is sufficient (https://www.activestate.com/activeperl/downloads).

2.2.2 Required

One of the following mass spectrometry search systems:

MaxQuant/Andomeda [5, 6]—freely downloadable from http://www.coxdocs.org — please follow installation instructions. For individual data files, where performance is not an issue, MaxQuant can run on a typical PC. For detailed hardware recommendations, see http://www.coxdocs.org/doku.php?id=maxquant:common:download_and_installation.

Morpheus 1.68 or later [7]—stand-alone freeware by Craig Wenger, available from SourceForge (https://sourceforge.net/projects/morpheus-ms/). Due to the efficiency of this algorithm, Morpheus searches are fast on virtually any modern PC.

Mascot—server software commercially available from Matrix Science Ltd. (http://matrixscience.com/). If you have access to a Mascot server, you will require administrative rights to change modifier models.

Others—in principle, the strategy outlined is implementable in any mass spectrometry search system, as long as one is able to define neutral losses on modifiers. Many are freely available for various computational platforms. We have tested it with the systems listed above.

3 Methods

3.1 Qualitative Assessment of Dataset ADPr Content (Optional)

Users with access to Xcalibur can perform an initial qualitative assessment of a dataset's ADPr content by looking for the diagnostic ions characteristic of ADPr illustrated in Fig. 1. To be sure that these signals are arising from modified peptides and not from polynucleotides, one should also look for the y1 ions characteristic of tryptic peptides.

1. Open a data file and apply one of the supplied Xcalibur layout files (*see* supplementary material) by looking under the File drop-down menu File → layout → apply and selecting the desired file (*see* **Note 3a**). The six chromatograms in this view (Fig. 5) represent the base peak intensity, followed by extracted ion chromatograms for the three MS/MS diagnostic ion masses labelled in Fig. 1, and finally the MS/MS masses characteristic of the terminal lysine or arginine of a tryptic peptide.
2. Global ADPr content: compare the MS/MS base peak intensity (BPI) chromatogram to the diagnostic peaks. When ADPr is present, the adenine diagnostic peak is generally the strongest (*see* **Note 3b**). The similarity of the BPI and ADN chromatograms reflects how ADPr-rich the sample is. In the Martello et al. dataset [2], the efficiency of the enrichment is easy to discern: most of the strongest MS/MS spectra contain the diagnostic masses.
3. Zoom in to examine regions of high diagnostic signals to see if they are arising from peptides. If they are, one of the y1 ions characteristic of a tryptic peptide should co-occur. In this example spectrum (Fig. 6), we see that fragmentation spectrum 8276 contains not only all three major diagnostic masses

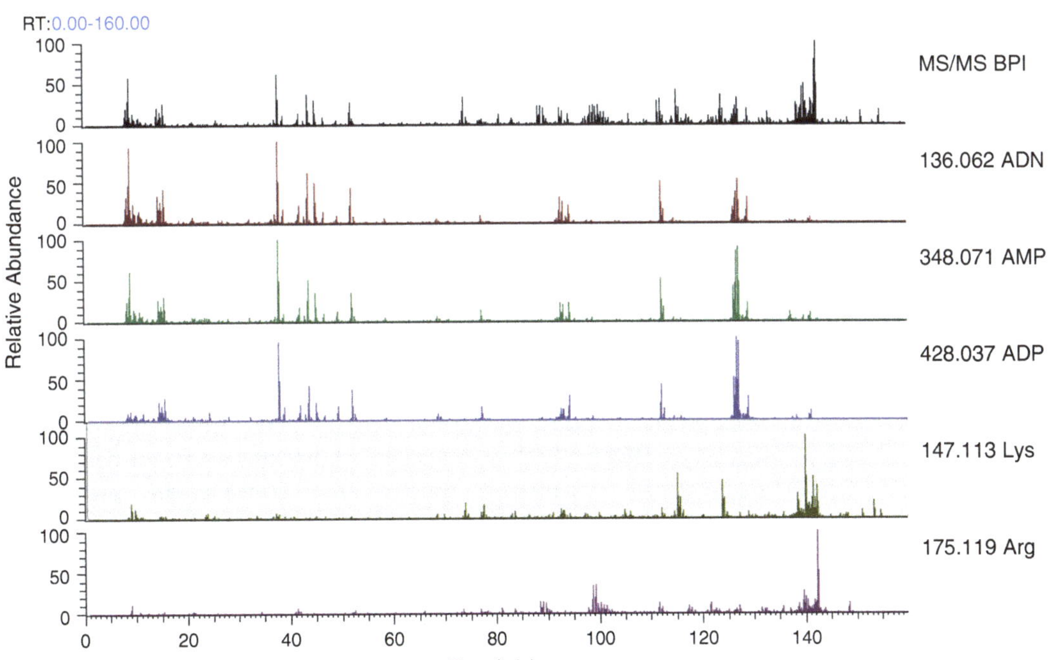

Fig. 5 Extracted ion chromatograms of the same example file including (top to bottom) the base peak intensity for MSMS and then the diagnostic masses for Adenine, AMP, ADP, lysine y1, and arginine y1. The view was created in Xcalibur using the layout available in the Supplementary Materials

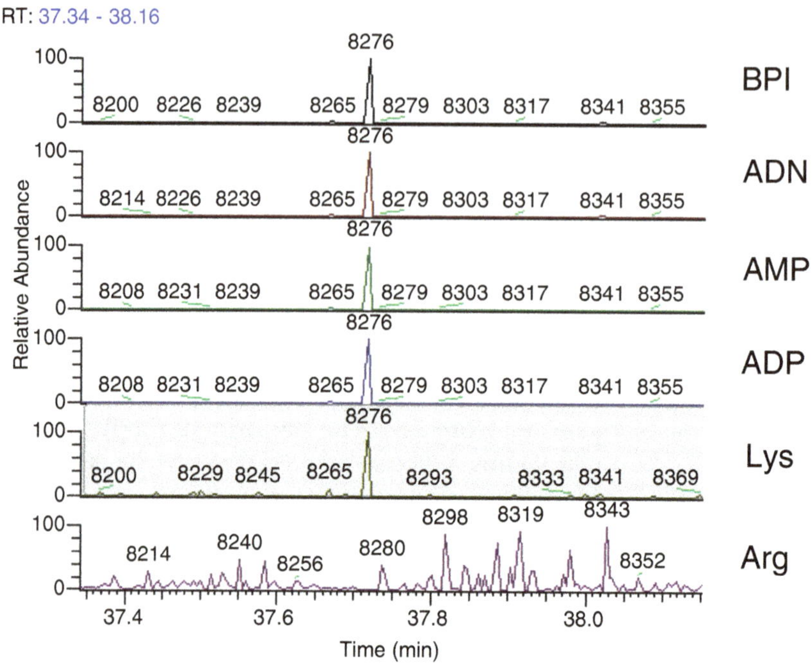

Fig. 6 A zoomed view similar to Fig. 5 showing the region of the chromatograms in the vicinity of the example scan from Figs. 1, 2, 3, and 4

but also the diagnostic mass for the C-terminal lysine of a tryptic peptide. Such co-occurrence of the diagnostic peaks is the hallmark of an ADP-ribosylated peptide spectrum (*see* **Note 3b and 3c**).

In a non-enriched sample, the overlap of peaks in the diagnostic ion chromatograms and the total MS/MS BPI would be much lower. Indeed, due to lower amounts of ADP-ribosylated peptides, only the strongest diagnostics (136.062 and 348.071) may be observable.

3.2 Post-Acquisition "Enrichment"

If the dataset you are examining appears to contain ADP-ribosylated peptides based on the qualitative assessment (Subheading 3.1), but is so large that it takes a long time to be searched, this step may help. While it does not strengthen the ADP-ribosylated peptide signals, it effectively "concentrates" them into a smaller dataset by automatically extracting spectra that contain the previously observed diagnostic ions.

1. Place the script Convert_Find_Extract_local.pl (from supplementary material) in a directory with the raw data files from which you would like to extract the likely ADPr peptide spectra.
2. Open the script in an editor and look for "SET CRITERIA HERE FOR THE SPECTRA TO BE EXTRACTED." The following lines offer different options for combining search

criteria in the order of increasing stringency. The "#" symbol comments out a line, so the script applies the criteria in the line without "#" at the beginning. Explanations of each criteria set follow the "###" midline. Place/delete your "#" accordingly. You will also need to edit the two lines containing "msconvert.exe" to reflect the location of the file on your computer (*see* **Notes 3d** and **3e**). Save your changes.

3. Double-click the script icon to run—this may take several minutes.
4. Your "enriched" mzML data files appear with "select_" preceding the filename. Use these if you are searching with Mascot or Morpheus.

3.3 Data Conversion

If you are using Morpheus or Mascot, you will need to convert your files from the Thermo rawfile (.raw) format, to something more generic like mzML or mgf.

1. If you have performed an in silico enrichment (Subheading 3.2)—your complete data are already available in mzML form in the directory in which you ran the conversion script.
2. To convert the files by hand, open a command console and move to the directory where your rawfiles are. You can then convert rawfiles to mzML with the ProteoWizard utility msconvert.exe with the following command:

   msconvert.exe --mzML --64 --filter "peakPicking true 2-2" --filter "msLevel 2–2" --filter "threshold count 1000 most-intense" *FILE.*
3. You can also modify the provided perlscript *Convert_MultiRawFiles.pl* to reflect the path to msconvert.exe on your computer. Once the script is edited (*see* **Notes 3d** and **3e**), run it by double-clicking.

3.4 Nonlocalized Searching

1. Setup—In each case we have to define the 100% labile ADPR modification.

A. Morpheus—Copy the table modifications.tsv to mod_original.tsv. Edit modifications.tsv and insert the line:

ADP ribosylation of C-term peptide C-terminus n/a 541.0611 541.304 541.0611 541.304 none X

Or simply copy the modification.tsv file found in the Morpheus folder of the supplementary materials.

B. MaxQuant—Under the *Andromeda Configuration* tab, select the subtab *Modifications* and press *Add*. Give the modifier a name (like *ADPR_Loss_Cterm*). Add a descriptor and enter the composition C(15) H(21) N(5) O(13) P(2). For *Position* select "any C-term" and set the *Type* as "Standard." For *Specificity* you will have to add K and R by pressing "+" and selecting the amino acids, otherwise MaxQuant will not annotate neutral losses. For each specificity, add the neutral loss with "+" and define the same composition as above. When done, press *save changes.*

C. Mascot—Log in as Mascot administrator and go to the *Configuration Editor* on the main Mascot page. On the configuration page, select *Modifications.* Next click *Add new modification* and give it a name. Enter the composition will be C(15) H(21) N(5) O(13) P(2). On the *Specificity* tab, select "C-term" for *Site* and "any C-term" for *position*. The *Classification* should be set to "Posttranslational" and *hidden* should NOT BE CHECKED. Most importantly you should add the complete neutral loss by clicking *New Neutral Loss* and entering the same composition as above. Make sure that *scoring* is marked. Save the modification.

2. Searching—Search the data versus the expected sequence database just as you normally would, but with the addition of the new form of ADPr as a variable modification. This should be the only form of ADPr allowed. Since many documented ADP-ribosylations appear in lysine-rich regions, you may wish to permit more mis-cleavages.

3. Interpreting results

The aim of our nonlocalized approach is to increase the sensitivity of identification of ADP-ribosylated peptides from HCD data and prevent mis-localization by more accurately modeling this labile modification. By assuming 100% lability, we choose to ignore all information localizing the modification on the peptide. In exchange we are able to fit all the strong mass signals arising from the "de-modified" peptide, which would otherwise contribute to mis-localization in conventional localized searches (*see* **Note 3h**).

The ADP-ribosylated peptides identified with this method are just that—they are identified, and according to their mass, they are modified. Both the identification and modification can be validated by the presence of diagnostic ions (*see* **Note 3i**).

Morpheus finds roughly 13% more unique ADP-ribosylated base peptides in the example data when a C-terminal ADPr model with complete neutral loss is used than when ADPr is permitted on D, E, R, K, and S with a neutral loss of AMP.

Plotting the highest Morpheus scores for the unique base peptides found with both models, most ADP-ribosylated peptides appear to be modeled better (more ions are matched) by the complete neutral loss than the loss of AMP (Fig. 7a). The fact that relatively few peptides score better in the AMP neutral loss model, and then only marginally, indicates that general use of a complete neutral loss model increases search sensitivity. A histogram of the distribution of score differences makes this point clear. The advantages of the complete neutral loss model far outweigh any disadvantages (Fig. 7b).

In addition, since the search space is much more limited (since every peptide can only have one possible ADP-ribosylated variant), the search runs faster. In this example, searching with the model of

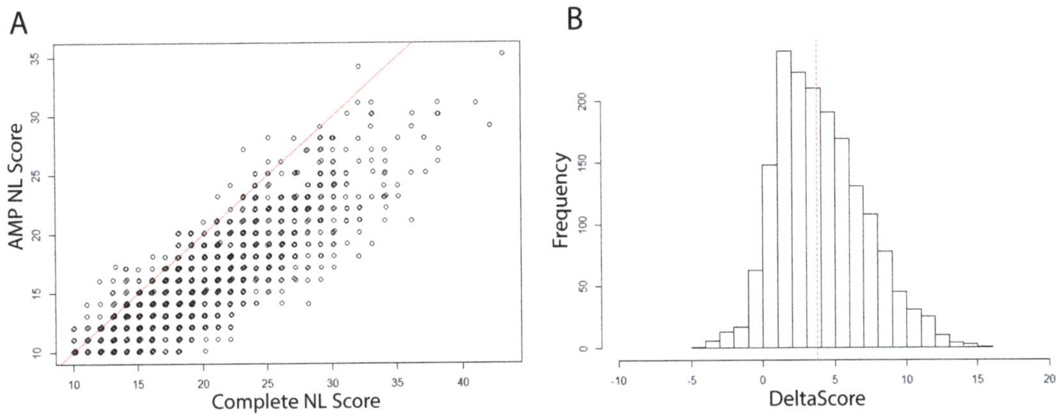

Fig. 7 (**a**) A 2-d plot of Morpheus scores for ADPr peptides identified in this dataset using both an AMP neutral-loss model with ADPr permitted on D, E, R, K, and S (vertical axis) versus a C-terminal complete loss model (horizontal). Peptides on the red line score equally well in both models. (**b**) A histogram of the differences in Morpheus scores (complete loss score—the AMP loss score) for ADPr peptides identified in this dataset with a c-terminal complete loss model and with an AMP neutral-loss model with ADPr permitted on D, E, R, K, and S. The mean difference of 3.7 is marked with a red line. This indicates that four more ions are matched on average by nonlocalized searching

ADPr on the peptide C-terminus (with a full neutral loss) took only **15% of the time** taken by a search that allowed ADPr on D, E, S, K, and R.

In short, nonlocalized searching of HCD data with a complete neutral loss model for ADPr yields significantly higher average scores for modified peptides in all three search engines tested, leading to the identification of more peptides that are modified. Not only does it appear to model the modifier more realistically, it does not deliver unreliable localizations, searches run much faster, and there is no need to assume any amino acid specificity. In our opinion, this is the best and most efficient means of using HCD data to identify ADP-ribosylated peptides.

4 Notes

1. Fragmentation behavior of ADP-rybosylated peptides
   - (a) The spectrum in this illustration is scan 8276 from the file 20140515_QE6_UPLC5_SCL_SA_Hela_PAR_PD_Rep1.raw (*see* **Note 2a**). This spectrum shows the strongest diagnostic signal for ADPr in this data file.
   - (b) This is the case for collision-induced dissociation, but other "softer" fragmentation modes like electron-transfer dissociation (ETD) preserve this particular modification [8].
   - (c) The example dataset from Martello et al. [2] was acquired with longer MS/MS injection time (250 ms) and higher MS/MS resolution (60 K) than are typical. 50–80 ms

injections and 15 K resolution are more usual settings, which would generally result in lower intensities and signal/noise.

(d) Search algorithms match spectra predicted for peptide models to measured spectra. The greater the overlap (matched peaks), the better the score. By default an ion corresponding to an unmodified peptide fragment is taken as evidence against the modification on that fragment. In light of the complete annotation of the unmodified ions in the spectrum in Fig. 2, there is a lot of evidence for the modifier not being anywhere. A localized search cannot score all of the annotated ions as matches, since it must place the modifier somewhere. Since the y-ion series is typically stronger than the b-ion series, search algorithms usually assign the ADPr to the permissible modification site closest to the peptide N-terminus, since this lets the intense y-ion series be matched. Thus the algorithm reports the modified peptide model that matches the most intense peaks and minimizes the contradiction, leading to arbitrary localization.

2. Data from Fusion or Lumos instruments can only be used when fragmentation spectra are collected in the orbitrap mass analyzer. Mixed-mode data collected on these tribrid instruments provide especially interesting test cases, since HCD and ETD spectra for the same samples can be compared. When using data from a mixed HCD/ETD run, you must be sure to separate the spectra based on the fragmentation. If you are converting the data prior to searching using MSConvert, be sure to filter based on the activation type. If you are using MaxQuant and raw files, you will have to look at the MSMS table and filter for HCD or ETD data. Viewed separately, these results show the relative merits of the two modes for identification and localization.

(a) Public datasets

ProteomeXchange:

PXD004245: files 20140515_QE6_UPLC5_SCL_SA_Hela_PAR_PD_Rep*.raw are the simplest to use. Other files require including SILAC labels in the search.

PXD004676: filter HCD spectra on conversion. *See* **Note 3d.**

PXD005462: filter HCD spectra on conversion. *See* **Note 3d.**

PXD005627: filter HCD spectra on conversion. *See* **Note 3d.**

Massive:

MSV000080334

3. Use ADPr_centroid.lyt for centroid MS/MS data or ADPr_profile.lyt for profile MS/MS data.

(a) If chromatograms are flat (i.e., no signal), check to see that the correct type of layout (profile or centroid) was chosen. If they are still flat, right click in the chromatogram window and move to the tab Ranges/Automatic processing and change the tolerance to 20 ppm. If coinciding peaks appear in all diagnostic chromatograms, the calibration of the dataset appears to be off. Increasing the tolerance beyond 20 ppm indicates severe miscalibration. Upon increasing the tolerance, peaks should appear in all three diagnostic chromatograms.

If peaks are conspicuously absent or rare in the 136.062 chromatogram, it could be that the low-mass cutoff in the data acquisition was set too high or not at all. In these cases the region in which this diagnostic ion appears was not even acquired.

(b) The following are diagnostic masses for ADPr—adenine (136.062), adenosine (250.094), AMP (348.071), ADP (428.037), and ADPR(H+) (542.069). When a precursor is ADP-ribosylated, these peaks are generally present. **CAUTION:** Due to their good ionization characteristics, these peaks may also appear in MSMS spectra for non-modified precursors in which a small amount of an ADP-ribosylated precursor has been co-isolated. This effect means that in "crowded" datasets many spectra selected on the basis of diagnostic ions will contain primarily ions from non-ADPr precursors.

(c) As stated in **Note 3a**, be certain that the MS/MS spectra include the mass region where the adenine ion appears, otherwise including it in the criteria list will result in empty datasets. If this low mass region is not acquired in every spectrum, use the AMP (348.071) mass to filter spectra instead. Please keep in mind the "leakage" of strong ion signals mentioned in **Note 3b**.

(d) Find msconvert.exe wherever ProteoWizard is installed on your machine. Once you've found this, you can drag the icon into the command window where you are working, and then add the parameters as listed in e. and a filename. If the data you are converting also contains ETD spectra add

--filter "activation HCD"

as an additional parameter on the command line.

(e) The line you need to modify in the script follows the comment

EDIT PATH TO MSCONVERT IN FOLLOWING LINE.

(f) A Morpheus only implements neutral loss scoring as of version 1.68. Unlike MaxQuant, Morpheus only scores the neutral loss product. It does not try to match the modified mass.

(g) B MaxQuant scores both the fully modified ion series and the neutral loss ion series for a given model but reports only the higher scoring option. If you like, you can add diagnostic peaks for adenine (C5 H5 N5), AMP (C10 H14 O7 N5 P), and ADP (C10 H15 O10 N5 P2). BE SURE THAT AFTER SAVING, THE NEUTRAL LOSS OF THE ENTIRE MODIFIER IS IN THE FIRST POSITION OF THE NEUTRAL LOSS LIST. MaxQuant only considers the first neutral loss in this list.

(h) The 100% lability model at the core of nonlocalized searching resolves the contradiction inherent in scoring a labile modifier by a localized model (*see* **Note 1d**). In a localized model, ions corresponding to unmodified peptide fragments are evidence against the modification, but with a labile modifier, the nonlocalized search can score the intense peaks in both series, and we have no localization instead of arbitrary localization.

(i) Diagnostic ions are NOT used as part of the scoring algorithm, but serve as an independent confirmation that the identified peptide is modified. Peptides identified by nonlocalized searches have a sequence and a precursor mass consistent with ADP-ribosylation. The presence of the diagnostic masses therefore constitutes an entirely independent indication of modification. MaxQuant provides a *Diagnostic Peak* column in the MS/MS tab of the viewer. For other search engines, the filtering of data by diagnostic ions in Subheading 3.2 ensures the presence of diagnostic peaks. Modified peptides identified from "enriched" datasets are thus already independently confirmed.

Acknowledgments

This work was funded by the Deutsche Forschungsgemeinschaft (Cellular Stress Responses in Aging-Associated Diseases) (grant EXC 229 to I.M.) and the European Union's Horizon 2020 research and innovation program (Marie Skłodowska-Curie grant agreement 657501 to J.J.B. and I.M.). Very special thanks to Craig Wenger for adding neutral loss searching features to the Morpheus system. Thanks as well to Dr. Ilian Atanassov for useful discussions.

References

1. Villén J, Beausoleil SA, Gygi SP (2009) Evaluation of the utility of neutral-loss-dependent MS3 strategies in large-scale phosphorylation analysis. Proteomics 8(21):4444–4452
2. Martello R, Leutert M, Jungmichel S, Bilan V, Larsen SC, Young C, Hottiger MO, Nielsen ML (2016) Proteome-wide identification of the endogenous ADP-ribosylome of mammalian cells and tissue. Nat Commun 7:12917. https://doi.org/10.1038/ncomms12917
3. Neuhauser N, Michalski A, Cox J, Mann M (2012) Expert system for computer-assisted annotation of MS/MS spectra. Mol Cell Proteomics 11(11):1500–1509
4. Myers SA, Daou S, Affar EB, Burlingame AL (2013) Electron transfer dissociation (ETD): the mass spectrometric breakthrough essential for O-GlcNAc protein site assignments – a study of the *O*-GlcNAcylated protein host cell factor C1. Proteomics 13(6):982–991
5. Cox J, Mann M (2008) MaxQuant enables high peptide identification rates, individualized p.p.b.-range mass accuracies and proteome-wide protein quantification. Nat Biotechnol 26:1367–1372
6. Cox J, Neuhauser N, Michalski A, Scheltema RA, Olsen JV, Mann M (2011) Andromeda: a peptide search engine integrated into the MaxQuant environment. J Proteome Res 10:1794–1805
7. Wenger CD, Coon JJ (2013) A proteomics search algorithm specifically designed for high-resolution tandem mass spectra. J Proteome Res 12(3):1377–1386
8. Leidecker O, Bonfiglio JJ, Colby T, Zhang Q, Atanassov I, Zaja R, Palazzo L, Stockum A, Ahel I, Matic I (2016) Serine is a new target residue for endogenous ADP-ribosylation on histones. Nat Chem Biol 12:998–1000. https://doi.org/10.1038/nchembio.2180

Chapter 19

Quantitative Determination of MAR Hydrolase Residue Specificity In Vitro by Tandem Mass Spectrometry

Robert Lyle McPherson, Shao-En Ong, and Anthony K. L. Leung

Abstract

ADP-ribosylation is a posttranslational modification that involves the conjugation of monomers and polymers of the small molecule ADP-ribose onto amino acid side chains. A family of ADP-ribosyltransferases catalyzes the transfer of the ADP-ribose moiety of nicotinamide adenine dinucleotide (NAD^+) onto a variety of amino acid side chains including aspartate, glutamate, lysine, arginine, cysteine, and serine. The monomeric form of the modification mono(ADP-ribosyl)ation (MARylation) is reversed by a number of enzymes including a family of MacroD-type macrodomain-containing mono(ADP-ribose) (MAR) hydrolases. Though it has been inferred from various chemical tests that these enzymes have specificity for MARylated aspartate and glutamate residues in vitro, the amino acid and site specificity of different family members are often not unambiguously defined. Here we describe a mass spectrometry-based assay to determine the site specificity of MAR hydrolases in vitro.

Key words ADP-ribosylation, Mass spectrometry, Macrodomain, ADP-ribosylhydrolase

1 Introduction

ADP-ribosylation is the posttranslational modification of protein substrates with monomers (MARylation) or polymers (PARylation) of the small molecule ADP-ribose [1]. The modification is enzymatically added to a range of chemically diverse amino acid side chains, including aspartate (D), glutamate (E), lysine (K), arginine (R), cysteine (C), and serine (S), by a class of enzymes called ADP-ribosyltransferases—the majority of which are limited to MAR transferase activity [1–5]. In addition, MARylation can be facilitated by other enzymes including sirtuins, extracellular membrane-associated ADP-ribosyltransferases, and bacterial toxins [4, 6–8]. In recent years, a number of studies have shown that proteins containing a macrodomain fold of the MacroD-type are enzymatic MAR hydrolases capable of reversing MARylation in vitro [9–22]. However, these assays often utilize MARylated substrates with undefined sites of modification; thus, the residue and, more

Paul Chang (ed.), *ADP-ribosylation and NAD+ Utilizing Enzymes: Methods and Protocols*, Methods in Molecular Biology, vol. 1813, https://doi.org/10.1007/978-1-4939-8588-3_19,

specifically, the site specificity of the studied MAR hydrolases, even in these simple vitro systems, are often unclear.

Recent advances in the field have led to the development and optimization of several proteomic techniques for identifying sites of ADP-ribosylation on individual substrates, as well as globally from cell lysates [5, 23–32]. Our lab and others developed a technique that uses phosphodiesterases to cleave protein-conjugated ADP-ribose to a phosphoribose tag that can be enriched and identified using established phosphoproteomic techniques [3, 5, 26, 27, 30, 32]. Recently we have used this technique to confidently identify the sites of modification of the MARylated catalytic domain of PARP10 (PARP10^{CD}) [15, 27], a MARylated substrate commonly used for in vitro MAR hydrolase assays [11–16]. We then quantified the intensity of these sites in samples incubated with or without our MAR hydrolase of choice (e.g., chikungunya viral macrodomain) to unambiguously determine the residue specificity of the MAR hydrolase activity. This experiment allowed us to determine that the MAR hydrolase of the chikungunya viral macrodomain has specificity for MARylated D and E but not K residues [15]. This protocol describes the steps necessary to carry out this experiment (the workflow summarized in Fig. 1) and should be applicable to study of a broad range of MARylated substrates and MAR hydrolases [21, 22, 33, 34].

2 Materials

2.1 Preparation of MARylated Substrate

1. Recombinant *H. Sapiens* PARP10 catalytic domain (residues 818–1025) (purified as in ref. [15]).
2. Buffer exchange system: we use Superdex™ 200 10/300 GL (GE) with a NGC Quest™ 10 FPLC system (BioRad).
3. Heat block that heats to 37 °C and shakes at 500 rpm.
4. 50 mM NAD^+.
5. 20× automodification (AM) buffer: 400 mM Tris–HCl pH 7.5, 1 M NaCl, 100 mM $MgCl_2$, 200 mM beta-mercaptoethanol.
6. Desalting buffer: 20 mM Tris–HCl pH 7.5, 200 mM NaCl, 10 mM beta-mercaptoethanol.
7. Protein quantification assay (Bradford or equivalent).
8. 10 K cutoff centrifugal spin concentrator.
9. Tabletop centrifuge.
10. 1.5 mL microcentrifuge tubes.
11. 15% tris-glycine SDS-PAGE gel and running apparatus.

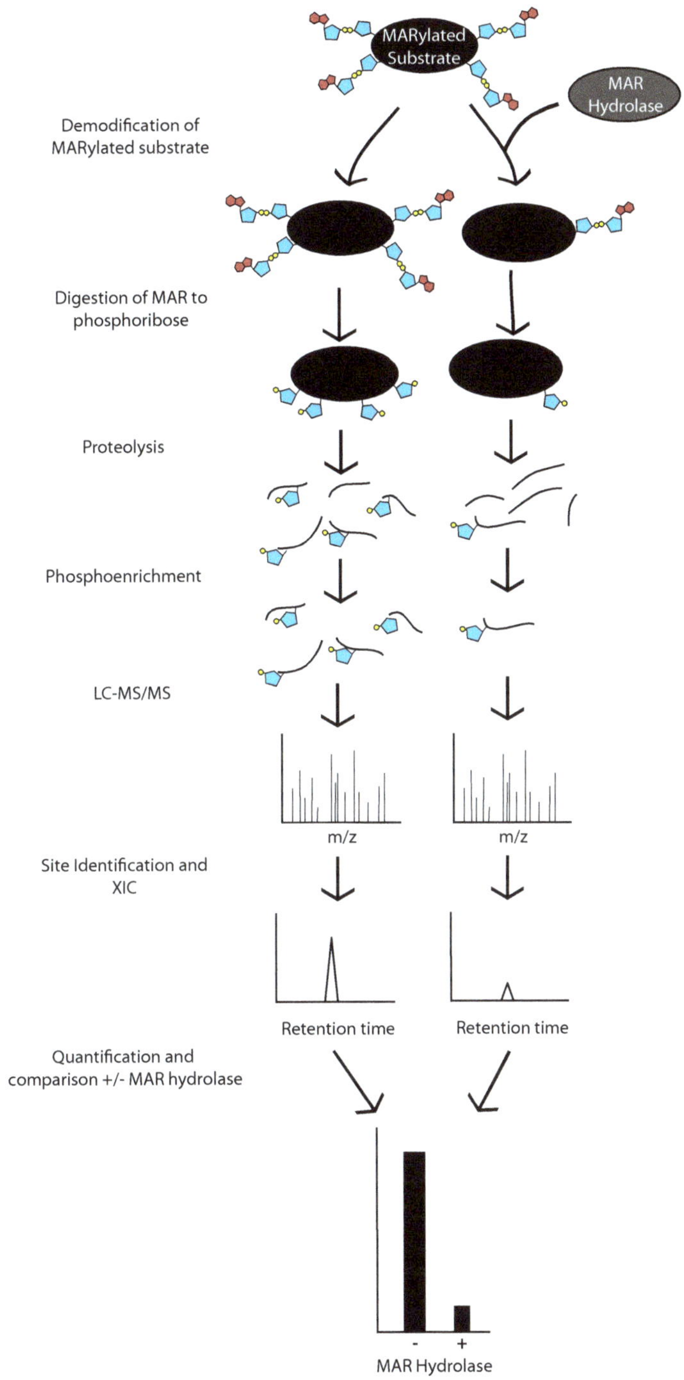

Fig. 1 Determining the site specificity of ADP-ribosylhydrolases. Workflow describing the steps necessary for (1) identifying sites of MARylation on in vitro modified substrates by LC-MS/MS and (2) quantifying their intensities under different conditions (e.g., incubation with MAR hydrolases)

12. Coomassie protein stain.
13. Nitrocellulose membrane and SDS-PAGE gel transfer apparatus.
14. Rabbit anti-PAN-ADP-ribose binding reagent (EMD Millipore).
15. Anti-rabbit secondary antibody.

2.2 MAR Hydrolase and Phosphodiesterase Treatment of MARylated Substrate

1. Recombinant MAR hydrolase (1 mg/mL) (e.g., chikungunya virus nsP3 macrodomain (residues 3–160) purified as in ref. [15]).
2. Recombinant *H. Sapiens* NudT16a phosphodiesterase A22V (full-length) (1 mg/mL) (purified as in ref. [30]).
3. Demodification (DM) buffer: 20 mM Tris–HCl pH 7.5, 200 mM NaCl, 10 mM beta-mercaptoethanol.
4. 20× NudT16 reaction buffer: 400 mM Tris–HCl pH 7.5, 1 M NaCl, 100 mM $MgCl_2$, 200 mM beta-mercaptoethanol.

2.3 Protein Digestion

1. Lysyl endopeptidase (LysC), stock solution 2 mg/mL in H_2O.
2. Modified trypsin, stock solution 2 mg/mL in 50 mM acetic acid.
3. Freshly made 3× denaturing buffer: 300 mM Tris–HCl pH 7.0, 4.5 M guanidine hydrochloride, 3 mM $CaCl_2$, 15 mM tris(2-carboxyethyl)phosphine, 30 mM chloroacetamide.
4. 100% trifluoroacetic acid (TFA).

2.4 Peptide Desalting

1. Reversed-phase C_{18} StageTips™ [35] (Thermo Scientific) (2 per sample).
2. 5 mL slip-tip syringe.
3. Peptide desalting buffer A: 5% acetonitrile, 0.1% trifluoroacetic acid made in Milli-Q (MQ) H_2O.
4. Peptide desalting buffer B: 80% acetonitrile, 0.1% trifluoroacetic acid made in MQ H_2O.

2.5 Phosphoribosylated Peptide Enrichment

1. PHOS-select™ Iron Affinity Gel (5 μL per sample) (Sigma Aldrich).
2. Peptide desalting buffer A: 5% acetonitrile, 0.1% trifluoroacetic acid made in MQ H_2O.
3. Peptide desalting buffer B: 80% acetonitrile, 0.1% trifluoroacetic acid made in MQ H_2O.
4. 1% formic acid (FA).
5. 0.5 M KH_2PO_4 pH 7.0.
6. Heat block that heats to 25 °C and shakes at 1000 rpm.
7. 1.5 mL microcentrifuge tubes.

2.6 Liquid Chromatography and Electrospray Tandem Mass Spectrometry

1. SpeedVac (evaporating centrifuge or equivalent).
2. MS buffer A: 0.1% acetic acid.
3. MS buffer B: 99.9% acetonitrile, 0.1% acetic acid.
4. Nanoliter-flow high performance liquid chromatography system.
5. 10 cm × 75 μm ID fused silica capillary column made in-house with laser puller and packed with 3 μm reversed-phase C_{18} beads.
6. Thermo Orbitrap Elite™ mass spectrometer (Thermo Electron).

2.7 Data Analysis and Phosphoribosylated Peptide Quantification

1. PC computer.
2. Xcalibur™ raw MS data analysis software.
3. MaxQuant [36] data analysis software.
4. Microsoft Office Excel.

3 Methods

3.1 Preparation of MARylated PARP10^{CD}

MARylation of the PARP10 catalytic domain followed by concentration and desalting by gel filtration.

Note: This protocol has been optimized for generating MARylated PARP10^{CD} but in principle can apply to any auto-MARylated substrate.

1. Automodify desired amount of PARP10^{CD} (*see* **Note 1**) with 1 mM NAD^+ in AM buffer at 37 °C shaking at 500 rpm for 2 h in a 1.5 mL microcentrifuge tube.
2. Optional: MARylation can be confirmed by SDS-PAGE of the reaction product. Total protein staining of the gel will show a shift in molecular weight of the MARylated product compared to its unmodified counterpart (Fig. 2a). Western blotting with anti-PAN-ADP-ribose binding reagent can also be used to detect MARylation (Fig. 2b).
3. Flow 30 mL of desalting buffer through a Superdex 200 10/300 column with an NGC Quest 10 FPLC to equilibrate the column. Inject 500 μL of MARylated PARP10^{CD} sample onto column, and flow sample through at a rate of 0.5 mL/min monitoring protein elution by measuring A_{280}. Collect 0.5 mL fractions, and combine fractions containing protein (fast-eluting A_{280} peak) but not NAD^+ (slow-eluting A_{280} peak) to maximize yield of MARylated protein and desalting of unincorporated NAD^+ (Fig. 2c) (*see* **Note 2**).
4. Measure protein concentration of desalted sample by Bradford assay, and if necessary dilute or concentrate with a 10 K cutoff centrifugal spin concentrator in a tabletop centrifuge to a final protein concentration of 0.5 mg/mL.

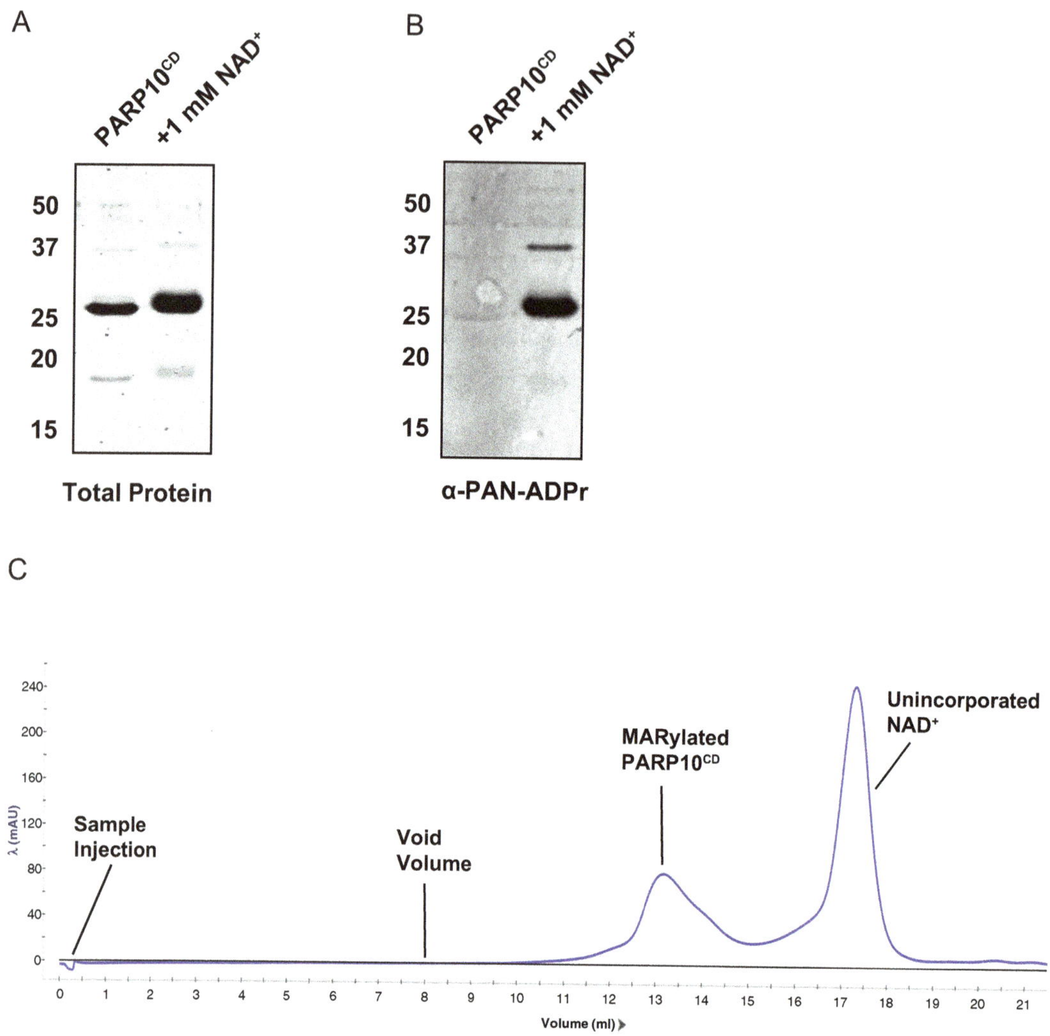

Fig. 2 Generating and desalting MARylated PARP10^{CD}. SDS-PAGE analysis of PARP10^{CD} incubated with or without 1 mM NAD^+ for 2 h at 37 °C analyzed by (**a**) total protein stain or (**b**) Western blotting with anti-PAN-ADP-ribose. (**c**) Gel filtration chromatography of PARP10^{CD} automodification reaction, blue trace is the absorbance of the sample at 280 nm over the course of the run

3.2 MAR Hydrolase and Phosphodiesterase Treatment of MARylated PARP10^{CD}

Demodification of MARylated PARP10 catalytic domain by MAR hydrolase and conversion of MAR to phosphoribose by phosphodiesterase treatment. Digestion of proteins to tryptic peptides.

1. Assemble the following demodification reaction and incubate at 37 °C for 1 h. For the negative control, recombinant MAR hydrolase is substituted with equal volume of demodification buffer:

   40 μL MARylated PARP10^{CD} (0.5 mg/mL)

   20 μL recombinant MAR hydrolase (1 mg/mL)

   15 μL DM buffer

2. Add 20 μL of recombinant *hs*NudT16a phosphodiesterase (1 mg/mL) and 5 μL of 20× NudT16 reaction buffer to each sample, and incubate at 37 °C for 2 h.
3. Add 50 μL of 3× denaturing buffer to each aliquot and incubate at 37 °C for 10 min in the dark.
4. Add 1 μL of lysyl endopeptidase (2 mg/mL) and 1 μL of modified trypsin (2 mg/mL) to each sample, and incubate at 37 °C shaking at 500 rpm overnight.
5. Add 1.5 μL of 100% TFA to each sample in the next morning.

3.3 Peptide Desalting

Desalting of peptides by C_{18} cleanup prior to phosphoenrichment.

Note: For subsequent steps, to apply liquid to StageTips, add liquid on top of the C_{18} plug and using a 5 mL slip-tip syringe apply air pressure to the top of the tip to push liquid through (*see* **Note 3**).

1. Activate StageTips™ (1 per sample) with 50 μL peptide desalting buffer B.
2. Equilibrate StageTips™ with 50 μL peptide desalting buffer A.
3. Load all of digested protein samples from Subheading 3.2, **step 5** onto StageTips.
4. Wash StageTips™ by applying three volumes 50 μL peptide desalting buffer A three times.
5. Samples on StageTips™ can be stored at 4 °C at this step, or proceed to phosphoenrichment (*see* Subheading 3.4).

3.4 Phosphoenrichment

Enrichment of phosphoribosylated peptides by immobilized metal affinity chromatography.

1. Wash PHOS-Select™ Iron Affinity Gel (5 μL per sample) with 500 μL peptide desalting buffer B three times in a clean 1.5 mL microcentrifuge tube, and resuspend in peptide desalting buffer B to a 50% resin/buffer slurry.
2. Make 10 μL aliquots of slurry for each sample in 1.5 mL microcentrifuge tubes.
3. Apply 90 μL of peptide desalting buffer B to each StageTip from Subheading 3.3, **step 5,** and elute directly into PHOS-Select Iron Affinity aliquots.
4. Incubate samples at 25 °C shaking at 1000 rpm for 1 h.
5. Allow resin to settle and aspirate supernatant in each sample, taking care to not disturb the resin, and then resuspend resin in 90 μL of Peptide desalting buffer B.
6. Activate new StageTips (1 per sample) as in Subheading 3.3, **step 1,** with 50 μL of peptide desalting buffer B.

7. Apply resuspended slurry samples to StageTips and push liquid through (*see* **Note 4**).
8. Wash StageTips by applying three 50 μL peptide desalting buffer B three times followed by 50 μL of 1% FA.
9. Elute sample from resin to C_{18} plug by applying 70 μL of 0.5 M KH_2PO_4 pH 7.0.
10. Wash once with 50 μL of 1% FA and 50 μL of peptide desalting buffer A.
11. Samples can be stored at 4 °C or proceed to LC-MS/MS (*see* Subheading 3.5).
12. Elute sample from StageTip with 50 μL of peptide desalting buffer B into a 1.5 mL microcentrifuge tube.
13. Dry samples in a SpeedVac, and resuspend in 7 μL of peptide desalting buffer A.

3.5 LC-MS/MS Analyses

Analysis of enriched peptides by liquid chromatography mass spectrometry.

1. Inject and load 5 μL of resuspended sample onto equilibrated analytical C_{18} column at 3% MS buffer B.
2. Apply a gradient of 3–30% MS buffer B over 30 min at a 300 nL/min flow rate.
3. Set the MS acquisition method to apply Top 5 HCD/CID (HCD, resolution = 30 K, 3E6 ion target; CID, Normal scan, 1E4 ion target) fragmentation methods per duty cycle over the course of the LC-MS run.

3.6 Data Searches

Computational searches of raw mass spectrometry data for peptide identification.

1. Load the .raw files into MaxQuant (e.g., version 1.5.3.8).
2. Set protein, peptide, and site FDRs to 0.01, and set a peptide score minimum of 40 with a delta score minimum of 17.
3. Define a new modification for phosphoribose (212.01 Daltons) within Andromeda with the chemical composition of $C_5H_9O_7P$ and residue specificities as D, E, K, R, C, and S.
4. Set the variable modifications as phosphoribosylation, phosphorylation (on STY), oxidation (on M), acetylation (at the protein N-terminus), and carbamidomethylation (on C). There are no fixed modifications.
5. Allow 5 peptide modifications, 2 missed cleavages, and a charge of up to 7. Set the enzyme as trypsin.
6. Choose an appropriate FASTA database that includes potential contaminants and the sequences of the recombinant proteins used in the experiment.
7. Run the searches.

3.7 Phosphoribosylated Peptide Quantification

Analysis and quantification of phosphoribosylated peptide intensity in different LC-MS runs.

1. After MaxQuant searches are complete, inspect the text file containing the phosphoribosylated peptide identifications to identify sites of modification on the MARylated substrate. For our study we only considered sites with >99% confidence for localization and MaxQuant scores >90 (Table 1).
2. For each site, inspect the MS/MS spectra of the phosphoribosylated peptide in MaxQuant viewer that contributed to the site identification. In general, the spectra should contain b and y ion peaks that fully cover the site of modification (e.g., Fig. 3a).

Table 1

Phosphoribose site identifications on PARP10^{CD} highlighting sites with high localization probabilities and MaxQuant scores (Data from ref. [15])

Positions within proteins	Leading proteins	Localization prob	Score
8	sp|Q53GL7|PAR10_HUMAN	1	140.45
99	sp|Q53GL7|PAR10_HUMAN	0.999999	109.16
22	sp|Q53GL7|PAR10_HUMAN	0.999997	159.34
15	sp|Q53GL7|PAR10_HUMAN	0.999911	151.16
49	sp|Q53GL7|PAR10_HUMAN	0.999127	111.31
207	sp|Q53GL7|PAR10_HUMAN	0.998717	142.19
12	sp|Q53GL7|PAR10_HUMAN	0.998087	145.81
204	sp|Q53GL7|PAR10_HUMAN	0.995927	138.31
121	sp|Q53GL7|PAR10_HUMAN	0.891376	48.376
124	sp|Q53GL7|PAR10_HUMAN	0.80551	47.64
114	sp|Q53GL7|PAR10_HUMAN	0.769917	61.524
54	sp|Q53GL7|PAR10_HUMAN	0.768854	176.42
52	sp|Q53GL7|PAR10_HUMAN	0.720788	137.52
53	sp|Q53GL7|PAR10_HUMAN	0.624452	73.279
113	sp|Q53GL7|PAR10_HUMAN	0.49998	61.524
37	sp|Q53GL7|PAR10_HUMAN	0.499965	46.41
38	sp|Q53GL7|PAR10_HUMAN	0.499965	46.41
25	sp|Q53GL7|PAR10_HUMAN	0.499781	41.014

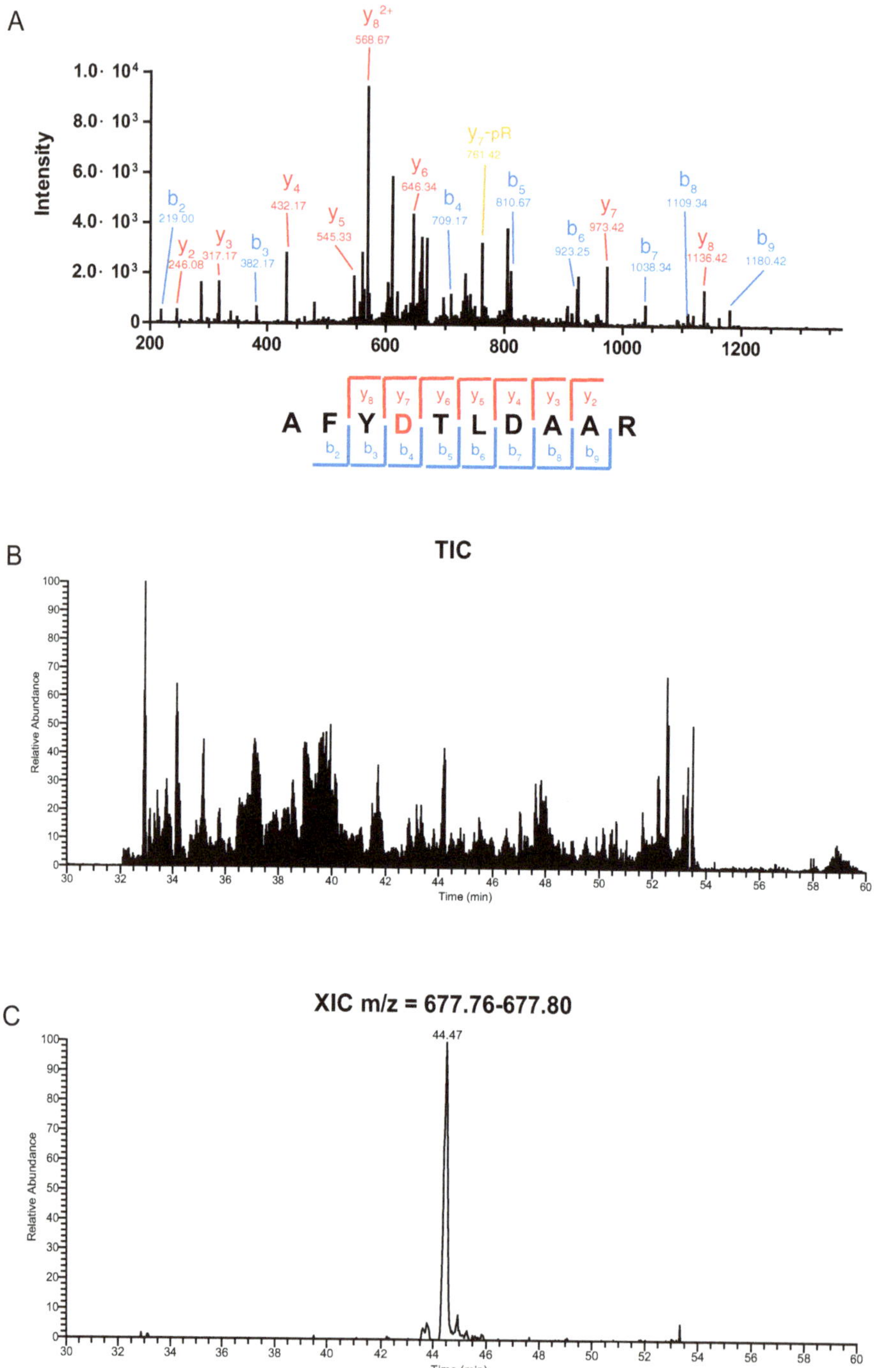

Fig. 3 Identification and quantification of phosphoribosylated sites on PARP10^{CD}. (**a**) MS/MS spectra of a PARP10^{CD} peptide containing a high-confidence phosphoribose modification at D839. (**b**) Total ion chromatogram (TIC) of LC-MS/MS run where peptide from (**a**) was identified. (**c**) Extracted ion chromatogram (XIC) of peptide from (**a**). (Data from ref. [15])

3. For each phosphoribosylated peptide that pass the criteria in Subheading 3.6, **steps 1** and **2**, open the .raw file above in Xcalibur. From the total ion current (TIC) of the run (e.g., Fig. 3b), set the mass range of the MS total ion chromatogram to the m/z of each phosphoribosylated peptide ±10 ppm to generate its extracted ion chromatogram (XIC) (e.g., Fig. 3c). Confirm that XIC contains a peak at the retention time that matches that of the MS/MS spectra that lead to the site identification. In general, the XIC can contain low-intensity peaks at other retention times, but these peaks ideally should not occur within 30 s of the peak of interest to avoid interference with quantification.
4. Export the raw data of the XIC into Excel or an equivalent spreadsheet program and sum the intensity of the peak of interest. We set the retention time window for quantification so that it encompasses the entire peak ±1 min.
5. Complete steps 3 and 4 for LC-MS runs of samples containing MAR hydrolases to determine the intensity of phosphoribosylated peptides in these samples. If the peptide of interest is not present in these samples, use the window for quantification from samples without MAR hydrolases.
6. Compare the intensities of phosphoribosylated peptides between replicates and different conditions to determine if incubation of the MAR hydrolase with MARylated substrate reduces their intensity.

4 Notes

1. We usually perform automodification in bulk with up to 1 mg of recombinant PARP10^{CD}.
2. This desalting step can also be performed on a desalting column (e.g., 5 mL HiTrap desalt column).
3. Take care to not allow the plug to dry out completely during peptide desalting as drying will reduce retention of peptides.
4. Resin will settle on top of the C_{18} plug; do not allow air to enter resin bed during enrichment.

References

1. Gupte R, Liu Z, Kraus WL (2017) PARPs and ADP-ribosylation: recent advances linking molecular functions to biological outcomes. Genes Dev 31:101–126. https://doi.org/10.1101/gad.291518.116
2. Vyas S, Matic I, Uchima L, Rood J, Zaja R, Hay RT et al (2014) Family-wide analysis of poly(ADP-ribose) polymerase activity. Nat Commun 5:4426. https://doi.org/10.1038/ncomms5426

3. Leidecker O, Bonfiglio JJ, Colby T, Zhang Q, Atanassov I, Zaja R et al (2016) Serine is a new target residue for endogenous ADP-ribosylation on histones. Nat Chem Biol 12(12):998–1000. https://doi.org/10.1038/nchembio.2180
4. Hottiger MO, Hassa PO, Lüscher B, Schüler H, Koch-Nolte F (2010) Toward a unified nomenclature for mammalian ADP-ribosyltransferases. Trends Biochem Sci 35:208–219. https://doi.org/10.1016/j.tibs.2009.12.003
5. Daniels CM, Ong S-E, Leung AKL (2015) The promise of proteomics for the study of ADP-ribosylation. Mol Cell 58:911–924. https://doi.org/10.1016/j.molcel.2015.06.012
6. Denu JM (2005) The Sir2 family of protein deacetylases. Curr Opin Chem Biol 9:431–440. https://doi.org/10.1016/j.cbpa.2005.08.010
7. Houtkooper RH, Pirinen E, Auwerx J (2012) Sirtuins as regulators of metabolism and healthspan. Nat Rev Mol Cell Biol 13:225–238. https://doi.org/10.1038/nrm3293
8. Di Girolamo M, Dani N, Stilla A, Corda D (2005) Physiological relevance of the endogenous mono(ADP-ribosyl)ation of cellular proteins. FEBS J 272:4565–4575. https://doi.org/10.1111/j.1742-4658.2005.04876.x
9. Gregor J, Rack M, Perina D, Ahel I (2016) Macrodomains: structure, function, evolution, and catalytic activities. Ann Rev Biochem 85:431–454. https://doi.org/10.1146/annurev-biochem-060815-014935
10. Palazzo L, Mikoč A, Ahel I (2017) ADP-RIBOSYLATION: new facets of an ancient modification. FEBS J 284(18):2932–2946. https://doi.org/10.1111/febs.14078
11. Jankevicius G, Hassler M, Golia B, Rybin V, Zacharias M, Timinszky G et al (2013) A family of macrodomain proteins reverses cellular mono-ADP-ribosylation. Nat Struct Mol Biol 20:508–514. https://doi.org/10.1038/nsmb.2523
12. Rosenthal F, Feijs KLH, Frugier E, Bonalli M, Forst AH, Imhof R et al (2013) Macrodomain-containing proteins are new mono-ADP-ribosylhydrolases. Nat Struct Mol Biol 20:502–507. https://doi.org/10.1038/nsmb.2521
13. Li C, Debing Y, Jankevicius G, Neyts J, Ahel I, Coutard B et al (2016) Viral *macro* domains reverse protein ADP-ribosylation. J Virol 90(19):8478–8486. https://doi.org/10.1128/JVI.00705-16
14. Fehr AR, Channappanavar R, Jankevicius G, Fett C, Zhao J, Athmer J (2016) The conserved coronavirus macrodomain promotes virulence and suppresses the innate immune response during severe acute. 7:1–12. https://doi.org/10.1128/mBio.01721-16.Editor
15. McPherson RL, Abraham R, Sreekumar E, Ong S-E, Cheng S-J, Baxter VK et al (2017) ADP-ribosylhydrolase activity of Chikungunya virus macrodomain is critical for virus replication and virulence. Proc Natl Acad Sci 114(7):1666–1671. https://doi.org/10.1073/PNAS.1621485114
16. Eckei L, Krieg S, Bütepage M, Lehmann A, Gross A, Lippok B et al (2017) The conserved macrodomains of the non-structural proteins of Chikungunya virus and other pathogenic positive strand RNA viruses function as mono-ADP-ribosylhydrolases. Sci Rep 7:41746. https://doi.org/10.1038/srep41746
17. Haikarainen T, Lehtiö L (2016) Proximal ADP-ribose hydrolysis in trypanosomatids is catalyzed by a macrodomain. Sci Rep 6:24213. https://doi.org/10.1038/srep24213
18. Zapata-Pérez R, Gil-Ortiz F, Martínez-Moñino AB, García-Saura AG, Juanhuix J, Sánchez-Ferrer Á (2017) Structural and functional analysis of Oceanobacillus iheyensis macrodomain reveals a network of waters involved in substrate binding and catalysis. Open Biol 7:160327. https://doi.org/10.1098/rsob.160327
19. Marjanovic MP, Palazzo L, Perina D, Sabljic I, Roko Z, Colby T et al (2016) Disruption of macrodomain protein SCO6735 increases antibiotic production in streptomyces coelicolor*. J Biol Chem 291:23175–23187. https://doi.org/10.1074/jbc.M116.721894
20. Rack JGM, Morra R, Barkauskaite E, Kraehenbuehl R, Ariza A, Qu Y et al (2015) Identification of a class of protein ADP-ribosylating sirtuins in microbial pathogens. Mol Cell 59:309–320. https://doi.org/10.1016/j.molcel.2015.06.013
21. Feijs KLH, Forst AH, Verheugd P, Lüscher B (2013) Macrodomain-containing proteins: regulating new intracellular functions of mono(ADP-ribosyl)ation. Nat Rev Mol Cell Biol 14:443–451. https://doi.org/10.1038/nrm3601
22. Barkauskaite E, Jankevicius G, Ladurner AG, Ahel I, Timinszky G (2013) The recognition and removal of cellular poly(ADP-ribose) signals. FEBS J 280:3491–3507. https://doi.org/10.1111/febs.12358
23. Zhang Y, Wang J, Ding M, Yu Y (2013) Site-specific characterization of the Asp- and Glu-ADP-ribosylated proteome. Nat Methods 10:981–984. https://doi.org/10.1038/nmeth.2603

24. Martello R, Leutert M, Jungmichel S, Bilan V, Larsen SC, Young C et al (2016) ADP-ribosylome of mammalian cells and tissue. Nat Commun 7:1–13. https://doi.org/10.1038/ncomms12917
25. Daniels CM, Ong SE, Leung AKL (2014) Phosphoproteomic approach to characterize protein mono- and poly(ADP-ribosyl)ation sites from cells. J Proteome Res 13:3510–3522. https://doi.org/10.1021/pr401032q
26. Palazzo L, Thomas B, Jemth A-S, Colby T, Leidecker O, Feijs KLH et al (2015) Processing of protein ADP-ribosylation by Nudix hydrolases. Biochem J 468:293–301. https://doi.org/10.1042/BJ20141554
27. Palazzo L, Daniels CM, Nettleship JE, Rahman N, McPherson RL, Ong S-E et al (2016) ENPP1 processes protein ADP-ribosylation in vitro. FEBS J 283(18):3371–3388. https://doi.org/10.1111/febs.13811
28. Rosenthal F, Nanni P, Barkow-Oesterreicher S, Hottiger MO (2015) Optimization of LTQ-Orbitrap mass spectrometer parameters for the identification of ADP-ribosylation sites. J Proteome Res 14(9):4072–4079. https://doi.org/10.1021/acs.jproteome.5b00432
29. Bilan V, Leutert M, Nanni P, Panse C, Hottiger MO (2017) Combining higher-energy collision dissociation and electron-transfer/higher-energy collision dissociation fragmentation in a product-dependent manner confidently assigns proteome wide ADP-ribose acceptor sites. Anal Chem. https://doi.org/10.1021/acs.analchem.6b03365
30. Daniels CM, Thirawatananond P, Ong S-E, Gabelli SB, Leung AKL (2015) Nudix hydrolases degrade protein-conjugated ADP-ribose. Sci Rep 5:18271. https://doi.org/10.1038/srep18271
31. Bonfiglio JJ, Fontana P, Zhang Q, Colby T, Gibbs-Seymour I, Atanassov I et al (2017) Serine ADP-ribosylation depends on HPF1. Mol Cell 65(5):932–940.e6. https://doi.org/10.1016/j.molcel.2017.01.003
32. Chapman JD, Gagné JP, Poirier GG, Goodlett DR (2013) Mapping PARP-1 auto-ADP-ribosylation sites by liquid chromatography-tandem mass spectrometry. J Proteome Res 12:1868–1880. https://doi.org/10.1021/pr301219h
33. Mashimo M, Kato J, Moss J (2014) Structure and function of the ARH family of ADP-ribosyl-acceptor hydrolases. DNA Repair (Amst) 23:88–94. https://doi.org/10.1016/j.dnarep.2014.03.005
34. Peterson FC, Chen D, Lytle BL, Rossi MN, Ahel I, Denu JM et al (2011) Orphan macrodomain protein (human C6orf130) is an O-acyl-ADP-ribose deacylase: solution structure and catalytic properties. J Biol Chem 286:35955–35965. https://doi.org/10.1074/jbc.M111.276238
35. Rappsilber J, Mann M, Ishihama Y (2007) Protocol for micro-purification, enrichment, pre-fractionation and storage of peptides for proteomics using StageTips Protocol for micro-purification, enrichment, pre-fractionation and storage of peptides for proteomics using StageTips. Nat Proto 2(8):1896–1906. https://doi.org/10.1038/nprot.2007.261
36. Cox J, Neuhauser N, Michalski A, Scheltema RA, Olsen JV, Mann M (2011) Andromeda: a peptide search engine integrated into the MaxQuant environment. J Proteome Res 10:1794–1805. https://doi.org/10.1021/pr101065j

Part VI

Mono-ADP-Ribose and Disease

Chapter 20

Detection of ADP-Ribosylating Bacterial Toxins

Chen Chen and Joseph T. Barbieri

Abstract

Many bacterial toxins catalyze the transfer of ADP-ribose from nicotinamide adenine dinucleotide (NAD) to a host protein. Greater than 35 bacterial ADP-ribosyltransferase toxins (bARTTs) have been identified. ADP-ribosylation of host proteins may be specific or promiscuous. Despite this diversity, bARTTs share a common reaction mechanism, three-dimensional active site structure, and a conserved active site glutamic acid. Here, we describe how to measure the ADP-ribosylation of host proteins as purified proteins or within a cell lysate.

Key words Bacterial toxins, Toxins, ADP-ribosyltransferase, NAD-glycohydrolase, ^{32}P-NAD, Biotin-NAD

1 Introduction

Mono-ADP-ribosylation is a common posttranslational modification catalyzed by bacterial toxins [1–3]. Bacterial ADP-ribosyltransferase toxins (bARTTs) are produced by Gram-negative and Gram-positive bacteria and catalyze the transfer of ADP-ribose from β-nicotinamide adenine dinucleotide (NAD) to one of several acceptor amino acids, including Arg, Asn, Thr, Cys, or Gln on a host protein(s). Often the action of the bARTT is responsible for the pathology of the disease elicited by the bacterial pathogen. Thus, vaccines targeting the bARTT are often generated against the bacterial pathogen, such as observed with diphtheria toxin.

bARTTs mono-ADP-ribosylation of host proteins can be specific or promiscuous for the host protein(s) targeted. Diphtheria toxin and exotoxin A of *Pseudomonas aeruginosa* (PA) ADP-ribosylate a single posttranslationally modified histidine (termed diphthamide) on Elongation Factor-2 (EF-2) [4], while ExoS of PA ADP-ribosylates numerous host proteins within a targeted cells or cell lysates [5]. Figure 1 shows the ADP-ribosylation catalyzed by ExoS of PA, which ADP-ribosylates numerous substrates in a HeLa cell lysate verses the ADP-ribosylation catalyzed by Certhrax

Paul Chang (ed.), *ADP-ribosylation and NAD+ Utilizing Enzymes: Methods and Protocols*, Methods in Molecular Biology, vol. 1813, https://doi.org/10.1007/978-1-4939-8588-3_20,

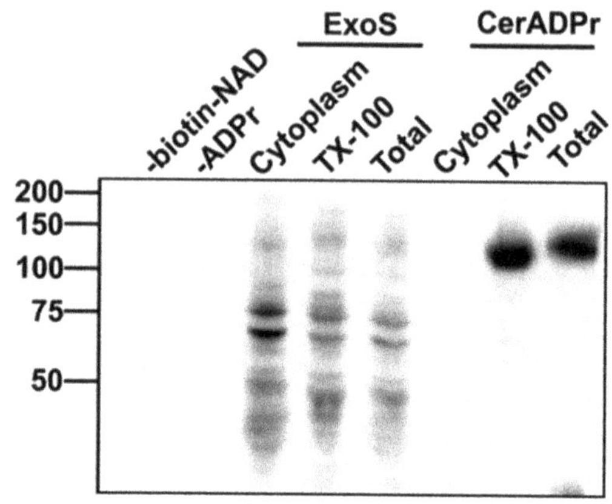

Fig. 1 Certhrax selectively ADP-ribosylates a high-molecular mass host protein (vinculin) in HeLa cell membranes. A HeLa cell lysate (total) was centrifuged to generate cytosol (cytoplasm) and membranes. Membranes were extracted with 0.1% Triton X-100 and centrifuged to remove the insoluble cell matrix (soluble membranes were termed Tx100). Cell lysates were incubated without the indicated bARTT (-ADPr) or with 50 ng of the catalytic domains of Certhrax toxin (CerADPr) or 20 ng of ExoS of PA (ExoS) for 1 h at 25 °C without substrate (-biotin-NAD) or with 4 μM biotin-NAD. The reactions were subjected to SDS–PAGE; proteins were transferred to a PVDF membrane, probed with streptavidin-conjugated HRP and SuperSignal (Pierce), and imaged with a digital camera system. The image is shown. Reprinted with permission from "Nathan C. Simon, James M. Vergis, Avesta V. Ebrahimi, Christy L. Ventura, Alison D. O'Brien, and Joseph T. Barbieri. Host Cell Cytotoxicity and Cytoskeleton Disruption by CerADPr, an ADP-Ribosyltransferase of Bacillus cereus G9241. Biochemistry, 2013, 52 [13], pp. 2309–2318. "Copyright (2013) American Chemical Society"

toxin of *Bacillus cereus*, which ADP-ribosylates only membrane bound vinculin. Despite this diversity, bARTTs share a common reaction mechanism, three-dimensional active site structure, and conserved active site amino acids [6–9]. While ADP-ribosylation often inhibits target protein action, occasionally ADP-ribosylation activates the targeted protein. For example, cholera toxin ADP-ribosylates Gs_{α}, yielding an active form of Gs_{α} [10]. One of the signature properties of members of the bARTTs family is the presence of an active site glutamic acid within the catalytic domain that can be photochemically cross-linked to the nicotinamide ring of NAD and contribute to catalysis [11]. Figure 2 shows the catalytic domain of diphtheria toxin, encoded by residues 1–193, with the active site glutamic acid (**Glu-148**) highlighted.

While the catalytic domains of the bARTT family are conserved for structure and function, the basis for binding host cell receptors and for the delivery of the catalytic into the host cell (translocation) differ. While diphtheria toxin and PA exotoxin A ADP-ribosylate diphthamide on EF-2, diphtheria toxin binds the heparin-binding

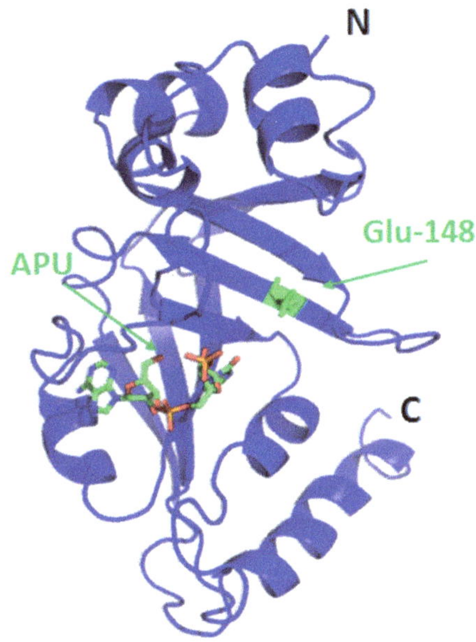

Fig. 2 The ADP-ribosyltransferase domain of diphtheria toxin. Shown is the catalytic domain of diphtheria toxin (amino acid residues 1–186) bound to the inhibitor adenylyl-3′,5′-uridine (APU) (PDB # 1DTP). Glu-148 is a conserved active site residue located within a series of antiparallel beta-strands common to members of the bARTT family

EGF-like growth factor precursor as a host receptor [12] and delivers the catalytic domain across the endosome membrane in response to a pH trigger [13], while exotoxin A binds the alpha 2-macroglobulin receptor/low density lipoprotein receptor-related protein as a host receptor [14]. After entering host cells, exotoxin A retrogrades traffics to the endoplasmic reticulum via a KDEL-like sequence [15–17], where the catalytic domain translocates into the cytosol. Detailed studies of other bARTTs, such as cholera toxin [18–22], show that upon entering the ER, the catalytic domain of cholera toxin unfolds and is recognized by the recycling machinery at the ER membrane and is translocated into the cytosol as an unfolded protein.

The bARTT family of protein toxins consists of more than 35 toxins, which are divided into four groups based on their **AB** domain organization, where **A** is the ADP-ribosyltransferase domain and **B** is the host cell binding domain. Diphtheria toxin-like toxins (diphtheria toxin and exotoxins A of PA) are single-chain proteins; cholera toxin-like toxins (cholera toxin, heat-labile enterotoxin, and pertussis toxin) are **AB5** toxins; C2-like toxins (C2 toxin, Iota toxin, *C. difficile* transferase, *C. spiroforme* toxin) are binary toxins where the **A** and **B** domains reside on independent proteins which associate upon **B** binding to the host cell membrane; and C3-like toxins are single **A** domains (C3bot, C3Stau/EDIN, C3cer, ExoS, ExoT).

Many laboratories measure the direct mono-ADP-ribosylation of host proteins with [^{32}P *adenylate phosphate*]-NAD where AD^{32}P-ribose is covalently transferred to host proteins, while Zhang J et al. used biotinylated-NAD as a substrate in diphtheria toxin-catalyzed ADP-ribosylation of EF-2 [23], and Takei Y et al. used digoxigenin-conjugated NAD as a substrate for pertussis toxin-catalyzed ADP-ribosylation of G proteins [24]. ^{32}P-labeled ADP-ribosylation proteins are detected by direct scintillation counting for β-emitting radiation, while biotinylated and digoxigenin-labeled ribosylation proteins are detected by Western blot or a conjugated ELISA-based assay. Diphtheria toxin-catalyzed eEF2 ribosylation, biotinylated-NAD provided similar sensitivity compared to [^{32}P]NAD [23]. One can also measure the amount of non-ADP-ribosylated host protein by an in vitro ADP-ribosylation reaction of a cell lysate prepared from cells previously incubated with an ADP-ribosylating toxin, as an indirect measurement of the action of the ADP-ribosylating toxin on cultured cells [25].

This chapter describes the use of [^{32}P]-*adenylate phosphate*-NAD and biotin-NAD to detect the ADP-ribosylation of host proteins within cell lysates for discovery or for the analysis of purified proteins. Using biotin-NAD, we identified vinculin as the protein ADP-ribosylated by Certhrax toxin from *Bacillus cereus* G9241 and then characterized the properties of ADP-ribosylated vinculin [26, 27].

2 Materials

2.1 Reagents

- Protease inhibitor cocktail for use with mammalian cells total cell lysates, membranes, and cytosol extracts (mPIC, Sigma # P8340, 100× in DMSO). Avoid repeated freeze thaw cycles, prepare aliquots, and store at −20 °C.
- Water. Prepare buffers with deionized water, sensitivity = 18 MΩ-cm at 25 °C.
- Reaction buffer: (5×) 50 mM Tris–HCl (pH 7.6), 100 mM NaCl (sterilize and store at 4 °C).
- HB buffer: Filter for sterilization and store at 4 °C.
- HB1 buffer = 3 mM imidazole, 250 mM sucrose, pH 7.4.
- HB2 buffer = HB1 buffer +0.5 mM EDTA +1% mPIC.
- Membrane solubilization buffer: HB2 buffer with 0.1% Triton X-100.
- Blocking solution: 50 mM Tris–HCl (pH 7.6) + 150 mM NaCl +2% BSA.
- Cells, adherent (HeLa) and suspension (HL-60), for cell lysate preparation.

(Biotin-NAD) ADP-ribosylation specific reaction:

- Substrate: 6-biotin-17-NAD (Trevigen, 250 μM stock in water, #4670-500-01) Avoid repeated freeze thaw cycles, prepare aliquots, and store at −80 °C.
- Pierce High Sensitivity Streptavidin-HRP (Pierce #21130).

(^{32}P-NAD) ADP-ribosylation specific reaction:

- [^{32}P]-*adenylate phosphate* NAD] (PerkinElmer, 800 Ci/mmol 5 mCi/mL, #NEG023X250UC) (^{32}P-NAD).
- TCA-impregnated Whatman #3 paper squares (¾" × ¾"). In a chemical hood, prepare 100% trichloroacetic acid-ether (shatterproof bottle). Place paper in a stainless-steel pan and soak paper with TCA-ether. Return unabsorbed TCA-ether to bottle. Paper "dries" in a few min and is stable for assay.

2.2 Disposables

- Pipette tips (10 to 1000 μL).
- Eppendorf tube 1.5 mL or ultracentrifugation tube.
- Six-well cell culture plate.
- Syringe (1 mL with 26–30 *g* needle).
- Filter.
- Polyvinylidene difluoride (PVDF) membrane.
- SDS-PAGE reagents.

2.3 Equipment

- Micropipettes.
- 37 °C tissue culture incubator.
- Bench-top microfuge (1.5 mL tube capacity).
- Ultracentrifuge.
- SDS-PAGE running system, transfer system.

3 Methods

Cell extracts are prepared according to substrates of ADPr proteins: cytosolic fraction, membrane fraction, or total cell lysate. Ribosylation action is prepared either with ^{32}P-NAD or biotin-NAD as described in the section.

3.1 Preparation of Cell Extracts

1. Adherent cells (HeLa cells): HeLa cells are seeded to ~70% confluence in 6-well plate and are cultured at 37 °C for various time. At desired time point, cells are washed with HB1 buffer, scraped into 1 mL HB2 buffer, and lysed by 25–30 passages through a 26-gauge needle (BD). Post-nuclear supernatant (PNS) is prepared by centrifuge at 400 × *g* for 5 min at 4 °C (total cell lysate). Cell membrane fraction is separated by ultracentrifugation at ~100,000 × *g* for 30 min at 4 °C (membrane and cytosol). Alternatively, membrane fraction can be separated

from cytosol by centrifuge at 16,000 × *g* for 60 min at 4 °C. Membrane fractions are solubilized with membrane solubilization buffer at 4 °C for 30 min followed by centrifugation at 20,000 × *g* for 10 min at 4 °C. Cell lysate will be aliquoted and stored at −80 °C.

2. Non-adherent cells (HL-60 cells): Centrifuge 2×10^6 to 1×10^7 HL-60 cells at 400 × *g* for 10 min at 4 °C. Discard the supernatant and wash the cell pellet in 1 mL of HB1 buffer and centrifuge at 400 × *g* for 10 min at 4 °C. Discard the supernatant and resuspend the pellet in 1 mL of HB2 buffer and pass the cells through a 30-gauge needle, 25–30 times. Separate the cytosolic fraction and membrane fraction as adherent cells described earlier.

3.2 ADP-Ribosylation Reaction

3.2.1 (^{32}P-NAD) ADP-Ribosylation Reaction [5]

Prepare the ADR-ribosylation reaction by mixing the following:

Cell lysate (*see* **Note 1**)	0–30 μL
H_2O	30–0 μL
^{32}P-NAD-(stock 250 μM) (*see* **Note 2**)	5 μL
ADPr protein (*see* **Note 3**)	5 μL (0–50 ng)
Reaction buffer (*see* **Note 4**)	10 μL
Total reaction	50 μL

Identification of host ADP-ribosylated substrates:

- Subject an aliquot of the ADRP reaction to SDS-PAGE gels, and transfer gel contents to PVDF membrane (*see* **Note 5**).
- Block the PVDF membrane with blocking solution.
- Incubate polyvinylidene fluoride (PVDF) membrane with Pierce High Sensitivity Streptavidin-HRP (1:8000) in TBS + 2% BSA at RT for 1 h.
- Wash PVDF membrane three times (5 min washes), incubate in SuperSignal for 5 min wash, and develop with X-ray film or by camera.

Determination of kinetic parameters of the ADP-ribosylation reaction:

- In a chemical hood, spot 50 μL of the reaction on TCA-saturated filter paper.
- Wash 3× with 10% TCA (5 min each).
- Wash 1× with ether and air dry.
- Place square into scintillation tube and count by Cherenkov radiation, without scintillation fluid.

3.2.2 (Biotin-NAD) ADP-Ribosylation Reaction [26]

Prepare the ADP-ribosylation reaction by mixing the following:

Cell lysate (*see* **Note 1**)	1–30 μL
H_2O	30–0 μL
Biotin-NAD (stock 250 μM) (*see* **Note 2**)	5 μL
ADPr protein (*see* **Note 3**)	5 μL (0–50 ng)
Reaction buffer (5×) (*see* **Note 4**)	10 μL
Total reaction	50 μL

Identification of host ADP-ribosylated substrates:

- Subject an aliquot of the ADPr reaction to SDS-PAGE (typically 12% acrylamide in running gel), and transfer (1 Amp for 1 h, at ~40 °C) gel contents to polyvinylidene fluoride (PVDF) membrane.
- Block the PVDF membrane with 2% BSA (20 min), and wash with TBS + 0.5% Tween 20.
- Incubate PVDF membrane with Pierce High Sensitivity Streptavidin-HRP (1:8000) in 2% BSA at RT for 1 h.
- Wash PVDF membranes three times (5 min) TBS +0.5% Tween 20 and develop with SuperSignal (Pierce).
- Develop on X-ray film or camera.

3.3 Assay Protocol

Incubate the reaction at room temperature for 5–60 min and at room temperature to 37 °C to establish optimal velocity. The extent of ADP-ribosylation should be time dependent. Therefore, this step can be extended if needed.

3.4 Controls

Controls include the ADP-ribosylation reaction without ADPr substrate protein or without cell lysate to provide a background determination and to measure possible auto-ADP-ribosylation of the ADPr substrate protein.

4 Notes

1. Use cell lysates (total, membrane, or cytosol) from adherent (ATCC # CCL-2, HeLa) or suspension (CCL-240, HL-60) cells. Titrate to achieve appropriate signal (+ lysate) to noise (- lysate). Note, cell lysate can be substituted with a known substrate protein. Establish the amount of target that yields a proportional amount of ADP-ribose incorporated, to allow the identification of the efficiency of ADP-ribosylation. This assay is useful when several substrates are ADP-ribosylated within

the cell lysate. Biotin-NAD protocol is used to identify ADP-ribosylated protein(s) in a cell lysate or to measure ADP-ribosylation of a purified protein substrate, but unlabeled β-NAD was used as a substrate to identify the amino acid ADP-ribosylated within a protein substrate [26].

2. Specific activity of ^{32}P-NAD or biotin-NAD is adjusted by dilution β-NAD (Sigma #N0632-1G). Titrate NAD to assess Km and k_{cat}. Linear velocity reactions are run at 10 μM NAD, utilizing <10% of NAD or substrate.
3. Titrate ADPr protein (toxin) to achieve <10% substrate incorporation into target protein.
4. Titrate NaCl and pH to optimize the reaction velocity.
5. This assay can also be developed into a solid phase assay by spotting the reaction mixture onto TCA-impregnated Whatman paper and assaying for the incorporation of ^{32}P-NAD into TCA precipitable material [28], as described for the "Determination of kinetic parameters of the ADP-ribosylation reaction" in Subheading 3.2.1.

Acknowledgments

JTB laboratory is supported by grants from the NIH: AI-030162 and AI-118389.

References

1. Aktories K, Barbieri JT (2005) Bacterial cytotoxins: targeting eukaryotic switches. Nat Rev Microbiol 3:397–410
2. Simon NC, Aktories K, Barbieri JT (2014) Novel bacterial ADP-ribosylating toxins: structure and function. Nat Rev Microbiol 12:599–611
3. Barbieri JT, Riese MJ, Aktories K (2002) Bacterial toxins that modify the actin cytoskeleton. Annu Rev Cell Dev Biol 18:315–344
4. Oppenheimer NJ, Bodley JW (1981) Diphtheria toxin. Site and configuration of ADP-ribosylation of diphthamide in elongation factor 2. J Biol Chem 256:8579–8581
5. Liu S, Yahr TL, Frank DW, Barbieri JT (1997) Biochemical relationships between the 53-kilodalton (Exo53) and 49-kilodalton (ExoS) forms of exoenzyme S of Pseudomonas aeruginosa. J Bacteriol 179:1609–1613
6. Wilson BA, Collier RJ (1992) Diphtheria toxin and Pseudomonas aeruginosa exotoxin A: active-site structure and enzymic mechanism. Curr Top Microbiol Immunol 175:27–41
7. Koch-Nolte F, Haag F (1997) Mono(ADP-ribosyl)transferases and related enzymes in animal tissues. Emerging gene families. Adv Exp Med Biol 419:1–13
8. Sixma TK, Pronk SE, Kalk KH, Wartna ES, van Zanten BA, Witholt B, Hol WG (1991) Crystal structure of a cholera toxin-related heat-labile enterotoxin from E. coli. Nature 351:371–377
9. Domenighini M, Magagnoli C, Pizza M, Rappuoli R (1994) Common features of the NAD-binding and catalytic site of ADP-ribosylating toxins. Mol Microbiol 14:41–50
10. Merritt EA, Hol WG (1995) AB5 toxins. Curr Opin Struct Biol 5:165–171
11. Carroll SF, Lory S, Collier RJ (1980) Ligand interactions of diphtheria toxin. III Direct photochemical cross-linking of ATP and NAD to toxin. J Biol Chem 255:12020–12024
12. Naglich JG, Metherall JE, Russell DW, Eidels L (1992) Expression cloning of a diphtheria toxin receptor: identity with a heparin-binding

EGF-like growth factor precursor. Cell 69:1051–1061

13. O'Keefe DO, Cabiaux V, Choe S, Eisenberg D, Collier RJ (1992) pH-dependent insertion of proteins into membranes: B-chain mutation of diphtheria toxin that inhibits membrane translocation, Glu-349----Lys. Proc Natl Acad Sci U S A 89:6202–6206
14. Kounnas MZ, Morris RE, Thompson MR, FitzGerald DJ, Strickland DK, Saelinger CB (1992) The alpha 2-macroglobulin receptor/low density lipoprotein receptor-related protein binds and internalizes Pseudomonas exotoxin A. J Biol Chem 267:12420–12423
15. Chaudhary VK, Jinno Y, FitzGerald D, Pastan I (1990) Pseudomonas exotoxin contains a specific sequence at the carboxyl terminus that is required for cytotoxicity. Proc Natl Acad Sci U S A 87:308–312
16. Hessler JL, Kreitman RJ (1997) An early step in Pseudomonas exotoxin action is removal of the terminal lysine residue, which allows binding to the KDEL receptor. Biochemistry 36:14577–14582
17. Teter K, Holmes RK (2002) Inhibition of endoplasmic reticulum-associated degradation in CHO cells resistant to cholera toxin, Pseudomonas aeruginosa exotoxin A, and ricin. Infect Immun 70:6172–6179
18. Taylor M, Burress H, Banerjee T, Ray S, Curtis D, Tatulian SA, Teter K (2014) Substrate-induced unfolding of protein disulfide isomerase displaces the cholera toxin A1 subunit from its holotoxin. PLoS Pathog 10:e1003925
19. Teter K, Allyn RL, Jobling MG, Holmes RK (2002) Transfer of the cholera toxin A1 polypeptide from the endoplasmic reticulum to the cytosol is a rapid process facilitated by the endoplasmic reticulum-associated degradation pathway. Infect Immun 70:6166–6171
20. Massey S, Banerjee T, Pande AH, Taylor M, Tatulian SA, Teter K (2009) Stabilization of the tertiary structure of the cholera toxin A1 subunit inhibits toxin dislocation and cellular intoxication. J Mol Biol 393:1083–1096
21. Chinnapen DJ, Hsieh WT, te Welscher YM, Saslowsky DE, Kaoutzani L, Brandsma E, D'Auria L, Park H, Wagner JS, Drake KR, Kang M, Benjamin T, Ullman MD, Costello CE, Kenworthy AK, Baumgart T, Massol RH, Lencer WI (2012) Lipid sorting by ceramide structure from plasma membrane to ER for the cholera toxin receptor ganglioside GM1. Dev Cell 23:573–586
22. Raghunathan K, Wong TH, Chinnapen DJ, Lencer WI, Jobling MG, Kenworthy AK (2016) Glycolipid crosslinking is required for cholera toxin to partition into and stabilize ordered domains. Biophys J 111:2547–2550
23. Zhang J, Snyder SH (1993) Purification of a nitric oxide-stimulated ADP-ribosylated protein using biotinylated beta-nicotinamide adenine dinucleotide. Biochemistry 32:2228–2233
24. Takei Y, Takahashi K, Kanaho Y, Katada T (1994) Pertussis toxin-catalyzed ADP-ribosylation of GTP-binding proteins with digoxigenin-conjugated NAD. Identification of the proteins in plasma membranes and nuclei. FEBS Lett 338:264–266
25. Gulke I, Pfeifer G, Liese J, Fritz M, Hofmann F, Aktories K, Barth H (2001) Characterization of the enzymatic component of the ADP-ribosyltransferase toxin CDTa from clostridium difficile. Infect Immun 69:6004–6011
26. Simon NC, Barbieri JT (2014) Bacillus cereus Certhrax ADP-ribosylates vinculin to disrupt focal adhesion complexes and cell adhesion. J Biol Chem 289:10650–10659
27. Simon NC, Vergis JM, Ebrahimi AV, Ventura CL, O'Brien AD, Barbieri JT (2013) Host cell cytotoxicity and cytoskeleton disruption by CerADPr, an ADP-ribosyltransferase of Bacillus cereus G9241. Biochemistry 52:2309–2318
28. Carroll SF, Collier RJ (1984) NAD binding site of diphtheria toxin: identification of a residue within the nicotinamide subsite by photochemical modification with NAD. Proc Natl Acad Sci U S A 81:3307–3311

Chapter 21

Preparation of Recombinant Alphaviruses for Functional Studies of ADP-Ribosylation

Rachy Abraham, Robert Lyle McPherson, Easwaran Sreekumar, Anthony K. L. Leung, and Diane E. Griffin

Abstract

Recently we characterized the mono(ADP-ribosyl) hydrolase (MAR hydrolase) activity of the macrodomain of nonstructural protein 3 ($nsP3^{MD}$) of chikungunya virus. Using recombinant viruses with targeted mutations in the macrodomain, we demonstrated that hydrolase function is important for viral replication in cultured neuronal cells and for neurovirulence in mice. Here, we describe the general cell culture and animal model infection protocols for alphaviruses and the technical details for biochemical characterization of the MAR hydrolase activity of $nsP3^{MD}$ mutants and the preparation of recombinant viruses incorporating those mutations through site-directed mutagenesis of an infectious cDNA virus clone.

Key words Alphavirus, Chikungunya virus, Sindbis virus, MAR hydrolase assay, Infectious cDNA clone, Reverse genetics, Site-directed mutagenesis, In vitro transcription

1 Introduction

Alphaviruses are mosquito-borne plus-strand, enveloped RNA viruses that cause a variety of human diseases ranging from rash and arthritis by the "Old World" viruses to encephalitis by the "New World" viruses. The encephalitic alphaviruses, like Venezuelan, western, and eastern equine encephalitis viruses, are endemic in the Americas, and recently introduced arthritic alphaviruses like chikungunya virus (CHIKV) can also cause atypical complications like encephalitis [1, 2]. Alphaviruses have an approximately 11.7 kilobase (kb) positive-sense RNA genome with a 5′ 7-methylguanosine cap and 3′ poly-A tail that encodes four nonstructural proteins (nsP1–4) and six structural proteins (C, E3, E2, 6K/TF, and E1) that are expressed from a subgenomic RNA [3]. Recently, we have shown that the macrodomain of viral nonstructural protein 3 (nsP3) of CHIKV enzymatically reverses the protein modification mono(ADP-ribosyl)ation in vitro. Macrodomains are a conserved protein fold found from archaea to

Paul Chang (ed.), *ADP-ribosylation and NAD+ Utilizing Enzymes: Methods and Protocols*, Methods in Molecular Biology, vol. 1813, https://doi.org/10.1007/978-1-4939-8588-3_21,

higher eukaryotes and are characterized by their binding to the small molecule ADP-ribose and its derivatives [4]. The viral macrodomain is part of the nonstructural protein in a subset of positive-strand RNA viruses, such as alphaviruses, coronaviruses, and hepatitis E virus [5, 6]. We identified and mutated residues in nsP3 critical for mono(ADP-ribosyl) binding and hydrolase (MAR hydrolase) activity, and using mutants with decreased MAR hydrolase activity, we demonstrated that reduction of this activity impairs viral replication in cultured mammalian and mosquito cells and virulence in neonatal mice [7].

Reverse genetic approaches have been extensively used in alphavirus research to understand virus biology and replication [8, 9], vector competence [10–13], attenuation and virulence [14, 15], and immunogenicity [16, 17]. The approach involves the construction of an infectious cDNA clone by incorporating a DNA copy of the viral RNA genome into a plasmid vector. With incorporation of the bacteriophage T7 or SP6 RNA polymerase promoters, the cDNA can be in vitro transcribed to produce full-length viral RNA transcripts. Transfection of the transcribed RNA into permissive cells results in production of infectious virus. To understand a loss or gain of gene function, variants or mutants are made in the infectious cDNA clone to produce mutant viruses in the same virus background for analysis.

In our studies of the MAR binding and hydrolase activities of the nsP3 macrodomain, we introduced mutations into the macrodomain region of nsP3 in the infectious cDNA clone of the 181/clone 25, the live attenuated vaccine strain of CHIKV. The 181/25 virus was derived from the Southeast Asian strain AF15561 by 18 plaque-to-plaque passages in MRC-5 human lung fibroblasts [18]. The infectious clone of 181/25 was prepared by employing site-directed mutagenesis to introduce five non-synonymous mutations into the parent clone 15,561 [10]. The parent clone 15,561 in a modified pSinRep5 plasmid was engineered to possess a SP6 RNA polymerase promoter at the 5′ terminus and a poly(A) tail, followed by a *NotI* restriction site at the 3'terminus [15] (Fig. 1). Because 181/25 is a vaccine strain, we reversed the attenuating mutations I12T and R82G [15] in the E2 glycoprotein to generate a recombinant CHIKV strain ($181/25^{E2\ I12T/R82G}$) that is virulent for newborn mice.

In this chapter, we describe the technical details of our strategies for biochemically characterizing the $nsP3^{MD}$ mutants with MAR hydrolase assays and for incorporating the mutations into the viral genome by site-directed mutagenesis. Here, the general protocols for cell culture of alphaviruses, infection of mice, and generation of recombinant mutant viruses using chikungunya virus 181/25 infectious cDNA clone are explained. Variations in the methodology for other alphaviruses, particularly the prototype Sindbis virus, are also mentioned. A summary of the methodology is outlined in Fig. 2.

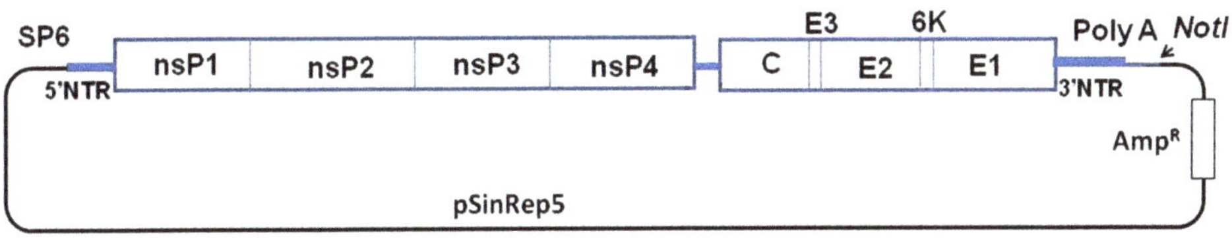

Fig. 1 Vector map of 181/25 infectious cDNA clone

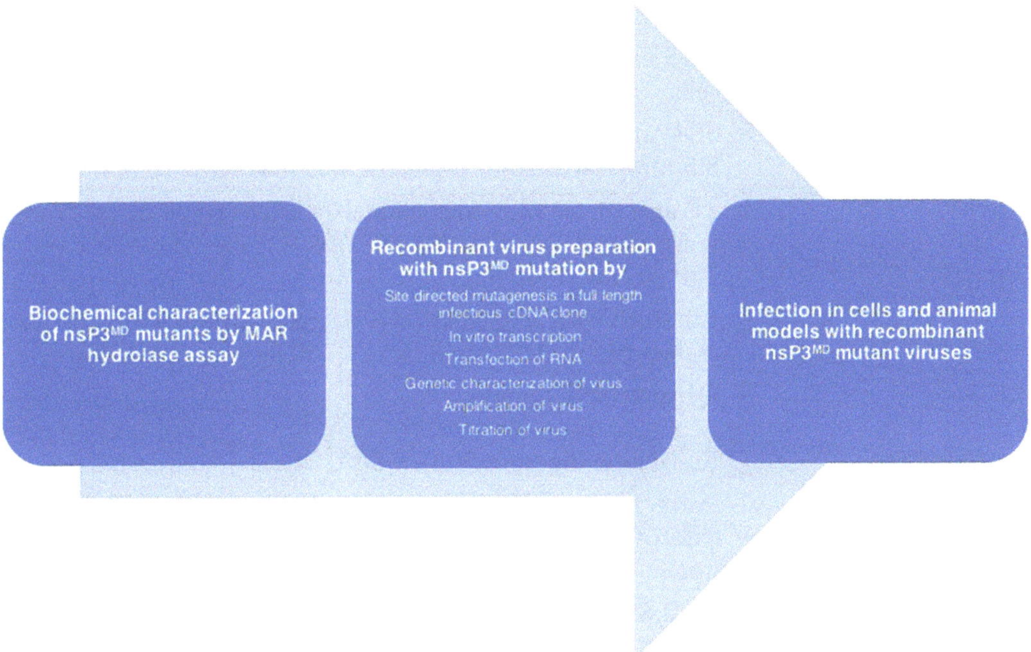

Fig. 2 Outline of the methodology

2 Materials

2.1 MAR Hydrolase Assay

1. Recombinant human PARP10 catalytic domain 818–1025 ($PARP10^{CD}$).
2. Recombinant MAR hydrolases (e.g., chikungunya viral macrodomain and human MacroD2).
3. ^{32}P-labeled β-nicotinamide adenine dinucleotide ($^{32}P\text{-}NAD^+$) (PerkinElmer).
4. Micro Bio-Spin 6 buffer exchange column.
5. Geiger counter.
6. Invitrogen Novex 14% Tris-glycine gel 10-well.
7. Automodification (AM) buffer: 20 mM Tris-HCl pH 7.5, 50 mM NaCl, 20 mM beta-mercaptoethanol, 5 mM $MgCl_2$.
8. Demodification (DM) buffer: 25 mM Tris-HCl pH 7.5, 200 mM NaCl, 20 mM beta-mercaptoethanol.
9. 4× SDS-PAGE sample loading buffer.

10. Novex SimplyBlue Safe Stain.
11. FujiFilm autoradiograph imaging plate.
12. FujiFilm imager.

2.2 Site-Directed Mutagenesis of cDNA and Transformation of Competent Bacteria

1. 10 ng/μL of full-length cDNA clone (e.g., CHIKV 181/25 cDNA clone).
2. QuikChange II site-directed mutagenesis kit.
3. 20 U *DpnI*.
4. One shot max efficiency chemically competent DH5α *E. coli* cells.
5. SOC medium: 2% tryptone (w/v), 1% yeast extract (w/v), 8.5 mM NaCl, 20 mM glucose, 2.5 mM KCl, 10 mM $MgCl_2$.
6. LB agar plates with ampicillin: 10% Bacto tryptone (w/v), 5% yeast extract (w/v), 10% NaCl (w/v), 2% Bacto agar(w/v), 100 μg/mL ampicillin.
7. LB broth with ampicillin: 10% Bacto tryptone (w/v), 5% yeast extract (w/v), 10% NaCl (w/v), 100 μg/mL ampicillin.
8. QIAprep spin miniprep kit.

2.3 Linearizing the Mutant Plasmid and In Vitro RNA Transcription

1. Restriction digestion buffer, 30 U *NotI*.
2. Purification of linearized plasmid: phenol-chloroform mix, chloroform, 3 M sodium acetate, 95% ethanol, nuclease-free water.
3. mMESSAGE mMACHINE SP6 transcription kit.
4. Purification of in vitro transcribed RNA: MEGAclear™ kit.

2.4 Transfection of RNA Transcripts into Susceptible Cells

1. BHK-21 cells (baby hamster kidney epithelial cells).
2. Growth media: Dulbecco's modified Eagle's medium (DMEM) supplemented with 10% heat-inactivated fetal bovine serum, L-glutamine (2 mM), penicillin (100 U/mL), and streptomycin (100 μg/mL).
3. Maintenance media: DMEM supplemented with 1% heat-inactivated fetal bovine serum, L-glutamine (2 mM), penicillin (100 U/mL), and streptomycin (100 μg/mL).
4. 1× Dulbecco's PBS (1.1 g Na_2HPO_4, 0.25 g KH_2PO_4, 0.2 g KCl, 8 g NaCl in IL distilled water).
5. 1× 0.5% trypsin EDTA.
6. OptiMEM serum-free media.
7. Lipofectamine 2000.

2.5 Viral RNA Extraction

1. Cell culture supernatant containing virus.
2. QIAamp® viral RNA mini kit.
3. RNaseZap.

2.6 One-Step Reverse Transcriptase Polymerase Chain Reaction (RT-PCR)

1. Superscript III one-step RT-PCR system: 1× reaction mix, 2 μL Superscript™ III RT/Platinum™ *Taq* Mix, 40 U ribonuclease inhibitor, 10 pmol of each oligonucleotide primer (*see* cf. **Note 11**).

2.7 Agarose Gel Electrophoresis and DNA Purification

1. 1% TAE-agarose gel (100 mL): 1 g agarose in 100 mL 1× TAE buffer (4.84 g Tris base, 0.58 g EDTA, and 1.1 mL glacial acetic acid in 1 L distilled water), 10 μL ethidium bromide (stock concentration 10 mg/mL).
2. 1 kb DNA ladder.
3. UV transilluminator.
4. QIAquick® gel extraction kit.
5. QIAquick® PCR purification kit.

2.8 Quantification of Virus by Plaque Assay

1. Vero cells (African green monkey kidney epithelial cells).
2. Growth media, maintenance media, 1× Dulbecco's PBS, 1× 0.5% trypsin EDTA.
3. 2× MEM supplemented with 1% heat-inactivated fetal bovine serum, penicillin (100 U/mL), and streptomycin (100 μg/mL).
4. 1.2% Bacto agar (w/v) in water.
5. 10% formaldehyde in PBS.
6. 1% crystal violet (w/v) in 20% ethanol in water.

2.9 Infection of Neuronal Cells

1. NSC34 cells (mouse neuronal cells) [19].
2. Growth media, maintenance media, 1× Dulbecco's PBS, 1× 0.5% trypsin EDTA.
3. RNeasy plus mini kit.
4. RIPA buffer (50 mM Tris pH 8, 150 mM NaCl, 0.1% (w/v) SDS, 1% (v/v) NP40, 0.5% (w/v) sodium orthovanadate, 1 mM EDTA pH 8), 1× protease inhibitor cocktail.

2.10 Infection of Newborn Mice

1. Timed pregnant CD-1 mice.
2. Inoculation of mice-29 G needle, anesthesia chamber, isoflurane.
3. Serum collection-1 mL 25 G needle, microtainer.
4. Ice cold PBS, 20 mL syringe, butterfly needle, dry ice.

2.11 Antiviral IgG and IgM Enzyme Immunoassay

1. 96-well flat immunoplates MaxiSorp.
2. 50 mM bicarbonate coating buffer pH 9.6 (1.58 g Na_2CO_3 and 2.92 g $NaHCO_3$ in 1 L distilled water).
3. 1× Dulbecco's PBS (1.1 g Na_2HPO_4, 0.25 g KH_2PO_4, 0.2 g KCl, 8 g NaCl in 1 L distilled water).

4. Washing buffer (0.05% Tween-20 in PBS).
5. Blocking buffer (10% fetal bovine serum in washing buffer).
6. Goat anti-mouse IgG-HRP antibody.
7. Goat anti-mouse IgM-HRP antibody.
8. BD OptEIA TMB substrate reagent kit.
9. Stop solution (2 M H_2SO_4).
10. Saranwrap.

3 Methods

3.1 MAR Hydrolase Assay

MAR hydrolase assay involves automodifying the substrate $PARP10^{CD}$ with radiolabeled ^{32}P-NAD^+ and incubating the substrate with recombinant proteins to be tested for MAR hydrolase activity followed by SDS electrophoresis and phosphoimaging.

1. Automodify desired amount of $PARP10^{CD}$ (*see* **Note 1**) with ^{32}P-NAD^+ at a concentration of 0.25 μCi/μg of $PARP10^{CD}$ in AM buffer at 30 °C for 30 min.
2. Add 1 mL of DM buffer to Bio-Rad Micro Bio-Spin column and allow gravity flow through the column. Repeat twice to equilibrate column in DM buffer. Add $PARP10^{CD}$ automodification reaction to the top of the bed (100–150 μL total volume) and allow it to enter the column. Add 600 μL of DM buffer to the column and collect 2-drop fractions using the Geiger counter to determine when the ^{32}P-labeled $PARP10^{CD}$ is eluted. Once the ^{32}P-labeled $PARP10^{CD}$ begins to elute, collect a 500 μL fraction to maximize yield and desalting of unincorporated ^{32}P-NAD^+.
3. For each reaction, make 1 μg aliquots of desalted ^{32}P-labeled $PARP10^{CD}$ (*see* **Note 2** about concentration of desalted sample from **step 2**). Add recombinant wild-type (WT), mutant MAR hydrolases, or DM buffer alone to reaction at concentrations equimolar to amounts of automodified $PARP10^{CD}$ (1 μg of $PARP10^{CD}$ is 43.1 pmol) and take reactions up to 15 μL total volume with DM buffer.
4. Incubate reactions at 37 °C for 1 h and stop reactions by adding 5 μL of 4× SDS-PAGE sample loading buffer.
5. Run samples by SDS-PAGE on Invitrogen Novex 14% Tris-glycine gel 10-well.
6. Rinse gel for 10 min with Milli-Q water followed by staining with Novex SimplyBlue Safe Stain for 1 h at room temperature. Once bands develop, rinse gel in Milli-Q water for 1 h at room temperature.

7. Image gel with a scanner to obtain total protein stain and expose the gel to a FujiFilm autoradiograph imaging plate overnight.
8. Image FujiFilm autoradiograph imaging plate on FujiFilm imager to acquire autoradiograph of gel.
9. For quantification, open the autoradiography file in ImageJ and quantify the signal intensity of each lane using the GelAnalyzer tool (*see* **Note 3**). Quantify the intensity of the ^{32}P signal from automodified PARP10^{CD} and obtain the percentage of signal removed using the below equation:

$$\%^{32}\text{P}-\text{signal removed} = \frac{((\text{Intensity from buffer alone})-(\text{Intensity from sample}))}{(\text{Intensity from buffer alone})}\times 100\%$$

To obtain % WT activity, use the below equation in combination with the above equation:

$$\%\text{WT activity} = \frac{(\%^{32}\text{P}-\text{signal removed from mutant MAR hydrolase})}{(\%^{32}\text{P}-\text{signal removed from WT MAR hydrolase})}\times 100\%$$

3.2 Site-Directed Mutagenesis of Infectious cDNA Clone and Transformation of Competent Cells

The infectious cDNA clone which is the double-stranded copy of the viral genome carried on a plasmid vector is modified for introducing mutation by PCR-based site-directed approach, followed by transformation to the *E. coli* cells. The mutant clone is then subjected to sequencing for checking the presence or absence of mutation.

1. Design oligonucleotide forward and reverse primers containing the desired mutation (*see* **Note 4**).
2. Prepare each PCR reaction in a 50 μL reaction volume:
   10× buffer 5 μL
   Pfu turbo polymerase (2.5 U/μL) 1 μL
   dNTP mix (10 mM) 1 μL
   Template plasmid (10 ng) 1 μL
   Forward primer (10 μM) 1.25 μL
   Reverse primer (10 μM) 1.25 μL
   Nuclease-free water Make up to 50 μL
3. Quickly spin the PCR tubes in a Mini Microcentrifuge and load them into a PCR thermal cycler. The thermal cycling conditions are 1 min of initial denaturation at 95 °C, followed by 18 cycles of the following: denaturation at 95 °C for 50 s, annealing at 60 °C for 15 s, extension at 68 °C for 16 min, and

a final extension at 68 °C for 7 min. After the reaction, cool the tubes on ice for 2 min.

4. To eliminate the template plasmid, digest with *DpnI* which cleaves only at methylated sites. Add 1 μL of *DpnI* (20 U/μL) endonuclease directly to the reaction, mix by pipetting. Quickly spin the PCR tubes in a Mini Microcentrifuge, and incubate at 37 °C for 1 h in the thermal cycler.
5. Transform competent DH5α cells with 2 μL of the *DpnI* digested product. Flick the tubes gently to mix and incubate the tubes on ice for 30 min. Heat shock the cells at 42 °C for 30 s, and immediately incubate the tubes on ice for 2 min. Add 250 μL of SOC media, and incubate at 37 °C for 1 h with continuous shaking at 200 rpm for recovery. Plate 100 μL of the transformation mixture onto LB agar plates with ampicillin, and incubate the plates at 37 °C for 16–18 h.
6. Pick individual *E.coli* colonies from the plate, and inoculate in 5 mL LB broth with ampicillin and incubate 16–18 h at 37 °C with continuous shaking at 200 rpm. Pellet the bacterial cells at 3000 rpm (1942 rcf) for 10 min at room temperature, and proceed with plasmid isolation using QIAprep spin mini plasmid prep kit or any other commercial mini prep kit (*see* **Note 5**).
7. The plasmids are sequenced and compared with the sequence of the parent clone (*see* **Note 6**).

3.3 Linearizing the Mutant Plasmid DNA and In Vitro RNA Transcription

In order to make the mutant viruses, the mutant full-length cDNA clones are linearized, purified, and in vitro transcribed to RNA.

Linearize 3 μg of plasmid containing the desired mutation in the full-length cDNA with *NotI* enzyme in a 100 μL reaction (*see* **Note 7**). Incubate the reaction for 16–18 h at 37°C. The component volumes for the reaction are:

10× restriction buffer 10 μL
Plasmid DNA 3 μg
NotI (20 U/μL) 1.5 μL
Nuclease-free water Make up to 100 μL

Analyze 5 μL of the digested reaction by gel electrophoresis to ensure that the plasmid is fully linearized.

1. Purification of the linearized plasmid (100 μL) is performed by phenol-chloroform extraction, followed by alcohol precipitation in an Eppendorf tube. Add an equal volume of phenol: chloroform to the digested product, mix well, and centrifuge at 15,700 rcf for 5 min. Transfer the aqueous phase to a new tube and add an equal volume of chloroform, mix well, and centrifuge at 15,700 rcf for 10 min. Harvest the aqueous phase and add 1/10 volume 3 M sodium acetate and 2 volumes 95% ethanol and incubate at −20 °C for 1 h. After 1 h, centrifuge at 13,000 rpm (15,700 rcf) for 15 min at 4 °C and discard super-

natant fluid completely and air-dry the tube. Resuspend the pellet in 10 μL nuclease-free water, and quantify the purified plasmid DNA with a NanoDrop spectrometer (*see* **Note 8**).

2. Transcribe 1.5 μg of purified plasmid DNA into viral RNA using mMESSAGE mMACHINE SP6 kit. The reaction mixture consists of:
   10× reaction buffer 3 μL
   GTP (20 mM) 3 μL
   2× NTP/CAP 10 μL
   Linear template 1.5 μg
   Enzyme mix 2 μL
   Nuclease-free water Make up to 30 μL

   Mix the reaction by pipetting thoroughly and incubate at 37 °C for 2 h. Add 1 μL of DNase to digest the template DNA, mix by pipetting, and incubate further at 37 °C for 15 min.

   Purify the RNA transcripts using the MEGAclear™ kit, and elute with 30 μL of RNase-free water (*see* **Note 9**). Generally, the expected concentration of the RNA synthesized from a standard 30 μL reaction starting from 1.5 μg of DNA is within a range 2–3 μg/μL.

3.4 Transfection of BHK-21 Cells to Recover Virus

This section deals with the transfection of the CHIKV RNA transcript into permissive cell lines to recover the clone-derived virus.

1. Transfer BHK-21 cells in growth medium into a T150 cm^2 flask, and incubate for 18–24 h in 37 °C 5% CO_2 incubator so that the cells will be 80–90% confluent at the time of transfection (*see* **Note 10**).
2. Transfect the in vitro transcribed RNA into cells using Lipofectamine 2000. Dilute 40 μL of Lipofectamine 2000 in 200 μL OptiMEM. Mix gently and keep at room temperature for 5 min. Also dilute 20–40 μg of in vitro transcribed RNA in 200 μL OptiMEM without serum and mix gently. Combine the diluted RNA with the diluted Lipofectamine (total volume is 400 μL). Mix by gentle pipetting, and incubate the mix for 20 min at room temperature to allow the RNA-Lipofectamine complex to form.
3. Remove the culture media from the T150 cm^2 flask of BHK-21 cells, wash once with PBS, and add 7 mL of OptiMEM without serum. Add the RNA-Lipofectamine 2000 complex drop by drop on top of the cells. Swirl gently to ensure equal distribution of the transfection mix to the cells.
4. Incubate the cells at 37 °C, 5% CO_2 for 6 h, with gentle rocking of the flask back and forth every 30 min. After 6 h, remove the transfection mix and add 12 mL DMEM 1% FBS to the flask. Incubate transfected cells at 37 °C, 5% CO_2 for 24–48 h until cytopathic effect (CPE) is visible (Fig. 3).

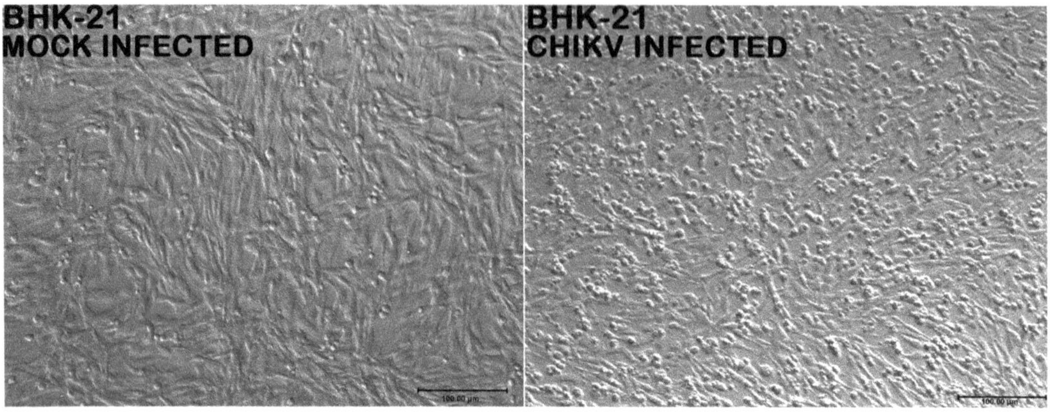

Fig. 3 Infection in BHK-21 cells showing cytopathic effect – Cultured BHK-21 cells were transfected with in vitro transcribed RNA using Lipofectamine and incubated for 24–48 h till visible cytopathic effects were observed

5. Collect the viral supernatant fluid from the flasks into a sterile 50 mL tube, and centrifuge at 1000 rpm (216 rcf) for 10 min at 4 °C. Virus should be stored as 1 mL aliquots in cryopreservation tubes at −80 °C (label as passage 0, P_0).

3.5 Viral RNA Extraction and Sequencing of the Mutated Macrodomain

The presence of the mutation in the clone-derived virus is verified by RNA isolation, followed by reverse transcriptase PCR, and finally sequencing.

1. Clean the work bench and pipettes with RNase Zap to remove all potential RNase contamination.
2. Extract viral RNA from viral supernatant fluid using QIAamp® viral RNA kit and store at −80 °C.
3. Perform one-step reverse transcriptase PCR using 5 μL of extracted RNA (*see* **Note 11**).

   2× reaction buffer 25 μL

   RNA template 5 μL

   Forward primer (10 μM) 1.25 μL

   Reverse primer (10 μM) 1.25 μL

   Superscript RT 2 μL

   RNaseOUT 1 μL

   Nuclease-free water Make up to 50 μL
4. Quickly spin the PCR tubes in a Mini Microcentrifuge and load them into a PCR thermal cycler. The thermal cycling conditions are 30 min of reverse transcription step at 50 °C and 2 min of initial denaturation at 94 °C, followed by 40 cycles of the following: denaturation at 94 °C for 15 s, annealing at 55 °C for 30 s, extension at 68 °C for 2 min, and a final extension at 68 °C for 5 min.

5. Perform agarose gel electrophoresis with a few microliters of the PCR product to determine whether the expected band is present. Mix 5 μL of the PCR reactions with 1 μL of loading dye, and load into individual wells. Load 1 kb DNA ladder into one of the wells to identify the expected band size. Run the agarose gel electrophoresis at 100 V for about 45 min or until the loading dye is nearing the bottom edge of the gel. Transfer the gel into a UV transilluminator to identify the expected band.
6. Purify the PCR product to remove excess primers and other PCR reaction mix components. If there is only a single band present, purify the remaining PCR product directly using QIAquick® PCR purification kit. If there are additional bands, excise the correct band from the gel and purify using QIAquick® gel extraction kit.
7. The purified PCR product will be used as a template for sequencing to confirm the presence or absence of the mutation in the clone-derived virus. The purified PCR product can be stored at −20 °C (*see* **Note 6** for the primer details).

3.6 Amplification of Recovered Recombinant Virus

1. Culture BHK-21 cells in growth medium in a T150 cm² flask the day before infection so that they are 80–90% confluent at the time of infection.
2. Infect the cells with a 1:10 dilution of P_0 virus (i.e., virus obtained after in vitro transcription; *see* Subheading 3.4, **step 5**). Incubate the cells for 1 h at 37 °C, with gentle rocking of the flask back and forth every 15 min. After 1 h, remove the unadsorbed virus, wash once with PBS, and add a fresh 12 mL of DMEM 1% FBS. Incubate the cells for 24–48 h until cytopathic effect is visible.
3. Collect supernatant containing the virus from the flasks into a sterile 50 mL tube, and centrifuge at 216 rcf for 10 min at 4 °C. Virus should be stored as 1 mL aliquots in cryopreservation tubes at −80 °C (label as passage 1, P_1).
4. Characterize the virus (P_1) by sequencing to confirm the presence of the desired mutation (follow Subheading 3.5), and proceed for virus quantitation by plaque assay titration.

3.7 Plaque Assay Titration for Viral Quantification (See Note 12)

Detecting the viral load or viral titer is essential for any virology-based study. Plaque assay allows the quantification of infectious virus particles which rely on the ability of viruses to cause cytopathic effects on the host cells because of infection. This section describes the steps involved: infection of cell monolayers with tenfold dilutions of viral supernatant, followed by overlaying with media having high molecular weight substrate like Bacto agar, agarose, or carboxymethylcellulose, which will limit the viral spread.

The plaques thus formed are visualized by staining cell monolayers with crystal violet or neutral red.

1. Day 1: Plate Vero cells in DMEM 10% FBS in a 6-well plate. A 100% confluent flask of Vero cells can be plated into five 6-well plates. Each plate will be for one sample; each well within this plate is for a tenfold dilution of that sample (Fig. 4). Incubate the plates overnight at 37 °C, 5% CO_2 (*see* **Note 13**).
2. Day 2: Infection
   (a) Preparation of overlay agar mixture – Melt 1.2% Bacto agar in the microwave and place in 42 °C water bath. Warm 2× MEM 1% FBS in 42 °C water bath. Mix equal amounts of 1.2% Bacto agar and 2× MEM and place in 42 °C water bath until use (*see* **Note 14**).
   (b) Sample preparation – Prepare tenfold serial dilutions of samples in 48-well plates.
   Add 450 μL of DMEM 1% FBS in each well of 48-well plate. Add 50 μL of viral stocks to be titrated to the wells in first row, mix well and transfer 50 μL from first row to second row, and thus continue serial dilutions as shown in Fig. 5.
   (c) Infection of Vero cells – Aspirate the media from the wells and wash briefly with PBS. Add 200 μL of serially diluted samples to the appropriate wells. Incubate at 37 °C in 5% CO_2 incubator for 1 h with gentle shaking every 15 min. After 1 h of incubation, aspirate the virus, wash with PBS, and add 1.5 mL overlay agar mixture to each well. Let agar solidify at room temperature, and incubate the plates at 37 °C, 5% CO_2, for 48 h.

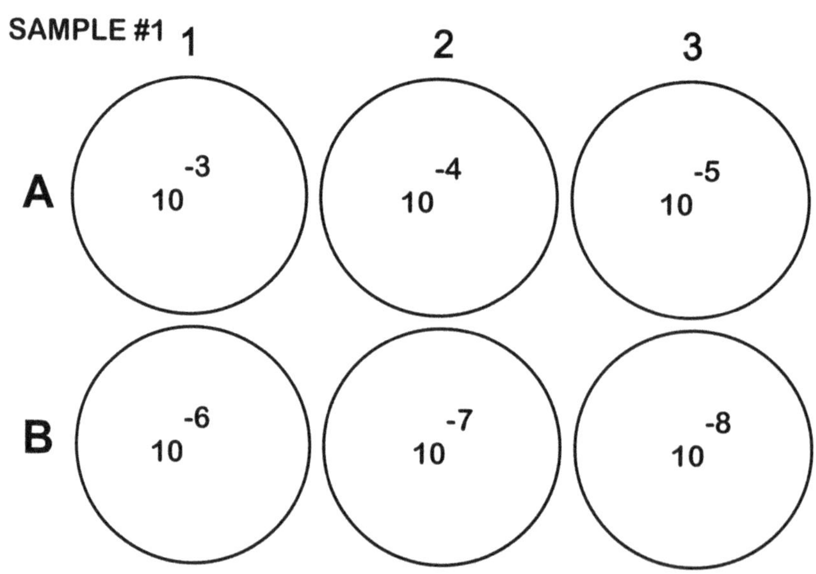

Fig. 4 Plate map of 6-well plate for the plaque assay

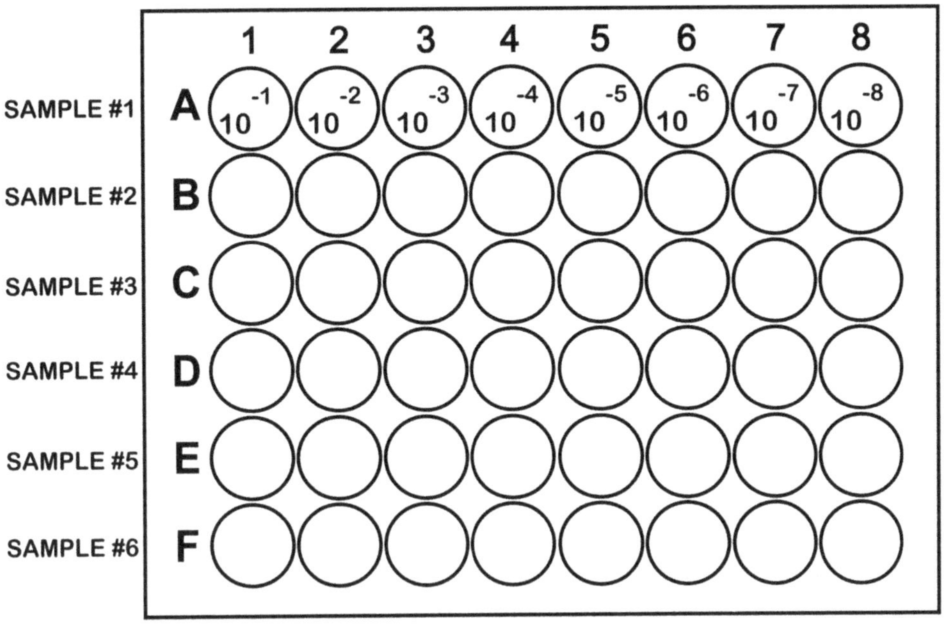

Fig. 5 Plate map of 48-well plate for serial diluting the viral stocks for the plaque assay

3. Day 4: Plaque staining and counting – Fix the cells by adding 1 mL of 10% formalin in PBS directly into the wells without removing the agar overlay. Incubate for 2 h to overnight (preferred). Aspirate the fixed overlay agar to an appropriate waste container, invert and tap the plate on a tissue paper pad to remove the agar overlay, and stain by adding 300 μL of 1% crystal violet in 20% ethanol in water. Keep for 10 min, and after cells are stained, wash in running tap water to remove excess stain solution, and air-dry the plates for an hour (Figs. 6 and 7).

$$\text{Calculate the PFU / mL} = \frac{(\text{plaque count}) \times (\text{dilutionfactor})}{(\text{amount of sample added in mL})}$$

3.8 Infection of Neuronal Cells to Assess Virus Growth

In vitro infection in cells involves plating cells, calculating the amount of the infectious virus to be added, infection of the cells, followed by incubation for the respective time points.

1. Seed 1×10^5 NSC34 cells per well in growth medium into 6-well plates a day before so the cells will be 80–90% confluent at the time of infection. Include one well for cell counts to calculate the multiplicity of infection (MOI). MOI is the ratio of plaque-forming units (PFU) to the total number of cells in the well. If one million virions are added to one million cells, the MOI is one.
2. Next day, trypsinize one of the wells in the plate, count the cells, and calculate the amount of virus needed for the desired

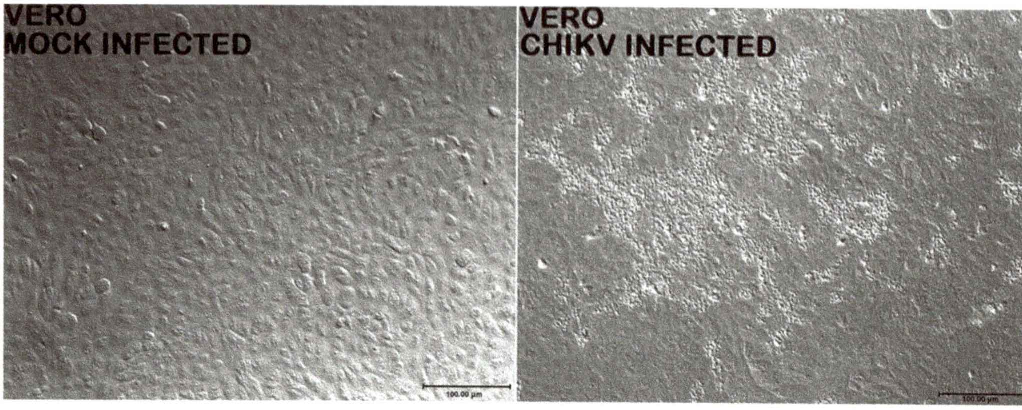

Fig. 6 Light microscopy image of plaque formation – Vero cells cultured in 6-well plate was infected with different dilutions of viral stocks for 1 h. After 1 h of incubation, the cells were washed with PBS, and overlay agar mixture was added and incubated for 48 h for plaque formation

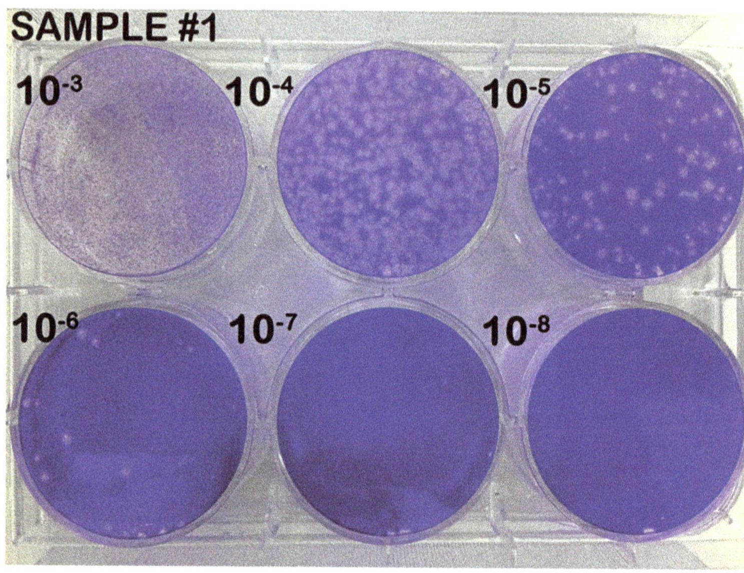

Fig. 7 Upon 48 h of incubation, the cells were fixed with 10% formaldehyde in PBS and stained with crystal violet to visualize the plaques formed

MOI. Prepare the diluted virus stock in DMEM 1% FBS for one experiment (if it involves two or more wells) in one tube to ensure homogeneity.

3. Remove the media from the cells, wash with PBS, and add 200 μL of diluted virus to the corresponding wells for infection. Add media alone to the mock infected wells. Incubate for 1 h with gentle swirling every 15 min at 37 °C, 5% CO_2. After 1 h of incubation, remove the unadsorbed virus, wash once with 1 mL PBS, and add 1.5 mL of DMEM 1% FBS.

4. Collect the supernatant at the desired time points and freeze at −80 °C. The supernatants can be further assayed for infectious virus titers using a plaque assay (follow Subheading 3.7) or for levels of secreted proteins such as cytokines by ELISA. The cells from each well can be collected either in buffer RLT (from Qiagen kit) for RNA isolation or in RIPA buffer for protein studies.

3.9 Infection of Mice to Assess Neurovirulence and Virus Growth (See Note 15)

Studies using mouse models to evaluate neurovirulence can be carried out by infecting the mice intracranially and monitoring them daily for clinical signs/mortality. The organs can be harvested to assess the viral replication by plaque assay or RNA quantification.

1. All animal experiments and protocols must be performed according to the guidelines approved by the Institutional Animal Care and Use Committee and carried out with strict adherence to the Guide for the Care and Use of Laboratory Animals by the National Institutes of Health and the Policy on Humane Care and Use of Laboratory Animals by the US Public Health Service.
2. The CHIKV mutant viruses in the virulent background $181/25^{E2\ I12TR82G}$ must be handled in the Biosafety level three facility and the animal experiments in the ABSL3 facility according to US regulations. All experiments with the biological samples after infection with CHIKV in the virulent background $181/25^{E2\ I12TR82G}$ must be handled in the BSL3 facility.
3. The mice should be bred under barrier conditions with a standard 12-h light/12-h dark cycle.
4. Timed pregnant CD-1 female mice are maintained in an ABSL3 facility, and the CHIKV infection experiments are conducted in 2-day old neonates.
5. The route of administration is intracranial in a 10 μL volume having 1000 PFU of virus diluted in PBS. Prepare the virus dilution for one experiment (if it involves more animals for same virus) in one tube to ensure homogeneity, and keep on ice until use to prevent thermal inactivation. The same mutant virus is administered to all pups of a litter.
6. Anesthetize each pup prior to infection using isoflurane (*see* **Note 16**).
7. Restrain the pup manually on a solid surface by firmly placing your thumb and index finger on either side of the head just at the nape, and insert the needle having 10 μL of virus into the right side of the cranium approximately halfway between the eye and the ear just off-center from the midline (*see* **Note 17**).
8. Monitor the mice twice daily for weight gain/loss and morbidity/mortality. If any animal shows signs of being moribund

or reaches experimental end points, euthanize immediately according to a method which has been approved by the ethics governing body. The 2-day old CD-1 mice infected with virulent strains of CHIKV will die without showing much evidence of illness, so clinical scoring is not possible for this model (*see* **Note 18**).

9. At the end of the morbidity/mortality experiment, the surviving pups and the mother are generally sedated with isoflurane and euthanized according to a method which has been approved by the ethics governing body.
10. For time course infection experiments to assess virus replication, at least three pups per group are euthanized at each time after infection, and blood and organs are collected. The pups are anesthetized using isoflurane, and blood is collected with a 1 mL syringe by cardiac puncture. The blood can be transferred to a microtainer tube and spun at 1500 rpm (200 rcf) for 15 min to collect serum.
11. Expose the heart, and perfuse with 20 mL ice cold PBS to remove the blood from the tissues before collecting the organs. When collecting the brain, one half brain can be homogenized in 1 mL QIAzol lysis reagent for RNA isolation using RNeasy lipid tissue mini kit. Genetic characterization of virus replicating in the brain is carried out by sequencing (*see* Subheading 3.5). The other half of the brain can be homogenized in PBS (*see* **Note 19**) and used for assessing the infectious virus titer using plaque assay (*see* Subheading 3.7). The viral titer is calculated using the formula:

$$\text{PFU/g} = \frac{[(\text{plaque count}) \times (\text{dilution factor}) / (\text{amount sample added in mL})]}{(\text{homogenate percent})}$$

3.10 Measurement of IgG and IgM Antibody Responses of Infected Mice

Levels of viral-specific IgM are quantified to evaluate the immediate, early humoral response following infection, while levels of viral-specific IgG are quantified to define the more specific humoral response that occurs later following infection and following reexposure. The section describes measuring and quantifying them through ELISA.

1. To coat plates, add 50 μL of coating buffer containing 10^6 PFU of virus to each well of a 96-well immunoplate (gently tap the plate against the bench to make sure the coating solution completely covers the entire well surface). Seal with saranwrap and incubate at 4 °C overnight.

2. Wash plate once with washing buffer, and add 200 μL of blocking buffer containing FBS. Incubate at 37 °C for 2 h or 4 °C overnight.
3. Wash plate twice with washing buffer, tap against paper towel to remove excess solution, and add 50 μL of sample in blocking buffer into each well. Cover with saranwrap and incubate at 4 °C overnight. The serum can be diluted 1:100 and brain homogenates 1:2 for 10% or 1:4 for 20% homogenates. Day 0 serum sample or blocking buffer alone must be included as a negative control.
4. Wash plate four times with washing buffer, and add 50 μL of secondary antibody (goat anti-mouse IgG-HRP/goat anti-mouse IgM-HRP) in blocking buffer in a dilution 1:1000 at room temperature for 2 h.
5. Wash plate four times with washing buffer, and add 100 μL of BD OptEIA TMB substrate reagent kit (*see* **Note 20**).
6. Stop the reaction by adding 50 μL stop solution to each well. A yellow color will develop, and read the optical density at 450 nm immediately.

4 Notes

1. We used 1 μg of automodified $PARP10^{CD}$ per reaction but performed the automodification in bulk up to 100 μg. The automodification and desalting procedure can be applied to other MARylated substrates.
2. Desalting on the Bio-Rad Micro Bio-Spin 6 column causes sample dilution and loss. We automodified 85 μg of $PARP10^{CD}$ in 100 μL for **step 1** and recovered 50 μg of ^{32}P-labeled $PARP10^{CD}$ in 500 μL after **step 2**. Concentration can be empirically determined by SDS-PAGE and coomassie blue staining with standards of known concentrations.
3. Extensive tutorials of this tool for quantifying band signals can be found online (e.g., http://lukemiller.org/index.php/2010/11/analyzing-gels-and-western-blots-with-image-j/).
4. Design the oligonucleotide primers as follows. The primers should be between 25 and 45 bases in length, with a melting temperature (T_m) of ≥78 °C. The desired mutation should be in the middle of the primer with ~10–15 bases of correct sequence on both sides. Adapted from quick change site-directed manual [https://www.agilent.com/cs/library/usermanuals/Public/200523.pdf].

5. Glycerol stock for the positive clones can be prepared and stored. But it is advisable to retransform the plasmid DNA and grow a bacterial culture from freshly grown culture to obtain a higher yield of plasmid DNA.
6. For sequencing the macrodomain region of CHIKV, CHK nsp3 MDR primer 5′- GGTACCCTCGAGTCATTAGG TCCGCATCTGTATGGCCTC -3′ was used. For sequencing the larger plasmid, at least 100 ng/μL is required. The sequencing primer anneals to the positive strand of the virus.
7. The restriction digestion enzyme for linearizing the plasmid is chosen depending on the restriction site engineered into the cDNA clone. For the 181/25 cDNA clone, Not I site is included at the 3'terminus (*see* Fig. 1).
8. For purification of the plasmid DNA, commercial PCR purification kits can be used. But for larger plasmids, the final yield will be much less for most of the kits, and therefore, the phenol-chloroform extraction followed by alcohol precipitation is advisable. However, the purity of the DNA must be ensured by finding A_{260}/A_{230} ratio which should be greater than 1.8. Lower A_{260}/A_{230} values indicate contamination from solvents (e.g., phenol).
9. Purification with the lithium chloride precipitation solution provided in the mMESSAGE mMACHINE SP6 kit will also yield the same quantity of RNA. Refer to mMESSAGE mMACHINE manual for further standardization [http://kirschner.med.harvard.edu/files/protocols/Ambion_mMessagemMachine.pdf].
10. Warm all cell culture media, PBS, and trypsin-EDTA in a water bath to 37 °C prior to adding to the cells.
11. The primers used for amplifying the nsP3 region of CHIKV in our study were:

    CHIKV nsP3 N1 forward primer

    (5′ AAGGCCGAATTCGGGCACCGTCGTACCGGGT AAAACGCA 3′)

    CHIKV nsP3 full reverse primer

    (5′ GGTACCCTCGAGTCATTACCCACCTGCCCTGT CTAGTC 3′)

    The PCR amplification with this primer set will yield an amplicon of 1.6 kb. The primers are designed to be highly specific and produce a single amplicon only. If other bands appear, further optimization can be carried out by gradient PCR or increasing the annealing temperature. Keep a nontemplate control for the PCR reaction.
12. The plaque assay for CHIKV uses Vero cells. For Sindbis virus, we use BHK-21 cells instead. The methodologies mentioned in **steps 1** and **2** are the same. The plaques in BHK-21 are

more clearly visible, when staining the live cells using 1:10 dilution of 0.33% neutral red in DPBS. Add 1 mL of diluted neutral red on the top of the agar overlay and incubate at 37 °C in a 5% CO_2 incubator for 2 h and count the plaques using a light box. For neutral red staining, the plaques should be counted immediately or/within a period of 2–3 h.

13. The cells should be 90–95% confluent with few spaces at the time of infection. Usually 1 × 10^5 cells were seeded overnight to obtain a monolayer the next day. The incubation time for culturing should not go beyond 24 h.
14. The 1.2% Bacto agar is autoclaved and stored at room temperature. The melted Bacto agar will solidify when cooled to room temperature. Therefore, take the overlay agar mixture from the 42 °C water bath just prior to use.
15. The mouse models differ for each strain of virus. To study the virulence of wild type versus mutants for CHIKV, we used 2-day old CD-1 pups. The same kind of virulence study for the TE strain of Sindbis virus can be done in 2-week old CD-1 mice.
16. Verify that the mice are adequately anesthetized by gently pinching the toes. If the animal withdraws their paw, anesthetize for few more minutes till a deep plane of sleep has been attained.
17. Place pups on a thermal mat to avoid a drop-in body temperature during anesthesia, and monitor until they recover.
18. Keep separate litters for morbidity/mortality studies and for a time course infection experiment involving the collection of organs and serum.
19. We use lysing matrix tubes with lysing matrix A (orange-capped tube having garnet matrix and ¼ ceramic sphere) for homogenizing the tissues for protein and tubes with lysing matrix D (green-capped tubes having 1.4 mm ceramic spheres) for homogenizing the tissues for RNA in QIAzol reagent using a MP FastPrep-24 homogenizer.
20. Make sure to take out the BD OptEIA TMB substrate kit to room temperature, at least 30 min prior to use. Combine equal amounts of reagent A and reagent B no more than 15 min prior to adding to the plates.

References

1. Schwartz O, Albert ML (2010) Biology and pathogenesis of chikungunya virus. Nat Rev Microbiol 8(7):491–500
2. Burt FJ, Chen W, Miner JJ, Lenschow DJ, Merits A, Schnettler E, Kohl A, Rudd PA, Taylor A, Herrero LJ, Zaid A et al (2017) Chikungunya virus: an update on the biology and pathogenesis of this emerging pathogen. Lancet Infect Dis 17(4):e107–e117
3. Strauss JH, Strauss EG (1994) The alphaviruses: gene expression, replication, and evolution. Microbiol Rev 58(3):491–562

4. Rack JG, Perina D, Ahel I (2016) Macrodomains: structure, function, evolution, and catalytic activities. Annu Rev Biochem 85:431–454
5. Gorbalenya AE, Koonin EV, Lai MM (1991) Putative papain-related thiol proteases of positive-strand RNA viruses. Identification of rubi- and aphthovirus proteases and delineation of a novel conserved domain associated with proteases of rubi-, alpha- and coronaviruses. FEBS Lett 288(1–2):201–205
6. Koonin EV, Gorbalenya AE, Purdy MA, Rozanov MN, Reyes GR, Bradley DW (1992) Computer-assisted assignment of functional domains in the nonstructural polyprotein of hepatitis E virus: delineation of an additional group of positive-strand RNA plant and animal viruses. Proc Natl Acad Sci U S A 89(17):8259–8263
7. McPherson RL, Abraham R, Sreekumar E, Ong SE, Cheng SJ, Baxter VK, Kistemaker HA, Filippov DV, Griffin DE, Leung AK (2017) ADP-ribosylhydrolase activity of Chikungunya virus macrodomain is critical for virus replication and virulence. Proc Natl Acad Sci 114(7):1666–1671
8. Kümmerer BM, Grywna K, Gläsker S, Wieseler J, Drosten C (2012) Construction of an infectious Chikungunya virus cDNA clone and stable insertion of mCherry reporter genes at two different sites. J Gen Virol 93(9):1991–1995
9. Scholte FE, Tas A, Martina BE, Cordioli P, Narayanan K, Makino S, Snijder EJ, van Hemert MJ (2013) Characterization of synthetic Chikungunya viruses based on the consensus sequence of recent E1-226V isolates. PLoS One 8(8):e71047
10. Tsetsarkin K, Higgs S, McGee CE, Lamballerie XD, Charrel RN, Vanlandingham DL (2006) Infectious clones of Chikungunya virus (La Reunion isolate) for vector competence studies. Vector-Borne Zoonotic Dis 6(4):325–337
11. Tsetsarkin KA, Vanlandingham DL, McGee CE, Higgs S (2007) A single mutation in chikungunya virus affects vector specificity and epidemic potential. PLoS Pathog 3(12):e201
12. Tsetsarkin KA, McGee CE, Volk SM, Vanlandingham DL, Weaver SC, Higgs S (2009) Epistatic roles of E2 glycoprotein mutations in adaption of chikungunya virus to Aedes albopictus and Ae. aegypti mosquitoes. PLoS One 4(8):e6835
13. Vanlandingham DL, Tsetsarkin K, Hong C, Klingler K, McElroy KL, Lehane MJ, Higgs S (2005) Development and characterization of a double subgenomic chikungunya virus infectious clone to express heterologous genes in Aedes aegypti mosqutioes. Insect Biochem Mol Biol 35(10):1162–1170
14. Plante K, Wang E, Partidos CD, Weger J, Gorchakov R, Tsetsarkin K, Borland EM, Powers AM, Seymour R, Stinchcomb DT, Osorio JE (2011) Novel chikungunya vaccine candidate with an IRES-based attenuation and host range alteration mechanism. PLoS Pathog 7(7):e1002142
15. Gorchakov R, Wang E, Leal G, Forrester NL, Plante K, Rossi SL, Partidos CD, Adams AP, Seymour RL, Weger J, Borland EM (2012) Attenuation of Chikungunya virus vaccine strain 181/clone 25 is determined by two amino acid substitutions in the E2 envelope glycoprotein. J Virol 86(11):6084–6096
16. Wang E, Weaver SC, Frolov I (2011) Chimeric Chikungunya viruses are nonpathogenic in highly sensitive mouse models but efficiently induce a protective immune response. J Virol 85(17):9249–9252
17. Tretyakova I, Hearn J, Wang E, Weaver S, Pushko P (2014) DNA vaccine initiates replication of live attenuated chikungunya virus in vitro and elicits protective immune response in mice. J Infect Dis 209:1882–1890
18. Levitt NH, Ramsburg HH, Hasty SE, Repik PM, Cole FE, Lupton HW (1986) Development of an attenuated strain of chikungunya virus for use in vaccine production. Vaccine 4(3):157–162
19. Cashman NR, Durham HD, Blusztajn JK, Oda K, Tabira T, Shaw IT, Dahrouge S, Antel JP (1992) Neuroblastoma× spinal cord (NSC) hybrid cell lines resemble developing motor neurons. Dev Dyn 194(3):209–221

Chapter 22

Monitoring the Sensitivity of T Cell Populations Towards NAD$^+$ Released During Cell Preparation

Björn Rissiek, Marco Lukowiak, Friedrich Haag, Tim Magnus, and Friedrich Koch-Nolte

Abstract

Mouse T cells express the toxin-related ecto-ADP-ribosyltransferase ARTC2 that catalyzes the posttranslational ADP-ribosylation of cell surface proteins by transferring the ADP-ribose group of its substrate nicotinamide adenine dinucleotide (NAD$^+$) to arginine residues of its target proteins. One well known target of ARTC2 is the ATP-gated P2X7 ion channel. ADP-ribosylation of P2X7 induces gating of the channel, calcium influx, ecto-domain shedding, phosphatidylserine externalization, and finally cell death. Previous studies have shown that the ARTC2 substrate NAD$^+$ is released during T cell preparation. Since P2X7 is differentially expressed among T cell subpopulations, preparation-related ADP-ribosylation has a strong impact on the vitality of T cells that express high levels of P2X7. With this chapter we provide a protocol to monitor the consequences of preparation-related P2X7 ADP-ribosylation on T cells using regulatory T cells as generic T cell subpopulation known to express high levels of P2X7. However, this protocol could be easily adapted to other T cell populations.

Key words ARTC2, P2X7, NAD-induced cell death, T cell preparation, s+16a, Nanobody

1 Introduction

Mouse T cells express the mammalian ecto-ADP-ribosyltransferase ARTC2 that utilizes extracellular nicotinamide adenine dinucleotide (NAD$^+$) as substrate to covalently modify arginine residues of various cell surface proteins by attaching an ADP-ribose group. In vivo, sites of inflammation are enriched in extracellular NAD$^+$, as dying cells can release NAD$^+$ as a danger molecule [1]. Interestingly, NAD$^+$ is also released during the preparation of primary mouse T cells from lymphatic organs leading to the ADP-ribosylation of multiple cell surface proteins. One well-characterized target protein of ARTC2 is the adenosine triphosphate (ATP)-gated P2X7 ion channel. Activation of P2X7 by binding of the soluble ligand ATP leads to the opening of the ion channel and the influx of Ca^{2+} ions. Similarly, ARTC2-mediated ADP-ribosylation

Paul Chang (ed.), *ADP-ribosylation and NAD+ Utilizing Enzymes: Methods and Protocols*, Methods in Molecular Biology, vol. 1813, https://doi.org/10.1007/978-1-4939-8588-3_22,

provides a covalent ligand that initiates P2X7 channel opening and induces a sustained influx of Ca^{2+} ions [2]. This further activates cell surface metalloproteases, which trigger ecto-domain shedding of various cell surface proteins such as CD27, CD62L, or ARCT2 itself [3–5] as well as the externalization of phosphatidylserine (PS) on the outer leaflet of the plasma membrane [6] within a few minutes after P2X7 activation. Long-term P2X7 activation leads to cell death that could be visualized by the uptake of propidium iodide (PI). Many of the cellular consequences of ARTC2-mediated P2X7 activation, such as ecto-domain shedding, externalization of PS, or the uptake of PI, can be easily monitored by flow cytometry.

ARTC2 is highly expressed on most splenic T cells in naïve mice. In contrast, P2X7 is differentially expressed among spleen T cell subpopulations [7]. Recently, we identified regulatory T cells (Tregs) and natural killer T cells (NKTs) to express high levels of P2X7, whereas conventional T helper cells or cytotoxic T cells express P2X7 at lower levels. Our previous studies have shown that NAD^+ released during cell preparation is sufficient to ADP-ribosylate P2X7 on Tregs and NKTs, even when cells are prepared on ice and are kept throughout at 4 °C [8, 9]. Activation of P2X7, however, needs temperatures above 30 °C. Therefore, ecto-domain shedding and PS flashing only occur after bringing cells back to 37 °C, e.g., for in vitro functional assays or adoptive transfer experiments.

In this chapter we provide a protocol to monitor the cellular consequences of P2X7 ADP-ribosylation by flow cytometry in order to identify T cell populations sensitive to NAD^+ released during cell preparation. We further demonstrate how ADP-ribosylation of T cell surface proteins can be prevented by injection of an ARTC2-blocking nanobody into mice prior to harvesting of T cell bearing organs.

2 Materials

2.1 Preparation of Mouse Spleen T Cells

1. Scissors.
2. Forceps.
3. Preparation board.
4. 70% ethanol.
5. 50 mL falcon tubes.
6. 15 mL falcon tubes.
7. 70 μM cell strainers
8. 10 cm petri dishes.
9. 5 mL syringe piston.
10. 10 mL serological pipettes.
11. Pipette boy.

12. Lab centrifuge.
13. ACK erythrocyte lysis buffer: 155 mM NH_4Cl + 10 mM $KHCO_3$ + 0.1 mM EDTA at pH 7.2.
14. Neubauer counting chamber.
15. Lab microscope.
16. FACS buffer: PBS pH 7.4 without calcium and magnesium, add 1 mM EDTA and 0.5% bovine serum albumin (BSA).
17. Multicolor flow cytometer: Use a BD FACSCanto II or any other flow cytometer that is equipped with sufficient channels to analyze the desired T cell population toward ecto-domain shedding or PS flashing/PI uptake.

2.2 Antibodies for Flow Cytometry

1. For the analyses of spleen $CD4^+$ regulatory T cells: anti-mouse CD4-Bv421 (clone RM4–4), anti-mouse CD25-PE (clone PC61), and anti-mouse CD27-APC (clone LG.3A10). Fluorochrome-conjugated antibodies usually come at a concentration of 0.2 mg/mL and are used at a final concentration of 2 μg/mL (1:100 dilution).
2. Alternatively, Foxp3-GFP reporter mice such as the DEREG mice [10] could be used to identify Tregs by the expression of GFP under control of the transcription factor Foxp3.
3. For detection of phosphatidylserine externalization and cell death: fluorochrome-labeled Annexin V (5 μL per sample, read manufactures instruction) can be used for the detection of PS externalization and propidium iodide (PI) at a final concentration of 1 μg/mL for the detection of cell death.

2.3 For 37 °C Incubations

1. Incubation Medium: RPMI + 10% fetal calf serum (FCS).
2. PS-buffer: 1.4 M NaCl + 25 mM $CaCl_2$ + 0.1 M HEPES 0.1 M at pH 7.4 in H_2O.

2.4 Blockade of ADP-Ribosylation during Cell Preparation

1. ARTC2-blocking nanobody clone s+16a as monomer or as Fc-fusion protein.
2. Solvent for s+16a injections: NaCl.
3. 1 mL syringe.
4. IR-light source.

3 Methods

The following protocol illustrates how to monitor the NAD^+-mediated activation of P2X7 on T cells by using CD27 ecto-domain shedding (Fig. 1a) or externalization of PS and propidium uptake (Fig. 2a) as readout for P2X7 activation. This approach can be used to identify T cell populations that are affected by NAD^+ released during cell preparation. Here, we use spleen regulatory

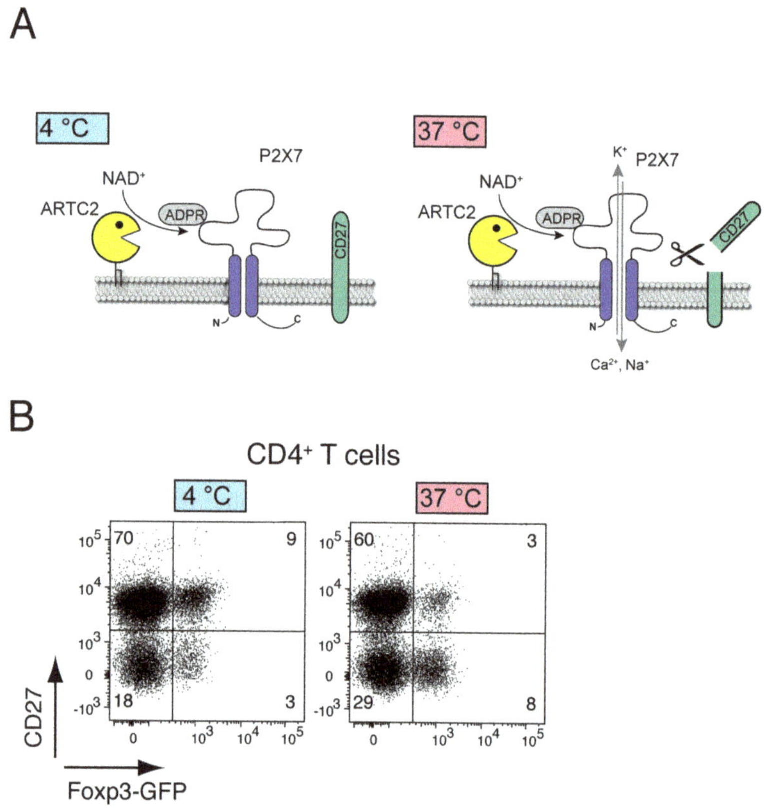

Fig. 1 ADP-ribosylation of P2X7 induces ecto-domain shedding at 37 ° C. (**a**) ARTC2 uses NAD+ release from dying cells during cell preparation to ADP-ribosylate P2X7 on spleen Tregs, even when cells are prepared on ice. Gating of P2X7 occurs when cells are brought back to 37 °C, triggering the shedding of CD27. (**b**) CD27 shedding on Tregs can be monitored by flow cytometry. Spleen cells from Foxp3-GFP reporter mice were incubated at 4 °C or 37 °C for 15 min and stained with anti-CD4 and anti-CD27. Total CD4+ T cells were included in the analyses. Incubation at 37 °C triggers shedding of CD27 on a large fraction of Foxp3-GFP+ Tregs

T cells (Tregs) as a known NAD+ sensitive T cell population. This protocol, however, could be easily adapted to other T cell populations from spleen, other lymphatic organs, or even nonlymphatic organs by adding fluorochrome-conjugated antibodies specific for other T cell subpopulations (*see* **Note 1**). Bring all buffers and solutions to 4 °C prior to use; otherwise ADP-ribosylation-induced P2X7 activation occurs immediately.

3.1 Preparation of Spleen T Cells

Lymphatic organs of the mouse, such as spleen and lymph nodes, are enriched in T cells and are therefore often used as T cell source for in vitro analyses. These cells are usually mechanically released from the lymphatic tissue, a process during which cell surface ADP-ribosylation occurs due to the release of NAD+ from dying cells.

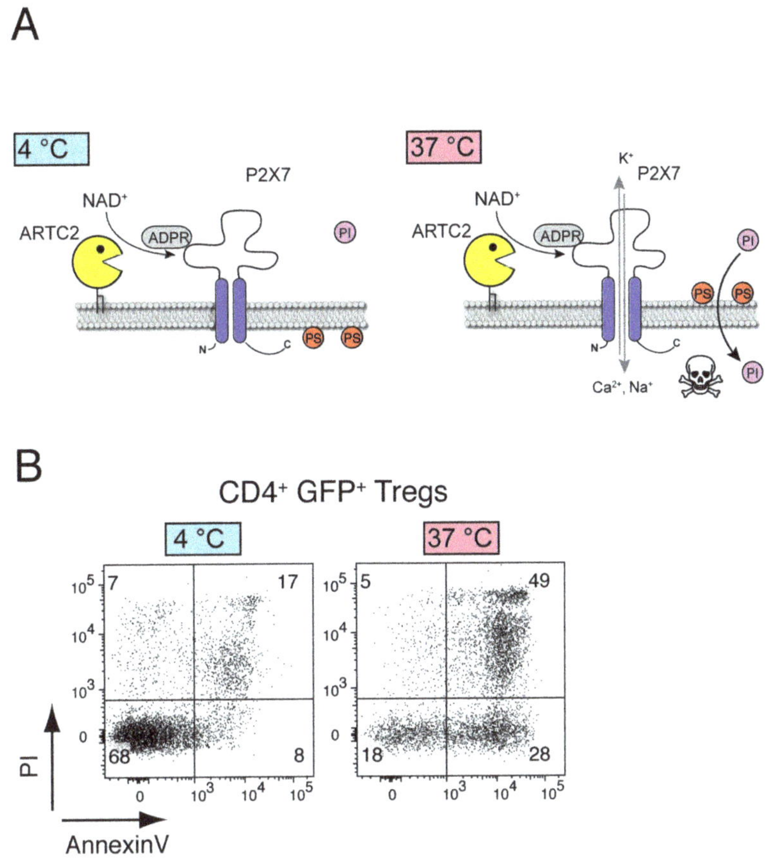

Fig. 2 ADP-ribosylation of P2X7 induces PS flashing and PI uptake at 37 °C. (**a**) ARTC2 uses NAD+ release from dying cells during cell preparation to ADP-ribosylate P2X7 on spleen Tregs, even when cells are prepared on ice. Gating of P2X7 occurs when cells are brought back at 37 °C, triggering the externalization of PS (PS flashing) and the uptake of PI. (**b**) PS flashing and PI uptake in Tregs can be monitored by flow cytometry. Spleen cells from Foxp3-GFP reporter mice were incubated at 4 °C or 37 °C for 60 min in the presence of 1 μg/mL PI. Cells were then stained with anti-CD4 and Annexin V. Total CD4+Foxp3-GFP+ Tregs were included in the analyses. Incubation at 37 °C triggers PS flashing and PI uptake in a large fraction of Tregs

1. Sacrifice the mouse by cervical dislocation, and clean the complete body with 70% ethanol. Fix the mouse head to the right and tail to the left on the preparation board, and use the scissors to make a 2 cm incision on the left flank. Open the abdominal cavity, and carefully take out the spleen, which is located behind the stomach.
2. Put the spleen in a 50 mL falcon tube with 10 mL of cold FACS buffer, and store it on ice. Place a 70 μm cell strainer into a 10 cm petri dish, and add 10 mL of FACS buffer. Take the spleen out of the falcon tube, and place it into the cell

strainer in the petri dish. Now, softly grind the spleen through the cell strainer by using the piston of a 3 mL syringe.

3. Use a 10 mL serological pipette and a pipette boy to collect the cell suspension, and filter it a second time through the cell strainer. Transfer the filtered cell suspension to a 15 mL falcon tube, and centrifuge at 300 × *g* in a lab centrifuge for 5 min at 4 °C.
4. Discard the supernatant, resuspend the cell pellet in 5 mL ACK erythrocyte lysis buffer, and incubate the cell solution for 5 min on ice. Add 10 mL of FACS buffer, and centrifuge the cells at 300 × *g* in a lab centrifuge for 5 min at 4 °C.
5. Resuspend the cell pellet in 5 mL FACS buffer, and count the cells using a Neubauer counting chamber and a lab microscope. The number of cells should be adjusted to 1×10^7/mL by adding FACS buffer.

3.2 Flow Cytometric Analysis of P2X7 Activation Induced ecto-domain Shedding on Regulatory T Cells

One immediate consequence of P2X7 activation on T cells is the mobilization of tumor necrosis factor-α-converting enzyme (TACE, ADAM17) to the plasma membrane. There, TACE cleaves off the ecto-domain of various cell surface proteins such as CD27 and CD62L. This process could be used as surrogate indicator of P2X7 ADP-ribosylation on T cells during cell preparation.

1. Transfer 1×10^6 cells (100 μL of your cell suspension) into two polystyrene FACS tube, and wash the cells with 2 mL FACS buffer by centrifugation at 300 × *g* for 5 min at 4 °C.
2. Resuspend the cells in 200 μL of cold incubation medium (RPMI + 10% FCS). Keep one FACS tube at 4 °C on ice, and incubate the other one for 15 min at 37 °C in a water bath.
3. Prepare an antibody cocktail to stain cells, including anti-CD4, anti-CD25, and anti-CD27 in FACS buffer, and keep it on ice in the dark.
4. Wash both samples once with 2 mL FACS buffer by centrifugation at 300 × *g* for 5 min at 4 °C.
5. Resuspend the two samples with 100 μL of antibody cocktail, and incubate them for 30 min in the fridge.
6. Wash both samples 2× with 2 mL FACS buffer by centrifugation at 300 × *g* for 5 min at 4 °C.
7. Resuspend the samples in 200 μL FACS buffer, and compare CD27 cell surface on $CD4^+CD25^+$ Tregs of the 4 °C and the 37 °C sample using a flow cytometer (*see* **Note 2**).
8. T cells sensitive to NAD^+ released during cell preparation, such as Tregs, exhibit a strong loss of CD27 upon 37 °C incubation (*see* Fig. 1b, **Note 3**).

3.3 Flow Cytometric Analysis of Phosphatidylserine (PS) Externalization and Propidium Iodide (PI) Uptake on Regulatory T Cells

A hallmark of early T cell apoptosis is the externalization of PS on the outer leaflet of the plasma membrane. Activation of P2X7 on T cells induces rapid PS externalization and ultimately cell death, which could be monitored by the uptake of PI. Like ecto-domain shedding, this process could be used as indicator to measure ADP-ribosylation of P2X7 on T cells during cell preparation (*see* **Note 4**).

1. Transfer 1×10^6 cells (100 μL of your cell suspension) into two polystyrene FACS tube, and wash the cells with 2 mL FACS buffer by centrifugation at $300 \times g$ for 5 min at 4 °C.
2. Resuspend the cells in 200 μL of cold incubation medium containing 1 μg/mL PI. Keep one FACS tube at 4 °C on ice, and incubate the other one for 1 h at 37 °C in a standard cell culture incubator.
3. Prepare a staining solution, containing anti-CD4, anti-CD25, and Annexin V in PS-buffer, and keep it on ice in the dark.
4. Wash both samples once with 2 mL PS-buffer by centrifugation at $300 \times g$ for 5 min at 4 °C.
5. Resuspend the two samples with 100 μL of antibody cocktail, and incubate them for 30 min in the fridge.
6. Wash both samples 2× with 2 mL PS-buffer by centrifugation at $300 \times g$ for 5 min at 4 °C.
7. Resuspend the samples in 200 μL PS-buffer, and compare the PS externalization and the PI uptake in $CD4^+CD25^+$ Tregs of the 4 °C and the 37 °C sample using a flow cytometer.
8. T cells sensitive to NAD^+ released during cell preparation, such as Tregs, exhibit strong PS externalization and PI uptake upon 37 °C incubation (*see* Fig. 2b, **Note 2**).

3.4 Blocking ARTC2-Mediated ADP-Ribosylation of P2X7 in T Cells During Cell Preparation by Injection of Nanobody S+16a

The ADP-ribosylation of cell surface proteins during T cell preparation leads to unwanted and often unrecognized side effects. ADP-ribosylation of P2X7, which is highly expressed on some T cell populations such as Tregs, can affect cell vitality, especially if cells are used in downstream applications that are carried out at 37 °C. To prevent this, the ARTC2-blocking nanobody s+16a could be injected into mice prior to harvesting of lymphatic organs (*see* Fig. 3a, **Note 5**).

1. Dilute 50 μg of nanobody s+16a monomer in 100 μL NaCl to prepare the injection solution.
2. Prewarm the mice by keeping them for 2–5 min under IR light to facilitate IV injection.
3. Inject 100 μL of the injection solution in the lateral tail vain of the mouse. Inject a second mouse with 100 μL NaCl as control. Allow the s+16a nanobody to circulate for 30 min. Then sacrifice the animals by cervical dislocation.

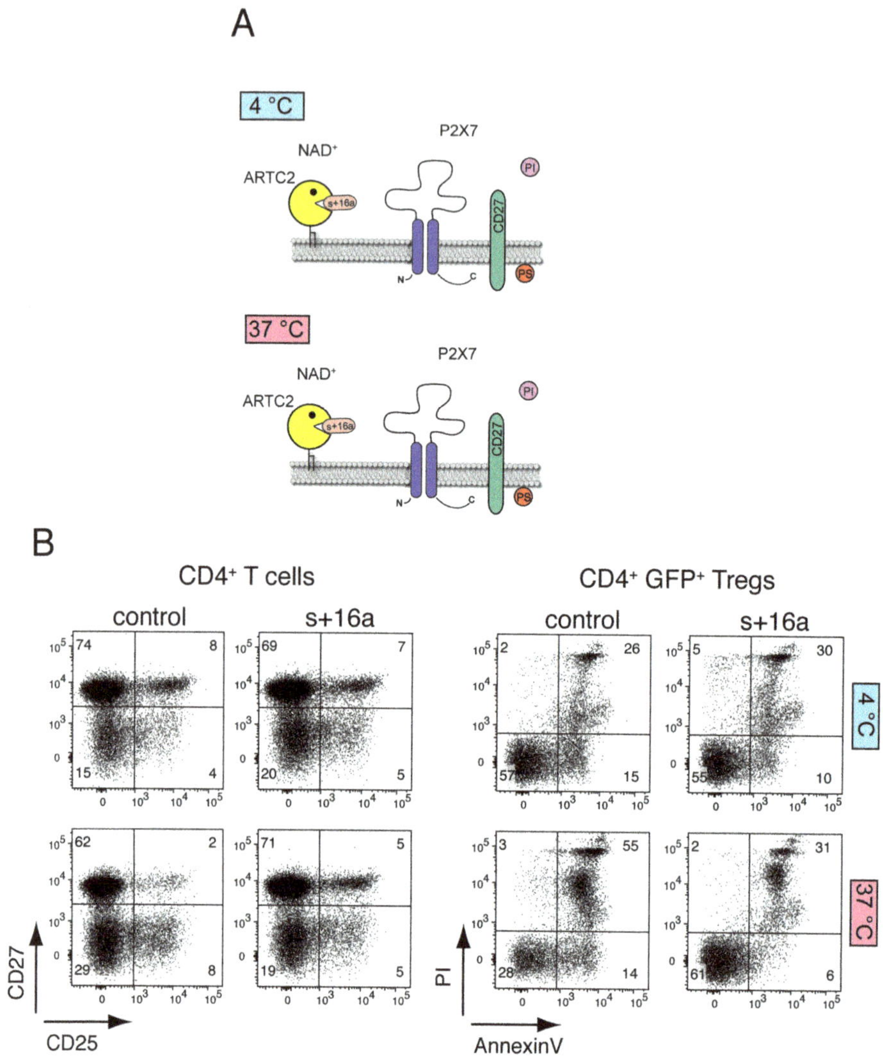

Fig. 3 Injection of the ARTC2 blocking nanobody s+16a prevents ADP-ribosylation of P2X7 during cell preparation. (**a**) Injection of nanobody s+16a into mice prior to sacrificing blocks the enzymatic activity of ARTC2 during cell preparation at 4 °C. Therefore, P2X7 is not ADP-ribosylated and stays in its closed state during 37 °C incubation. (**b**) Injection of s+16a prevents CD27 shedding on $CD4^+CD25^+$ Tregs as well as PS flashing and PI uptake, when cells are incubated at 37 °C for 15 min or 1 h, respectively

4. Isolate the spleen, as described under Subheading 3.1.
5. Perform ecto-domain shedding and/or PS-flashing/PI uptake analyses as described under Subheadings 3.2 and 3.3, respectively, to monitor P2X7 activation on spleen Tregs upon 37 °C incubation.
6. Injection of s+16a prevents CD27 shedding and PS-flashing/PI uptake during 37 °C incubation and thereby maintains the vitality of spleen Tregs after cell preparation (Fig. 3b).

4 Notes

1. Some phenotypic cell surface markers, e.g., chemokine receptors used to identify other T cell populations, might be internalized during incubation at 37 °C. To monitor NAD+ sensitivity in this case, cells can be isolated by flow cytometric cell sorting and ecto-domain shedding, PS flashing or PI uptake can be monitored directly on the sorted cell populations.
2. Activation of P2X7 can lead to cell shrinkage; therefore, cells might "migrate" out of the FSC/SSC lymphocyte gate. Be aware of that!
3. Similar to CD27, CD62L is a target for P2X7-mediated ecto-domain shedding. CD62L is often used as marker for naïve T cells. ADP-ribosylation of P2X7 and its subsequent activation at 37 °C, especially during the liberation of T cells from solid organs using collagenase digestion, can affect the phenotypic appearance of T cells. Be aware of that!
4. Ecto-domain shedding and PS flashing can be measured in the same tube when using PS-buffer for FACS staining.
5. Isolation of T cells from solid organs, e.g., the intestine or the kidney, usually involves a collagenase digestion step at 37 °C for 15–30 min. This might be sufficient to already induce ecto-domain shedding, PS flashing, or even cell death during the digestion procedure. To test whether this is the case, s+16a nanobody can be injected prior to sacrificing the mouse, and a direct comparison of the s+16a treated mouse and an untreated mouse will give more insight into the NAD^+ sensitivity of the desired T cell population.

References

1. Adriouch S, Hubert S, Pechberty S et al (2007) NAD+ released during inflammation participates in T cell homeostasis by inducing ART2-mediated death of naive T cells in vivo. J Immunol 179:186–194
2. Schwarz N, Fliegert R, Adriouch S et al (2009) Activation of the P2X7 ion channel by soluble and covalently bound ligands. Purinergic Signal 5:139–149. https://doi.org/10.1007/s11302-009-9135-5
3. Gu B, Bendall LJ, Wiley JS (1998) Adenosine triphosphate-induced shedding of CD23 and L-selectin (CD62L) from lymphocytes is mediated by the same receptor but different metalloproteases. Blood 92:946–951
4. Moon H, Na H-Y, Chong KH, Kim TJ (2006) P2X7 receptor-dependent ATP-induced shedding of CD27 in mouse lymphocytes. Immunol Lett 102:98–105. https://doi.org/10.1016/j.imlet.2005.08.004
5. Menzel S, Rissiek B, Bannas P et al (2015) Nucleotide-Induced Membrane-Proximal Proteolysis Controls the Substrate Specificity of T Cell Ecto-ADP-Ribosyltransferase ARTC2.2. J Immunol 195(5):2057–2066. https://doi.org/10.4049/jimmunol.1401677
6. Adriouch S, Ohlrogge W, Haag F et al (2001) Rapid induction of naive T cell apoptosis by ecto-nicotinamide adenine dinucleotide: requirement for mono(ADP-ribosyl)transferase 2 and a downstream effector. J Immunol 167:196–203
7. Rissiek B, Danquah W, Haag F, Koch-Nolte F (2014) Technical advance: a new cell prepara-

tion strategy that greatly improves the yield of vital and functional Tregs and NKT cells. J Leukoc Biol 95:543–549. https://doi.org/10.1189/jlb.0713407

8. Rissiek B, Haag F, Boyer O et al (2015) ADP-ribosylation of P2X7: a matter of life and death for regulatory T cells and natural killer T cells. Curr Top Microbiol Immunol 384:107–126. https://doi.org/10.1007/82_2014_420

9. Hubert S, Rissiek B, Klages K et al (2010) Extracellular NAD+ shapes the Foxp3+ regulatory T cell compartment through the ART2-P2X7 pathway. J Exp Med 207:2561–2568. https://doi.org/10.1084/jem.20091154

10. Lahl K, Loddenkemper C, Drouin C et al (2007) Selective depletion of Foxp3+ regulatory T cells induces a scurfy-like disease. J Exp Med 204:57–63. https://doi.org/10.1084/jem.20061852

Chapter 23

Identifying Target RNAs of PARPs

Florian J. Bock and Paul Chang

Abstract

Posttranscriptional regulation of RNA is an important component of gene expression by controlling the total amount of mRNA available for translation into protein. It involves multiple pathways including nuclear processing of mRNA and its precursors, RNA silencing, and regulation of RNA decay. Poly(ADP-ribose) polymerases (PARPs), enzymes that modify target proteins with ADP-ribose, play important roles in several RNA-regulatory pathways. RNA-binding PARPs target specific transcripts for regulation, and multiple PARPs ADP-ribosylate RNA-regulatory proteins to alter their localization, activity, or RNA binding. Additionally, RNA-binding proteins can bind directly to poly(ADP-ribose) with various effects on their function. Here we describe methods to identify and confirm specific transcripts that are regulated by PARPs.

Key words ADP-ribosylation, RNAseq, RNA, PARP

1 Introduction

The defining feature of the poly(ADP-ribose) polymerase (PARP) family of proteins is the presence of a PARP domain, a protein domain that catalyzes the transfer of ADP-ribose onto target proteins [1, 2]. Enzymatically active PARPs use this domain to modify target proteins with ADP-ribose, using NAD+ as substrate. The PARP family can be further divided into mono- and poly(ADP-ribose) polymerases, the former attaching a single ADP-ribose, whereas the latter attaches multiple units of ADP-ribose, giving rise to poly(ADP-ribose) [3, 4]. Modification of target proteins often alters their function; therefore direct modification is a major mechanism of PARP activity. In addition to a PARP domain, most PARPs contain additional protein domains that have well-characterized functions [2, 3]. These can, for example, target the PARPs to the correct location in the cells, promote protein-protein interaction, or regulate the enzymatic activity of the PARP domain. One category of protein domains found in PARPs is RNA-binding domains, which are present in several PARPs, PARP7, 12, 13, 10, and 14, thus defining them as RNA-binding proteins (RBPs) [2, 5].

Paul Chang (ed.), *ADP-ribosylation and NAD+ Utilizing Enzymes: Methods and Protocols*, Methods in Molecular Biology, vol. 1813, https://doi.org/10.1007/978-1-4939-8588-3_23,

Several of these RNA-binding PARPs are involved in RNA processing; however they have not been shown to directly ADP-ribosylate or otherwise modify their bound RNAs, but instead recruit RNA-regulatory proteins to act on the RNAs. Thus, an additional mechanism of PARP activity functions through direct binding of RNA [5].

Two distinct RNA-binding domains are found in PARPs: the CCCH zinc finger domain and the RNA recognition motif (RRM). PARPs that contain CCCH domains include PARP7, PARP12, and PARP13, while PARP10 and PARP14 contain RNA recognition motifs [5–8]. Proteins containing CCCH domains have been shown to play important roles in RNA processing including splicing, translation, RNA export, transcription, and RNA degradation [9]. For example, PARP13 can target viral as well as cellular RNAs for degradation [10–13], therefore protecting cells from viral infection or modulating sensitivity to programmed cell death. PARP7, 12, and 13 have all been implicated in regulation of the translation of specific transcripts, and PARP7 and 12 have also been shown to play roles in the antiviral response [8, 14–17].

There are two splicing isoforms of PARP13 found in humans, PARP13.2 that is devoid of a PARP domain and PARP13.1 which contains an enzymatically inactive PARP domain lacking key residues for catalysis. PARP13.2 has been shown to play important roles in the cellular antiviral response and is an interferon response gene – its expression becomes highly upregulated upon viral infection and activation of interferon beta [18]. In contrast PARP13.1 is constitutively expressed, and its expression levels are not interferon regulated [18]. For the purpose of this chapter, PARP13 will refer to full-length PARP13.1.

Although PARP13.1 likely plays some role in the antiviral response, its primary functions appear to involve the regulation of cellular RNAs. PARP13.1 has been shown to regulate the stability of cellular RNAs including TRAILR4, playing an important role in the cellular response to TRAIL-mediated apoptosis [8, 13, 19]. It has also been shown to regulate LINE1 mRNA expression and hence is involved in LINE1-mediated transposition [12, 20]. PARP13.1 regulates the stability of TRAILR4 mRNA by binding to its 3'UTR and recruiting decay factors such as the exosome complex [8, 13]. Although how PARP13.1 regulates the stability of other mRNA targets identified has not been studied, it likely includes this mechanism of decay complex recruitment [10, 21].

One major obstacle to our understanding of the function of RNA-binding PARPs is that very few direct RNA targets have been identified. In this chapter we describe a method to identify RNA targets of RNA-binding PARPs. Although we focus on PARP13 and HeLa cells, these methods are applicable to any PARP that binds RNA and any cell line in which knockout clonal cell lines can be generated.

This approach uses high-throughput RNA sequencing to generate multiple transcriptome profiles from different samples which are then compared using pair-wise analysis to identify specific transcripts that are up- or downregulated (Fig. 1) [22–26]. This transcriptome analysis is performed on wild-type parental HeLa cells, PARP13 knockout cell lines derived from parental HeLa cells, and PARP13 knockout cell lines where PARP13 expression is rescued (Fig. 2) [5, 13]. These rescue cell lines reexpress either PARP13 or a PARP13 mutant where key amino acid residues required for RNA binding have been mutated (Y72A, Y108A, F144A, H176A, R189A) (PARP13^{q}) [13]. The levels of PARP13 RNA targets will be upregulated in PARP13 knockout cells relative to wild-type cells since PARP13 is no longer targeting them for degradation. However, not all RNAs that are upregulated upon PARP13 knockout are direct targets of the protein since indirect targets can become misregulated as a downstream consequence of PARP13 knockout. To identify direct targets, we use a rescue approach (Fig. 2). Reexpression of wild-type PARP13 results in the rescue of wild-type levels of the protein, and therefore the expression of direct RNA targets of PARP13 decreases relative to knockout cell lines and is more similar to wild-type cellular levels. In contrast expression of PARP13^{q} should not affect the levels of these PARP13 RNA targets relative to PARP13 knockout cells since these direct RNA targets require binding to the PARP13 RNA-binding domain (Fig. 2). Thus, using this approach, transcripts whose expression levels are regulated by direct binding to PARP13 are identified by pair-wise comparison of the transcriptomes of the wild-type, knockout, and rescue cell lines (Figs. 1 and 2). Direct targets will be upregulated in PARP13 knockouts relative to wild-type cells, downregulated in PARP13 rescue cells relative to PARP13 knockouts, and unchanged in PARP13 RNA-binding mutants relative to PARP13 knockouts (Fig. 2).

2 Materials

2.1 Cell Lines and Cell Culture Reagents

1. Exponentially growing human HeLa cells that were used to generate PARP13 knockout cells.
2. Clonal PARP13 knockout cell lines generated from HeLa cells (*see* **Note 1**).
3. Humidified 37°C incubator with 5% CO_2.
4. Complete media; Dulbecco's Modified Eagle Medium (with 4.5 g/L glucose) supplemented with 10% fetal bovine serum and 2 mM glutamine.
5. Phosphate-buffered saline (PBS).
6. Trypsin-EDTA.

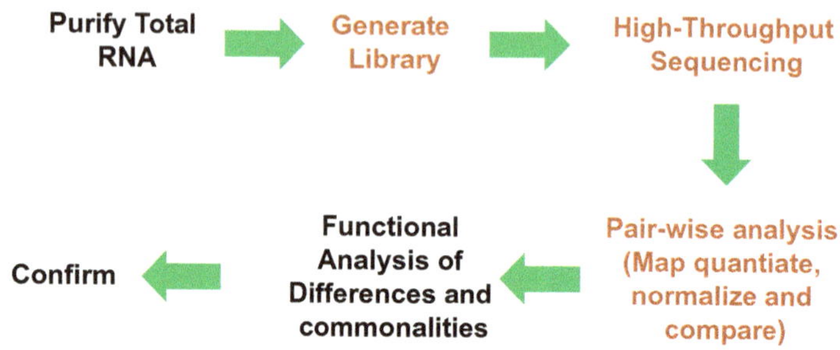

Fig. 1 General outline of RNA sequencing approach. Steps performed by bioinformatics and sequencing facilities are shown in orange

7. Cell counting device of choice.
8. Trypan blue.
9. 10 cm tissue culture plates.
10. 6-well tissue culture plates.

2.2 Transfections

1. DNA expression constructs for mammalian cells encoding the following: an N-terminal fusion of streptavidin-binding protein (SBP) to full-length PARP13 (SBP-PARP13); an N-terminal fusion of SBP to full-length PARP13 containing 5 mutations in the CCCH domain that prevent RNA binding (Y72A, Y108A, F144A, H176A, R189A) (PARP13^{q}) (*see* **Note 2**).
2. Lipofectamine 2000 Transfection Reagent.
3. Opti-MEM media.
4. Hoechst 33342 (10 mg/mL).

2.3 Immunoblot

1. NuPAGE Novex Bis-Tris mini gel (*see* **Note 3**).
2. NuPAGE MOPS SDS running buffer.
3. NuPAGE LDS sample buffer (4×).
4. NuPAGE reducing agent (10×).
5. Protein sample buffer; PBS containing NuPAGE LDS sample buffer diluted 4× and NuPAGE reducing agent diluted 10×. NuPAGE reducing agent should be added last minute.
6. XCell SureLock mini-cell electrophoresis system gel box.
7. XCell II Blot module.
8. Broad range protein molecular weight standards.
9. NuPAGE transfer buffer (20×).
10. Methanol.
11. Novex precut membrane/filter paper sandwiches.
12. TBS-T: 0.05% (v/v) Tween-20 in TBS.
13. Ponceau S solution: 0.05% (w/v) Ponceau S, 1% acetic acid, in H_2O.

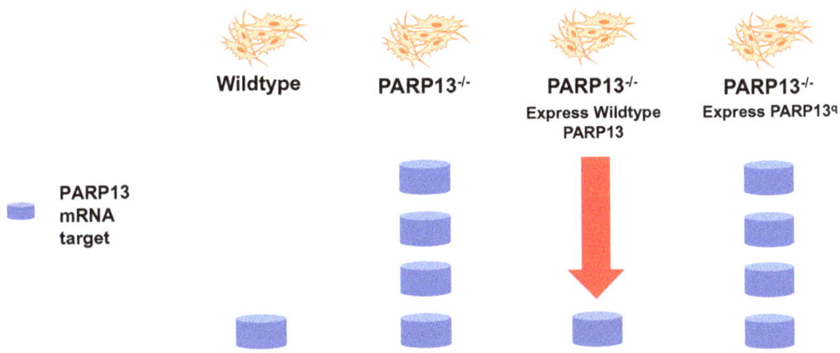

Fig. 2 The rescue approach to identify PARP13 RNA targets. The levels of PARP13 RNA targets, shown in blue, are upregulated in PARP13 knockout cells (PARP13$^{-/-}$) relative to wild-type cells. Reexpression of wild-type PARP13 in PARP13 knockout cells results in decreased expression of direct RNA targets relative to PARP13$^{-/-}$. Expression of PARP13^{q} does not affect the expression levels of PARP13 RNA targets relative to PARP13$^{-/-}$

14. Blocking solution: 5% nonfat dry milk powder (w/v) in TBS-T.
15. Antibodies diluted in milk: anti-SBP antibody, anti-PARP13 antibody, loading control of choice.
16. ECL Western blotting substrate.
17. Luminescence imaging device.

2.4 RNA Purification

1. RNeasy total RNA purification kit or similar kit, alternatively TRIzol or similar products, can be used.
2. Nuclease-free ddH_2O.
3. Spectrophotometer to measure absorbance at 230 nm, 260 nm, and 280 nm.

2.5 Transcriptome Analysis

An RNA sequencing facility with the ability to perform multiplexed high-throughput RNA sequencing with a minimum of ten million reads and QC steps including analysis of RNA integrity with a bioanalyzer for quality control of the input RNA. Furthermore, a specialist familiar with the analysis of RNA sequencing data.

3 Methods

The purpose of this protocol is to identify RNA targets of RNA-binding PARPs using PARP13 as an example. A general outline of the approach is shown in Fig. 1.

Key steps involve the expression of wild-type and RNA-binding mutant PARP13 (rescue) in PARP13 knockout cells, confirming that expression levels of the reexpressed proteins are similar in each sample and confirming that a known target of PARP13, TRAILR4 mRNA, is changed accordingly in response to PARP13 expression. Once these steps are verified, the total RNA is used to generate a library for high-throughput sequencing, the library is sequenced, the sequences mapped to a known transcriptome, and the sequence reads quantitated. The data is now ready for pair-wise analysis to identify specific transcripts that are up or down regulated between two samples. Although pathway analysis is a critical component of this, we will not describe these steps here since many excellent reviews are available [27–29].

3.1 Cell Culture

RNA sequencing provides a wealth of quantitative data. One major challenge involves identifying which data is relevant as this technique is so sensitive that minor differences between samples irrelevant to an experiment can be detected. This can have a major impact on the interpretation of the generated data, particularly during the process of pathway analysis post sequencing. Thus irrelevant pathways can be misidentified as biologically relevant, and often these minor experimental differences arise from tissue culture technique or flawed experimental design. It is therefore critical to minimize differences in the treatment of the cells wherever possible. For example, the media used to passage the cells should always be the same, the frequency of passaging the cells identical, and if cells are transfected, control transfections should be used.

It is also critical to start with high-quality, early split cell lines and to practice rigorous tissue culture techniques. When generating knockout cell lines, it is a good practice to save multiple early splits of the knockouts and the parental cell lines and pair the splits so that each passage of the knockout has a matched parental split. When performing the experiments thaw early splits of all cells at least 3–4 passages prior to collecting RNA.

Cells must be healthy and not contaminated, in log-linear growth, and should have no less than 90% viability as assayed by trypan blue at the time of RNA purification.

1. For all steps requiring media, prewarm to 37 C. Make sure to use standard sterile technique.
2. Thaw PARP13 knockout cells and matched split parental HeLa cell lines into pre-complete media in 10 cm tissue culture plates.
3. Allow cells to adhere, and then replace with fresh complete media.
4. 24 h later passage cells by aspirating media and washing with PBS.
5. Replace PBS with trypsin-EDTA. Add enough prewarmed trypsin-EDTA to cover the cells.
6. Observe cells under tissue culture light microscope. When cells are rounded, tap the dish to detach.

7. Resuspend cells in fresh complete medium. Add at least double the volume of the trypsin-EDTA solution to inactivate trypsin.
8. Transfer cells to a fresh plate and add complete media to a total volume of 10 mL.
9. Passage cells in this manner twice. Confirm viability of the cells using a cell counting device and trypan blue viability assay (*see* **Note 4**).
10. For each cell line, generate growth curves at different seeding densities to identify optimal passaging conditions for log-linear growth.
11. Passage each cell line at the correct seeding density every 2 days for a minimum of three passages prior to RNA purification.

3.2 Transfection of Rescue Constructs

As noted in Subheading 3.1, it is critical to choose a transfection protocol that is least disruptive to the cell in order to minimize changes to the transcriptome that are transfection dependent. The protocol described below is designed to minimize the amount of time the transfection mixture is on the cells while maximizing transfection efficiency and protein expression. This protocol was developed for HeLa cells and for expression of PARP13 protein. When using other cell types and expression vectors, the optimal transfection conditions need to be determined for that cell type prior to commencing an RNA purification and sequencing project. Transfection controls need to be included, as they are of critical importance and should always be performed in parallel for knockout and parental cell lines.

In our experience, the transfection efficiency of Lipofectamine 2000 appears to vary from batch to batch, and each DNA prep can result in differences in protein expression. In addition, cell lines obtained from different sources often exhibit differences in transfection efficiency. It is therefore important to titrate the amount of DNA and Lipofectamine 2000 needed for ideal protein expression for each expression vector whenever new transfection reagent is purchased, a new DNA prep is made, or a new aliquot of cells thawed. Protocols for DNA purification are described on the manufacturer's website and are dependent on the specific kit being used.

1. For all steps requiring media, prewarm to 37 C. Make sure to use standard sterile technique.
2. First determine the amount of DNA required to express optimal amounts of PARP13 for each expression vector in PARP13 knockout cells and wild-type HeLa parental cell lines.
3. Use one 6-well plate for each transfection reaction and cell type. Confirm that the cells are in logarithmic growth phase

and that viability is 90% or above using trypan blue the day before transfection.

4. Split the cells into a 6-well plate containing fresh complete media the evening prior to transfection. Cells should be at 75% confluence the day of transfection.
5. In this protocol we test expression of SBP, SBP-PARP13 and SBP-PARP13q, requiring a total of 12 reactions per cell line. SBP serves as an empty vector control.
6. For each transfection reaction, prepare the following in batch: dilute 5 μL Lipofectamine 2000 reagent into 245 μL of Opti-MEM per reaction. Mix gently.
7. For each expression vector, assemble 4 separate dilutions of DNA. Use 250 μL Opti-MEM and add 0, 1, 2.5, or 5 μg DNA. The 0 DNA dilution serves as a transfection reagent only control (*see* **Note 5**).
8. Incubate reactions at room temperature for 5 min.
9. To each DNA dilution, add 250 μL of the Lipofectamine 2000 Opti-MEM mixture to assemble the transfection liposome complexes. Incubate for 20 min and no more than 25 min. Total volume should be 500 μL.
10. While waiting for the reaction, replace media in each well of 6-well plate of cells with 1.5 mL fresh media. Confirm that the cells are at 75% confluence and healthy on a tissue culture microscope.
11. Add 500 μL of each transfection liposome complex solution to each well dropwise. Gently rock plates back and forth to mix. Incubate for 6 h in 37 C incubator.
12. Replace the media containing transfection mixture with 2 mL fresh complete media twice in each well. Incubate overnight.
13. 24 h after transfection, split cells into a fresh 6-well plate. Cells should be split to 75% confluence.
14. 48 h after transfection, wash plates with PBS. Add 250 μL protein sample buffer per well. Boil for 5 min 95 C.
15. Run 5 μL of each sample including empty vector controls and no vector controls. Process each sample for immunoblot as described in Subheading 3.3.
16. Determine which concentration of DNA is required to obtain expression levels that are similar to wild-type PARP13 in the no vector transfection samples of the HeLa parental cell lines. Make sure to normalize total protein concentrations as per Subheading 3.3.
17. Using the concentration of DNA determined in **step 14**, transfect wild-type and PARP13 knockout cell lines with SBP, SBP-PARP13, and SBP-PARP13q in 2 identical sets of 6-well

plates as described above. One set will be used to confirm protein levels as described in Subheading 3.3, and the other set will be used to purify RNA as described in Subheading 3.4. Make sure to include a no vector transfection control for each cell line. It is important to transfect both wild-type and knockout cells with each vector because general expression of protein vectors can result in changes in the transcriptome in a manner independent of the function of the protein itself.

18. Make sure to perform a minimum of two separate and independent transfection reactions for each DNA expression plasmid (two sets of plates for each replicate). Each independent transfection reaction will serve as an experimental replicate. It is important to always have experimental replicates whenever an RNA sequencing reaction is performed (*see* **Note 6**).

3.3 Immunoblot

1. Assemble XCell SureLock mini-cell electrophoresis system with a NuPAGE Novex Bis-Tris mini gel in a density suitable to detect the protein of interest according to the provided protocol.
2. Fill the tank with NuPAGE MOPS SDS running buffer.
3. Load 5 μL of sample as prepared in Subheading 3.2, including a protein ladder for size determination.
4. Run gel at 200 V for around 50 min until dye front reaches the bottom of the gel.
5. Remove gel from cassette and assemble XCell II Blot module according to the manufacturer's guidelines.
6. Transfer at 30 V for 60 min.
7. Remove the membrane and wash once in distilled water, and then wash in Ponceau solution until lanes become visible.
8. Wash twice in distilled water to remove membrane background and check quality of the run and transfer.
9. Wash off Ponceau with TBST and block for 60 min in milk.
10. Incubate in primary antibody overnight at 4 degrees under agitation.
11. Wash three times 5 min with TBST, and incubate at RT for 60 min with the appropriate secondary antibody in milk.
12. Wash three times with TBST and develop using available systems (e.g., ECL, Licor).

3.4 Total RNA Purification and Sequencing

Several key variables need to be considered in conducting experiments to identify mRNA targets: The amount of time after transfection to collect the samples for RNA analysis is important since a key PARP13 function involves RNA decay and specific targets have different mRNA amounts and half-lives. A too short rescue window for protein expression might not impact the mRNA amounts significantly resulting in false negatives, whereas expression of even small amounts of PARP13 for longer time periods results in the assembly of stress granules which will confound results.

It is also critical that all experimental samples for each specific *n* are prepared at the same time using the same reagents in independent setups and that all experimental samples are collected at the same time. Also, parental wild-type cell lines used to generate the knockout cell lines need to be included.

As with any RNA sequencing experiment, the data is only as good as the experimental design and the samples used. Therefore, considerable attention must be taken to ensure that the cells are healthy and the RNA is of high quality to be sequenced.

For preparation of RNA, proper handling under RNAse-free conditions needs to be employed (*see* **Note 7**). Although we suggest the use of commercial kits for total purification of RNA, traditional RNA purification techniques such as TRIzol extraction work equally well.

Library synthesis is now mainly automated, and sequencing facilities will perform synthesis and quality control steps, so we will not describe those steps in detail. Instead we will focus on how to interpret the quality control analysis of the RNA prior to proceeding with library synthesis. Nowadays multiple different ways and methods exist to investigate different aspects of transcriptome analysis using sequencing. Since the majority of isolated RNA will actually be ribosomal RNA, before sequencing the mRNA fraction has to be enriched. This can be achieved by positive selection for poly(A) RNA or depletion of rRNA, with poly(A) RNA requiring a higher amount of good quality RNA.

Another point to consider is read length or the number of bases sequenced in each individual fragment as well as single-end (SE) or paired-end (PE) sequencing. Whereas in SE sequencing the transcript is sequenced in one direction, PE sequencing includes the reverse direction as well. Getting the sequencing data from both ends of the fragment facilitates sequencing of repetitive sequencing, splicing analysis, or de novo assembly of unknown sequences. Since gene expression analysis of known genomes generally requires shorter read lengths, a standard size of 40 or 50 nucleotides SE should be sufficient in most cases.

Lastly and probably most importantly, the sequencing depth has to be taken into consideration. Different sequencing platforms have different limitations in the number of individual fragments they can sequence; current standard machines range between 5 and 500 million reads. Whereas highly expressed genes can be readily detected even with a low number of total reads, rare transcripts require a deeper sequencing depth. Additionally, for economical and temporal considerations multiple samples can be sequenced in the same run, a process called multiplexing. In this case the different samples are "barcoded" with unique adapters during library preparation, which allows the discrimination of individual transcripts in the subsequent analysis. However, each sample pooled together in this way decreases the read depth of the individual samples. Typically, for broad gene expression analysis of mRNA (like the experiments detailed in this chapter), 10–30 million reads per sample should be sufficient.

1. Make sure to use RNAse-free technique from this point forward.
2. The night prior to collecting RNA, split cells of interest into a fresh 6 well at 75% confluence. This is sufficient to harvest enough RNA for both RNA sequencing as well as prior control experiments (e.g., determine expression levels of genes regulated under the experimental conditions) or subsequent validation.
3. Remove culture medium and wash cells once in PBS.
4. Adherent cells can be directly lysed in the well. Depending on the preferred method of RNA extraction, add TRIzol or lysis buffer contained in RNA extraction kit.
5. Follow the instruction of the RNA extraction kit. Elute RNA in ~50 μL RNase-free water.
6. Measure RNA concentration and purity on a spectrophotometer using a fresh, RNAse-free tip. Concentration is determined as the absorbance at 260 nm, which corresponds to 40 μg/mL ssRNA. Contaminations are assessed using the 260 nm/280 nm ratio, which should be close to 2.0.for RNA, and the 260 nm/230 nm ratio, which should be 2.0–2.2.
7. If desired, RNA can be reverse transcribed into cDNA to test the expression of genes of interest prior to RNA sequencing, e.g., to check if the experimental conditions have the desired effect on transcripts known to be regulated accordingly. For PARP13.1 confirm that TRAILR4 mRNA levels are high in PARP13 knockout cells and low in knockout cells rescued with wild-type PARP13.
8. If not submitted for sequencing immediately, RNA should be stored at −80 °C until further use.
9. Once submitted to a genomics facility equipped to perform the desired RNA sequencing, the samples are subjected to a number of quality control steps followed by preparation of the library.
10. RNA quality will be assessed using a bioanalyzer to electrophoretically determine the RNA integrity number (RIN). This allows the detection of RNA degradation or the presence of anomalies. RINs have values between 0 (bad) and 10 (good).

11. Ask your bioinformatics facility and computational expert if the RIN value is high enough to proceed. Generally, they are most familiar with their systems and have the best sense of what leads to a good quality library and what does not.
12. The accurate concentration of the submitted RNA will be determined by qPCR, before the RNA is subjected to library preparation to make it suitable for sequencing.

3.5 Transcriptome Analysis

As data analysis of RNA sequencing requires some expertise, they are often performed by bioinformatics experts associated with the sequencing facility for labs with limited experience. However, this may vary between institutions and should be clarified in advance to conduct the experiments in accordance with facility requirements.

1. Before the actual data analysis, a number of quality control steps are conducted. These include the detection of read duplicates, GC content, adapter presence, sequence quality, and the number of reads mapping to the genome.
2. Confirm with your bioinformatics facility that these variables are within norms and that the data is of sufficient quality to proceed with analysis.
3. If the sequencing data is of sufficient quality, the adaptors attached to the 5′ and 3′ end are removed to get a clean sequence of the individual fragments.
4. Subsequently the reads are mapped to their individual location of the (known) reference genome or transcriptome, identifying the transcript from which the read originated.
5. To quantify the amount of transcripts present in each sample, the reads mapping to each individual transcript are counted. Normalization, e.g., reads per kilobase (RPKM), is used to be able to compare different samples.
6. Make sure to completely understand how the normalization is performed, and discuss this with your bioinformatics experts before the analysis begins. The normalization must occur with experimentally relevant sample databases, for example, from the same cell types, since many cell types express different transcripts. This is of particular importance for stressed samples since the transcriptome of a stressed sample is quite different than a non-stressed sample from the same cell type; therefore normalization to non-stress cells can result in loss of relevant information.
7. Comparing RPKMs for individual genes between different samples finally enables the determination of changes in expression level.

Transcripts that are up- or downregulated in PARP13-deficient cells compared to wild-type cells, which are rescued by reexpression of wild-type PARP13 but not an RNA-binding mutant, are likely candidates to be direct and specific targets of PARP13. It should however be noted that additional, more direct assays like EMSA or CLIP-qPCR need to be conducted to confirm direct regulation. Once a set of direct PARP13 targets has been established, these data can be used to perform pathway analysis (e.g., using DAVID or GSEA analysis) to gain insights into the modulated biological function.

4 Notes

1. Knockouts were originally generated using Sigma zinc finger nuclease constructs; however CRISPR is currently the most attractive approach.
2. For identification of the RNA targets of PARPs, other than PARP13 generate rescue constructs for that PARP. For example, for other CCCH PARPs, analogous mutations for each residue mutated in the PARP13 zinc finger can be generated, or each of the three cysteines and the histidine required to bind RNA can be mutated. These positions are well characterized within CCCH domains, and multiple reviews have been written to identify them.
3. Although we describe the use of Novex gel system, other commercial types of pre-poured SDS PAGE gels will work as well as lab-made gels.
4. Any commercial viability assay will work. Trypan blue is convenient, economical, and compatible with most tissue culture microscopes or cell counting devices.
5. Use maxiprep DNA that is endotoxin-free with a 260/280 ratio >2.8. For SBP-PARP13 expression, the range of DNA we test for transfection varies between 1, 2.5, and 5 μg for a 10 cm plate, and additional expression was not observed above 5 μg.
6. It is important to perform experimental replicates for each RNA sequencing project at the same time if possible. This reduces artifacts due to technical error and minimizes nonexperimental variabilities in the transcriptome of cells between experiments. Due to the sensitivity of RNA sequencing, these variabilities that are not relevant to the actual experiments can affect the ability to analyze the data effectively.
7. RNAses are everywhere and can cause degradation of your RNA samples. Purchase RNAse ZAP or another similar prod-

uct that inactivates RNAses, and spray work surfaces and pipettes. Follow common RNAse-free procedures, and purchase RNAse-free filter tips and tubes. Keep a separate stock of reagents, tips, and tubes for RNA-relevant experiments, and if possible keep a separate work area that is restricted for RNA work.

Acknowledgments

This work was partially supported by Cancer Center Support (core; grant P30-CA14051) and RO1GM087465 from the National Institutes of Health to PC. FJB was funded by a Ludwig Postdoctoral Fellowship.

References

1. Gibson BA, Kraus LW (2012) New insights into the molecular and cellular functions of poly(ADP-ribose) and PARPs. Nat Rev Mol Cell Bio 13:411–424
2. Bock F, Chang P (2016) New directions in poly(ADP-ribose) polymerase biology. FEBS J 283:4017–4031
3. Vyas S, Chang P (2014) New PARP targets for cancer therapy. Nat Rev Cancer 14(7):502–509
4. Vyas S, Matic I, Uchima L et al (2014) Family-wide analysis of poly(ADP-ribose) polymerase activity. Nat Commun 5:4426
5. Bock F, Todorova T, Chang P (2015) RNA regulation by poly(ADP-ribose) polymerases. Mol Cell 58:959–969
6. Chen S, Xu Y, Zhang K et al (2012) Structure of N-terminal domain of ZAP indicates how a zinc-finger protein recognizes complex RNA. Nat Struct Mol Biol 19(4):430–435
7. Iqbal BM, Johns M, Cao J et al (2014) PARP-14 combines with tristetraprolin in the selective posttranscriptional control of macrophage tissue factor expression. Blood 124:3646–3655
8. Todorova T, Bock FJ, Chang P (2015) Poly(ADP-ribose) polymerase-13 and RNA regulation in immunity and cancer. Trends Mol Med 21:373–384
9. Hall TM (2005) Multiple modes of RNA recognition by zinc finger proteins. Curr Opin Struct Biol 15:367–373
10. Zhu Y, Chen G, Lv F et al (2011) Zinc-finger antiviral protein inhibits HIV-1 infection by selectively targeting multiply spliced viral mRNAs for degradation. Proc Natl Acad Sci 108:15834–15839
11. Mao R, Nie H, Cai D et al (2013) Inhibition of hepatitis B virus replication by the host zinc finger antiviral protein. PLoS Pathog 9:e1003494
12. Goodier JL, Pereira GC, Cheung LE et al (2015) The broad-Spectrum antiviral protein ZAP restricts human retrotransposition. PLoS Genet 11:e1005252
13. Todorova T, Bock FJ, Chang P (2014) PARP13 regulates cellular mRNA post-transcriptionally and functions as a pro-apoptotic factor by destabilizing TRAILR4 transcript. Nat Commun 5:5362
14. Atasheva S, Akhrymuk M, Frolova EI, Frolov I (2012) New PARP gene with an anti-alphavirus function. J Virol 86:8147–8160
15. Atasheva S, Frolova EI, Frolov I (2014) Interferon-stimulated poly(ADP-ribose) polymerases are potent inhibitors of cellular translation and virus replication. J Virol 88:2116–2130
16. Yamada T, Horimoto H, Kameyama T et al (2016) Constitutive aryl hydrocarbon receptor signaling constrains type I interferon-mediated antiviral innate defense. Nat Immunol 17:687–694
17. Lee H, Komano J, Saitoh Y et al (2013) Zinc-finger antiviral protein mediates retinoic acid inducible gene I–like receptor-independent antiviral response to murine leukemia virus. Proc Natl Acad Sci 110:12379–12384
18. Hayakawa S, Shiratori S, Yamato H et al (2010) ZAPS is a potent stimulator of signaling mediated by the RNA helicase RIG-I during antiviral responses. Nat Immunol 12:37–44
19. Johnstone RW, Frew AJ, Smyth MJ (2008) The TRAIL apoptotic pathway in cancer onset,

progression and therapy. Nat Rev Cancer 8(10):782–798

20. Moldovan JB, Moran JV (2015) The zinc-finger antiviral protein ZAP inhibits LINE and Alu retrotransposition. PLoS Genet 11:1–34
21. Guo X, Ma J, Sun J, Gao G (2007) The zinc-finger antiviral protein recruits the RNA processing exosome to degrade the target mRNA. Proc Natl Acad Sci 104:151–156
22. Wang Z, Gerstein M, Snyder M (2009) RNA-Seq: a revolutionary tool for transcriptomics. Nat Rev Genet 10(1):57–63
23. Oshlack A, Robinson MD, Young MD (2010) From RNA-seq reads to differential expression results. Genome Biol 11:220
24. Ramsköld D, Kavak E, Sandberg R (2012) How to analyze gene expression using RNA-sequencing data. Methods Mol Biol 802:259–274
25. Finotello F, Camillo B (2015) Measuring differential gene expression with RNA-seq: challenges and strategies for data analysis. Brief Funct Genomics 14:130–142
26. Kukurba KR, Montgomery SB (2015) RNA sequencing and analysis. Cold Spring Harb Protoc 11:951–969
27. Subramanian A, Tamayo P, Mootha VK et al (2005) Gene set enrichment analysis: a knowledge-based approach for interpreting genome-wide expression profiles. Proc Natl Acad Sci USA 102:15545–15550
28. Huang D, Sherman BT, Lempicki RA (2008) Systematic and integrative analysis of large gene lists using DAVID bioinformatics resources. Nat Protoc 4(1):44–57
29. Frank E-S, Glazko GV (2011) Pathway analysis of expression data: deciphering functional building blocks of complex diseases. PLoS Comput Biol 7:e1002053

Part VII

NAD+ Analogue Approaches and Chemical Synthesis

Chapter 24

ADPr-Peptide Synthesis

Hans A. V. Kistemaker, Jim Voorneveld, and Dmitri V. Filippov

Abstract

Synthetic mono-ADPr-peptides are useful for structural, biochemical, and proteomics studies. We describe here a protocol for the preparation of mono-ADPr-peptides based on a fairly standard Fmoc-based solid-phase synthesis. Phosphoribosylated precursor building blocks are introduced into the peptide chain on solid-phase and subsequently converted to ADPr-sites by chemical phosphorylation with adenosine phosphoramidite. Suitably protected phosphoribosylated glutamine, asparagine, and citrulline building blocks described in this protocol allow introduction of ADP-Gln, ADPr-Asn, and ADPr-Cit into peptide chains as demonstrated for three peptides. Trifunctional amino acids, for which base-sensitive side-chain protection is available, can be accommodated in the sequences flanking the ADPr-cites.

Key words Mono-ADPr-peptides, Solid-phase synthesis, Pyrophosphate, Glycosylation, Phosphitylation, Phosphoribosylated amino acids

Abbreviations

ACN	Acetonitrile
AcOH	Acetic acid
CSO	(1S)-(+)-(10-camphorsulfonyl)-oxaziridine
DBU	1,8-diazabicycloundec-7-ene
DCM	Dichloromethane
DiPEA	Diisopropylethylamine
Dmab	4-(N-[1-(4,4-dimethyl-2,6-dioxocyclohexylidene)-3-methylbutyl]amino) benzyl ester
DMF	Dimethylformamide
Et_2O	Diethyl ether
EtOH	Ethanol
HCTU	O-(1H-6-Chlorobenzotriazole-1-yl)-1,1,3,3-tetramethyluronium hexafluorophosphate
HFIP	1,1,1,3,3,3-hexafluoro-2-propanol
NMP	*N*-methylpyrrolidone
TBDPS	tert-butyldiphenylsilyl

Paul Chang (ed.), *ADP-ribosylation and NAD+ Utilizing Enzymes: Methods and Protocols*, Methods in Molecular Biology, vol. 1813, https://doi.org/10.1007/978-1-4939-8588-3_24,

TBSOTf	tert-butyldimethylsilyl triflate
TEA	Triethylamine
Tfa	Trifluroacetamide (protective group)
TFA	Trifluoroacetic acid

1 Introduction

Although the biological importance of protein ADP-ribosylation in general is now well established, the details of the physiological function of mono-ADPr-proteins need further clarification [1, 2]. Mono-ADP-ribosylation is widespread in higher organisms and can occur both extra- and intracellularly [3, 4]. Synthetic well-defined mono-ADPr-peptides could play a significant role in structural and biochemical studies, and it has been shown already that synthetically prepared mono-ADPr-peptides can be important for the studies of the binding properties of macrodomains [5] and in the investigation of ADP-ribosylome [6]. ADP-ribosylation has been reported to occur on a wide variety of nucleophilic amino acid side chains [7]. However, the acidic amino acids (Asp and Glu) were most often reported as the abundant modification sites of intracellular ADP-ribosylation, while Arg is the most important modified amino acid when it comes to extracellular mono-ADP-ribosylation [8]. Although it seems straightforward to target these three amino acids as the modification sites when developing the chemical synthesis of ADPr-peptides, one has to take into account the extreme chemical lability of glycosylic ester linkages in case of ADPr-Glu and ADPr-Asp and a very high basicity of arginine side chain in case of ADPr-Arg. Therefore, on the outset, we elected to develop the solid-phase synthesis of mono-ADPr-peptides that contained more synthetically tractable bioisosteres of the ADPr-amino acids mentioned above, and so we substituted glutamine for glutamic acid, asparagine for aspartic acid, and citrulline for arginine as the acceptors of ADP-ribose. In this chapter, we describe the synthesis of mono-ADPr-peptides that contain these three amino acids (Asn, Gln, and Cit) as the sites of mono-ADP-ribosylation. The solid-phase protocol we developed over the years [5, 9, 10] uses as the basis the well-established Fmoc chemistry to build up the peptide chain. The sites of ADP-ribosylated amino acids are introduced in two stages; first the phosphoribosyl amino acid building blocks (**18**, **19**, **20**) are prepared in solution (Fig. 1) and incorporated into a peptide chain via the solid-phase synthesis (Fig. 2). Subsequently the immobilized phosphoribosyl peptides, thus obtained, are converted on-resin into protected ADPr-peptides (Fig. 2) via a phosphoramidite-based method that we originally developed for the synthesis of sugar nucleotides [11]. This method proved to be also amenable for the solid-phase synthesis of pyrophosphates of various nature as shown by us [12, 13]

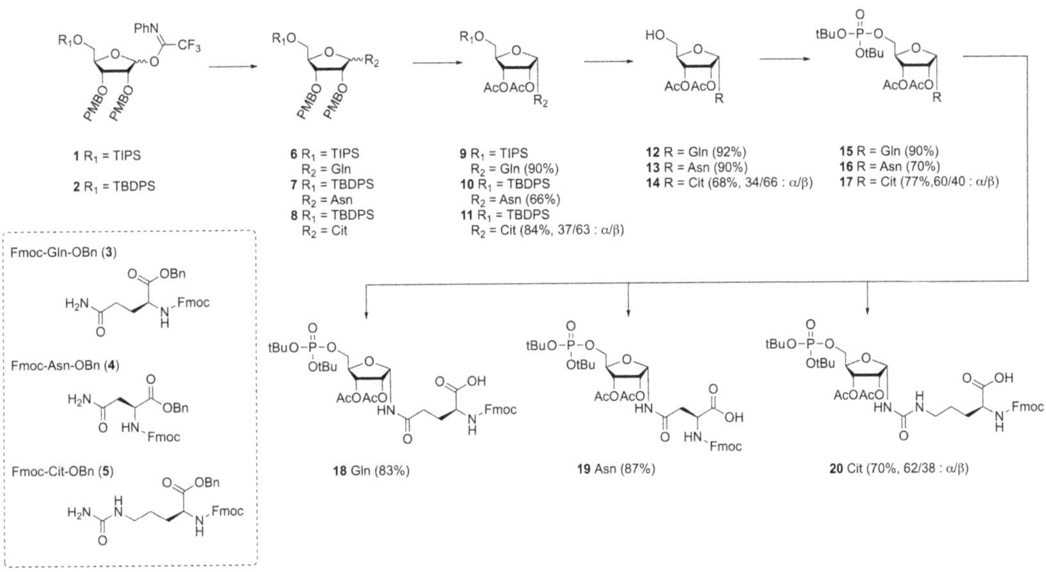

Fig. 1 A schematic outline of the solution synthesis of phosphoribosylated Fmoc-amino acids **18**, **19**, and **20**. The reagents, conditions, and exact description of the synthetic steps depicted are provided in "Methods," Subheadings 3.1.1–3.1.3

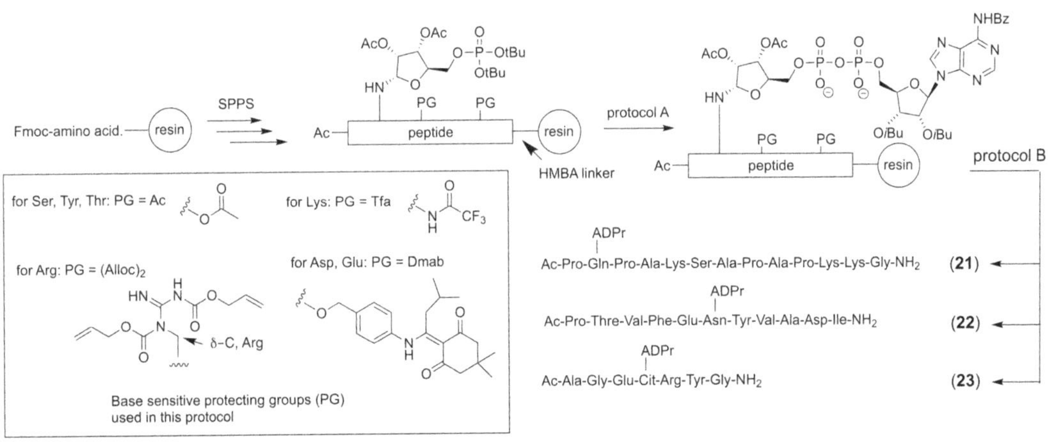

Fig. 2 A schematic outline of the solid-phase synthesis of ADPr-peptides **21** ("Gln"), **22** ("Asn"), and **23** ("Cit"). The detailed description is provided in "Methods," Subheading 3.2. *PG* protecting group

and by others [14]. The important feature of the synthetic protocol we present here is the use of base-sensitive semipermanent protective groups for amino acid side chain in the sequences flanking the ADPr-site. Our procedure at this stage of development cannot accommodate all native trifunctional amino acids in the flanking sequences. In particular peptides containing His, Asn, Gln, Cys, and Trp residues cannot be prepared by our current protocol. In addition, incorporation of Asp and Glu with our chemistry is somewhat problematic at this stage of development due to insufficient compatibility of the available hydrazine-sensitive Dmab protection [15] with the linker we use.

It is to be expected that the protocol presented in this article can be adapted to the synthesis of ADPr-peptides containing ADPr-amino acids other than explicitly described here, provided the suitable ribosylated building blocks are available. For instance, the synthesis of ADPr-serine peptides could be readily accomplished, when the synthesis of Fmoc-phosphoribosylserine is developed [16].

2 Materials

2.1 General

1. All solvents used under anhydrous conditions were stored over 4 Å molecular sieves except for methanol which was stored over 3 Å molecular sieves. Solvents used for workup and column chromatography were of technical grade and used directly. NMP and DCM used for peptide synthesis were of peptide synthesis grade.
2. 20% sulfuric acid in EtOH.
3. 25 g/L ammonium molybdate tetrahydrate in water.
4. Ammonium cerium(IV) sulfate hydrate (10 g/L) in 10% sulfuric acid.
5. Ninhydrin (3 g/L) in EtOH/AcOH (20/1 v/v).
6. TLC plates.
7. UV indicator.
8. Silica gel (0.04–0.063 mm).
9. Rotary evaporator equipped with a membrane pump or a water aspirator.
10. High-field NMR-spectrometer, for example, Brüker AV-400 (400 MHz) or Brüker AV-500 (500 MHz).
11. Optional: a system to automate silica gel chromatography, for example, **Biotage® Isolera™ Spektra Four** machine.

2.2 Solution-Phase Synthesis of the Fmoc-Protected Building Blocks

1. 1-Methyl-imidazole.
2. 1-Methyl-imidazolium chloride.
3. Acetic anhydride.
4. *tert*-Butylhydroperoxide (5.5 M solution in nonane).
5. Celite.
6. Di-*tert*-butyl-*N*,*N*-diisopropylphosphoramidite.
7. Et_3.HF.
8. Fmoc-Asn-OBn (**4**).

9. Fmoc-Gln-OBn (**3**).
10. Fmoc-Cit-OBn (**5**).
11. 0.1 M solution of HCl in HFIP (prepared out of 37% HCl).
12. HF.pyridine.
13. HFIP.
14. Hydrogen (g).
15. Pd/C (10 w/t%).
16. $HClO_4.SiO_2$ (0.4 mmol/g loading), prepared as described [17].
17. TEA.
18. 5'-*O*-TIPS-ribosyltrifluoroacetimidate donor **1** (Fig. 1) synthesized as described in [9].
19. 5'-*O*-TBDPS-ribosyltrifluoroacetimidate donor **2** (Fig. 1) synthesized as described in [5].

2.3 Solid-Phase Peptide Synthesis

1. All reagents and the resin used in the peptide synthesis are commercially available, were of peptide synthesis grade, and were purchased from the vendors specialized in the supplies for peptide synthesis.
2. Reagents and solvents used in the "on-resin" introduction of ADP-ribosyl moiety (*General protocol A*) were either of DNA-synthesis grade (ACN, DCM, CSO, ETT) or of highest available purity (DBU, HFIP, HCl).
3. Saturated solution of ammonia in TFE: Pass ammonia gas through precooled TFE (0 °C) till saturation is reached. The solution should be kept in a tightly closed vessel and can be stored for at least 1 month in a freezer (−20 °C).
4. Standard analytical chromatography and mass spectroscopy equipment, for example, Thermo Finnigan LCQ Advantage MAX ion-trap mass spectrometer with an electrospray ion source coupled to Surveyor HPLC system (LC-MS analysis) and Thermo Finnigan LTQ Orbitrap (HRMS analysis).
5. A system for preparative liquid chromatography suitable for reverse-phase and ion-exchange columns.
6. Analytical and semi-preparative reverse-phase C18 and anion-exchange columns suitable for HPLC-systems.
7. TSKgel Boronate-5PW 10 μm stainless-steel column for boronate affinity chromatography. The purification on boronate is done according to the protocol supplied by the manufacturer.
8. Optional: automated peptide synthesizer (we used ABI-433A from Applied Biosystems).

3 Methods

The synthetic strategy described in Subheading 3 can be subdivided in four preparative stages that can be outlined as follows. First, the pre-phosphoribosylated Fmoc-amino acids **18**, **19**, and **20** (Fig. 1, Subheadings 3.1–3.3) are prepared in the quantities sufficient to operate solid-phase peptide synthesis of protected and immobilized phosphoribosyl peptides (Fig. 2, Subheading 3.4), which constitutes the next stage of the synthesis. Subsequent step consists of chemoselective cleavage of tert-butyl groups with diluted acid followed by phosphorylation of the liberated phosphomonoester via phosphoramidite approach (*see* Fig. 3, Subheading 3.4). The forth, and the final, stage of the synthesis comprise the cleavage of the protection groups and the linker with ammonia treatment (*see* Subheading 3.4). The step-by-step description of the procedures involved is provided in the following sections.

3.1 Synthesis of Glutamine Building Block (18) Suitable for Fmoc-Based Solid-Phase Peptide Synthesis

Well-known ribofuranosyl trifluoroacetimidate donor **1** is prepared as previously described [9] and used in the glycosylation of Fmoc-Gln-OBn (**3**) to give ribofuranosylated glutamine **6** with high α-selectivity as shown in Fig. 1. Upon purification by silica gel column chromatography to obtain pure α-anomer of compound **6**, the para-methoxybenzyl protections (PMB) at the positions 2 and 3 are replaced by base-labile acetyl groups (from **6** to **9**) in order to maintain the compatibility with the protecting scheme of the solid-phase peptide synthesis (*see* Subheading 3.4). Desilylation of 5-OH in **9** with a fluoride source, followed by two-step one-pot phosphorylation of thus obtained **12**, gives fully protected phosphoribosylated building block **15**. Compound **15** is converted into Fmoc-Gln derivative **18** by rapid hydrogenolysis of **15** that cleaves benzyl ester but leaves Fmoc group intact. Phosphoribosylated Fmoc-glutamine **18** is used for the assembly of the phosphoribosyl precursor of ADPr-Gln peptide **21** as described in Subheading 3.4.1.

3.1.1 Glycosylation with Ribose Donor (See Note 1)

Activate 3 Å molecular sieves in a 25 mL round bottom flask by flame-heating in vacuo (oil pump) for 2 min, and let the molecular sieves cool to room temperature while maintaining the vacuum.

1. Co-evaporate in a 25 mL round bottom flask a mixture of 215 mg (0.30 mmol) of the trifluoroacetimidate donor (**1**) and 92 mg (0.2 mmol) Fmoc-Gln-OBn (**3**) thrice with 1,4-dioxane using a rotary evaporator.
2. Dissolve the residue in a 1:1 v/v mixture of dry DCM/1,4-dioxane (4.5 mL), and add the activated molecular sieves. Leave the flask to stand for 1 h to remove traces of water.
3. Add 50 mg of $HClO_4.SiO_2$ and stir for 2 h.

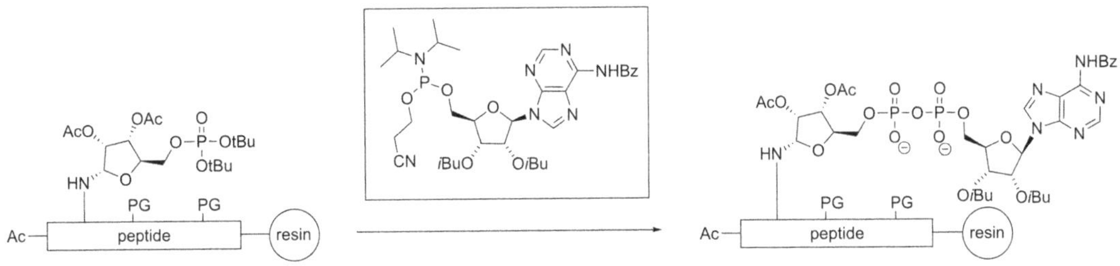

Fig. 3 A generic overview of the chemical transformations involved in the conversion of phosphoribosyl precursor into ADP-ribosyl amino acid. Details are provided in *General protocol A*

4. Filter the reaction through a pad of Celite on a glass filter and concentrate the reaction mixture in vacuo.
5. Purify the compound by flash column chromatography using silica gel and a linear gradient (EtOAc/pentane, 10/90 – 40/60).

$N^δ$-(5-triisopropylsilyl-2,3-di-*O*-(4-methoxybenzyl)-α-Dribosyl)-$N^α$-fluorenylmethoxycarbonyl glutamine benzyl ester (**6**) is yielded (137 mg, 139 μmol, 69% yield).

^{1}H-NMR (400 MHz, MeOD-$CDCl_3$) δ: 7.72 (d, *J* = 7.5 Hz,2H, Fmoc arom.), 7.63 – 7.55 (m, 2H, Fmoc arom.), 7.35 (t, *J* = 7.3 Hz, 2H, arom.), 7.31 – 7.11 (m,10H, arom.), 6.83 (t, *J* = 8.1 Hz, 4H, PMB arom.), 5.73 (d, *J* = 5.3 Hz, 1H, H1'-α), [5.62 (d, *J* = 4.4 Hz,H1'-β)], 5.18 – 5.06 (m, 2H, CH2 Bn), 4.58 – 4.49 (m, 2H, CH2 PMB), 4.48 – 4.41 (m, 2H, CH2 PMB),4.39 – 4.27 (m, 2H. CH2 Fmoc), 4.27 – 4.21 (m, 1H. CH α-Gln), 4.14 (t, *J* = 6.9 Hz, 1H, CH Fmoc),4.09 (bs, 1H, H4'), 4.07 – 3.99 (m, 2H, H2', H3'), 3.73 (s, 6H, CH3 PMB), 3.64 – 3.56 (m, 2H, H5'), 2.22(t, *J* = 7.3 Hz, 2H, CH2 γ-Gln), 2.18 – 2.05 (m, 1H, CH2 β-Gln), 2.01 – 1.88 (m, 1H, CH2 β-Gln), 0.98 (s,21H, TIPS).

^{13}C-NMR (100 MHz, MeOD-$CDCl_3$) δ: 174.03, 172.77 (CO α-Gln, CO δ-Gln), 160.35 (Cq. PMB), 157.79 (CO Fmoc), 144.68, 144.48, 142.01, 136.27 (Cq. arom.), 130.49 (arom.), 130.31 (Cq. arom.), 130.25 (arom.), 130.03 (Cq. arom.), 129.16 – 127.75 (arom.), 125.84, 125.79, 120.51 (Fmoc arom.), 114.54 (PMB arom.), 83.95 (C4'), 79.52 (C1'), 77.83 (C3'), 76.86 (C2'), 72.90, 72.89 (CH2 PMB), 67.71 (CH2 Bn), 67.59 (CH2 Fmoc), 64.43 (C5'), 55.58, 55.56 (CH3 PMB), 54.49 (CH α-Gln), 47.85 (CH Fmoc), 33.26 (CH2 γ-Gln), 27.67 (CH2 β-Gln), 18.33, 18.30, 12.54 (CH, CH3,TIPS).

HRMS $[C_{57}H_70N_2O_{11}Si + H]^+$: 987.4830 found, 987.4822 calculated.

3.1.2 Deprotection the 2′ and 3′ OH and Subsequent Acetylation

1. Dissolve 197 mg (0.2 mmol) of compound (**8**) in 4 mL HFIP in a 50 mL round bottom flask, and add 0.2 mL of a 0.1 M stock solution of HCl in HFIP. Stir for 15 min at room temperature.

2. Quench the reaction by adding 0.3 mL pyridine to the reaction mixture, and co-evaporate the reaction thrice with toluene in vacuo.
3. Dissolve the residue in 2 mL pyridine, and add 378 μL (4 mmol) acetic anhydride. Let the reaction stir overnight.
4. Concentrate the mixture in vacuo and dissolve the residue in DCM.
5. Wash the organic layer with a solution of saturated aqueous $NaHCO_3$. Collect the organic layer and dry with $MgSO_4$. Filtrate the organics and concentrate in vacuo.
6. Purify the compound by flash column chromatography using silica gel (EtOAc/pentane 40/60).

N^{δ}-(5-triisopropylsilyl-2,3-di-*O*-acetyl-α-D-ribosyl)-N^{α}-fluorenylmethoxycarbonyl glutamine benzyl ester (**9**) is yielded as a white foam (150 mg, 0.18 mmol, yield 90%).

3.1.3 Removal of the Silyl Protecting Group

1. Dissolve 1.02 g (1.23 mmol) of compound (**9**) in 12 mL THF. in a 50 mL round bottom flask, and cool the flask to 0 °C. Add 5.3 mL (32.5 mmol) Et_3.HF, and stir for 16 h while allowing the temperature to raise to room temperature.
2. Carefully quench the reaction with saturated aqueous $NaHCO_3$ until no bubbles are generated from the reaction mixture.
3. Transfer the reaction mixture into a separating funnel, and extract the water layer thrice with EtOAc.
4. Combine the organics and dry them with $MgSO_4$. Filtrate and concentrate the organics in vacuo.
5. Purify the compound by flash column chromatography using silica gel and a linear gradient (EtOAc/pentane, 50/50 – 70/30 – 80/20).

N^{δ}-(2,3-di-*O*-acetyl-α-D-ribosyl)-N^{α}-fluorenylmethoxycar bonyl glutamine benzyl ester (**12**) is yielded as a white foam (654 mg, 1.1 mmol, 92%).

3.1.4 Phosphorylation of the 5′ OH

1. Co-evaporate 674 mg (1.0 mmol) of compound **12** in a 100 mL round bottom flask thrice with 1,4-dioxane using a rotary evaporator equipped with a membrane pump. Apply a dry nitrogen atmosphere.
2. Dissolve the compound in 10 mL of a mixture of 0.2 M 1-methyl-imidazole and 0.3 M 1-methyl-imidazolium chloride in dry DMF. Add 473 μL (1.5 mmol) di-*tert*-butyl-*N,N*-diisopropylphosphoramidite, and stir for 15 min at room temperature.
3. Add 3.6 mL (19.8 mmol) of a 5.5 M *t*BuOOH solution in nonane, and stir the reaction mixture for 1 h.

4. Quench the reaction mixture by adding saturated aqueous $NaHCO_3$, and transfer the mixture into a separating funnel.
5. Extract the water layer thrice with Et_2O. Combine the organic layer and dry the organics with $MgSO_4$. Filtrate the organics and concentrate in vacuo.
6. Purify the compound by flash column chromatography using silica gel and a linear gradient (EtOAc/toluene, 50/50 – 60/40 – 70/30).

N^{δ}-(2,3-di-*O*-acetyl-5-*O*-(di-tert-butyl)-phosphoryl-α-D-ribosyl)-N^{α}-fluorenylmethoxycarbonyl glutamine benzyl ester (**15**) is yielded (781 mg, 0.90 mmol, 90%).

3.1.5 Hydrogenolysis of the Benzyl Protecting Group

1. Dissolve 700 mg (0.81 mmol) of compound **15** in 16 mL of a 4/4/1 mixture of *t*BuOH/1,4-dioxane/H_2O in a 50 mL round bottom flask. Add 17 mg Pd/C and let the suspension stir.
2. Fill a balloon with hydrogen gas, and allow the hydrogen gas to bubble through the suspension by using a long needle. Let the reaction mixture stir for 1 h under hydrogen atmosphere.
3. Replace the hydrogen gas for nitrogen, and purge the round bottom flask with nitrogen to remove all the hydrogen. Filter the reaction mixture over a pad of Celite using a glass filter, and concentrate in vacuo.
4. Purify the compound by flash column chromatography using silica gel (acidify the silica gel by using 1 v/v% AcOH in the eluens while preparing the column). Use a linear gradient (MeOH/DCM/AcOH 0/99/1 – 5/94/1 – 10/89/1).

N^{δ}-(2,3-di-*O*-acetyl-5-*O*-(di-tert-butyl)-phosphoryl-α-D-ribosyl)-N^{α}-fluorenylmethoxycarbonyl glutamine (**18**) is yielded as a white foam (521 mg, 0.67 mmol, 83%).

^{1}H-NMR (500 MHz, MeOD-$CDCl_3$) δ 7.74 (d, J = 7.6 Hz, 2H, Fmoc), 7.62 (dd, J = 11.0, 7.7 Hz, 2H, Fmoc), 7.36 (t, J = 7.4 Hz, 2H, Fmoc), 7.28 (t, J = 7.4 Hz, 2H, Fmoc), 6.01 (d, J = 3.8 Hz, 1H, H1`), 5.38 – 5.28 (m, 2H, H2', H3'), 4.34 (AB, J = 10.2, 7.3 Hz, 1H, CH_2 Fmoc), 4.32 – 4.27 (m, 1H, CH_2 Fmoc), 4.27 – 4.23 (m, 1H, H4'), 4.23 – 4.15 (m, 2H, CH α-Gln, CH Fmoc), 4.10 – 4.05 (m, 1H, H5'), 4.05 – 3.99 (m, 1H, H5'), 2.38 (t, J = 7.5 Hz, 2H, CH_2 γ-Gln), 2.23 – 2.12 (m, 1H, CH_2 β-Gln), 2.09 (s, 3H, CH_3 Ac), 2.06 (s, 3H, CH_3 Ac), 2.03 – 1.95 (m, 1H, CH_2 β-Gln), 1.47 (s, 18H, CH_3 tBu).

^{13}C-NMR (126 MHz, MeOD-$CDCl_3$) δ 175.05, 174.64 (CO α-Gln, CO δ-Gln), 171.05, 170.83 (CO Ac), 158.05 (CO Fmoc), 144.86, 144.69, 142.15 (Cq. arom.), 128.49, 127.87, 125.97, 125.90, 120.65 (arom.), 84.60, 84.57, 84.54, 84.51 (Cq. tBu), 80.09 (C1'), 79.89, 79.82 (C4'), 72.13, 71.24 (C2', C3'), 67.74 (CH_2 Fmoc), 67.02, 66.97 (C5'), 54.58 (CH Fmoc), 47.98 (CH α-Gln), 33.15 (CH_2 γ-Gln), 30.10, 30.07 (CH_3 tBu), 28.63 (CH_2 β-Gln), 20.58, 20.48 (CH_3 Ac).

^{31}P NMR (162 MHz, MeOD-$CDCl_3$) δ –9.50.

HRMS [$C_{37}H_{49}N_2O_{14}P$ + H]$^+$: 777.2996 found, 777.2994 calculated.

3.2 Synthesis of Asparagine Building Block (19) Suitable for Fmoc-Based Solid-Phase Peptide Synthesis

Phosphoribosyl asparagine building block **19** is synthesized following a procedure similar to the one described in Subheading 3.1 for the glutamine counterpart **18** (Fig. 1). Here a known ribofuranosyl trifluoroacetimidate donor **2** prepared as described [5] is applied to stereoselectively glycosylate Fmoc-Asn-OBn (**4**) providing the α-anomer of **7** that is further processed as described for glutamine derivative **6** (*see* Subheading 3.1, Fig. 1). Phosphoribosylated Fmoc-Asn building block **19**, thus obtained, is used subsequently to prepare the phosphoribosylated precursor of ADPr-Asn peptide **22** as described in Subheading 3.4.2.

3.2.1 Glycosylation with Ribose Donor (See Note 1)

1. Activate 3 Å molecular sieves in a 50 mL round bottom flask by flame-heating in vacuo (oil pump) for 2 min, and let the molecular sieves cool to room temperature while maintaining the vacuum.
2. Co-evaporate in a 100 mL round bottom flask a mixture of 2.82 g (3.6 mmol) of the trifluoroacetimidate donor (**2**) and 1.33 g (3.0 mmol) Fmoc-Asn-OBn (**4**) thrice with 1,4-dioxane using a rotary evaporator.
3. Dissolve the residue in a 1:1 v/v mixture of dry DCM/1,4-dioxane (30 mL), and add the activated molecular sieves. Stir for 1 h at room temperature to remove traces of water.
4. Cool the solution to –10 °C (use an iced water bath mixed with sodium chloride or a cryostat), and add 41 μL (0.18 mmol) TBSOTf. Let the temperature rise to room temperature by removing the ice bath or cryostat, and stir for 30 min.
5. Quench the reaction by adding a few drops of TEA, and filter the reaction through a pad of Celite on a glass filter. Concentrate the reaction mixture in vacuo.
6. Purify the compound by flash column chromatography using silica gel and a linear gradient (EtOAc/pentane, 20/80 – 30/70 – 60/40).

N^γ-(2,3-di-*O*-(4-methoxybenzyl)-5-tert-butyldiphenylsilyl-α-D-ribosyl)-*N*-α-fluorenylmethoxycarbonyl asparagine benzyl ester (**7**) is yielded as a white foam (2.50 g, 2.37 mmol, 79%, 97/3 α/β).

^{1}H-NMR (500 MHz, MeOD-$CDCl_3$) δ 7.71 (d, *J* = 7.6 Hz, 2H, arom.), 7.62 – 7.50 (m, 6H, arom.), 7.46 – 7.10 (m, 19H, arom.), 6.80 (t, *J* = 8.0 Hz, 4H, arom.), 5.77 (d, *J* = 5.3 Hz, 1H, H1'-α), [5.63 (d, *J* = 4.4 Hz, H1'-β)], 5.18 – 5.08 (m, 2H, CH_2 Bn), 4.62 (t, *J* = 5.9 Hz, 1H, CH α-Asn), 4.54 – 4.40 (m, 4H, CH_2 PMB), 4.31 (AB, *J* = 10.4, 7.2 Hz, 1H, CH_2 Fmoc), 4.21

(AB, J = 10.0, 7.7 Hz, 1H, CH_2 Fmoc), 4.14 – 4.08 (m, 1H, CH Fmoc), 4.08 – 4.04 (m, 1H, H2'), 4.04 – 3.98 (m, 2H, H3', H4'), 3.76 – 3.66 (m, 6H, CH_3 PMB), 3.54 (AB, J = 11.3, 2.7 Hz, 1H, H5'), 3.53 – 3.47 (m, 1H, H5'), 2.78 (d, J = 5.9 Hz, 2H, CH_2 β-Asn), 0.94 (s, 9H, CH_3 TBDPS).

^{13}C-NMR (100 MHz, MeOD-$CDCl_3$) δ 172.24, 171.75 (CO α-Asn, CO γ-Asn), 160.60 (Cq. PMB), 157.85 (CO Fmoc), 144.81, 144.76, 142.19, 136.57 (Cq. arom.), 136.46, 136.38 (arom.), 133.93, 133.72 (Cq. arom.), 130.75, 130.71, 130.61, 130.51 (arom.), 130.38 (Cq. arom.), 129.29, 129.00, 128.86, 128.65, 128.61, 128.50, 127.91, 126.03, 125.97, 120.66, 114.75 (arom.), 83.72 (C4'), 80.02 (C1'), 77.99 (C3'), 77.32 (C2'), 73.33, 73.10 (CH_2 PMB), 68.09, 68.01 (CH_2 Bn, CH_2 Fmoc), 64.79 (C5'), 55.66, 55.61 (CH_3 PMB), 52.02 (CH α-Asn), 47.97 (CH Fmoc), 38.67 (CH_2 β-Asn), 27.30 (CH_3 TBDPS), 19.83 (Cq. TBDPS).

HRMS $[C_{63}H_{66}N_2O_{11}Si + H]^+$: 1055.4513 found, 1055.4509 calculated.

3.2.2 Deprotection of the 2′ and 3′ OH and Subsequent Acetylation

1. Dissolve 770 mg (0.73 mmol) of compound (7) in 15 mL HFIP in 50 mL round bottom flask, and add 730 μL of a 0.1 M stock solution of HCl in HFIP. Stir for 15 min at room temperature.
2. Quench the reaction by adding 1 mL pyridine to the reaction mixture, and co-evaporate the reaction thrice with toluene in vacuo.
3. Dissolve the residue in 7.3 mL pyridine, and add 1.4 mL (14.6 mmol) acetic anhydride. Let the reaction stir overnight.
4. Concentrate the mixture in vacuo and dissolve the residue in 15 mL DCM.
5. Wash the organic layer with a solution of saturated aqueous $NaHCO_3$. Collect the organic layer and dry with $MgSO_4$. Filtrate the organics and concentrate in vacuo.
6. Purify the compound by flash column chromatography using silica gel and a linear gradient (EtOAc/pentane, 30/70 – 35/65).

N^{γ}-(2,3-di-*O*-acetyl-5-tert-butyldiphenylsilyl-α-D-ribosyl)-*N*-α-fluorenylmethoxycarbonyl asparagine benzyl ester (**10**) is yielded as a white foam (328 mg, 0.36 mmol, 49%).

3.2.3 Removal of the Silyl Protecting Group

1. Dissolve 198 mg (0.22 mmol) of compound (**10**) in 2.2 mL pyridine in a 50 mL round bottom flask. Add 0.22 mL HF.pyridine and stir for 1 h at room temperature.
2. Carefully quench the reaction with saturated aqueous $NaHCO_3$ until no bubbles are generated from the reaction mixture.

3. Transfer the reaction mixture into a 100 mL separating funnel, and extract the water layer thrice with 30 mL DCM.
4. Combine the organics and dry them with $MgSO_4$. Filtrate and concentrate the organics in vacuo.
5. Purify the compound by flash column chromatography using silica gel and a linear gradient (EtOAc/pentane, 50/50 – 75/25 – 100/0).

N$^{\gamma}$-(2,3-di-*O*-acetyl-α-D-ribosyl)-*N*$^{\alpha}$-fluorenylmethoxycarbonyl asparagine benzyl ester (**13**) is obtained as a white foam (135 mg, 0.20 mmol, 90%).

3.2.4 Phosphorylation of the 5′ OH

1. Co-evaporate 126 mg (0.19 mmol) of compound **13** in a 25 mL round bottom flask thrice with 1,4-dioxane using a rotary evaporator equipped with a membrane pump. Apply a dry nitrogen atmosphere.
2. Dissolve the compound in 1.9 mL of a mixture of 0.2 M 1-methyl-imidazole and 0.3 M 1-methyl-imidazolium chloride in dry DMF. Add 90 μL (0.29 mmol) di-*tert*-butyl-*N,N*-diisopropylphosphoramidite, and stir for 15 min at room temperature.
3. Add 0.35 mL (1.9 mmol) of a 5.5 M *t*BuOOH solution in nonane, and stir the reaction mixture for 1 h.
4. Quench the reaction mixture by adding saturated aqueous $NaHCO_3$ and transfer the mixture into a separating funnel.
5. Extract the water layer thrice with Et_2O (15 mL). Combine the organic layer and dry the organics with $MgSO_4$. Filtrate the organics and concentrate in vacuo.
6. Purify the compound by flash column chromatography using silica gel and a linear gradient (EtOAc/pentane, 40/60 – 50/50 – 60/40).

N$^{\gamma}$-(2,3-di-*O*-acetyl-5-*O*-(di-tert-butyl)-phosphoryl-α-D-ribosyl)-*N*$^{\alpha}$-fluorenylmethoxycarbonyl asparagine benzyl ester (**16**) is yielded (113 mg, 0.13 mmol, 70%).

3.2.5 Hydrogenolysis of the Benzyl Protecting Group

1. Dissolve (0.55 mmol) of compound **16** in 11 mL of a 4/4/1 mixture of *t*BuOH/1,4-dioxane/H_2O in a 50 mL round bottom flask. Add 83 mg Pd/C and let the suspension stir.
2. Fill a balloon with hydrogen gas, and allow the hydrogen gas to bubble through the suspension by using a long needle. Let the reaction mixture stir for 1 h under hydrogen atmosphere.
3. Replace the hydrogen gas for nitrogen, and purge the round bottom flask with nitrogen to remove all the hydrogen. Filter the reaction mixture over a pad of Celite using a glass filter, and concentrate in vacuo.

4. Purify the compound by flash column chromatography using silica gel (acidify the silica gel by using 1 v/v% AcOH in the eluens while preparing the column). Use a linear gradient (EtOH/DCM/AcOH 0/99/1 – 3/96/1 – 5/94/1).

N^{γ}-(2,3-di-*O*-acetyl-5-*O*-(di-tert-butyl)-phosphoryl-α-D-ribosyl)-N^{α}-fluorenylmethoxycarbonyl asparagine (**19**) is obtained as a white foam (390 mg, 0.48 mmol, 87%).

^{1}H-NMR (500 MHz, MeOD-$CDCl_3$) δ 7.76 (d, J = 7.5 Hz, 2H, Fmoc), 7.63 (d, J = 7.4 Hz, 2H, Fmoc), 7.37 (t, J = 7.5 Hz, 2H, Fmoc), 7.29 (t, J = 7.4 Hz, 2H, Fmoc), 6.00 (d, J = 2.3 Hz, 1H, H1'), 5.34 (d, J = 3.1 Hz, 2H, H2', H3'), 4.55 (t, J = 6.0 Hz, 1H, CH α-Asn), 4.35 (AB, J = 10.3, 7.3 Hz, 1H, CH_2 Fmoc), 4.32 – 4.26 (m, 1H, CH_2 Fmoc), 4.25 – 4.18 (m, 2H, CH Fmoc, H4'), 4.10 – 3.98 (m, 2H, H5'), 2.87 (AB, J = 15.8, 5.2 Hz, 1H, CH_2 β-Asn), 2.81 (AB, J = 15.7, 6.9 Hz, 1H, CH_2 β-Asn), 2.10 (s, 3H, CH_3 Ac), 2.07 (s, 3H, CH_3 Ac), 1.47 (s, 18H, CH_3 tBu).

^{13}C-NMR (126 MHz, MeOD-$CDCl_3$) δ 174.26 (CO α-Asn), 172.12, 171.20, 171.00 (CO γ-Asn, CO Ac), 157.99 (CO Fmoc), 144.90, 144.85, 142.26 (Cq. arom.), 128.58, 127.96, 126.06, 126.03, 120.72 (arom.), 84.74, 84.71, 84.68, 84.64 (Cq. tBu), 80.25 (C1'), 80.03, 79.96 (C4'), 72.23, 71.38 (C2', C3'), 68.01 (CH_2 Fmoc), 67.10, 67.06 (C5'), 51.81 (CH α-Asn), 48.06 (CH Fmoc), 38.61 (CH_2 β-Asn), 30.13, 30.10 (CH_3 tBu), 20.57, 20.49 (CH_3 Ac).

^{31}P NMR (162 MHz, MeOH) δ –9.24.

HRMS $[C_{36}H_{47}N_2O_{14}P + H]^+$: 763.2839 found, 763.2838 calculated.

3.3 Synthesis of Citrulline Building Block (20) Suitable for Fmoc-Based Solid-Phase Peptide Synthesis

The citrulline building block (**20**) is synthesized via the same sequence as described for Fmoc-asparagine derivative **19** (*see* Subheading 3.2, Fig. 1). Glycosylation of Fmoc-Cit-OBn with trifluoroacetimidate donor **2**, subsequent protective group manipulations (**8** into **14**), phosphorylation (**14** into **17**), and, finally, the cleavage of benzyl ester in **17** provided phosphoribosylated Fmoc-Cit building block **20**. Note, however, that phosphoribosyl citrulline proved to be configurationally unstable at the anomeric center of the ribofuranosyl moiety even under neutral conditions and has to be used in the solid-phase synthesis of ADPr-Cit peptide **23** (*see* Subheading 3.4.3) as a mixture of anomers of the riboside of the citrulline side chain.

3.3.1 Glycosylation with Ribose Donor (See Note 1)

1. Activate 3 Å molecular sieves in a 100 mL round bottom flask by flame-heating in vacuo (oil pump) for 2 min, and let the molecular sieves cool to room temperature while maintaining the vacuum.

2. Co-evaporate in a 250 mL round bottom flask a mixture of 5.0 g (6.4 mmol) of the trifluoroacetimidate donor (**2**) and 2.93 g (6.0 mmol) Fmoc-Asn-OBn (**4**) thrice with 1,4-dioxane using a rotary evaporator.
3. Dissolve the residue in a 1:1 v/v mixture of dry DCM/1,4-dioxane (60 mL), and add the activated molecular sieves. Apply a dry nitrogen atmosphere, and leave the flask to stand for 16 h to remove traces of water.
4. Cool the solution to −15 °C (use an iced water bath mixed with sodium chloride or a cryostat), and add 438 μL (1.92 mmol) TBSOTf. Leave the reaction to stir for 13 h at −15 °C.
5. Quench the reaction by adding 1 mL TEA, and filter the reaction through a pad of Celite on a glass filter. Concentrate the reaction mixture in vacuo.
6. Purify the compound by flash column chromatography using silica gel and a linear gradient (EtOAc/pentane, 40/60 – 50/50 – 60/40).

N^{ω}-(2,3-di-*O*-(4-methoxybenzyl)-5-tert-butyldiphenylsilyl-α-D-ribosyl)-N^{α}-fluorenylmethoxycarbonyl citrulline benzyl ester (**8**) is yielded as a white foam (2.64 g, 2.41 mmol, 40%, 78/22 α/β).

^{1}H-NMR (400 MHz, MeOD-$CDCl_3$) δ 7.72 (d, J = 7.5 Hz, 2H, arom.), 7.63 – 7.57 (m, 6H, arom.), [7.57 – 7.52 (m, 2H, arom-β)], 7.45 – 7.13 (m, 19H, arom.), 6.86 – 6.75 (m, 4H), 5.73 (d, J = 4.3 Hz, 1H, H1'-α), [5.54 (d, J = 5.9 Hz, H1'-β)], 5.14 – 5.06 (m, 2H, CH_2 Bn), 4.55 – 4.38 (m, 4H, CH_2 PMB), 4.35 – 4.24 (m, 2H, CH_2 Fmoc), 4.24 – 4.15 (m, 1H, CH α-Cit), 4.15 – 4.07 (m, 1H, CH Fmoc), 4.07 – 3.98 (m, 3H, H2', H3', H4'), [3.96 (dd, J = 5.2, 3.9 Hz, H3'-β)], 3.74 – 3.71 (m, 3H, CH_3 PMB), 3.70 (s, 3H, CH_3 PMB), [3.62 (d, J = 4.1 Hz, H5'-β)], 3.59 – 3.48 (m, 2H, H5'), 3.15 – 2.99 (m, 2H, CH_2 δ-Cit), 1.89 – 1.75 (m, 1H, CH_2 β-Cit), 1.75 – 1.60 (m, 1H, CH_2 β-Cit), 1.58 – 1.43 (m, 2H, CH_2 γ-Cit), [0.98 (s, CH_3 TBDPS-β)], 0.94 (s, 9H, CH_3 TBDPS).

^{13}C-NMR (100 MHz, MeOD-$CDCl_3$) δ 173.66 (CO α-Cit), 160.72 (Cq. PMB), 159.69 (CO Fmoc), 158.43 (CO ω-Cit), 145.10, 144.94, 142.39, 136.93 (Cq. arom.), 136.61, 136.52 (arom.), 134.14, 133.97, 130.94 (Cq. arom.), 130.86, 130.81, 130.76, 130.72, 130.68, 130.57, 129.42, 129.15, 129.04, 128.80, 128.77, 128.73, 128.64, 128.05, 128.03, 126.15, 126.11, 120.78, 114.78, 114.77, 114.73, 114.69 (arom.), 84.90 (C1'-β), 83.22 (C4'-α), 82.73 (C4'-β), 81.70 (C1'-α), 80.70 (C2'-β), 78.25 (C3'-α), 77.24 (C2'-α), 76.66 (C3'-β), 73.17, 73.10 (CH_2 PMB-α), 72.82, 72.39 (CH_2 PMB-β), 67.84, 67.77 (CH_2 Bn, CH_2 Fmoc), 65.21 (C5'-β), 65.06 (C5'-α), 55.68,

55.66, 55.64 (CH_3 PMB), 55.35 (CH α-Cit), 48.21 (CH Fmoc), 40.23 (CH_2 δ-Cit), 29.70 (CH_2 β-Cit), 27.50 (CH_2 γ-Cit), 27.43, 27.34 30 (CH_3 TBDPS), 20.00, 19.92 (Cq. TBDPS).

HRMS $[C_{65}H_{71}N_3O_{11}Si + H]^+$: 1098.4944 found, 1098.4931 calculated.

3.3.2 Deprotection of the 2′ and 3′ OH and Subsequent Acetylation

1. Dissolve 2.60 g (2.37 mmol) of compound (**8**) in 50 mL HFIP in a 250 mL round bottom flask, and add 2.37 mL of a 0.1 M stock solution of HCl in HFIP. Stir for 15 min at room temperature.
2. Quench the reaction by adding 3 mL pyridine to the reaction mixture, and co-evaporate the reaction thrice with toluene in vacuo.
3. Dissolve the residue in 24 mL pyridine, and add 4.5 mL (47.6 mmol) acetic anhydride. Let the reaction stir overnight.
4. Concentrate the mixture in vacuo and dissolve the residue in 50 mL DCM.
5. Wash the organic layer with a solution of saturated aqueous $NaHCO_3$. Collect the organic layer and dry with $MgSO_4$. Filtrate the organics and concentrate in vacuo.
6. Purify the compound by flash column chromatography using silica gel and a linear gradient (EtOAc/pentane, 30/70 – 35/65 – 40/60).

N^{ω}-(2,3-di-*O*-acetyl-5-tert-butyldiphenylsilyl-α-D-ribosyl)-N^{α}-fluorenylmethoxycarbonyl citrulline benzyl ester (**11**) is yielded as a white foam (602 mg, 0.64 mmol, 27%) separated from the β-anomer (1.00 g, 1.07 mmol, 44%) along with a mixture of anomers (303 mg, 0.32 mmol, 13%).

3.3.3 Removal of the Silyl Protecting Group

1. Dissolve 707 mg (0.75 mmol) of compound (**11**) in 7.5 mL pyridine in a 50 mL round bottom flask. Add 0.75 mL HF.pyridine and stir for 1 h at room temperature.
2. Carefully quench the reaction with saturated aqueous $NaHCO_3$ until no bubbles are generated from the reaction mixture.
3. Transfer the reaction mixture into a separating funnel, and extract the water layer thrice with DCM.
4. Combine the organics and dry them with $MgSO_4$. Filtrate and concentrate the organics in vacuo.
5. Purify the compound by flash column chromatography using silica gel and a linear gradient (EtOAc/pentane, 50/50 – 60/40 – 100/0).

N^{ω}-(2,3-di-*O*-acetyl-α-D-ribosyl)-N^{α}-fluorenylmeth oxycarbonyl citrulline benzyl ester (**14**) is yielded as the α-anomer (125 mg, 0.18 mmol, 24%) and the β-anomer (231 mg, 0.33 mmol, 44%) separately.

3.3.4 Phosphorylation of the 5′ OH

1. Co-evaporate 190 mg (0.27 mmol) of compound **14** (α-anomer) in a 25 mL round bottom flask thrice with 1,4-dioxane using a rotary evaporator equipped with a membrane pump. Apply a dry nitrogen atmosphere.
2. Dissolve the compound in 2.7 mL of a mixture of 0.2 M 1-methyl-imidazole and 0.3 M 1-methyl-imidazolium chloride in dry DMF. Add 129 μL (0.41 mmol) di-*tert*-butyl-*N,N*-diisopropylphosphoramidite, and stir for 15 min at room temperature.
3. Add 0.49 mL (2.7 mmol) of a 5.5 M *t*BuOOH solution in nonane, and stir the reaction mixture for 1 h.
4. Quench the reaction mixture by adding saturated aqueous $NaHCO_3$, and transfer the mixture into a separating funnel.
5. Extract the water layer thrice with Et_2O (15 mL). Combine the organic layer, and dry the organics with $MgSO_4$. Filtrate the organics and concentrate in vacuo.
6. Purify the compound by flash column chromatography using silica gel and a linear gradient (EtOAc/pentane, 60/40 – 70/30 – 80/20).

N^{ω}-(2,3-di-*O*-acetyl-5-*O*-(di-tert-butyl)-phosphoryl-α-D-ribosyl)-N^{α}-fluorenylmethoxycarbonyl citrulline benzyl ester (**17**) is yielded as the α-anomer (104 mg, 0.12 mmol, 45%) and the β-anomer (75 mg, 0.08 mmol, 32%) separately.

3.3.5 Hydrogenolysis of the Benzyl Protecting Group

1. Dissolve 99 mg (0.11 mmol) of compound **17** in 2.2 mL of a 4/4/1 mixture of *t*BuOH/1,4-dioxane/H_2O in a 50 mL round bottom flask. Add 17 mg Pd/C and let the suspension stir.
2. Fill a balloon with hydrogen gas, and allow the hydrogen gas to bubble through the suspension by using a long needle. Let the reaction mixture stir for 1 h under hydrogen atmosphere.
3. Replace the hydrogen gas for nitrogen and purge the round bottom flask with nitrogen to remove all the hydrogen. Filter the reaction mixture over a pad of Celite using a glass filter, and concentrate in vacuo.
4. Purify the compound by flash column chromatography using silica gel (acidify the silica gel by using 1 v/v% AcOH in the eluens while preparing the column). Use a linear gradient (EtOH/DCM/AcOH 0/99/1 – 5/94/1 – 7/92/1).

N^{ω}-(2,3-di-*O*-acetyl-5-*O*-(di-tert-butyl)-phosphoryl-α/β-D-ribosyl)-N^{α}-fluorenylmethoxycarbonyl citrulline (**20**) is obtained as a mixture of anomers (62 mg, 77 μmol, 70%, α/β: 62/38).

^{1}H-NMR (500 MHz, MeOD-$CDCl_3$) δ 7.77 (d, J = 8.2 Hz, 4H, arom. Fmoc), 7.65 (t, J = 8.0 Hz, 4H, arom. Fmoc), 7.37 (t, J = 7.4 Hz, 4H, arom. Fmoc), 7.30 (t, J = 7.4 Hz, 4H, arom.

Fmoc), 5.92 (d, J = 5.1 Hz, 1H, H1'-α), 5.58 (d, J = 6.5 Hz, 1H, H1'-β), 5.36 (t, J = 5.2 Hz, 1H, H3'-α), 5.32 (dd, J = 5.5, 3.6 Hz, 1H, H3'-β), 5.29 (t, J = 5.3 Hz, 1H, H2'-α), 5.04 (t, J = 6.1 Hz, 1H, H2'-β), 4.35 (d, J = 6.9 Hz, 4H, CH_2 Fmoc), 4.23 – 4.15 (m, 5H, CH Fmoc, CH α-Cit, H4'-α), 4.13 – 4.09 (m, 1H, H4'-β), 4.08 – 4.00 (m, 4H, H5′-β, H5′-α), 3.20 – 3.13 (m, 4H, CH_2 δ-Cit), 2.11 (s, 3H, CH_3 Ac), 2.09 (s, 6H, CH_3 Ac), 2.05 (s, 3H, CH_3 Ac), 1.94 – 1.82 (m, 2H, CH_2 β-Cit), 1.76 – 1.64 (m, 2H, CH_2 β-Cit), 1.64 – 1.52 (m, 4H, CH_2 γ-Cit), 1.50 – 1.47 (m, 36H, CH_3 tBu).

^{13}**C-NMR** (126 MHz, MeOD-$CDCl_3$) δ 175.51 (CO α-Cit), 171.38, 171.23, 171.17, 170.85 (CO Ac), 159.42, 159.12 (CO Fmoc), 158.35 (CO ω-Cit), 145.04, 144.87, 142.33 (Cq. arom. Fmoc), 128.58, 127.98, 126.06, 120.72 (arom. Fmoc), 84.17 (C1'-β), 81.45 (C1'-α), 80.35, 80.28 (C4'-β), 79.59, 79.52 (C4'-α), 74.23 (C2'-β), 72.51 (C3'-α), 71.93 (C3'-β), 71.33 (C2'-α), 67.79 (CH_2 Fmoc), 67.49, 67.44 (C5′-β), 67.31, 67.26 (C5′-α), 54.91 (CH α-Cit), 48.20 (CH Fmoc), 40.28, 40.19 (CH_2 δ-Cit), 30.15, 30.12 (CH_3 tBu), 29.94, 29.89 (CH_2 β-Cit), 27.49, 27.43 (CH_2 γ-Cit), 20.58, 20.53, 20.44 (CH_3 Ac).

^{31}**P-NMR** (162 MHz, MeOD-$CDCl_3$) δ –9.20, –9.48.

HRMS $[C_{38}H_{52}N_3O_{14}P + H]^+$: 806.3258 found, 806.3260 calculated.

3.4 Solid-Phase Peptide Synthesis of Peptides 21, 22, and 23

The solid-phase methodology toward ADPr-peptides **21, 22,** and **23** containing, respectively, ADPr-Gln, ADPr-Asn, and ADPr-Cit as the ADP-ribosylated amino acids described in this section (Fig. 2) starts with the assembling of phosphoribosylated precursor peptides. The key aspect of this synthetic approach is the use of pre-phosphoribosylated Fmoc-amino acids (**18**, **19**, and **20**, Subheading 3.1) to avoid potentially difficult phosphorylation of the 5-hydroxyl group of the ribofuranose residue of peptide resins [10]. The solid-phase synthesis described in this section is different from standard Fmoc-based peptide synthesis protocol in that it uses the side-chain protections and the linker sensitive to nucleophilic bases, such as alkali and ammonia. Upon the completion of otherwise a fairly standard stepwise solid-phase synthesis of phosphoribosylated peptide chain, the peptide resin is subjected to General protocol A (as detailed in Fig. 3) designed to introduce the pyrophosphate bridge between the 5-OH of ribofuranosyl moiety at the peptide and the 5'-OH of adenosine. To cleave the ADPr-peptides from the resin and to remove the base-sensitive side-chain protections, General protocol B (Fig. 2) is employed. It is important to execute this protocol as a two-step procedure. First, the linker is cleaved with a solution of ammonia in trifluoroethanol to give N-terminal carboxamide exclusively. Subsequently, the cleavage of trifluoroacetyl protections of lysine side chains and the *N*-benzoyl protecting group at the exocyclic amino group of the adenine base is effected by aqueous ammonia.

General Protocol A: On-Resin t-Butyl Cleavage and Pyrophosphate Formation

This protocol (Fig. 3) is based on 50 μmol peptide resin and is applicable for the synthesis of peptides **21**, **22**, and **23**. The resin can be best handled in a fritted syringe.

1. Add HFIP (2 mL) and HCl (12.3 M, 1 eq.) to the resin, and shake for 1 h at room temperature.
2. Filter and wash the resin with pyridine and subsequently repeatedly with DCM.
3. Analyze the resin with ^{31}P-NMR spectroscopy to determine the extent of t-Bu cleavage (*see* **Note 2**). If ^{31}P-NMR indicates incomplete cleavage of t-Bu, repeat the treatment with HFIP (2 mL) and HCl (1 eq.) for 30 min.
4. Wash the resin with pyridine and DCM, and place it under an argon atmosphere.
5. Wash with dry ACN extensively to remove traces of water (*see* **Note 3**).
6. Add a solution of ETT (6 eq., 0.25 M in ACN) and a solution of adenosine amidite (= *N*^{6}-benzoyl-9-[5-*O*-(*N*,*N*-diisopropylamino-*O*-cyanoethyl)phosphoramidite-2,3-di-*O*-isobutyryl-β-D-ribofuranosyl]-adenine (**24**), prepared as described [17, 18] (3 eq., 0.3 M in ACN), and shake the resin for 30 min.
7. Filter the resin off, wash it with anhydrous ACN, and add a solution of CSO (2 mL, 0.5 M in ACN). Shake the resin for 10 min and filter and wash with anhydrous ACN.
8. Add a solution of DBU (2 mL, 0.5 M in DMF) and shake the mixture for 10 min and filter and wash sequentially with DMF, ACN, and DCM.

General Protocol B: Cleavage and Deprotection Protocol

This protocol is based on 50 μmol peptide resin and is applicable for the synthesis of peptides **21**, **22**, and **23**.

1. Place the resin in a vessel with a screw cap (15–20 mL), add a saturated solution of NH_3 in trifluoroethanol (5 mL), wait for 4 h to cleave the peptide from the resin, and partially remove the base-labile protecting groups (*see* **Note 4**.)
2. Add NH_4OH (5 mL), and shake the resin for 16 h to complete the deprotection.
3. Filter off the resin, wash it with H_2O, and concentrate the combined filtrates under reduced pressure.
4. Dissolve the crude peptide in a minimal amount of MeOH, add a few drops of AcOH, precipitate it in ice-cold Et_2O, centrifuge (5.0 min at 4000 rpm [2000 × *g*]), and decant the

supernatant to obtain the precipitate of ADPr-peptide that is ready for purification. The purification protocol to be followed is dependent on the nature of peptide and described separately for each peptide.

3.4.1 Ac-Pro-Gln(N^{δ}-[5'-O-Adenosine Diphosphate-α-D-Ribosyl])-Pro-Ala-Lys-Ser-Ala-Pro-Ala-Pro-Lys-Lys-Gly-NH_2 (21)

Synthesis of Gln-ADPr peptide **21** is relatively straightforward and usually high-yielding. The phosphoribosyl peptide is first assembled up to Pro-3 (counting from the N-terminus) starting from TentaGel® resin equipped with HMBA linker and loaded with Fmoc-Gly-OH. Next, the Ser(Trt) is converted to Ser(Ac) on-resin via cleavage of the trityl with diluted acid and subsequent acetylation, then Fmoc group is cleaved from the amino group of the proline residue, and the synthesis of the peptide chain can be completed. It is not advisable to use Fmoc-Ser(Ac)-OH as a building block in peptide coupling (even if it is available) due to possible risk of side reactions such as elimination of acetate or epimerization of the activated serine. After the completion of the stepwise synthesis of the peptide chain, the immobilized phosphoribosyl precursor is subjected to General protocol A (pyrophosphate formation) and General protocol B (cleavage and deprotection).

1. Synthesize the intermediate peptide first using a standard SPPS Fmoc-protocol either on an automated peptide synthesizer or manually starting from TentaGel® S HMBA resin (in this example 225 mg, 50 μmol is used) preloaded with Fmoc-Gly-OH. The resin preloaded with Fmoc-Gly can be either obtained commercially or prepared using standard carbodiimide-/DMAP-based method. The loading in this example (0.22 mmol/g) was determined by UV-VIS spectroscopy as described.

   Use the following amino acid building blocks: Fmoc-Lys(Tfa)-OH, Fmoc-Pro-OH, Fmoc-Ser(Trt)-OH, and Fmoc-Ala-OH (all commercially available). Abbreviated solid-phase protocol we followed: Fmoc cleavage using 20% piperidine in DMF, coupling of the appropriate amino acid applying a fivefold excess, activation by 5 eq. HCTU in NMP (0.25 M), and 12.5 eq. DiPEA in NMP (1.25 M) for 1 h (*see* **Note 5**).

2. Transfer the resin with immobilized precursor peptide to a fritted syringe (10 mL) before coupling of phosphoribosylated glutamine **18**.
3. Treat the resin repeatedly with 5% TFA in DCM until no yellow color appears upon addition of the fresh TFA solution indicating complete removal of trityl from serine residue.
4. Wash the resin with DCM and NMP repeatedly.
5. Add Ac_2O (10% $^v/_v$) and DMAP (cat.) in NMP (2 mL), wait for 20 min, and then wash with NMP and DCM consecutively to protect the serine residue with acetyl group.

6. Add compound **18** (116 mg, 150 μmol), HCTU (150 μmol), and DIPEA (300 μmol) in NMP (1.5 mL), and shake for 16 h.
7. Wash with NMP and DCM, and couple the last amino acid (Fmoc-Pro-OH), using the standard cycle (either automated or manual).
8. Cleave Fmoc, cap (with acetic anhydride), and wash (NMP, DCM) the resin using standard peptide chemistry protocols to obtain fully protected phosphorylribosyl peptide.
9. Apply *General protocol A* to the resin to obtain the crude immobilized ADPr-peptide.
10. Apply *General protocol B* to the resin to obtain crude peptide **21**.
11. Purify the crude peptide obtained after **step 10** by RP-HPLC using a gradient of 5–20% B (eluent: A. 10 mM NH_4OAc in H_2O, B. MeCN). Pool the semi-pure fractions containing the product which were applied to a boronate affinity column to yield the title compound. The combined fractions were lyophilized to yield the title compound (**21**) as a white solid (3.3 mg, 1.8 μmol, 4% based on initial loading).

^{1}H NMR (500 MHz, D_2O) δ 8.55 (s, 1H, H2), 8.31 (s, 1H, H8), 6.17 (d, J = 6.1 Hz, 1H, H1'-adenosine), 5.67 (d, J = 4.5 Hz, 1H, H1'-ribose).

^{31}P-NMR (162 MHz, D_2O): δ −10.34, −10.47, −10.63, −10.76. (EDTA was added to sharpen up the signals).

LC-MS (5–20% MeCN [1% TFA] in 12.5 min), Rt = 6.38.

HRMS $[C_{74}H_{121}N_{23}O_{29}P_2 + 2H]^{2+}$: 929.9165 found, 929.9161 calculated, $[C_{74}H_{121}N_{23}O_{29}P_2 + 3H]^{3+}$: 620.2800 found, 620.2798 calculated.

3.4.2 Ac-Pro-Thr-Val-Phe-Glu-Asn(N^t-[5'-O-Adenosine Diphosphate-α-D-Ribosyl])-Tyr-Val-Ala-Asp-Ile-NH_2 (22)

Preparation of this particular of Asn-ADPr peptide (**22**) is still somewhat problematic in terms of yield (*see* **Note** 7). The phosphoribosylated peptide is assembled here completely (up to the N-terminal proline) starting from TentaGel® resin equipped with HMBA linker and loaded with Fmoc-Ile-OH. The Fmoc-Tyr-OH and Fmoc-Thr-OH are introduced without side-chain protection and acetylated on-resin "post-synthetically." After the completion of the stepwise synthesis of the peptide chain, the immobilized phosphoribosyl precursor is subjected to General protocol A (pyrophosphate formation) as described for ADPr-Gln peptide in Subheading 3.4.1. The protective groups (Dmab) are cleaved next (*see* **Note** 7), and the immobilized partially protected ADPr-peptide is submitted to General protocol B (cleavage and deprotection).

1. Synthesize the intermediate peptide first using a standard SPPS Fmoc-protocol exactly as described for ADPr-peptide **21** start-

ing from TentaGel® S HMBA resin (in this example 225 mg, 50 μmol is used) preloaded with Fmoc-Ile-OH (0.22 mmol/g).

Use the following amino acid building blocks: Fmoc-Asp(Dmab)-OH, Fmoc-Glu(Dmab)-OH, Fmoc-Pro-OH, Fmoc-Tyr-OH, Fmoc-Ala-OH, Fmoc-Val-OH, Fmoc-Thr-OH, Fmoc-Phe-OH (all commercially available), and Fmoc-phosphoribosylated building block **19** (*see* **Note 6**).

2. Transfer the resin with immobilized precursor peptide to a fritted syringe (10 mL); wash with NMP.
3. Transfer the resin to a fritted syringe and cap it with Ac_2O (50 eq.) and DIPEA (50 eq.) in NMP (2 mL) for 1 h, and then wash it with NMP.
4. Apply *General protocol A* to the resin to obtain the crude immobilized ADPr-peptide.
5. Add 2%$^v/_v$ hydrazine in DMF (2 mL) to remove the Dmab groups, and shake the resin for 5 min. Repeat this treatment two times more.
6. Wash the resin with NMP and DCM repeatedly. Dry the resin in vacuo to remove the residual solvent.
7. Apply *General protocol B* to the resin to obtain crude peptide **22**.
8. Purify the crude peptide obtained after **step 7** by ion-exchange HPLC using a strong anion exchanger (we used source 15Q 10 mm × 10 cm column) applying a gradient elution with NH_4OAc (20 mM to 1.0 M), and lyophilize the product fractions repeatedly to obtain ADPr-peptide **22** as a white solid (0.98 mg, 0.53 μmol, 1.1% based on initial loading) (*see* **Notes 7** and **8**).

^{1}H NMR (500 MHz, D_2O) δ 8.48 (s, 1H, H2), 8.21 (s, 1H, H8), 7.29 – 7.19 (m, 3H, arom. Phe), 7.15 (d, J = 7.3 Hz, 2H, arom. Phe), 7.02 (d, J = 8.5 Hz, 2H, arom. Tyr), 6.75 (d, J = 8.4 Hz, 2H, arom. Tyr), 6.08 (d, J = 5.8 Hz, 1H, H1'-adenosine), 5.64 (d, J = 4.4 Hz, 1H, H1'-ribose).

^{31}P-NMR (202 MHz, D2O): δ −10.38, −10.49, −10.64, −10.74. (EDTA was added to sharpen signals).

LC-MS (5–50% MeCN [1% TFA] in 12.5 min), Rt = 7.90.

HRMS $[C_{76}H_{110}N_{18}O_{32}P_2 + H]^+$: 1849.7084 found, 1849.7082 calculated, $[C_{76}H_{110}N_{18}O_{32}P_2 + 2H]^{2+}$: 925.3572 found, 925.3577 calculated, $[C_{76}H_{110}N_{18}O_{32}P_2 + 2Na]^{2+}$: 936.8469 found, 936.3487 calculated.

3.4.3 Ac-Ala-Gly-Glu-Cit(N^ω-[5'-O-Adenosine Diphosphate-α-D-Ribosyl])-Arg-Tyr-Gly-NH_2 (23)

Synthesis of ADPr-Arg peptide **23** suffers from poor yield for the same reason as ADPr-Asn counterpart **22** (*see* **Note 7**). In addition, due to the presence of arginine residue bearing a bis-Alloc protection at its side chain, DBU should be used, at least, from the moment the Fmoc-Arg(Alloc)$_2$-OH has been coupled to the resin, in order to preserve the δ-Alloc group known to be labile toward piperidine, which is the standard reagent in Fmoc chemistry. The phosphoribosyl peptide is first assembled up to Tyr-6 (counting from the N-terminus) starting from TentaGel® resin equipped with HMBA linker and loaded with Fmoc-Gly-OH. Next, the tyrosine side chain is acetylated on-resin, Fmoc group is cleaved from the amino group of the tyrosine residue, and the synthesis of the peptide chain is completed. Next, the immobilized phosphoribosyl precursor is subjected to General protocol A (pyrophosphate formation) as described for ADPr-Gln peptide in Subheading 3.4.1. Subsequently, the Alloc groups cleaved using Pd(0) in solution, and hydrazine hydrate is applied to cleave Dmab group from the glutamic acid side chain (*see* **Note 7**). Finally, the immobilized partially protected ADPr-peptide is submitted to General protocol B (cleavage and deprotection).

1. Start the synthesis from TentaGel® S HMBA resin, preloaded with Fmoc-Gly-OH (0.22 mmol/g, 550 mg, 100 μmol) and placed in a fritted syringe, as described for ADPr-Gln peptide **21**. First cleave Fmoc-protection with DBU (2%$^v/_v$) in DMF (*see* **Note 9**).
2. Couple Fmoc-Tyr-OH via a standard peptide synthesis cycle.
3. Cap the resin with Ac_2O (80 eq.), DIPEA (80 eq.) and DMAP (cat.) in NMP (4 mL) for 1 h and then wash it with NMP. Cleave Fmoc group.
4. Complete the synthesis of the intermediate peptide using Fmoc-protocol with DBU (2%$^v/_v$ in DMF) for Fmoc cleavage as follows. Couple sequentially Fmoc-Arg(Alloc)$_2$-OH (standard coupling), citrulline building block **20** (3 eq.16 h), and then Fmoc-Glu(Dmab)-OH, Fmoc-Gly-OH, and Fmoc-Ala-OH.
5. Cap the peptide resin with Ac_2O (0.5 mL) and DIPEA (0.8 mL) in NMP (4 mL) for 10 min. Repeat the capping once more.
6. Apply *General protocol A* to the resin to obtain the crude immobilized ADPr-peptide.
7. Remove Alloc protection from arginine side chain by shaking the resin for 15 min with N,N′-dimethylbarbituric acid (46 mg) and $Pd(PPh_3)_4$ (20 mg) in DCM (5 mL).
8. Repeat **step 7** treatment two times more.
9. Add 2%$^v/_v$ hydrazine in DMF (2 mL) to remove the Dmab groups, and shake the resin for 5 min. Repeat this treatment two times more.
10. Wash the resin with NMP and DCM repeatedly. Dry the resin in vacuo to remove the residual solvent.

11. Apply *general cleavage and deprotection protocol* to the resin to obtain crude peptide **23**.
12. Purify the crude peptide obtained after **step 11** by ion-exchange HPLC using a strong anion exchanger as described for ADPr-peptide **22** to obtain **23** as a white solid (0.31 mg, 0.2 μmol, 60/40;α/β, 0.2% based on initial loading) (*see* **Notes** 7 and **10**).

^{1}H NMR (500 MHz, D_2O) δ 8.48 (s, 1H, H2), 8.46 (s, 1H, H2), 8.22 (s, 1H, H8), 8.21 (s, 1H, H8), 7.06 (d, J = 8.6 Hz, 2H, arom. Tyr), 7.03 (d, J = 8.6 Hz, 2H, arom. Tyr), 6.76 (d, J = 8.6 Hz, 3H, arom. Tyr), 6.74 (d, J = 8.5 Hz, 2H, arom. Tyr), 6.12 – 6.07 (m, 2H, H1'α-, H1'β-adenosine), 5.54 (d, J = 4.4 Hz, 1H, H1'α-ribose), 5.36 (d, J = 6.3 Hz, 1H, H1'β-ribose).

^{31}P-NMR (202 MHz, D2O): δ −10.17, −10.28, −10.32, −10.43, −10.60, −10.62, −10.70, −10.72. (EDTA was added to sharpen signals).

LC-MS (10–50% MeCN [1% TFA] in 12.5 min), Rt = 4.47.

HRMS $[C_{50}H_{76}N_{18}O_{25}P_2 + H]^+$: 1391.4777 found, 1391.4777 calc, $[C_{50}H_{76}N_{18}O_{25}P_2 + 2H]^{2+}$: 696.2428.

4 Notes

1. Glycosylation step has to be performed under exclusion of moisture. Always keep the reaction mixture under an atmosphere of dry nitrogen or argon.
2. To perform "on-resin" NMR measurement, swell the resin in DMF, and use an external lock, for example, by inserting a capillary with aceton-d_6 into the NMR-tube.
3. It is imperative here to perform **steps 5–8** (under exclusion of moisture).
4. Treatment of NH_3 in TFE ensures that peptide carboxamide is formed exclusively when peptide is cleaved from HMBA linker, and this procedure is high-yielding [19].
5. The required base-labile O-acetyl protection is introduced at serine residue "post-synthetically" to exclude a potential elimination of acetate from activated Fmoc-Ser(Ac)-OH; Ser is therefore coupled as Fmoc-Ser(Trt)-OH.
6. Please note that Thr and Tyr building blocks are introduced here without side-chain protection and acetylated later on-resin.
7. The poor yield is probably due to premature cleavage of the peptide from the resin during the hydrazine treatment and appears unavoidable.
8. The same synthesis of **22** (100 μmol) and final deprotection using Tesser's base (dioxane, methanol, and aqueous NaOH)

followed by strong anion-exchange chromatography on source 15Q (10 mm × 10 cm) using gradient elution with NH_4OAc (20 mM to 1.0 M), and repeated lyophilization gave a complex mixture of ADPr-peptide-related products (11.8 mg, 6.4 μmol, 6% based on initial loading).

9. Use of DBU instead of piperidine is optional in **steps** 1 and **3** but compulsory in **step 4** to avoid cleavage of δ-Alloc group off the side chain of the arginine residue [20].
10. ADPr-citrulline loses the stereochemical purity of its glycosidic bond in course of the synthesis. The implications of this are discussed by us previously [5].

Acknowledgments

This work was supported by the Netherlands Organization for Scientific Research (NWO).

References

1. Gupte R, Liu Z, Kraus WL (2017) PARPs and ADP-ribosylation: recent advances linking molecular functions to biological outcomes. Genes Dev 31(2):101–126. https://doi.org/10.1101/gad.291518.116
2. Ryu KW, Kim DS, Kraus WL (2015) New facets in the regulation of gene expression by ADP-ribosylation and poly(ADP-ribose) polymerases. Chem Rev 115(6):2453–2481. https://doi.org/10.1021/cr5004248
3. Hassa PO, Haenni SS, Elser M, Hottiger MO (2006) Nuclear ADP-ribosylation reactions in mammalian cells: where are we today and where are we going? Microbiol Mol Biol Rev 70(3):789–829. https://doi.org/10.1128/Mmbr.00040-05
4. Vyas S, Chesarone-Cataldo M, Todorova T, Huang YH, Chang P (2013) A systematic analysis of the PARP protein family identifies new functions critical for cell physiology. Nat Commun 4:2240. https://doi.org/10.1038/ncomms3240
5. Kistemaker HA, Nardozza AP, Overkleeft HS, van der Marel GA, Ladurner AG, Filippov DV (2016) Synthesis and macrodomain binding of mono-ADP-ribosylated peptides. Angew Chem Int Ed Engl 55(36):10634–10638. https://doi.org/10.1002/anie.201604058
6. Bilan V, Selevsek N, Kistemaker HA, Abplanalp J, Feurer R, Filippov DV, Hottiger MO (2017) New quantitative mass spectrometry approaches reveal different ADP-ribosylation phases dependent on the levels of oxidative stress. Mol Cell Proteomics 16(5):949–958. https://doi.org/10.1074/mcp.O116.065623
7. Daniels CM, Ong SE, Leung AK (2015) The promise of proteomics for the study of ADP-ribosylation. Mol Cell 58(6):911–924. https://doi.org/10.1016/j.molcel.2015.06.012
8. Laing S, Unger M, Koch-Nolte F, Haag F (2011) ADP-ribosylation of arginine. Amino Acids 41(2):257–269. https://doi.org/10.1007/S00726-010-0676-2
9. Kistemaker HAV, van der Heden van Noort GJ, Overkleeft HS, van der Marel GA, Filippov DV (2013) Stereoselective ribosylation of amino acids. Org Lett 15(9):2306–2309. https://doi.org/10.1021/Ol400929c
10. van der Heden van Noort GJ, van der Horst MG, Overkleeft HS, van der Marel GA, Filippov DV (2010) Synthesis of mono-ADP-ribosylated oligopeptides using ribosylated amino acid building blocks. J Am Chem Soc 132(14):5236–5240. https://doi.org/10.1021/Ja910940q
11. Gold H, van Delft P, Meeuwenoord N, Codee JDC, Filippov DV, Eggink G, Overkleeft HS, van der Marel GA (2008) Synthesis of sugar nucleotides by application of phosphoramidites. J Org Chem 73(23):9458–9460. https://doi.org/10.1021/jo802021t
12. Kistemaker HAV, Lameijer LN, Meeuwenoord NJ, Overkleeft HS, van der Marel GA, Filippov DV (2015) Synthesis of well-defined adenosine

diphosphate ribose oligomers. Angew Chem Int Ed Engl 54(16):4915–4918. https://doi.org/10.1002/anie.201412283

13. Kistemaker HAV, Meeuwenoord NJ, Overkleeft HS, van der Marel GA, Filippov DV (2015) On the synthesis of oligonucleotides interconnected through pyrophosphate linkages. Eur J Org Chem 2015(27):6084–6091. https://doi.org/10.1002/ejoc.201500911
14. Cremosnik GS, Hofer A, Jessen HJ (2014) Iterative synthesis of nucleoside oligophosphates with phosphoramidites. Angew Chem Int Ed Engl 53(1):286–289. https://doi.org/10.1002/Anie.201306265
15. Conroy T, Jolliffe KA, Payne RJ (2009) Efficient use of the Dmab protecting group: applications for the solid-phase synthesis of N-linked glycopeptides. Org Biomol Chem 7(11):2255–2258. https://doi.org/10.1039/b821051a
16. Bonfiglio JJ, Fontana P, Zhang Q, Colby T, Gibbs-Seymour I, Atanassov I, Bartlett E, Zaja R, Ahel I, Matic I (2017) Serine ADP-ribosylation depends on HPF1. Mol Cell 65(5):932–940 e936. https://doi.org/10.1016/j.molcel.2017.01.003
17. Kistemaker HAV, Meeuwenoord NJ, Overkleeft HS, van der Marel GA, Filippov DV (2016) Solid-phase synthesis of oligo-ADP-ribose. Curr Protoc Nucl Acid Chem 64:4.68.61–64.68.27. https://doi.org/10.1002/0471142700.nc0468s64
18. van der Heden van Noort GJ, Verhagen CP, van der Horst MG, Overkleeft HS, van der Marel GA, Filippov DV (2008) A versatile one-pot procedure to phosphate monoesters and pyrophosphates using Di(p-methoxybenzyl)-N,N-diisopropylphosphoramidite. Org Lett 10(20):4461–4464. https://doi.org/10.1021/Ol801608j
19. van der Heden van Noort GJ, van Delft P, Meeuwenoord NJ, Overkleeft HS, van der Marel GA, Filippov DV (2012) Fully automated sequential solid phase approach towards viral RNA-nucleopeptides. Chem Commun 48(65):8093–8095. https://doi.org/10.1039/C2cc33477a
20. Filippov D, Kuyl-Yeheskiely E, van der Marel GA, Tesser GI, van Boom JH (1998) Synthesis of a nucleopeptide fragment from poliovirus genome. Tetrahedron Lett 39(21):3597–3600

Chapter 25

Identifying Genomic Sites of ADP-Ribosylation Mediated by Specific Nuclear PARP Enzymes Using Click-ChIP

Ryan A. Rogge, Bryan A. Gibson, and W. Lee Kraus

Abstract

Nuclear poly(ADP-ribose) polymerases (PARPs), including PARPs 1, 2, and 3 and the Tankyrases, belong to a family of enzymes that can bind to chromatin and covalently modify histone- and chromatin-associated proteins with ADP-ribose derived from nuclear NAD^+. The genomic loci where the nuclear PARPs bind and covalently modify chromatin are a fundamental question in PARP biology. Chromatin immunoprecipitation coupled with deep sequencing (ChIP-seq) has become an essential tool for determining specific sites of binding and modification genome-wide. Few methods are available, however, for localizing PARP-specific ADP-ribosylation events across the genome. Here we describe a variation of ChIP-seq, called Click-ChIP-seq, for identifying sites of ADP-ribosylation mediated by specific PARP family members. This method uses analog-sensitive PARP (asPARP) technology, including asPARP mutants and the alkyne-containing "clickable" NAD^+ analog 8-Bu(3-yne)T-NAD^+. In this assay, nuclei from cells expressing an asPARP protein of interest are incubated with 8-Bu(3-yne)T-NAD^+, which is incorporated into ADP-ribose modifications mediated only by that specific asPARP protein. The nuclei are then subjected to cross-linking with formaldehyde, and the protein-linked analog ADP-ribose is clicked to biotin using copper-catalyzed alkyne-azide "click" chemistry. The chromatin is fragmented, and the fragments containing analog ADP-ribose are enriched using streptavidin-mediated precipitation. Finally, the enriched DNA is analyzed by qPCR or deep-sequencing experiments to determine which genomic loci contain ADP-ribose modifications mediated by the specific PARP protein of interest. Click-ChIP-seq has proven to be a robust and reproducible method for identifying chromatin-associated, PARP-specific ADP-ribosylation events genome-wide.

Key words ADP-ribosylation, Analog sensitivity, Automodification, Chromatin, Chromatin immunoprecipitation (ChIP), Click chemistry, Cross-link, Mono(ADP-ribosyl)ation (MARylation), Mutation, NAD^+ analog, Nucleus, Nucleosome, Poly(ADP-ribose) polymerase (PARP), Poly(ADP-ribosyl)ation (PARylation), Posttranslational modification (PTM)

1 Introduction

Chromatin, the physiological template for nuclear processes, is bound by a host of nuclear proteins. Many of these proteins possess enzymatic activities that covalently modify nucleosomal histones, as well as other nonhistone chromatin-associated proteins [1, 2]. These posttranslational modifications can alter the structure

Paul Chang (ed.), *ADP-ribosylation and NAD+ Utilizing Enzymes: Methods and Protocols*, Methods in Molecular Biology, vol. 1813, https://doi.org/10.1007/978-1-4939-8588-3_25,

and function of chromatin by altering nucleosomes locally, as well as higher-order chromatin structures globally [1, 3]. Nuclear poly(ADP-ribose) polymerases (PARPs), including PARPs 1, 2, and 3 and the Tankyrases, belong to a family of enzymes that can bind to chromatin and covalently modify histones and chromatin-associated proteins with ADP-ribose derived from nuclear NAD^+ [4–6]. PARP-mediated ADP-ribosylation can occur as mono- or poly-modifications, yielding mono(ADP-ribose) (MAR) or poly(ADP-ribose) (PAR), respectively [7–9]. The unique biologies of MAR and PAR are poorly understood.

Nuclear PARPs possess mono- or poly(ADP-ribosyl)transferase activity. For example, PARPs 1 and 2 and the Tankyrases are polyenzymes [i.e., catalyzes poly(ADP-ribosyl)ation], while PARP-3 is a monoenzyme [i.e., catalyzes mono(ADP-ribosyl) ation] [4, 9]. PARP-1 in particular is a highly abundant protein, and the vast majority of PAR in the nucleus is associated with the automodification of PARP-1 [6, 8, 10]. Much less is known about the biology of other nuclear PARP enzymes. While first implicated as a facet of the cellular response to DNA damage [11], nuclear ADP-ribosylation (both MARylation and PARylation) has now been shown to play key roles in transcription, as well as RNA processing and stability [5, 6, 12]. Identifying the location of ADP-ribosylation across the genome is important for understanding the role of PARPs and PAR in chromosome biology. In addition to core histones, other chromatin-associated proteins that are ADP-ribosylated include the PARP enzymes (through automodification) [13], linker histones [14–16], and transcription-related factors such as NELF-E, C/EBPβ, and CTCF [14, 17, 18]. Although identification of these and other substrates has been informative, identifying PARP-specific ADP-ribosylation events at specific genomic loci, however, has proven challenging.

The Kraus Lab recently developed a method for the isolation of chromatin fragments that have been modified by a specific PARP enzyme [14]. This method, which we call Click-ChIP, is a variation of chromatin immunoprecipitation (ChIP) [19–21]. It uses recently developed analog-sensitive PARP (asPARP) technology, which relies on asPARP mutant enzymes (Fig. 1a, b), and the alkyne-containing "clickable" NAD^+ analog 8-Bu(3-yne)T-NAD^+ [14, 22] (Fig. 1c). The use of 8-Bu(3-yne)T-NAD^+ for nuclear ADP-ribosylation is made possible by an engineered analog-sensitive mutant of the PARP enzyme of interest [14, 22] (Fig. 1a, b). Mutation of a "gatekeeper" residue in the adenine-binding pocket of the PARP enzyme of interest to glycine or alanine (e.g., L877A in PARP-1) (Fig. 1b) creates a "hole" that is filled by the alkyne substituent added at the eighth position of the adenine ring in 8-Bu(3-yne)T-NAD^+ (Fig. 1c), allowing for its incorporation into ADP-ribosylated proteins [14].

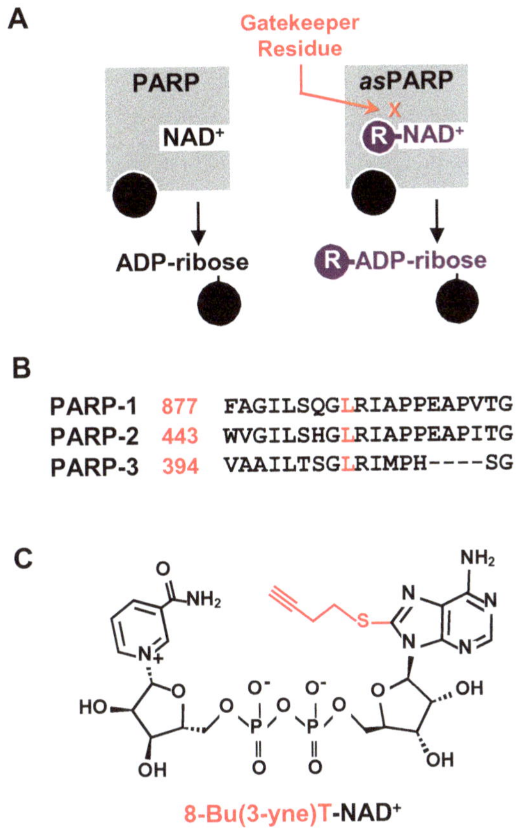

Fig. 1 Analog-sensitive PARP (asPARP) methodology used in Click-ChIP. (**a**) Schematic representation of the asPARP methodology used in Click-ChIP, which allows detection of genomic loci ADP-ribosylated by a specific PARP family member. In the asPARP approach, an asPARP protein (*gray square*) uses an alkyne-containing NAD^+ analog (e.g., 8-Bu(3-yne)T-NAD^+ (*shown in purple*) to catalyze the transfer of 8-Bu(3-yne)T-ADP-ribose (*also shown in purple*) onto a target protein *iblack circle*). (**b**) Multiple sequence alignment of a portion of the catalytic domains from PARPs 1, 2, and 3. The gatekeeper residues that confer NAD^+ analog sensitivity are highlighted in red. (**c**) Chemical structure of the clickable NAD^+ analog 8-Bu(3-yne)T-NAD^+. Adapted from Gibson and Kraus, 2017 [14]

In Click-ChIP, nuclei from cells expressing an asPARP mutant of interest are incubated with the NAD^+ analog, which is incorporated into ADP-ribose modifications mediated only by that specific asPARP protein [14] (Fig. 2). After cross-linking the nuclei with formaldehyde, the protein-linked analog ADP-ribose is clicked to biotin using copper-catalyzed alkyne-azide "click" chemistry [23] (Fig. 2). The chromatin is fragmented, and the fragments containing analog ADP-ribose are enriched using streptavidin-mediated precipitation [14] (Fig. 2). The genomic DNA enriched by Click-ChIP can be analyzed by qPCR or deep sequencing to

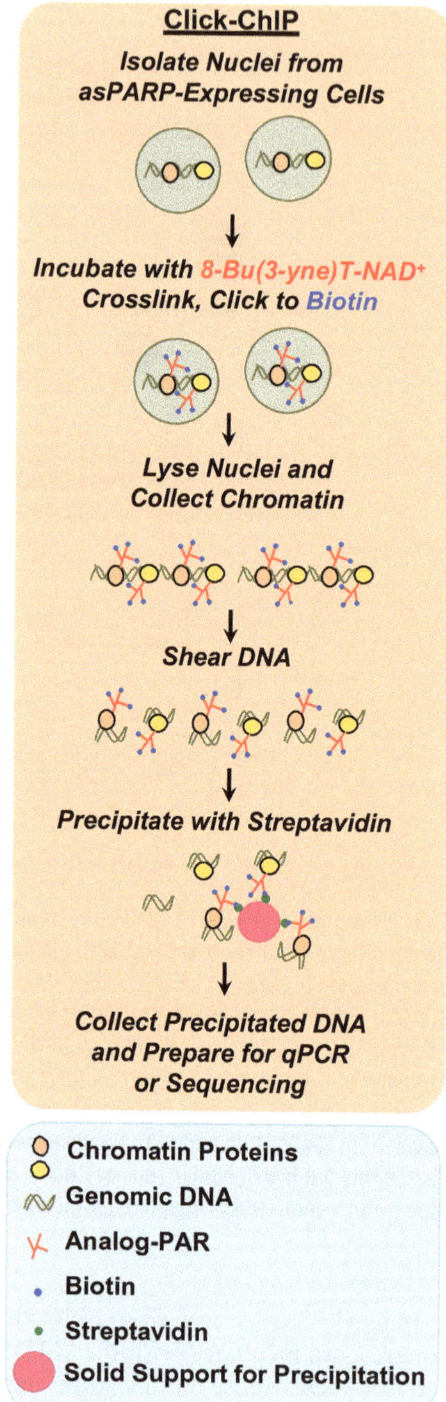

Fig. 2 The Click-ChIP-seq workflow. In Click-ChIP, nuclei from cells expressing an asPARP mutant of interest are incubated with the NAD^+ analog, which is incorporated into ADP-ribose modifications mediated only by that specific asPARP protein. After cross-linking the nuclei with formaldehyde, the protein-linked analog ADP-ribose is clicked to biotin using copper-catalyzed alkyne-azide "click" chemistry. The chromatin is fragmented, and the fragments containing analog ADP-ribose are enriched using streptavidin-mediated precipitation. The genomic DNA enriched by Click-ChIP can be analyzed by qPCR or deep sequencing to determine which genomic loci contain ADP-ribose modifications mediated by the specific PARP protein of interest. Adapted from Gupte et al., 2017 [12]

determine which genomic loci contain ADP-ribose modifications mediated by the specific PARP protein of interest [14]. The protocol described herein describes how to react isolated asPARP-containing nuclei with the NAD^+ analog, click the analog to biotin, and enrich the fragmented chromatin. A brief description and references are provided for further analysis of the genomic DNA associated with enriched ADP-ribosylated chromatin by deep sequencing (Click-ChIP-seq) (Fig. 2).

2 Materials

All reagents should be the highest-quality molecular biology and biochemistry grade.

2.1 Analog NAD^+ Labeling Reaction

1. 18 MΩ double-distilled H_2O (ddH_2O).
2. Protease inhibitor cocktail (e.g., Roche cOmplete™ EDTA-free; tablets provided in glass vials, which can be dissolved to 1×) (*see* **Note 1**).
3. Nuclei reaction buffer: 30 mM Tris-HCl (pH 7.5), 5 mM $MgCl_2$, 5 mM CaCl2, 10 mM KCl, 0.01% IGEPAL CA-630, 0.05 mM EDTA, 20% glycerol, 1 mM DTT, 1× protease inhibitor cocktail (*see* **Note 1**).
4. 8-Bu(3-yne)T-NAD^+ solution: 10 mM 8-Bu(3-yne)T-NAD^+ (analog NAD^+; (Biolog Life Science Institute) in ddH_2O.
5. β-NAD^+ solution: 10 mM in ddH_2O. Store in small, single-use aliquots at −80 °C. Dilute right before use.
6. Cross-linking buffer: 10 mM HEPES (pH 8.0), 10 mM NaCl, 10 mM $MgCl_2$, 0.1% IGEPAL CA-630, 1 mM DTT, 1× protease inhibitor cocktail (*see* **Note 1**).
7. Formaldehyde, 16% (methanol-free).
8. Glycine solution: 2.5 M in ddH_2O.

2.2 Click Reaction

1. Click reaction buffer: 10 mM HEPES (pH 8.0), 10 mM NaCl, 10 mM spermine, 0.1% IGEPAL CA-630, 1× protease inhibitor cocktail (*see* **Note 1**).
2. Biotin-azide solution: 100 mM in ddH_2O.
3. Copper sulfate ($CuSO_4$) solution: 100 mM dissolved in ddH2O (*see* **Note 1**).
4. Tris(3-hydroxypropyltriazolylmethyl)amine (THPTA) solution: 500 mM in ddH_2O (*see* **Note 2**).
5. Sodium ascorbate solution: 500 mM in ddH_2O (*see* **Note 1**).
6. 25× click catalyst mix: 125 mM THPTA, 25 mM $CuSO_4$, 125 mM sodium ascorbate (*see* Subheading 3.2, **step 2,** and **Note 1**).

2.3 Chromatin Enrichment

1. ChIP lysis buffer: 50 mM Tris-HCl (pH 7.9), 1% SDS, 10 mM EDTA, 1× protease inhibitor cocktail (*see* **Note 1**).
2. Water bath sonicator (Diagenode).
3. 1× TBE: 89 mM Tris-HCl (pH 7.6), 89 mM boric acid, 2 mM EDTA.
4. 10× TBE loading solution: 400 mg/mL sucrose, 2 mg/mL Orange G.
5. ChIP dilution buffer: 20 mM Tris-HCl (pH 7.9), 2 mM EDTA, 150 mM NaCl, 0.5% Triton X-100, 1 mM DTT, 1× protease inhibitor cocktail.
6. Nitrocellulose membrane.
7. HRP-conjugated streptavidin.
8. Enhanced chemiluminescent (ECL) reagent.
9. Streptavidin-coated beads (e.g., Dynabeads MyOne Streptavidin T1).
10. Magnetic isolation stand.
11. 2% sodium dodecyl sulfate.
12. Low-salt wash buffer: 20 mM Tris-HCl (pH 7.9), 2 mM EDTA, 150 mM NaCl, 0.5% Triton X-100, 1 mM DTT, 1× protease inhibitor cocktail (*see* **Note 1**).
13. High-salt wash buffer: 20 mM Tris-HCl (pH 7.9), 2 mM EDTA, 500 mM NaCl, 0.05% SDS, 1% Triton X-100, 1× protease inhibitor cocktail (*see* **Note 1**).
14. LiCl wash buffer: 10 mM Tris-HCl (pH 7.9), 1 mM EDTA, 250 mM LiCl, 1% IGEPAL, 1% sodium deoxycholate, 1× protease inhibitor cocktail (*see* **Note 1**).
15. TE buffer: 10 mM Tris-HCl (pH 7.9), 0.25 mM EDTA.

2.4 Decross-Linking the Chromatin and Purifying the Genomic DNA

1. Decross-linking buffer: 100 mM sodium bicarbonate, 100 mM NaCl, 1% SDS.
2. 500 μg/mL RNase, DNase-free.
3. 20 mg/mL proteinase K.
4. Phenol/chloroform/isoamyl alcohol mixture (25:24:1).
5. Chloroform.
6. Glycogen solution: 10 mg/mL in ddH_2O.
7. Sodium acetate (NaOAc) solution: 3 M in ddH_2O.
8. Isopropanol.
9. Ethanol solution: 75% (v/v) in ddH_2O.
10. DNA purification resin (e.g., AMPure XP beads; Beckman Coulter).

2.5 Genomic Analyses

1. BOWTIE short read aligner (http://bowtie-bio.sourceforge.net/index.shtml) [24] or another appropriate short read aligner.
2. Custom scripts for analyzing Click-ChIP-seq data, as described in Gibson et al. (2016) [14].

3 Methods

3.1 In Nuclei Labeling with Analog NAD+ and Formaldehyde Cross-Linking

The goal of this section of the protocol is to label PARP-specific sites of ADP-ribosylation across the genome using an analog-sensitive mutant of a nuclear PARP of interest and a "clickable" NAD^+ analog. Since the cell membrane is impermeable to the NAD^+ analog, the reactions are carried out in isolated nuclei from the cells of interest. The nuclei are incubated with the NAD^+ analog, which is incorporated into posttranslational modifications of chromatin proteins by the asPARP. The nuclei are then cross-linked with formaldehyde to preserve protein-DNA interactions in a standard chromatin immunoprecipitation (ChIP) protocol. After cross-linking, the nuclei contain proteins cross-linked to genomic DNA and other chromatin proteins, some of which are modified by clickable ADP-ribosylation from the activity of the asPARP.

1. For each experimental condition, resuspend approximately five million nuclei isolated from cells expressing an asPARP enzyme of interest in 500 μL of room temperature nuclei reaction buffer (*see* **Note 3**).
2. Set up both experimental and control reactions using analog NAD^+ and natural β-NAD^+. Add β-NAD^+ or analog NAD^+ to their respective reactions of five million nuclei to a final concentration of 250 μM (i.e., add 12.5 μL of 10 mM β-NAD^+ or analog NAD^+ to each 500 μL reaction) (*see* **Notes 4** and **5**).
3. Allow the ADP-ribosylation reactions to proceed by incubating the nuclei at room temperature for 30 min with gentle mixing.
4. Collect the nuclei by centrifugation at 2000 × *g* for 1 min at room temperature in a microcentrifuge.
5. Wash the nuclei by gentle resuspension in 500 μL of nuclei reaction buffer without analog NAD^+ or natural β-NAD^+, with subsequent centrifugation at 2000 × *g* for 1 min at room temperature in a microcentrifuge.
6. Resuspend the nuclei in 750 μL of cross-linking buffer.
7. To cross-link the chromatin within the nuclei, add 25 μL of 16% formaldehyde to 750 μL of cross-linking buffer. Incubate the samples at room temperature for 10 min with gentle mixing to prevent the nuclei from settling.

8. After the incubation, quench the formaldehyde cross-linking reaction by adding 100 μL of 2.5 M glycine. Mix gently and incubate on ice for 10 min.
9. Collect the nuclei by centrifugation at 500 × *g* for 10 min at room temperature in a microcentrifuge. Aspirate the supernatant.
10. Wash the nuclei by resuspending in 100 μL of click reaction buffer.
11. Collect the nuclei by centrifugation at 500 × *g* for 5 min at room temperature in a microcentrifuge. Resuspend the nuclei in 100 μL of click reaction buffer.
12. Collect the nuclei by centrifugation at 500 × *g* for 5 min at room temperature in a microcentrifuge (*see* **Note 6**).

3.2 Clicking the ADP-Ribosylated and Cross-Linked Chromatin Proteins to Biotin

The goal of this section of the protocol is to click the ADP-ribosylated and cross-linked chromatin proteins to biotin using copper-catalyzed alkyne-azide click chemistry. The click reaction is carried out in intact, analog NAD^+-labeled nuclei.

1. For the click reaction, resuspend the nuclei from Subheading 3.1, **step 12**, in 100 μL of click reaction buffer containing 50 μM of biotin-azide. (e.g., add 1 μL of 100 mM azide to 2 mL of click reaction buffer. Prepare a sufficient amount for two sequential click reactions).
2. Prepare 25× click catalyst mix by adding the following stepwise to a tube: 1 part ddH_2O, 1 part 100 mM $CuSO_4$ (at this point the solution should be faint blue), 1 part 500 mM THPTA (at this point the solution should be intensely blue), and 1 part 500 mM sodium ascorbate (at this point the solution should be clear) (*see* **Note 7**).
3. To initiate the click reaction, add 4 μL of 25× click catalyst mix (from **step 2**) and mix gently.
4. Let the click reaction proceed at room temperature with gentle mixing using a rotisserie-style mixer (or a nutator) at a low speed (~10–15 rpm) for 1 h (*see* **Note 8**).
5. Collect the nuclei by centrifugation at 500 × *g* for 5 min at room temperature in a microcentrifuge. Repeat the click reaction by resuspending the nuclei in fresh click reaction buffer containing 50 μM biotin-azide with subsequent addition of 4 μL of 25× click reaction buffer. Incubate at room temperature for 1 h with gentle mixing using a rotisserie-style mixer.
6. Collect the nuclei by centrifugation at 500 × *g* for 5 min at room temperature in a microcentrifuge. Wash the nuclei by resuspending them in click reaction buffer (without biotin-azide), and then collect them by centrifugation at 500 × *g* for 5 min at room temperature in a microcentrifuge (*see* **Note 9**).

3.3 Enrichment of ADP-Ribosylated Chromatin Fragments Clicked to Biotin

At this point in protocol, the chromatin within the nuclei contains ADP-ribosylated proteins that have been cross-linked to genomic DNA and clicked to biotin. This section of the protocol describes how to disrupt the nuclei, fragment the chromatin to a size that is appropriate for the generation of deep-sequencing libraries, and isolate the fragmented chromatin using streptavidin beads, which bind to the biotinylated, cross-linked chromatin proteins.

1. Resuspend the nuclei from Subheading 3.2, **step 4**, in 300 μL of ChIP lysis buffer using a Bioruptor bath sonicator (Diagenode). As a starting point, sonicate on high with 1 min pulses on a 50% duty cycle (i.e., 30 s on, 30 s off), for a total time period of 3 min. Mix the tubes by inversion, and repeat the previous sonication program twice (i.e., run the 3 min program three times). The goal of sonication is to generate DNA fragments with an average size of 200–300 bp (Fig. 3a) (*see* **Note 10**).

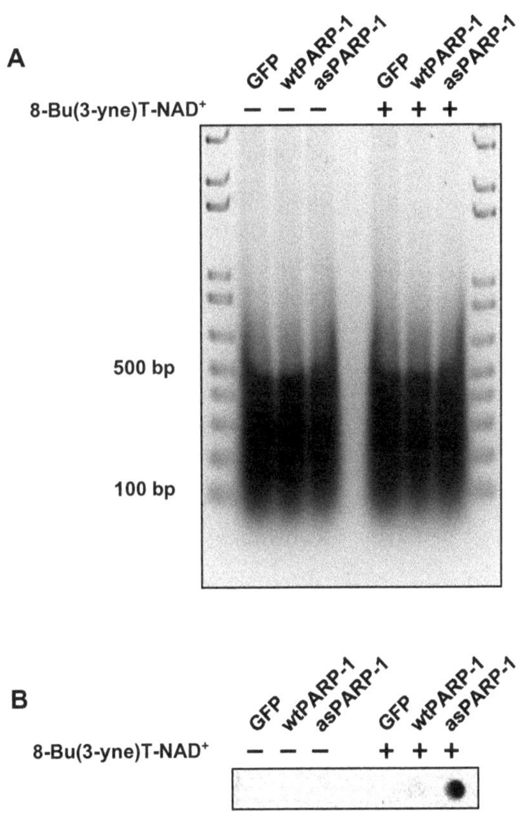

Fig. 3 Quality control assays of asPARP-labeled chromatin. (**a**) Agarose gel analysis of sonicated chromatin from asPARP-1-labeled or control labeled nuclei. Nuclei from mouse embryo fibroblasts (MEFs) expressing asPARP-1, wtPARP-1, or GFP were incubated ±8-Bu(3-yne)T-ADP-ribose. The nuclei were then cross-linked, and the protein-linked analog ADP-ribose was clicked to biotin using copper-catalyzed alkyne-azide "click" chemistry. The chromatin was then fragmented by sonication and analyzed on an ethidium bromide-stained 1.5% agarose gel run in 1× TBE. (**b**) HRP-streptavidin dot blot on nitrocellulose for biotin detection of the same sonicated chromatin samples shown in (**a**)

2. Perform a quality control check of the chromatin fragmentation. Deproteinize a small aliquot (10 μL) of each sample by adding 1 μL of 20 mg/mL proteinase K and incubating at 65 ° C for 2 h. Add 1 μL of 10× TBE loading solution to each aliquot, mix, and run on a 1.5% agarose gel in 1× TBE. Stain the gel with ethidium bromide and visualize (Fig. 3a).
3. Perform a quality control check of the biotin content of the chromatin. Adsorb 1 μL of each sample onto dry nitrocellulose membrane by pipetting. Let the spots dry. Perform dot blotting using an HRP-streptavidin detection reagent (*see* **Note 11**). Biotin incorporation should only be observed in samples in which asPARP-containing nuclei were incubated with analog NAD^+ (Fig. 3b) (*see* **Note 4**).
4. Add ChIP dilution buffer to the remaining sonicated chromatin samples from **step 1** above to a final volume of 1 mL, and mix thoroughly but gently.
5. Remove an aliquot from each sonicated chromatin sample as "input." A 50 μL aliquot will represent 5% of the total input. Do not add streptavidin-coated beads to the input sample; this sample will not be enriched for biotin-containing chromatin.
6. The samples are now ready for enrichment using streptavidin-coated magnetic beads. Resuspend the beads evenly, remove 50 μL for each sample, and wash the total amount using 1 mL of ChIP dilution buffer in a microcentrifuge tube (*see* **Note 12**). Place the tube in magnetic stand, and allow the magnet to collect the beads, aspirating the supernatant, and resuspend the beads in the original volume of ChIP dilution buffer. Add the 50 μL of beads to each of your samples. Incubate overnight with gentle mixing at 4 ° C (*see* **Note 13**).
7. Place each sample tube in a magnetic stand, and allow the magnet to collect the beads. Aspirate the supernatant.
8. Wash the beads by repeated magnetic isolation, aspiration of supernatant, and addition of 1 mL of wash buffer as described below, followed by a 5 min incubation at 4 °C with gentle mixing. The washes should be as follows:
   (a) Twice with 2% SDS.
   (b) Once with low-salt wash buffer.
   (c) Three times with high-salt wash buffer.
   (d) Once with LiCl wash buffer.
   (e) Twice with TE.

3.4 Decross-Linking the Chromatin and Purifying the Genomic DNA

In order to analyze the genomic DNA in the chromatin fragments, the chromatin must be decross-linked, eluted from the streptavidin beads, and purified. The steps below describe the protocol for doing so.

1. Decross-link the bead-bound chromatin from Subheading 3.3, **step 8**, by resuspending the beads in 200 μL of decross-linking buffer. Incubate at 65 ° C overnight. Decross-link the input samples from Subheading 3.3, **step 5**, as well by bringing the volume to 200 μL with decross-linking buffer and incubating alongside the enriched samples at 65 ° C overnight.
2. Isolate the beads by placing the tubes in a magnetic stand. Collect and save the supernatants in fresh tubes.
3. Wash the beads with 300 μL of TE, and add the wash (containing additional decross-linked genomic DNA released from the beads) to the cognate supernatants that were collected for each sample. Bring the input samples to a final volume of 500 μL with TE as well (to match the enriched samples).
4. Digest the RNA in the samples by adding RNase to a final concentration of 3 μg/mL (by adding 3 μL of a 500 μg/mL stock solution), mixing well, and incubating at 37 ° C for 30 min.
5. Digest the protein in the samples by adding proteinase K to a final concentration of 50 μg/mL (by adding 1.25 μL of 20 mg/mL stock solution), mixing well, and incubating at 55 ° C for 2 h.
6. Extract the DNA samples with phenol/chloroform/isoamyl alcohol (25:24:1) as follows:
   (a) Add an equal volume of phenol/chloroform/isoamyl alcohol and vortex the samples for 20 s. The solution should turn cloudy.
   (b) Separate the phases by centrifuging the samples at full speed (21,000 × *g*) in a microcentrifuge for 5 min at room temperature.
   (c) Remove, transfer to a new tube, and save the aqueous phase, which should be the top layer containing the enriched genomic DNA.
7. Extract the samples in a similar manner using an equal volume of chloroform, and collect the aqueous phase in a new tube.
8. Precipitate the DNA from the aqueous phase by the addition of 50 μL of 3 M NaOAc, 6 μL of 10 mg/mL glycogen, and 500 μL of isopropanol. Mix thoroughly, and collect the precipitate by centrifuging the samples at full speed in a microcentrifuge for 25 min at 4 °C.
9. Carefully remove the supernatant, making sure not to disturb the pellet, which will be translucent. Wash the pellet carefully with 500 μL of 75% ethanol.

10. Pellet the DNA again by centrifuging the samples at full speed in a microcentrifuge for 10 min at 4 °C. Carefully remove the supernatant, making sure not to disturb the pellet, which will be white at this point.
11. Dry the pellet for 5 min at room temperature to evaporate the alcohol by leaving the tubes open and inverted. Do not let the pellets dry completely, or they will become difficult to dissolve. The enriched genomic DNA pellets can be stored at −20 °C this point until further analysis.
12. Dissolve the DNA pellet in 100 μL of ddH_2O, TE, or your buffer of choice. At this point, the enriched genomic DNA can be stored at −20 °C, analyzed to be qPCR, or subjected to other analytical methods (such as deep sequencing; *see* Subheading 3.4 below).
13. Purify the DNA further using AMPure XP beads (or a similar bead product) according to the manufacturer's instructions.
14. Quantify the DNA from both the enriched and input fractions using a Qubit fluorimeter.
15. If desired, check the enrichment at specific genomic loci of interest using qPCR, as shown in Fig. 4.

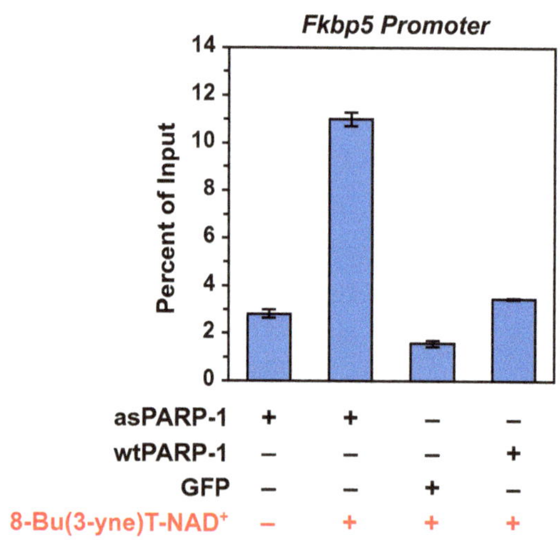

Fig. 4 Gene-specific enrichment of PARP-1 ADP-ribosylated chromatin using Click-ChIP-qPCR in MEFs. Bar graph showing Click-ChIP-qPCR results for a genomic locus (i.e., the promoter of the *Fkbp5* promoter) enriched in PARP-1 ADP-ribosylated chromatin in MEFs. The results are shown for nuclei expressing asPARP-1 with or without incubation with analog NAD^+, as well as nuclei expressing wild-type PARP-1 or GFP incubated with analog NAD^+. Each bar represents the mean ± SEM from three separate determinations

3.5 Genomic Analyses of Genomic DNA Enriched Using Click-ChIP

The DNA isolated from the Click-ChIP protocol described above can be processed into genomic libraries for deep sequencing as described below, allowing a genome-wide view of PARP-specific ADP-ribosylation events.

1. Prepare Click-ChIP-Seq libraries from the enriched and input DNA samples using methods that we described previously [14] (*see* **Note 14**).
2. Sequence the libraries using an appropriate next-generation sequencing technology (e.g., Illumina HiSeq or NextSeq), and collect the data (*see* **Note 15**).
3. Align the reads to a reference genome using the standard settings in BOWTIE [24].
4. Analyze the data and call peaks. Present the enrichment data in browser track representations, heatmaps, and boxplots (*see*, e.g., Fig. 5) (*see* **Note 16**).

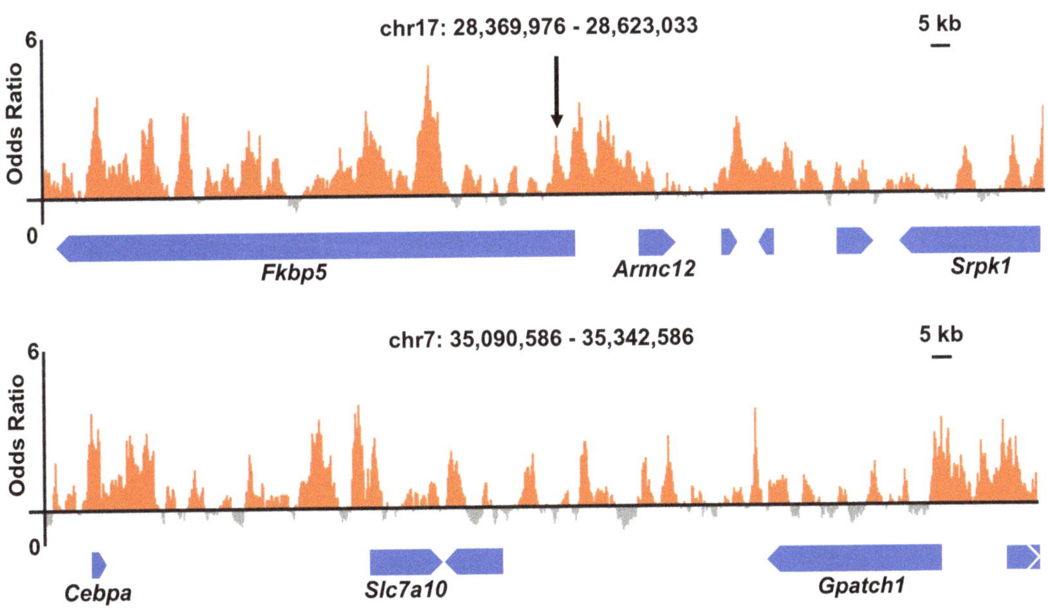

Fig. 5 Global enrichment of PARP-1 ADP-ribosylated chromatin using Click-ChIP-seq in MEFs. Browser tracks showing Click-ChIP-seq results for two different genomic regions enriched in PARP-1 ADP-ribosylated chromatin in MEFs. Enrichment is displayed as the odds ratio (reads per bin treatment/reads per bin input, normalized by the ratio of their respective total read counts). The Click-ChIP-seq data are from Gibson et al., 2016 [14], and the browser track was generated using the Integrative Genomic Viewer (IGV) [25, 26]. Gene locations are displayed as horizontal blue block arrows. The region analyzed by qPCR in Fig. 4 is marked by a black arrow

4 Notes

1. Make fresh for each use.
2. Make THPTA in ddH_2O and store in single-use aliquots at −20 °C.
3. Isolate nuclei from cells expressing an asPARP of interest using a method appropriate for those cells in an amount appropriate for all downstream assays. The Click-ChIP protocol described herein has been used successfully with one to ten million nuclei isolated from mouse embryonic fibroblasts (MEFs) and human MCF-7 cells.
4. Major factors that we have tested that affect the final enrichment and fold enrichment over control include (1) the number of nuclei, (2) the concentration of analog NAD^+ with which the nuclei are reacted (0.001 to 250 μM has been a useful range), and (3) the concentration of biotin-azide used in the click reaction (10 to 100 μM has been a useful range). The conditions listed in the protocol were used in the published experiments from Gibson et al. (2016) [14]. Low concentrations of analog NAD^+ for the labeling of the isolated nuclei may result in low-signal dot blots but still yield good enrichment detectable by qPCR. Adjustment of the ratios noted above is a good place to start when optimizing for new biological systems or new experimental systems.
5. Additional controls include the same cells expressing GFP or wild-type version of the PARP enzyme. These controls will provide information about the background levels of analog NAD^+ incorporation in the absence of a functional asPARP enzyme.
6. After this step, the nuclei are labeled with analog NAD^+, cross-linked with formaldehyde, and ready for the alkyne-azide click reaction.
7. In order for the reaction to proceed at an appreciable rate, a copper catalyst is required, in particular Cu(I) ions. In this protocol, the Cu(I) ions are provided by $CuSO_4$ in the 25× click buffer. Sodium ascorbate is added to reduce Cu(II) to Cu(I) ions, and THPTA binds to and stabilizes Cu(I) ions. Prepare the 25× click buffer fresh for each use, especially the sodium ascorbate stocks. The particular order of addition used for the components in the 25× click catalyst mix is to ensure binding of the copper ions to the copper ligand THPTA. If altering the mix, always premix the $CuSO_4$ and the THPTA.
8. If altering the methods described herein, some important factors to keep in mind concerning the click reaction are as follows: (a) The click reaction is sensitive to dissolved oxygen,

from which reactive oxygen species are produced. The addition of THPTA and sodium ascorbate helps to scavenge reactive oxygen species, and (b) the oxidized form of the reducing agent, sodium ascorbate, can cross-link some amino acids. While unnecessary for this particular application, aminoguanidine hydrochloride can be added to prevent the cross-linking (c).

9. After this step, the nuclei are labeled with analog NAD^+, cross-linked with formaldehyde, and clicked to biotin. The chromatin is now ready to be fragmented by sonication. Alternatively, the nuclei can be collected by centrifugation and flash frozen. The specifics of the freezing process may affect the amount of sonication necessary for fragmentation of genomic DNA in the subsequent steps.
10. The sonication requires optimization for each cell type used. The parameters will vary based on the sonicator used. The goal of sonication is to generate DNA fragments with an average size of 200–300 bp. You may have to vary the intensity setting, duty cycle, duration of sonication, and the number of cycles. When working out the optimal sonication conditions, collect analytical samples between cycles and analyze by gel electrophoresis (as in Subheading 3.3, **step 2**; Fig. 3a).
11. The choice of blocking agent for the dot blotting can affect the biotin detection results. We recommend BSA for blocking and for diluting the HRP-streptavidin. Milk contains biotin, which can lead to an increased background if used as a blocking agent.
12. When pipetting the streptavidin-coated magnetic beads, use a wide bore plastic pipette tip, which can be prepared by severing a small portion at the end of the tip with a clean razor blade.
13. If the background is too high after chromatin enrichment (i.e., a low signal-to-background ratio), consider preclearing samples using uncoated magnetic beads (or bead with a different affinity tag, e.g., Protein A Dynabeads), which can strip away non-specific interactions. For example, add 5 μL of washed Protein A Dynabeads, incubate at 4 °C with gentle mixing (e.g., on a nutator) for 1 h, collect the beads using a magnetic stand, transfer the supernatant to a fresh tube, and continue as described in Subheading 3.3, **step 6**.
14. Most standard ChIP-seq library preparation protocols will work [19–21].
15. Single-end 50 base reads should be sufficient.
16. For the browser tracks shown in Fig. 5, Gibson et al. [14] developed and used a custom script. The reference genome

was divided into 2500 bp windows with a 250 bp step size. The "intersectBed" tool in the bedTools package (http://bedtools.readthedocs.io/en/latest/) was used to count the number of input and treatment reads in each bin. The ratio of these read counts was calculated and then normalized by the total number of reads for each. The bins were then reduced to non-overlapping 250 bp segments [14, 27, 28].

Acknowledgements

The PARP-related research in the Kraus Lab is supported by grants from the National Institutes of Health, NIDDK (DK069710), and the Cancer Prevention and Research Institute of Texas (CPRIT) (RP160319).

References

1. Bannister AJ, Kouzarides T (2011) Regulation of chromatin by histone modifications. Cell Res 21:381–395
2. Desjarlais R, Tummino PJ (2016) Role of histone-modifying enzymes and their complexes in regulation of chromatin biology. Biochemistry 55:1584–1599
3. Chen P, Li G (2010) Dynamics of the higher-order structure of chromatin. Protein Cell 1:967–971
4. Ame JC, Spenlehauer C, De Murcia G (2004) The PARP superfamily. BioEssays 26:882–893
5. Kraus WL, Hottiger MO (2013) PARP-1 and gene regulation: progress and puzzles. Mol Asp Med 34:1109–1123
6. Krishnakumar R, Kraus WL (2010) The PARP side of the nucleus: molecular actions, physiological outcomes, and clinical targets. Mol Cell 39:8–24
7. Gibson BA, Kraus WL (2012) New insights into the molecular and cellular functions of poly(ADP-ribose) and PARPs. Nat Rev Mol Cell Biol 13:411–424
8. Ryu KW, Kim DS, Kraus WL (2015) New facets in the regulation of gene expression by ADP-ribosylation and poly(ADP-ribose) polymerases. Chem Rev 115:2453–2481
9. Vyas S, Matic I, Uchima L et al (2014) Family-wide analysis of poly(ADP-ribose) polymerase activity. Nat Commun 5:4426
10. Vyas S, Chesarone-Cataldo M, Todorova T et al (2013) A systematic analysis of the PARP protein family identifies new functions critical for cell physiology. Nat Commun 4:15972
11. Skidmore CJ, Davies MI, Goodwin PM et al (1979) The involvement of poly(ADP-ribose) polymerase in the degradation of NAD caused by gamma-radiation and N-methyl-N-nitrosourea. Eur J Biochem 101:135–142
12. Gupte R, Liu Z, Kraus WL (2017) PARPs and ADP-ribosylation: recent advances linking molecular functions to biological outcomes. Genes Dev 31:101–126
13. Ogata N, Ueda K, Kawaichi M et al (1981) Poly(ADP-ribose) synthetase, a main acceptor of poly(ADP-ribose) in isolated nuclei. J Biol Chem 256:4135–4137
14. Gibson BA, Zhang Y, Jiang H et al (2016) Chemical genetic discovery of PARP targets reveals a role for PARP-1 in transcription elongation. Science 353:45–50
15. Messner S, Altmeyer M, Zhao H et al (2010) PARP1 ADP-ribosylates lysine residues of the core histone tails. Nucleic Acids Res 38:6350–6362
16. Ogata N, Ueda K, Kagamiyama H et al (1980) ADP-ribosylation of histone H1. Identification of glutamic acid residues 2, 14, and the COOH-terminal lysine residue as modification sites. J Biol Chem 255:7616–7620
17. Farrar D, Rai S, Chernukhin I et al (2010) Mutational analysis of the poly(ADP-ribosyl)ation sites of the transcription factor CTCF provides an insight into the mechanism of its regulation by poly(ADP-ribosyl)ation. Mol Cell Biol 30:1199–1216
18. Luo X, Ryu KW, Kim DS et al (2017) PARP-1 controls the adipogenic transcriptional pro-

gram by PARylating C/EBPβ and modulating its transcriptional activity. Mol Cell 65:260–271

19. Carey MF, Peterson CL, Smale ST (2009) Chromatin immunoprecipitation (ChIP). Cold Spring Harb Protoc 2009(9):pdb prot5279
20. Furey TS (2012) ChIP-seq and beyond: new and improved methodologies to detect and characterize protein-DNA interactions. Nat Rev Genet 13:840–852
21. Milne TA, Zhao K, Hess JL (2009) Chromatin immunoprecipitation (ChIP) for analysis of histone modifications and chromatin-associated proteins. Methods Mol Biol 538:409–423
22. Gibson BA, Kraus WL (2017) Identification of protein substrates of specific PARP enzymes using analog-sensitive PARP mutants and a 'clickable' NAD+ analog. Meth Mol Biol 1608:111–135
23. Presolski SI, Hong VP, Finn MG (2011) Copper-catalyzed azide-alkyne click chemistry for bioconjugation. Curr Protoc Chem Biol 3:153–162
24. Langmead B, Trapnell C, Pop M et al (2009) Ultrafast and memory-efficient alignment of short DNA sequences to the human genome. Genome Biol 10:R25
25. Robinson JT, Thorvaldsdottir H, Winckler W et al (2011) Integrative genomics viewer. Nat Biotechnol 29:24–26
26. Thorvaldsdottir H, Robinson JT, Mesirov JP (2013) Integrative genomics viewer (IGV): high-performance genomics data visualization and exploration. Brief Bioinform 14:178–192
27. Quinlan AR, Hall IM (2010) BEDTools: a flexible suite of utilities for comparing genomic features. Bioinformatics 26:841–842
28. Ramirez F, Ryan DP, Gruning B et al (2016) deepTools2: a next generation web server for deep-sequencing data analysis. Nucleic Acids Res 44:W160–W165

Part VIII

NAD+ Sensors

Chapter 26

Methods for Using a Genetically Encoded Fluorescent Biosensor to Monitor Nuclear NAD+

Michael S. Cohen, Melissa L. Stewart, Richard H. Goodman, and Xiaolu A. Cambronne

Abstract

Free nicotinamide adenine dinucleotide (NAD^+) serves as substrate for NAD^+-consuming enzymes. As such, the local concentration of free NAD^+ can influence enzymatic activities. Here we describe methods for using a fluorescent, genetically-encoded sensor to measure subcellular NAD^+ concentrations. We also include a discussion of the limitations and potential applications for the current sensor. Presented in this chapter are (1) guidelines for calibrating instrumentation and experimental setups using a bead-based method, (2) instructions for incorporating required controls and properly performing ratiometric measurements in cells, and (3) descriptions of how to evaluate relative and quantitative fluctuations using appropriate statistical methods for ratio-of-ratio measurements.

Key words NAD^+, Nicotinamide adenine dinucleotide, Biosensor, Metabolite, Fluorescent sensor, Circularly permuted fluorescent protein, ARTD, PARP

1 Introduction

The key advantage of using a sensor to measure oxidized β-nicotinamide adenine dinucleotide (NAD^+) is its ability to provide measurements of free NAD^+ in relevant physiological contexts. Because NAD^+ can have distinct roles in different compartments and additionally be regulated on a subcellular level, it is important to assess and differentiate its local concentrations. Whole-cell measurements from lysates are at a disadvantage in this regard because compartmentalized information is typically lost and it is technically challenging to differentiate free and bound NAD^+. A genetically encoded sensor can be (1) subcellularly localized or expression-restricted to specific cell or tissue types, (2) used on live cells for dynamic monitoring, and (3) used in context to distinguish free NAD^+, which is the fraction relevant to cellular signaling.

Paul Chang (ed.), *ADP-ribosylation and NAD+ Utilizing Enzymes: Methods and Protocols*, Methods in Molecular Biology, vol. 1813, https://doi.org/10.1007/978-1-4939-8588-3_26,

NAD^+ has a well-known role as an electron acceptor in reductive and oxidative (redox) reactions. In addition, NAD^+ is utilized as a substrate for NAD^+-consuming enzymes, such as PARP (also known as ARTD) family members [1, reviewed 3–5] and sirtuins [2, reviewed in 3–5]. To mediate enzymatic ADP-ribosylation, PARP/ARTD family members transfer the ADP-ribose unit from NAD^+ to target proteins and release nicotinamide. As such, one NAD^+ molecule is consumed for each posttranslational ADP-ribosylation modification. The activities of PARP/ARTD family members thus rely on the local availability of free NAD^+. We describe here an approach to directly monitor this free NAD^+ pool in specific subcellular compartments.

In cells, a significant proportion of NAD(H) exists as a protein-bound fraction that is predominantly engaged in redox reactions [6–8]. Importantly, steady-state levels of this NAD(H) pool are not thought to fluctuate due to the nonconsuming nature of redox reactions. In contrast, there exists a free NAD^+ fraction—the fraction available for consumption by NAD^+-consuming enzymes—which is susceptible to steady-state changes. This free pool is estimated to comprise ~30% of total NAD content [8]. Total NAD^+ measurements from lysates typically represent a combination of the bound and free fraction. Therefore, data obtained from this approach may not accurately represent the amount of free NAD^+ available to NAD^+-consuming enzymes.

Free NAD^+ is further compartmentalized subcellularly, and its levels may be differentially regulated among cell types and tissues. The enzymes regulating NAD^+ biosynthesis and consumption localize to specific subcellular compartments and have varying expression levels in different cell types [4, 5, 9–13]. This provides a mechanism for how steady-state, free NAD^+ levels are differentially regulated within and among cells [14–20]. The subcellular compartmentalization of NAD^+ further highlights a limitation of whole-cell, lysate measurements for monitoring the specific pool of NAD^+ available to a particular NAD^+-consuming enzyme.

Unlike NADH, NAD^+ does not have intrinsic fluorescence to facilitate its detection in cells. NADH measurements or measurements of the NADH/NAD^+ ratio are unreliable indicators of free intracellular NAD^+ concentrations because of the high ratio of free NAD^+ to NADH [8]. Instead, fluctuations in NADH or ratiometric NADH/NAD^+ sensors predominantly reflect cellular redox state. In the nucleocytoplasm, the ratio of free NAD^+ to NADH is estimated at ~700:1 and in the mitochondria ~7:1 [8]. Furthermore, current ratiometric NADH/NAD^+ sensors are prone to saturation for NAD^+ at physiological concentrations, in effect being overly sensitive for most *in-cell* NAD^+ measurements [21–23]. Magnetic resonance scanning is helpful for in vivo patient diagnosis and can distinguish NAD^+ from NADH, but it is limited with regard to subcellular resolution and distinguishing the bound from free fractions [24]. Lastly, while indirect readouts such as the PARAPLAY assay have been extremely informative [15], we sought to develop a quantitative approach for direct detection of NAD+ in intact cells. The advance we have developed is a genetically encoded fluorescent sensor for free NAD+ that can be localized to specific subcellular compartments for real-time, direct measurements in live cells [17].

The NAD+ sensor is based on a circularly permutated fluorescent protein, cpVenus, cleaved between its original amino acids 144 and 145 to position new N- and C-termini in close proximity to the central fluorophore; the original termini were joined with a small linker [25, 26]. Circularly permutated fluorescent proteins have been used in a variety of single-fluorescent protein sensors by connecting the fluorescent protein to a peptide-binding domain for the target molecule of interest [reviewed in 27, 28]. In this case, we attached a specific NAD+-binding pocket modeled after the NAD+-binding domain from LigA *E. faecalis* to the new termini and engineered it to reversibly bind and not consume NAD+ (*see* Fig. 1) [17, 29, 30]. The sensor was further confirmed to specifically recognize free NAD+ (and not NADH or other analogous molecules) and was optimized for its recognition of NAD+ within the physiological range [17]. This general design for a single-fluorescent protein sensor is analogous to what has been described by Tsien and Miyawaki groups, in their pioneering calcium sensors [25, 31], and also what has been subsequently utilized in many other single-fluorescent protein biosensor designs, including those used for the GCaMP and GECO sensor series and the NADH/NAD+ sensors [21–23, 32–35]. Because the sensor is genetically encoded, we can incorporate subcellular localization sequences to target the sensor to specific subcellular compartments for local NAD+ measurements [17].

The NAD+ sensor has a major fluorescence peak that is excitable at 488 nm and emits at ~520 nm and a weaker secondary peak that is excitable at 405 nm and emits at ~510 nm [17]. The fluorescence from 488 nm excitation decreases in the presence of free NAD+ and represents the readout of NAD+ concentration. In contrast, the 405 nm fluorescence is not significantly altered by NAD+ concentration, tracks with the expression level of the sensor, and can be used to normalize for varying sensor concentrations in cells [17]. Thus, the sensor uses ratiometric 488/405 nm measurements to follow how NAD+ fluctuates in cells. This type of ratiometric measurement is similarly used by other sensors and

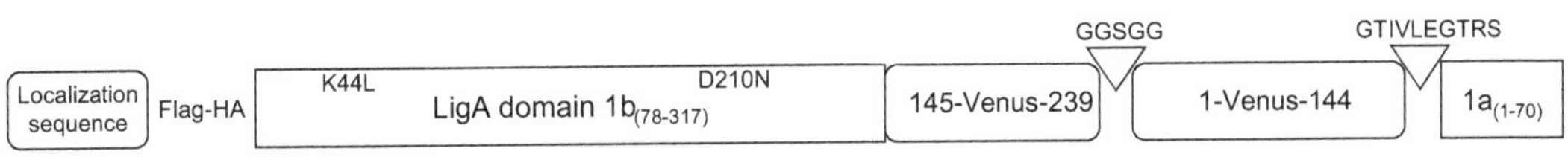

Fig. 1 Schematic of NAD+ sensor

probes, such as Pericam, HyPer, and Fura-2, to normalize for probe or sensor concentration or expression level [31, 36, 37]. There is an inverse relationship between the 488 nm fluorescence from the sensor and local NAD^+ concentrations; when NAD^+ concentrations are increased, the sensor's 488/405 nm fluorescence dims, and when NAD^+ concentrations are lowered, this fluorescence ratio brightens.

An additional important control for all experiments is cpVenus alone, without the NAD^+-binding pocket. NAD^+ has no measurable effect on the 488/405 nm fluorescence of cpVenus in vitro. However, we found that pH influences the fluorescence intensities of both the sensor and cpVenus control to comparable degrees between pH 6.6 and 8.0 [17]. Thus, a parallel analysis of fluorescent changes in cpVenus is required for normalization of pH effects or any other non-specific changes in fluorescence.

This leads us to a discussion of the current limitations of the sensor. One of the limitations is its pH dependency, most likely related to the pKa of cpVenus that was used in construction of the sensor. Nevertheless, because cpVenus fluorescence is similarly affected by pH, one can use cpVenus 488/405 nm fluorescence to normalize for NAD^+-independent changes in the sensor. We found that normalizing with cpVenus across this pH range minimizes the influence of pH and exposes the NAD^+-dependent changes [17]; similar strategies have been previously successful [38, 39]. The limitation is that these dual measurements require parallel analysis of the sensor and cpVenus under similar experimental conditions. Nevertheless, if it is initially established that pH does not change over the course of an experiment, NAD^+ fluctuations can be monitored in single cells with the sensor's 488 nm fluorescence.

The presence of high NAD^+ concentrations maximally lowers the 488 nm fluorescence by ~50% but does not eliminate it. Thus, cellular conditions where fluorescence is eliminated both from the sensor and cpVenus—e.g., pH $\leq$6.5 or exogenous H_2O_2 treatment—are incompatible with measurements. Only a few subcellular compartments—namely, the *trans*-Golgi network, secretory granules, endosomes, and lysosomes—exhibit pH <6.5, however. Most intracellular compartments fall within the measurable pH range, including the nucleus, cytosol, mitochondria, peroxisomes, endoplasmic reticulum, and *cis*-Golgi network [reviewed in 40].

Another limitation of the sensor is that this version may not be sufficiently sensitive to reliably detect all modest NAD^+ changes; this sensitivity may also depend on how well the instrumentation can monitor its fluorescence.

We describe here methods for expressing the sensor and cpVenus control in mammalian cells, as well as for generating a standard curve to quantify nuclear measurements using sensor-coated and cpVenus-coated beads and a buffered NAD^+ stock (*see* Materials). We have found that this bead-based approach coupled to flow cytometry is amenable for quantifying depleted NAD^+ concentrations, as could be expected in scenarios with unbalanced rates of NAD^+ consumption.

2 Materials

2.1 Plasmid DNA, Cell Culture, Transfection, and Small-Molecule Treatments

1. Plasmid DNA: Follow manufacturer's instructions for column-prep plasmid maxiprep. Resuspend plasmid DNA in 10 mM Tris pH 8.0 at a concentration of 1 mg/mL.
2. HEK293T cells: Maintain and propagate in a humidified incubator at 37 °C in 5% CO_2, and routinely passage with sterile technique in a biosafety cabinet.
3. Complete DMEM growth media: DMEM, 10% fetal bovine serum, and 25 mM HEPES pH 7.4.
4. Complete MEM media for small-molecule treatments and analysis: MEM, 10% fetal bovine serum, and 25 mM HEPES pH 7.4.
5. Sterile water.
6. 2.5 M $CaCl_2$ stock.
7. 2× HBS pH 7.1: 40 mM HEPES, 10 mM KCl, 270 mM NaCl, 1.5 mM Na_2HPO_4, and 10 mM dextrose; adjust to pH 7.1 with NaOH and filter sterilize (*see* **Note 4**).
8. Lipofectamine 2000™.
9. Serum-free Opti-MEM™ media.
10. Inverted light microscope with phase contrast, a broad LED excitation source (Lumencor Sola SMII Light Engine), and fluorescence filter sets for 525 ± 25 nm.
11. PVDF low protein-binding filter syringe, 0.45 μm.
12. Puromycin: 10 mg/mL aliquoted stocks in sterile water can be store up to a year at −20 °C. Puromycin is freshly diluted in media prior to use.
13. FK866 (N-[4-(1-benzoyl-4-piperidinyl)butyl]-3-(3-pyridinyl)-2ε-propenamide): 50 mM stocks in DMSO, stored in aliquots at −20 °C. Immediately before use, serially dilute FK866 with complete MEM media to a final concentration of 10 nM.
14. Nicotinamide riboside chloride: freshly resuspend powder as a 200 mM stock solution in 10 mM citrate buffer pH 3, aliquot, and store at −20 °C for use within 2 weeks. Immediately before use, dilute the nicotinamide riboside stock to 500 μM in complete MEM media. An equivalent volume of citrate buffer is added to control cells.

2.2 Generation of Sensor-/cpVenus-Coated Beads

1. 0.5% NETN buffer: 20 mM Tris pH 7.4, 150 mM NaCl, 0.5% NP-40, 1 mM EDTA, and protease inhibitors. To make 40 mL of 0.5% NETN buffer, dissolve two EDTA-free protease inhibitor tablets (Sigma) into 36 mL ultrapure water. Add 800 μL of 1 M Tris pH 7.4, 1.2 mL of 5 M NaCl, 2 mL of 10% NP-40, and 80 μL of 0.5 mM EDTA pH 8. Assemble buffer and prechill the same day you plan to perform the cell lysis.
2. Anti-Flag M2 magnetic beads (Sigma); average diameter 50 μm.
3. Magnetic stand compatible with 1.5 mL snap-cap tubes and 15 mL conical tubes.

2.3 Buffered 50 mM NAD^+ Stock Solution

1. Assay buffer: 100 mM Tris pH 7.4 and 150 mM NaCl.
2. 50 mM NAD^+ stock: Resuspend 25 mg of NAD^+ (>99% pure, molecular weight 663.4 g/mol) in 500 μL of assay buffer. This buffer composition is important, as NAD^+ is unstable in phosphate buffers [41], and addition of salt helps maintain proper protein conformation when incubated with sensor or cpVenus proteins. Initially increase the pH with ~25 μL of 1 N NaOH to approximately pH 6.5. Then slowly add 0.1 N NaOH (~10 μL) to increase to pH 7.5. Mix well and use pH strips to confirm pH. Total volume is now approximately 535 μL. Add the remaining volume of buffer up to 753.6 μL total for a 50 mM stock. Mix well. Aliquot and store at −20 °C or −80 °C.

2.4 Ratiometric Fluorescent Measurements

1. Flow cytometer: BD LSR II™ (or similar) with filter set Violet-2A (e.g., 405 nm; Em. 525 ± 25 nm) and filter set FL 1-A (e.g., 488 nm, Em. 530 ± 15 nm).
2. FlowJo™ data analysis software (or similar).
3. 5 mL polypropylene round bottom tubes.

3 Methods

3.1 Expression of the Sensor and cpVenus Control in Mammalian Cells

The approach described below is generally accessible to a wide scientific audience and does not require a BL2-approved workspace, lentivirus helper plasmids, or viral packaging. However, it requires micrograms of maxi-prepped DNA. For some researchers, it may be more cost-effective to generate stably expressing cell lines via viral transduction. We briefly describe that protocol in Subheading 3.2. Once cells are either transiently or stably expressing the sensor, the researcher can advance directly to Subheading 3.3.

1. Begin with a healthy HEK293T cell culture that has been passaged at regular intervals. Prepare a total of 8 × 10 cm dishes, where 6×10^6 cells have been seeded for each 10 cm dish. Four of the dishes will be used for expressing and purifying the sensor, and the remaining four will be used for cpVenus (*see* **Note 1**).

2. The next day, confirm with light microscopy that HEK293T cells have attached and are evenly distributed. Following a standard calcium phosphate protocol, transfect high-quality maxi-prepped plasmid DNA expressing the sensor or cpVenus control (20 μg per dish) (*see* **Notes 2** and **3**). For simplicity, DNA stocks can be prepared at a concentration of 1 mg/mL.

   In summary, for each condition, combine the following together in a 50 mL conical tube and vortex to mix well: a total of 80 μg plasmid DNA, 3.52 mL of sterile water, and 400 μL of 2.5 M $CaCl_2$. While aerating the mixture with a 5 mL pipette, slowly add 4 mL of 2× HBS in a dropwise manner (*see* **Note 4**). This mixture is incubated for 15 min at room temperature before being added dropwise to cells, 2 mL of mixture per 10 cm plate.

3. Forty-eight hours post transfection, using an epifluorescence microscope with an inverted stage that can accommodate dishes confirms that the cells are expressing either sensor or cpVenus with a transfection efficiency >80%. Cells should fluoresce at ~520 nm after either broad excitation or with a 488 nm light source and are expected to express predominantly in the cytoplasm.

4. Proceed to the generation of sensor-coated beads. Alternatively, cell pellets can be collected, flash frozen in liquid nitrogen, and stored at -80 °C for up to several months.

3.2 Viral Transduction to Generate a Stably Expressed Cell Line

The section describes an alternative method for generating stably expressing sensor and cpVenus cell lines that can be expanded to obtain 4 × 10 cm dishes of cells or used for measurement, if generated using targeted versions of the sensor and cpVenus. This approach requires a biosafety level 2 (BL2)-approved workspace, and steps are performed in a biosafety cabinet, using sterile technique and reagents. Standard BL2 viral handling safety precautions must be followed and performed in a BL2-approved workspace. These versions of the sensor and cpVenus are encoded in lenti-compatible plasmids, and their expression is driven by a CMV promoter. Their coding sequences are followed by an internal ribosome entry site (IRES) and puromycin selection marker. We describe transfection with the Lipofectamine 2000™ reagent, but analogous approaches for introducing plasmids into mammalian cells work equally well. Co-transfection of the sensor plasmids with second-generation helper plasmids will produce infective lentiviral particles.

1. Begin with a healthy HEK293T cell culture that has been passaged at regular intervals. Split one million cells per well in a 6-well (35 mm) plate.
2. The next day, confirm with light microscopy that HEK293T cells have attached and are evenly distributed in the 35 mm

dish or well. Transfect these cells with Lipofectamine 2000™, following manufacturer's directions. Optimal transfections are achieved using high-quality, maxi-prepped plasmid DNA for the sensor/cpVenus and helper vectors.

In summary, for each well, we premix the sensor/cpVenus (2 μg) with the helper plasmids (2 μg) in a sterile 1.5 mL snap-cap tube (total DNA is 4 μg) and then mix in 200 μL of neutral-pH serum-free Opti-MEM™ media. In a separate snap-cap tube, first dilute 10 μL of Lipofectamine 2000™ reagent into 200 μL of Opti-MEM™, and then transfer this Lipofectamine 2000™/Opti-MEM™ mixture to the DNA/Opti-MEM™ mixture. Vortex together to mix and incubate at room temperature for 20 min. After the incubation, add this mixture dropwise to the pre-seeded HEK293T cells.

3. After 4–6 h, gently remove media on cells and replace with fresh growth media.
4. Allow cells to express the sensor and cpVenus control from the transfected plasmids for 48–72 h. Confirm this expression and a >80% transfection efficiency with epifluorescence widefield microscopy. Successful viral packaging can often be observed at this point by a slightly round and swelling morphology of the packaging cell. Very gently transport your plate to and from the incubator, as the cells are liable to detach very easily at this stage.
5. Collect the virus-containing growth medium from the transfected cells between 48 and 72 h post transfection; do not disturb or collect the cell monolayer. Spin the virus-containing growth medium at 1500 × *g* for 5 min to pellet cellular debris. Carefully remove the supernatant without disturbing the pellet, and pass this supernatant through a low protein-binding 0.45 μm PVDF syringe filter. Approximately 1.5 mL of viral supernatant is typically recovered from 2 mL of media. Viral titer or infectivity units can be calculated at this point, if desired.
6. Trypsinize and count HEK293T cells from a healthy culture. Mix 500×10^3 cells with 0.5 mL of filtered viral supernatant in a 12-well plate. Add 0.5 mL of fresh growth media and incubate overnight.
7. The next day, gently remove viral supernatant and growth media from the transduced cells that are now adhered to the 12-well plate. Replace with fresh growth media.
8. After waiting 48 h to allow for expression of the puromycin resistance gene, trypsinize cells from the 12-well plate and replate in a 10 cm plate containing growth media with 1 μg/mL puromycin for selection of transduced cells. As a control, an untransduced 12-well plate of cells is also split into growth

media containing puromycin. The cells in this control should not survive the puromycin selection because they are not transduced, and in contrast the transduced cells should fluoresce green and proliferate under the same puromycin selection conditions.

9. Check cells undergoing puromycin selection daily. If there are a lot of unhealthy or unattached cells in the culture, gently wash away the debris from attached growing cells by removing the media and replacing with fresh growth media containing puromycin.
10. Cell lines should be expanded and grown under puromycin selection for a minimum of 1–2 weeks. At this point, cells can be frozen in 10% DMSO as freezer stocks for storage at −80 °C or in liquid nitrogen.

3.3 Generation of Sensor-Coated and cpVenus-Coated Beads

All steps are performed at 4 °C or on ice. Store sensor-coated beads at 4 °C, and do not freeze-thaw. We recommend using these beads the same day for best results. If stored in buffer with protease inhibitors, we have not noticed substantial differences for up to 3 days. We do not recommend using protein-coated beads that have been stored for over a week, due to the chance of proteolysis or denaturation.

1. On ice, remove growth media from 10 cm plate of cells, and collect cells in 5 mL of PBS per 10 cm plate with gentle scraping using a cell scraper. Pool cells from the same condition into a 50 mL conical tube.
2. To collect cells as a pellet, centrifuge tubes at 4 °C at 1000 × *g* for 5 min. Carefully remove and discard PBS supernatant from cells.
3. To extract the sensor and cpVenus proteins, lyse cells by resuspending them in 0.5% NETN buffer, and incubate suspension on ice for 10 min. Use 2.5 mL of buffer per 10 cm plate; for 4 × 10 cm plates, this volume is 10 mL.
4. Clarify the lysate by centrifugation at 12000 × *g* for 20 min at 4 °C (*see* **Note 5**).
5. While the lysates are spinning, prepare the anti-Flag magnetic beads with an average diameter of 50 μm. Use 20 μL of slurry per 10 cm dish. For eight dishes, we need 160 μL slurry of beads and therefore prepare 175 μL slurry to provide excess to account for any pipetting error. Use a cut tip or one with a wide orifice to pipette to a new 1.5 mL tube. Place the tube on a magnetic stand for 2 min, and wash beads twice with 0.5% NETN buffer. Resuspend the washed beads in 600 μL of 0.5% NETN, and distribute 300 μL of the resuspended beads into 2 × 15 mL conical tubes. Place conical tubes on magnetic stand at 4 °C until lysate is prepared.

6. Following centrifugation of the lysate, the sensor and cpVenus proteins will remain in the supernatant. Remove this supernatant to a new tube.
7. Quantitate the amount of total protein in the lysate obtained from each sample using a Bradford assay or similar (*see* **Note 6**). With this procedure, we typically obtain values between 2 and 5 mg/mL for total protein concentration in the lysate. Calculate the required volumes from each sample such that relative total protein concentration is normalized between samples (*see* **Note** 7).
8. The washed beads in the conical tubes have now been magnetically separated. Carefully remove the buffer, and resuspend beads with the appropriate volume of clarified lysate. Perform this step sequentially for the sensor and cpVenus so that the beads do not dry out when the buffer is removed.
9. Secure the cap on the conical tubes, and incubate beads with lysate for 2 h at 4 °C on a rotator. This step coats the beads with sensor and cpVenus protein (*see* Fig. 2 and **Note 8**).
10. After incubation, place conical tubes back on the magnetic stand to separate the beads that are now coated with either the sensor or cpVenus protein.
11. While keeping the tube on the stand, gently wash beads twice with 1 mL of 0.5% NETN buffer.
12. It is highly recommended to proceed directly to calibration on the same day that the beads are generated. However, if beads need to be saved for the next day, they can be stored in 0.5% NETN at 4 °C following the wash steps.

3.4 Analyzing Sensor-Coated and cpVenus-Coated Beads on Flow Cytometer

1. NAD^+ is acidic, and it must be prepared as a buffered stock prior to use and storage in solution. Instructions for preparing a buffered 50 mM NAD^+ stock are found in Subheading 2.3. Using the buffered NAD^+ stock, prepare 1 mL of 2× NAD^+ buffers 2 mM and 6 mM NAD^+ in assay buffer (100 mM Tris pH 7.4, 150 mM NaCl). Using assay buffer, serially dilute the 2 mM and 6 mM stocks to prepare the series of 2× NAD^+ concentrations for the calibration curve (*see* Table 1).
2. Place tubes on the magnetic stand, gently remove 0.5% NETN, and resuspend each set of sensor-coated and cpVenus-coated beads in 1.05 mL of assay buffer. Using a cut tip, distribute 100 μL of each of the bead suspension evenly into 10 × 1.5 mL tubes. As each tube will correspond to different NAD^+ assay conditions, label each tube appropriately.
3. Add 100 μL of the appropriate 2× NAD^+ buffer to each tube. Keep beads cold during this process.

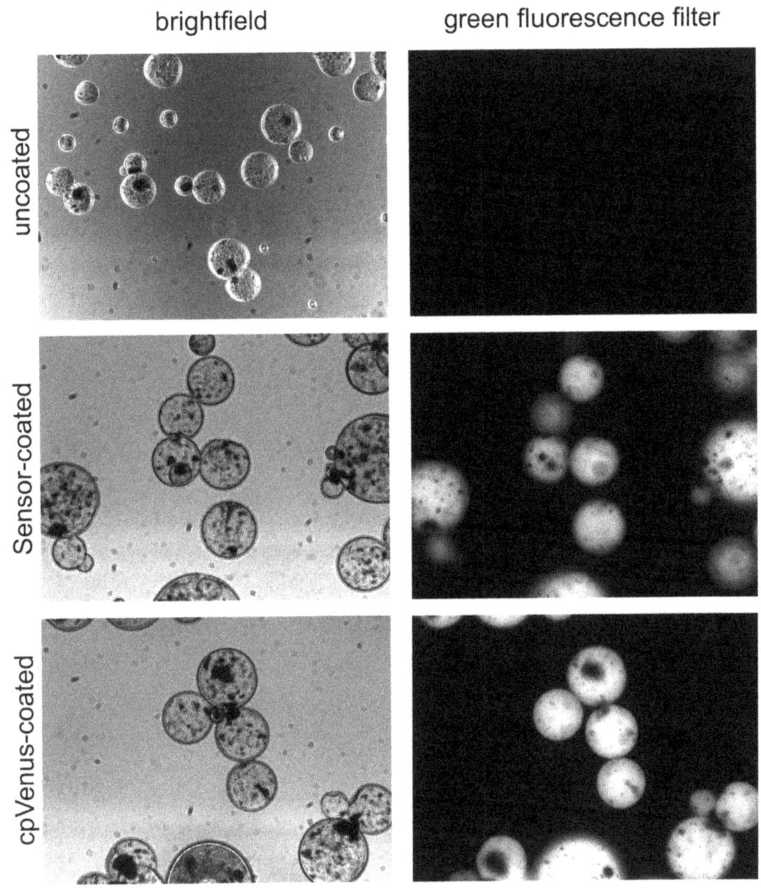

Fig. 2 Brightfield and epifluorescence images of uncoated (top), sensor-coated (middle), and cpVenus-coated (bottom) beads in suspension. Fluorescence was excited by an LED Sola SMII Light Engine, and emission was viewed through a 525 ± 25 nm filter, 30 ms exposure time

4. Ensure that the lasers on the flow cytometer have been warmed up and the detectors are stabilized. If needed, immediately prior to analysis, pass the resuspended beads through a 100 μm nylon mesh filter into the 5 mL round-bottom tube that will be used with the flow cytometer. This can help prevent clogging of the instrument from clumped or particularly large beads.
5. The first sample to run is uncoated beads (no fluorescence). This identifies a relatively uniform bead size (G1) based on forward and size scatter (*see* Fig. 3a). Generate a plot, and set the *x*-axis to FSC-A (log scale) and the *y*-axis to SSC-A (log scale) for forward and side scatter, respectively. Adjust voltages on the instrument to center the nonfluorescent bead popula-

Table 1
Conditions for bead-based calibration

cpVenus-coated beads		Sensor-coated beads		2× NAD+ stock
Condition	[NAD+] μM	Condition	[NAD+] μM	[NAD+] μM
1	0.01	11	0.01	0.02
2	0.1	12	0.1	0.2
3	1	13	1	2
4	3	14	3	6
5	10	15	10	20
6	30	16	30	60
7	100	17	100	200
8	300	18	300	600
9	1000	19	1000	2000
10	3000	20	3000	6000
21 uncoated beads (nonfluorescent)				

tion onto the lower left quadrant of this plot (*see* Fig. 3a and **Note 9**).

6. The sensor and cpVenus proteins externally coat the beads, and thus coated beads are expected to be irregularly shaped and have increased complexity, reflected by increased forward and side scatter values (G2).
7. To define fluorescent beads (FL), generate a second plot that displays events from G1 and G2. Set the *x*-axis to Violet-2A for the BD LSRII (e.g., 405 nm, filter set 525/50 nm) and the *y*-axis to FL1-A (e.g., 488 nm, filter set 530/30) (*see* **Notes 9** and **10**). Detectable fluorescence is defined by exclusion of the nonfluorescent beads (*see* Fig. 3a). Draw a gate (FL) to define the fluorescent beads above and to the right of the nonfluorescent bead sample.
8. Analyze the 20 total samples representing the effects of varying NAD+ concentration on the fluorescence of cpVenus- or sensor-coated beads. Collect a minimum of 500 events from the FL gate, and save data from all events (*see* **Note 11**).

3.5 Generating a Standard Curve

1. Upload the data files (FCS files) obtained from the flow cytometer for post-capture analysis in FlowJo™ software or similar.
2. Open the file corresponding to the nonfluorescent beads, and define gates similarly as previously performed on the cytometer, with the FL gate being defined from the G1/G2 subsets.

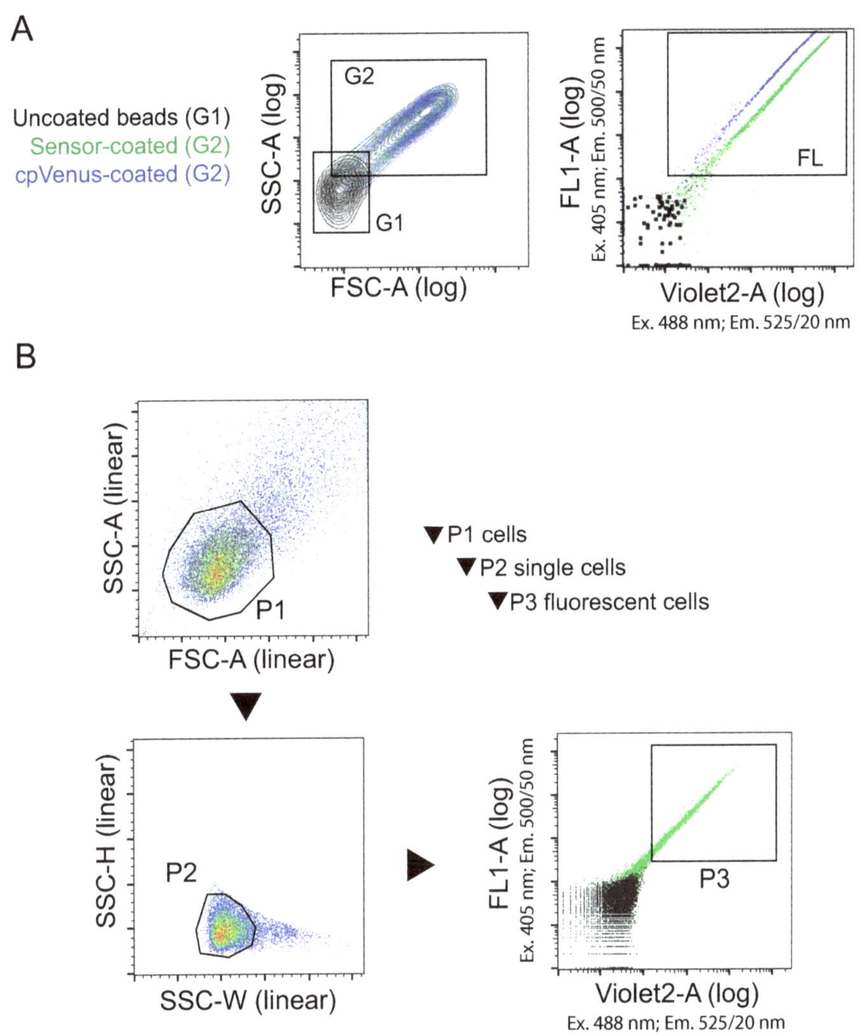

Fig. 3 (**a**) Example of flow cytometry analysis of sensor-coated (green) and cpVenus-coated (blue) beads, relative to uncoated beads (gray). Approximately 500 fluorescent events are analyzed in FL. (**b**) Hierarchical analysis of the size, granularity, and fluorescence of cells as measured through flow cytometry and represented as dot plots. Approximately 10×10^3 fluorescent cells are analyzed in P3. Nonfluorescent cells are overlaid in gray

Use exclusion of the nonfluorescent beads as boundaries for the FL population. We suggest viewing the data as a density plot or in pseudocolor. After defining these gates, drag these gate names to "All Samples" under the "group" heading in the workspace to ensure that identical gating is performed on all samples (*see* **Notes 12** and **13**).

3. To determine the fluorescence ratio 488 nm/405 nm per bead, select "derive parameters" under the "Tools" tab. Enter the "formula" by clicking "Insert Reference" to select the FL1-A value; click the ÷ key and "Insert Reference" to select Violet-2A (*see* **Note 14**).

4. Apply this "derived" parameter to "all samples." A histogram of the derived value should result in a defined peak. This represents the response of the sensor or cpVenus to the NAD^+ concentration, normalized for the amount of protein coating the bead. The geometric mean values of the derived parameter from the fluorescent FL group can be exported using the "Table Editor" function (*see* **Note 15**).
5. Divide the geometric mean of the sensor by the geometric mean of cpVenus for each NAD^+ concentration. We refer to this value as "the ratio of ratios" (*see* Table 2). The 488/405 nm fluorescence of cpVenus should not dramatically change across these conditions, and it is used in the denominator to normalize for any experimental technicalities that may non-specifically alter fluorescence. This can include slight pH differences across samples. If the NAD^+ stock is not sufficiently buffered, the cpVenus values will also trend as a decrease with its addition.
6. To obtain the standard curve, plot the ratio-of-ratio values in the *y*-axis against the log μM on the *x*-axis (*see* Fig. 4). This is expected to yield a sigmoidal curve and represents one replicate. Additional replicates can provide 95% confidence intervals for the curve.

Table 2

Example of ratio-of-ratio calculations for calibration curve

	Geo. mean 488/405 nm			Ratio of ratios
NAD^+ μM	cpVenus	Sensor	Log [NAD^+] μM	$(Sensor_{488/405nm})/(cpVenus_{488/405nm})$
0.01	5.18	1.95	−2.00	0.376
0.1	5.19	1.98	−1.00	0.382
1	5.2	1.97	0.00	0.379
3	5.09	1.93	0.48	0.379
10	5.12	1.84	1.00	0.359
30	5.17	1.75	1.48	0.338
100	5.17	1.53	2.00	0.296
300	4.97	1.33	2.48	0.268
1000	5.08	1.14	3.00	0.224
3000	5.08	1.01	3.48	0.199

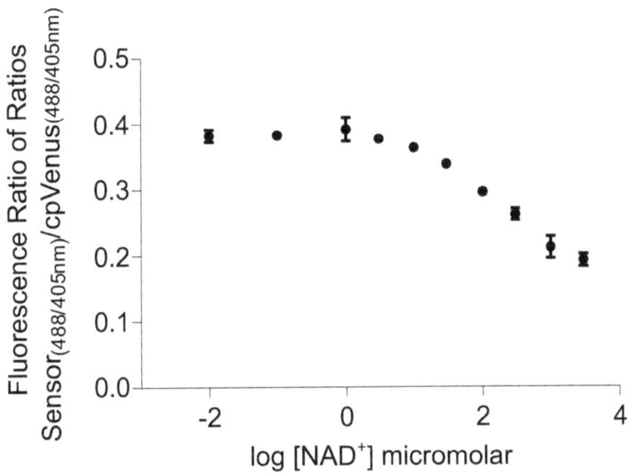

Fig. 4 Fluorescence data obtained with sensor- and cpVenus-coated beads, graphed against the log of NAD^+ concentrations. Mean ± SD, $n = 2$

3.6 Nuclear Expression of the Sensor in Cells Using Transient Transfection

To obtain measurements for intracellular NAD^+, the localized sensor and cpVenus control need to be expressed in your cell of interest. Because the presence of NAD^+ only decreases but doesn't eliminate the sensor's fluorescence, monitoring fluorescence and confirming nuclear localization of the sensor and cpVenus should be straightforward using fluorescence microscopy. The expression level of both the sensor and cpVenus can be normalized by the fluorescence at ~515 nm, following excitation at 405 nm; thus with sufficient sample size, measurements can be obtained from transiently transfected cells or non-clonal lines. Here we describe a basic protocol for examining NAD^+ depletion and increases in the nuclei of transiently transfected HEK293T cells (*see* **Note 16**).

1. Day 0: Starting with a healthy HEK293T cell culture, seed 5×10^5 cells in a 35 mm well. For this protocol, a minimum of 6×35 mm dishes or a 6-well plate is required, plus an extra well as an untransfected, nonfluorescent control for the flow cytometry data capture (*see* Table 3).
2. Day 1: The next day the cells are ready for transfection after confirming with light microscopy that they have adhered and are evenly distributed, ~30–50% confluent.
3. Under sterile conditions, and following manufacturer's directions, transfect 2 μg of nuclear cpVenus or nuclear sensor plasmid DNA using 5 μL of Lipofectamine 2000™ in 400 μL total of serum-free Opti-MEM™ media per well (*see* **Notes 17** and **18**).

   First prepare and vortex to mix well a master mix of 1.2 mL Opti-MEM™ media and 30 μL of Lipofectamine 2000™. Meanwhile, in separate tubes, dilute and vortex to mix well 6 μg of each plasmid DNA (nuclear cpVenus or sensor) in 600 μL of Opti-MEM™ media. Add 600 μL of the Lipofectamine 2000™/Opti-MEM™ mixture to each of the tubes containing

Table 3

Suggested experimental conditions for assessing NAD$^+$ sensor changes in cells

Condition	Transfected plasmid	Treatment
1 *Nonfluorescent control*	–	–
2	Nuclear cpVenus	–
3	Nuclear sensor	–
4	Nuclear cpVenus	10 nM FK866, 6–8 h
5	Nuclear sensor	10 nM FK866, 6–8 h
6	Nuclear cpVenus	500 μM nicotinamide riboside, 24 h
7	Nuclear sensor	500 μM nicotinamide riboside, 24 h

FK866 is a specific inhibitor of nicotinamide phosphoribosyltransferase (NAMPT), the rate-limiting enzyme in the NAD$^+$ salvage biosynthesis pathway that generates nicotinamide mononucleotide (NMN). NMN is a direct precursor for intracellular NAD$^+$ Treatment with exogenous nicotinamide riboside can also influence steady-state NAD$^+$ levels through NMN. Nicotinamide riboside serves as substrate for nicotinamide riboside kinase (NRK) to synthesize NMN

diluted plasmid DNA. Vortex to mix. Incubate for 20 min at room temperature. Add 400 μL of each transfection mixture to 3 × 35 mm wells.

4. Day 2: After allowing cells ~20–24 h to express the plasmid, use fluorescent microscopy to confirm nuclear expression of the sensor and cpVenus. Transfected cells should fluoresce green and localize to the nucleus (*see* **Note 19**).

3.7 Small-Molecule Cell Treatments

After confirmation of a successful transfection, cells are now ready for experimental treatment conditions 24 h post transfection. Here, we will use FK866 and nicotinamide riboside treatments to deplete or increase, respectively, nuclear NAD$^+$ concentrations (*see* Table 3).

1. Serially dilute FK866 in complete MEM to a final concentration of 10 nM (5 mL). Separately, also prepare 500 μM nicotinamide riboside in complete MEM (5 mL) (*see* Table 3, **Note 20**, and Subheading 2).
2. Gently remove growth media completely from each well, and carefully replace with 2 mL complete MEM media containing

either buffer, 10 nM FK866, or 500 μM nicotinamide riboside.

3. Return to incubator for 4–16 h for nuclear FK866 treatment (*see* **Note 21**) and 16–24 h for nicotinamide riboside treatment.

3.8 Obtaining Nuclear Ratio-of-Ratio Measurements in Cells

1. Cells are ready for flow cytometry analysis at 48 h post transfection and 4–24 h after treatment (*see* **Note 22**). The time taken to prepare cells for flow cytometry analysis must be within 5 and 10 min from start to finish due to the analysis being performed on live cells out of the incubator. If you have more samples than can be handled within this timeframe, trypsinize, collect cells, and analyze in small batches (i.e., one 6-well plate at a time). The media used to resuspend cells (complete MEM growth media) must be made fresh such that it is within near-neutral pH. If it contains phenol red, the medium should be orange-red in color without any traces of pink or magenta. Ideally media is made immediately before cell collection. Media that is not near-neutral pH is unable to sufficiently buffer the cell suspension for the duration of the flow analysis (*see* **Notes 23** and **24**).
2. Prepare and label flow cytometry sample tubes on ice. Remove growth media from cells and gently wash 1× with PBS.
3. Add 100 μL of 0.05% EDTA-trypsin to each well of cells, and incubate at 37 °C until you can easily see cells come off the plate (approximately 3–5 min).
4. Add 400 μL freshly prepared complete MEM media to each well to quench the trypsin. Triturate with a 1 mL pipette tip to obtain single-cell suspension, and immediately transfer to a labeled sample tube on ice.
5. Prepare the flow cytometer similarly to what had been described for evaluating the beads. Use the untransfected control cells to adjust voltages to center the cells on a forward and side scatter plot; the *x*-axis is set to *linear* FSC-A, and the *y*-axis is set to *linear* SSC-A (*see* Fig. 3b). To identify the relatively uniform HEK293T cell size, adjust voltages on the instrument to center the population (*see* **Note 25**). Draw a gate to define this population as P1 (*see* Fig. 3b).
6. To select for individual cells (P2) and minimize evaluation of cells that have remained clumped together, generate a new plot for doublet exclusion as evaluated by wider widths. Select the P1 population for viewing, and set the *x*-axis to linear SSC-W and the *y*-axis to linear SSC-H. Draw the gate (P2) around the population on the left (*see* Fig. 3b).
7. To define the fluorescent cells (P3), generate a third plot that displays events from P2. Set the *x*-axis to log Violet-2A for the

BD LSRII (e.g., 405 nm, filter set 525/50 nm) and the *y*-axis to log FL1-A (e.g., 488 nm, filter set 530/30). The P3 population is derived from the P2 population and is defined by exclusion of the nonfluorescent cells (*see* Fig. 3b). Draw the P3 gate to the right and above the cell population observed in the untransfected control sample, ensuring that you do not include any events in this P3 gate (*see* **Note 26**).

8. After setting up these gates, you are now ready to evaluate the remaining samples. Each batch of samples to be analyzed should be completed within 15–20 min. It is necessary to collect at least 1×10^4 cells in P3 to rigorously evaluate the fluorescence of each sensor or cpVenus population. Save data from all events, and ensure that collected data includes fluorescence following 488 nm excitation at ~525 ± 25 nm and fluorescence following 405 nm excitation at ~515 ± 10 nm.
9. Perform a post-capture analysis with the data using FlowJo software, or similar, to obtain the geometric mean of the 488/405 nm fluorescence ratio from each cell in P3. This is done with the derived parameter function, akin to what was described for analyzing coated-bead fluorescence in Subheading 3.5.
10. For each condition (control, 10 nM FK866, or 500 μM nicotinamide riboside), divide the geometric mean of the 488/405 nm ratio of the sensor by the geometric mean of the 488/405 nm ratio of cpVenus. This is a normalization step that accounts for any non-specific changes independent of NAD^+ that may have influenced fluorescence intensities during the experiment (*see* Table 4).
11. Interpolate this value onto the standard curve to obtain values for nuclear NAD^+ concentrations under each condition. Confidence intervals are obtained from the curve.

3.9 REML Statistics Compared to Control to Evaluate Significance of Relative Changes

It is also possible to evaluate relative changes without generating a standard curve by comparing the change in the treated conditions relative to the untreated samples.

1. We begin by dividing the geometric mean 488/405 fluorescence value for the sensor by the corresponding value for the cpVenus control subjected to the same condition as described above (*see* Table 4).
2. We next use this normalized value to compare the treated condition to the control condition. This is performed by dividing treated normalized value by the value of normalized untreated cells. Because the sensor's fluorescence is inversely proportional to NAD^+ concentrations, it is expected that the sensor's fluorescence will increase in brightness with FK866 treatment

Table 4
Measuring relative changes with the nuclear sensor

Treatment	Nuclear sensor	Nuclear cpVenus	Ratio of ratios Sensor$_{(488/405nm)}$/cpVenus$_{(488/405nm)}$	Relative change (treated/untreated)
0 nM FK866	0.93	3.00	*0.93/3.00* = 0.31	
10 nM FK866, 7 h	1.14	2.94	*1.14/2.94* = 0.39	*0.39/0.31* = **1.25**
Buffer	1.01	3.13	*1.01/3.13* = 0.32	
500 μM nicotinamide riboside, 24 h	0.91	3.34	*0.91/3.34* = 0.27	*0.27/0.32* = **0.84**

Steady-state NAD^+ levels are depleted following inhibition of its biosynthetic pathway with FK866 treatment. This is indicated by a ratio >1 when treated conditions are compared to control. Increases in steady-state NAD^+ levels are indicated by a ratio <1

and decrease in brightness with nicotinamide riboside treatment, compared to the untreated sensor. A relative increase in fluorescence is denoted by a value >1; a relative decrease in fluorescence will result in a value <1 (*see* Table 4).

3. To estimate the reliability of the experimental results and evaluate the likelihood that any observed fluorescence changes occurred by chance, we recommend analyzing the data using a mixed-effect model in which the experimental replicate is regarded as a random factor and the selective conditions of treated/control and sensor/cpVenus are both considered as fixed factors. This approach is most appropriate for evaluating the ratio of ratios, compared to other statistical evaluations, e.g., t-tests, and requires biological triplicates.
4. To stabilize variance and limit the impact of outliers, we first log transform the geometric mean fluorescence intensity value for 488/405 nm. The requirement for this transformation is based on our observations that the untransformed data violates the model's assumptions of normality of the error distribution and homogeneity of error variances; log-transformed data showed no such violations.
5. We estimate variance by performing a residual maximum likelihood (REML) analysis. This is performed using the 488 nm/405 nm geometric mean values for each experimental condition and requires a minimum of three biological replicates for statistical analysis. The *p*-value obtained for the interaction between the treated and control sensors, compared to the treated and untreated cpVenus, represents the *p*-value for the experimental replicates.

4 Notes

1. The exact number of required dishes will depend on the transfection efficiency and expression level. The sensor and cpVenus are expressed from CMV promoters on high-copy plasmids. When expressed in HEK293T cells, an efficient transfection of 4 × 10 cm dishes typically yields enough sensor or cpVenus protein for purification and calibration. Alternatively, stable cell lines can be generated (*see* Subheading 3.2).
2. We have successfully used lipid, calcium phosphate, and nucleofection approaches for transient transfection in HEK293T cells. We have found that high-quality, maxi-prepped plasmid DNA is optimal for obtaining high-efficiency transfections.
3. For the generation of sensor- and cpVenus-coated beads, it is most straightforward to express either the cytoplasmic or unlocalized plasmid versions. Sensor and cpVenus expressed without specific localization sequences predominantly reside in the cytoplasm.
4. The pH of the 2× HBS is critical for achieving a high rate of transfection. We have had success by adjusting to pH 7.1. We recommend preparing 3–4 solutions that range from pH 7.05 to 7.15 and first testing each one on a small scale with an eGFP plasmid to determine the optimal pH for transfection.
5. The lysate may need to be transferred to a shatter-proof centrifuge tube, depending on composition of the conical tube material. If using a fixed-angle rotor, an adaptor for conical tubes may be used as well.
6. We usually perform a Bradford assay [42] using the 5× BioRad Protein Assay™ following manufacturer's directions and BSA standard curve. In summary, the 5× BioRad reagent is freshly diluted to 1× in water, and 1 mL is added to a spectrophotometer cuvette. Add 2 μL of lysate, or 0.5% NETN buffer for the blank, to the reagent in the cuvette. Invert with parafilm to mix and immediately read absorbance at 595 nm. Interpolate absorbance measurement to BSA standard curve to obtain mg/mL of total protein in the lysate.
7. Typically, we obtain 20–50 mg of total protein from each lysate.
8. The sensor and cpVenus proteins encode N-terminal Flag-HA epitope tags and so will bind to the anti-Flag beads. These epitope tags follow the localization sequences and are connected to the sensor or cpVenus coding region with a flexible linker.

9. The exact values used for voltages are specific to every instrument and may take some time and sample to figure out for each instrumental setup. We suggest defining the setup prior to each new sample type, e.g., cell type, beads, etc. Suggested starting values for analyzing the anti-Flag magnetic beads include FSC-A (log), 50; SSC-A (log), 100; FL1-A (log), 250; and Violet-2A (log), 275.
10. The specific fluorescence filters denoted here for detection of emission are critical for taking advantage of the two distinct fluorescence peaks in the sensor and for ratiometric measurements.
11. Depending on instrumentation, it may be necessary to frequently clean the nozzle between samples to prevent clogs in the flow cell.
12. Confirm by checking for each condition that the gate is transferred and appropriately drawn such that it includes the correct cell population.
13. Confirm that a relatively equal number of beads (~500 events) are being evaluated in FL across all experimental conditions, excluding the nonfluorescent control sample.
14. The derived parameter function is used to obtain the 488/405 nm measurement per event, which is mathematically distinct from taking the average 488 nm fluorescence of the population and dividing by the average 405 nm fluorescence of the same population.
15. It is appropriate to use the geometric mean of this derived ratio as the fluorescence value for each experimental condition because fluorescence measurements are log-amplified.
16. This protocol represents a general framework that can be adapted to different cell types and for measurements in various subcellular compartments.
17. We describe a protocol for Lipofectamine 2000™, but many other methods to express exogenous DNA in mammalian cells will work here including other lipid-based approaches, calcium phosphate, viral transduction, electroporation, and nucleofection.
18. An N-terminal fusion of the SV40 nuclear localization sequence (PKKKRKV), connected via a flexible linker, targets the sensor and cpVenus proteins to the nucleus.
19. Do not leave cells out of the incubator for over 10 min as this may negatively affect their health.
20. We have found that MEM media works most consistently with nicotinamide riboside treatments. FK866 treatments can work in either MEM- or DMEM-based media.

21. Incubation longer than 18 h with FK866 results in dramatic NAD^+ depletion and compromised health of cells. Once cells are unhealthy, it is difficult to properly evaluate sensor responses and not advised.
22. From here onward, contamination of the cell culture is no longer a consideration, and cells can be prepared for analysis outside of the biosafety cabinet if needed.
23. HEPES is necessary in the media to buffer the cell suspension during flow analysis, which is typically performed on the bench outside of an incubator.
24. Phenol red will not interfere with the flow analysis because cells are diluted in sheath buffer during the analysis. Including serum in the media and keeping the cells on ice help to prevent cells from clumping.
25. For HEK293T cells, our suggested starting voltages are FSC-A (linear), 250; SSC-A (linear), 300; FL1-A (log), 250; and Violet-2A (log), 275.
26. For the untransfected sample, confirm that cells appear in P1 and P2 but not in P3 (Fig. 3b, gray population).

Acknowledgments

This work was supported by funding from the Hillcrest Committee Pilot Project Award and NIH DP2GM126897 and P30CA069533 to X.A.C. and The Pew Charitable Trusts to M.S.C. NAD^+ sensors which are available to the academic research community under a material transfer agreement with Oregon Health & Science University (OHSU). The authors are listed as inventors on patent PCT/US15/62003 for the NAD^+ sensor.

References

1. Chambon P, Weill JD, Mandel P (1963) Nicotinamide mononucleotide activation of new DNA-dependent polyadenylic acid synthesizing nuclear enzyme. Biochem Biophys Res Commun 11:39–43
2. Imai S, Armstrong CM, Kaeberlein M, Guarente L (2000) Transcriptional silencing and longevity protein Sir2 is an NAD-dependent histone deacetylase. Nature 403:795–800
3. Hottiger MO, Hassa PO, Lüscher B, Schüler H, Koch-Nolte F (2010) Toward a unified nomenclature for mammalian ADP-ribosyltransferases. Trends Biochem Sci 35:208–219
4. Gupte R, Liu Z, Kraus WL (2017) PARPs and ADP-ribosylation: recent advances linking molecular functions to biological outcomes. Genes Dev 31:101–126
5. Yang Y, Sauve AA (2016) NAD^+ metabolism: bioenergetics, signaling and manipulation for therapy. Biochim Biophys Acta 1864:1787–1800
6. Holzer H, Lynen F, Schultz G (1956) Determination of diphosphopyridine nucleotide/reduced diphosphopyridine nucleotide quotient in living yeast cells by analysis of constant alcohol and acetaldehyde concentrations. Biochem Z 328:252–263

7. Bucher T, Klingenberg M (1958) Wege des Wasserstoffs in der lebendigen Organisation. Angew Chem 70:552–570
8. Williamson DH, Lund P, Krebs HA (1967) The redox state of free nicotinamide-adenine dinucleotide in the cytoplasm and mitochondria of rat liver. Biochem J 103:514–527
9. Berger F, Lau C, Dahlmann M, Ziegler M (2005) Subcellular compartmentation and differential catalytic properties of the three human nicotinamide mononucleotide adenylyltransferase isoforms. J Biol Chem 280:36334–36341
10. Mori V, Amici A, Mazzola F, Di Stefano M, Conforti L, Magni G, Ruggieri S, Raffaelli N, Orsomando G (2014) Metabolic profiling of alternative NAD biosynthetic routes in mouse tissues. PLoS One 9:e113939
11. Zhang X, Kurnasov OV, Karthikeyan S, Grishin NV, Osterman AL, Zhang H (2003) Structural characterization of a human cytosolic NMN/NaMN adenylyltransferase and implication in human NAD biosynthesis. J Biol Chem 278:13503–13511
12. Lau C, Dölle C, Gossmann TI, Agledal L, Niere M, Ziegler M (2010) Isoform-specific targeting and interaction domains in human nicotinamide mononucleotide adenylyltransferases. J Biol Chem 285:18868–18876
13. Di Stefano M, Galassi L, Magni G (2010) Unique expression pattern of human nicotinamide mononucleotide adenylyltransferase isozymes in red blood cells. Blood Cells Mol Dis 45:33–39
14. Yang H, Yang T, Baur JA, Perez E, Matsui T, Carmona JJ, Lamming DW, Souza-Pinto NC, Bohr VA, Rosenzweig A, de Cabo R, Sauve AA, Sinclair DA (2007) Nutrient-sensitive mitochondrial NAD+ levels dictate cell survival. Cell 130:1095–1107
15. Dölle C, Niere M, Lohndal E, Ziegler M (2010) Visualization of subcellular NAD pools and intra-organellar protein localization by poly-ADP-ribose formation. Cell Mol Life Sci 67:433–443
16. Pittelli M, Formentini L, Faraco G, Lapucci A, Rapizzi E, Cialdai F, Romano G, Moneti G, Moroni F, Chiarugi A (2010) Inhibition of nicotinamide phosphoribosyltransferase: cellular bioenergetics reveals a mitochondrial insensitive NAD pool. J Biol Chem 285:34106–34114
17. Cambronne XA, Stewart ML, Kim D, Jones-Brunette AM, Morgan RK, Farrens DL, Cohen MS, Goodman RH (2016) Biosensor reveals multiple sources for mitochondrial NAD^+. Science 352:1474–1477
18. VanLinden MR, Dölle C, Pettersen IK, Kulikova VA, Niere M, Agrimi G, Dyrstad SE, Palmieri F, Nikiforov AA, Tronstad KJ, Ziegler M (2015) Subcellular distribution of NAD+ between cytosol and mitochondria determines the metabolic profile of human cells. J Biol Chem 290:27644–27659
19. Nikiforov A, Dölle C, Niere M, Ziegler M (2011) Pathways and subcellular compartmentation of NAD biosynthesis in human cells: from entry of extracellular precursors to mitochondrial NAD generation. J Biol Chem 286:21767–21778
20. Hikosaka K, Ikutani M, Shito M, Kazuma K, Gulshan M, Nagai Y, Takatsu K, Konno K, Tobe K, Kanno H, Nakagawa T (2014) Deficiency of nicotinamide mononucleotide adenylyltransferase 3 (nmnat3) causes hemolytic anemia by altering the glycolytic flow in mature erythrocytes. J Biol Chem 289:14796–14811
21. Hung YP, Albeck JG, Tantama M, Yellen G (2011) Imaging cytosolic NADH-NAD(+) redox state with a genetically encoded fluorescent biosensor. Cell Metab 14:545–554
22. Bilan DS, Matlashov ME, Gorokhovatsky AY, Schultz C, Enikolopov G, Belousov VV (2014) Genetically encoded fluorescent indicator for imaging NAD(+)/NADH ratio changes in different cellular compartments. Biochim Biophys Acta 1840:951–957
23. Zhao Y, Hu Q, Cheng F, Su N, Wang A, Zou Y, Hu H, Chen X, Zhou HM, Huang X, Yang K, Zhu Q, Wang X, Yi J, Zhu L, Qian X, Chen L, Tang Y, Loscalzo J, Yang Y (2015) SoNar, a highly responsive NAD+/NADH sensor, allows high-throughput metabolic screening of anti-tumor agents. Cell Metab 21:777–789
24. Zhu XH, Lu M, Lee BY, Ugurbil K, Chen W (2015) In vivo NAD assay reveals the intracellular NAD contents and redox state in healthy human brain and their age dependences. Proc Natl Acad Sci U S A 112:2876–2881
25. Baird GS, Zacharias DA, Tsien RY (1999) Circular permutation and receptor insertion within green fluorescent proteins. PNAS 96:11241–11246
26. Nagai T, Ibata K, Park ES, Kubota M, Mikoshiba K, Miyawaki A (2002) A variant of yellow fluorescent protein with fast and efficient maturation for cell-biological applications. Nat Biotechnol 20:87–90
27. Sanford L, Palmer A (2017) Recent advances in development of genetically encoded fluorescent sensors. Methods Enzymol 589:1–49
28. Rodriguez EA, Campbell RE, Lin JY, Lin MZ, Miyawaki A, Palmer AE, Shu X, Zhang J, Tsien RY (2017) The growing and glowing toolbox of fluorescent and photoactive proteins. Trends Biochem Sci 42:111–129

29. Gajiwala KS, Pinko C (2004) Structural rearrangement accompanying NAD^+ synthesis within a bacterial DNA ligase crystal. Structure 12:1449–1459
30. Lahiri SD, Gu RF, Gao N, Karantzeni I, Walkup GK, Mills SD (2012) Structure guided understanding of NAD^+ recognition in bacterial DNA ligases. ACS Chem Biol 7:571–580
31. Nagai T, Sawano A, Park ES, Miyawaki A (2001) Circularly permuted green fluorescent proteins engineered to sense Ca2+. PNAS 98:3197–3202
32. Tian L, Hires SA, Mao T, Huber D, Chiappe ME, Chalasani SH, Petreanu L, Akerboom J, McKinney SA, Schreiter ER, Bargmann CI, Jayaraman V, Svoboda K, Looger LL (2009) Imaging neural activity in worms, flies and mice with improved GCaMP calcium indicators. Nat Methods 6:875–881
33. Akerboom J, Chen TW, Wardill TJ, Tian L, Marvin JS, Mutlu S, Calderón NC, Esposti F, Borghuis BG, Sun XR, Gordus A, Orger MB, Portugues R, Engert F, Macklin JJ, Filosa A, Aggarwal A, Kerr RA, Takagi R, Kracun S, Shigetomi E, Khakh BS, Baier H, Lagnado L, Wang SS, Bargmann CI, Kimmel BE, Jayaraman V, Svoboda K, Kim DS, Schreiter ER, Looger LL (2012) Optimization of a GCaMP calcium indicator for neural activity imaging. J Neurosci 32:13819–13840
34. Chen TW, Wardill TJ, Sun Y, Pulver SR, Renninger SL, Baohan A, Schreiter ER, Kerr RA, Orger MB, Jayaraman V, Looger LL, Svoboda K, Kim DS (2013) Ultrasensitive fluorescent proteins for imaging neuronal activity. Nature 499:295–300
35. Zhao Y, Araki S, Wu J, Teramoto T, Chang YF, Nakano M, Abdelfattah AS, Fujiwara M, Ishihara T, Nagai T, Campbell RE (2011) An expanded palette of genetically encoded Ca^{2+} indicators. Science 333:1888–1891
36. Belousov VV, Fradkov AF, Lukyanov KA, Staroverov DB, Shakhbazov KS, Terskikh AV, Lukyanov S (2006) Genetically encoded fluorescent indicator for intracellular hydrogen peroxide. Nat Methods 3:281–286
37. Grynkiewicz G, Poenie M, Tsien RY (1985) A new generation of Ca2+ indicators with greatly improved fluorescence properties. J Biol Chem 260:3440–3450
38. Sakaguchi R, Endoh T, Yamamoto S, Tainaka K, Sugimoto K, Fujieda N, Kiyonaka S, Mori Y, Morii T (2009) A single circularly permuted GFP sensor for inositol-1,3,4,5-tetrakisphosphate based on a split PH domain. Bioorg Med Chem 17:7381–7386
39. Zhao Y, Jin J, Hu Q, Zhou HM, Yi J, Yu Z, Xu L, Wang X, Yang Y, Loscalzo J (2011) Genetically encoded fluorescent sensors for intracellular NADH detection. Cell Metab 14:555–566
40. Casey JR, Grinstein S, Orlowski J (2010) Sensors and regulators of intracellular pH. Nat Rev Mol Cell Biol 11:50–61
41. Anderson BM, Anderson CD (1963) The effect of buffers on nicotinamide adenine dinucleotide hydrolysis. J Biol Chem 238:1475–1478
42. Bradford MM (1976) A rapid and sensitive method for the quantitation of microgram quantities of protein utilizing the principle of protein-dye binding. Anal Biochem 72:248–254

INDEX

Paul Chang (ed.), *ADP-ribosylation and NAD+ Utilizing Enzymes: Methods and Protocols*, Methods in Molecular Biology, vol. 1813, https://doi.org/10.1007/978-1-4939-8588-3, © Springer Science+Business Media, LLC, part of Springer Nature 2018

G

H

I

L

M

N

O

P

Lightning Source UK Ltd.
Milton Keynes UK
UKHW05n1217160818
327221UK00005B/101/P

9 781493 985876